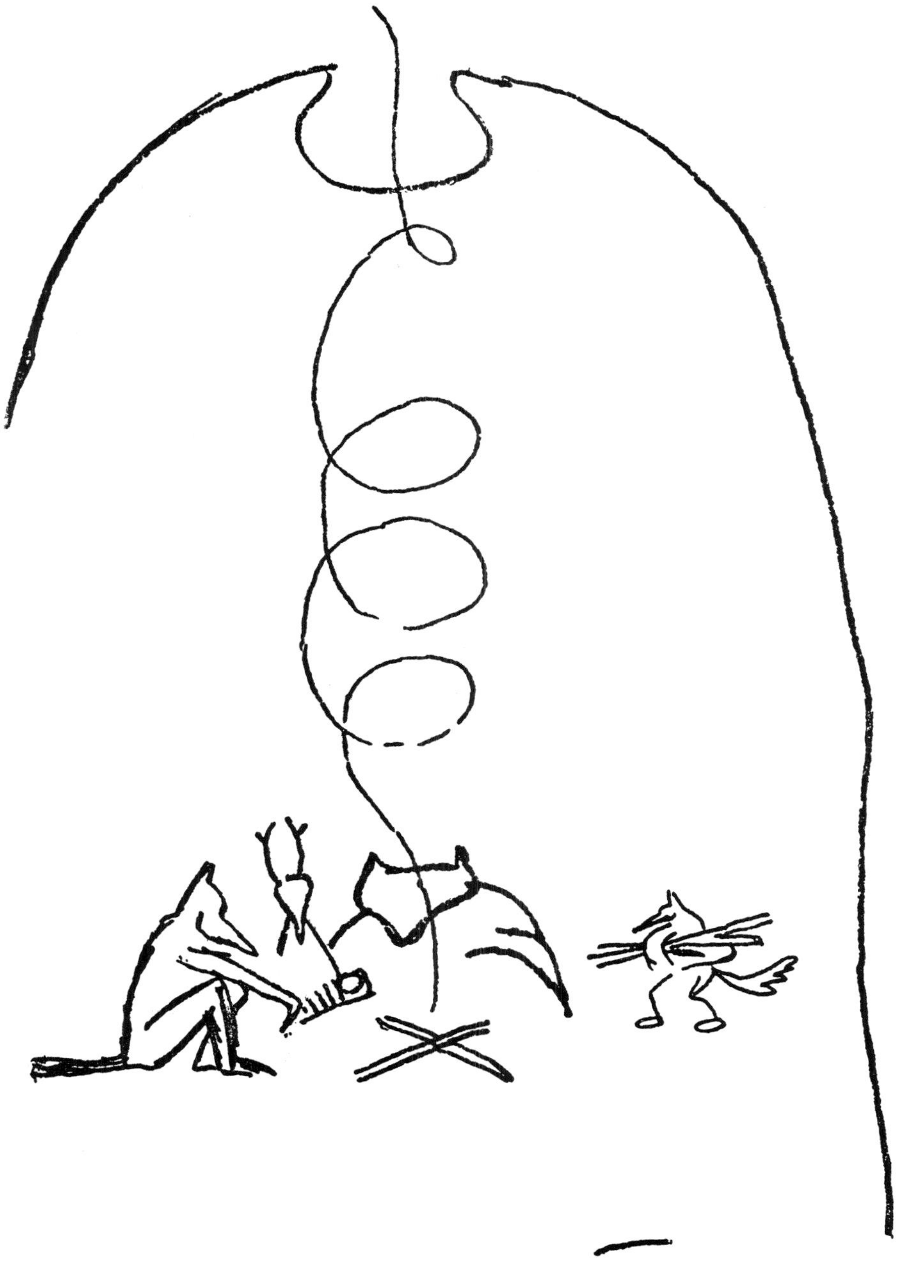

Fox was the only living man. There was no earth. The water was everywhere. "What shall I do," Fox asked himself. He began to sing in order to find out.

"I would like to meet somebody," he sang to the sky.

Then he met Coyote.

"I thought I was going to meet someone," Fox said.

"Where are you going?" Coyote asked.

"I've been wandering all over trying to find someone. I was worried there for a while."

"Well it's better for two people to go together... that's what they always say."

"O.K. But what will we do?"

"I don't know."

"I got it! Let's try to make the world."

"And how are we going to do that?" Coyote asked.

"Sing!" said Fox.

Jaime de Angulo

CONTENTS

CAVES, HUTS, TENTS

4 Shelter
5 Caves
6 Huts
7 Dogon *Aldo van Eyck*
8 Masai, Ethopia, Kabre
9 Iron Age Huts
10 Tin and Thatch in Togo *Kelly Jon Morris*
11 Tents
12 Tuareg *Johannes Nicolaisen*
13 Bedouin *Dale McLeod*
14 North Africa *Rich Storek*
14 Tekna
16 Yurts

NATIVE AMERICANS

17 Black Elk
18 Tipis, Pomo, Mandan, Miwok
19 Navajo, Hopi, Wichita, Pima
20 Coyote and Silver Fox *Jaime de Angulo*

EUROPEAN TIMBER

21 Early Timber *James Acland*
22 English Cottage Frame
24 Yugoslavia, Eastern Europe
25 Norway
26 Kizhi, Russia

THE NEW WORLD

27 The New World
28 Adobe, Baled Straw, Soddies
29 Studs

BARNS

30 Great Timber Barns
31 North American Barns

BUILDING

37 Building
40 Shed
41 Gable
42 Hip
43 Adobe
44 Hexagon
45 Barn
46 Floors and Footings
47 Concrete Floors
48 Windows and Doors
49 Roofing and Skylights
50 Tools and Tips
51 Japanese Homes *Edward S. Morse*
52 Residence Renaissance *Eric Park*
55 Dr. Tinkerpaw
56 Inside

MATERIALS

60 Materials and Methods
60 Materials and Animal Energy *Peter Warshall*
61 Shakes and Shingling
62 The Nature of Wood
62 Eucalyptus Lumber
63 Saplings
64 Mortise and Tenon *John Welles*
66 Dirt
68 Cinva-Ram *Kelly Jon Morris*
69 Stone
70 Baled Hay *Roger L. Welsch*
70 Plaster
71 Sod *Roger L. Welsch*
72 Canvas
72 Winter Tent *Keith Jones*
73 Hawaiian Lashing *Te Rangi Hiroa*
74 Reed
75 Bamboo
76 Thatching
78 Craftsmen of Necessity *Christopher and Charlotte Williams*
80 Wrecking and Salvage
82 Demolition Addict *Eric Park*
82 Basho Demolition *Martin Bartlett*
83 Captain Bill
85 Earth Shelter *Peter Warshall*

NOMAD LIVING

88 No-Mad Living *Ben Eagle*
89 Housecar *Kelly Hart*
89 Tin Lizzies *Jaime de Angulo*
90 Joaquin and Gypsy's Housetruck
93 Houseboats, Junks

DWELLING

94 Treehouse *Hugh Brown*
96 Towers
98 The Cones of Cappadocia *Paul Oliver, Herbert A. Feuerlicht*
100 New York City *Ned Cherry*
100 London Squatters *Graham Wells*
101 Medieval NYC *Herbert Muschamp*
102 Barrios *Charles Jencks*
103 Banani, A Dogon Village
104 $40 a Month, $40 a Day
105 Carpenter Gothic *Phil Palmer*
105 100 More Years *Michael Geraghty*
106 Libre

DOMES *Domebook 3*

108 The Dome
109 Introduction to *Domebook 3*
110 The Wonder of Jena
112 Smart But Not Wise
115 Technology Review
118 Drop City Revisited
119 Pacific High School Revisited
120 Bill Woods
120 Sealing Wood Domes
122 Ferro Cement
123 Tao Foam *Charles Harker*
123 Tensegrity Mast
124 Raw Stuff *Bill Bennett*
125 House of the Century *Ant Farm*
126 Divine Proportion
126 Chord Factors, Models
127 Log Dome *Bob Lander*
127 Adobe Dome
128 Crystal Windows *Kim Hick*
129 Bindu Dome
130 Decoding Arabic Design *David Saltman*
131 Polygons *Ananda*
132 Zarch *Geoffrey Bornemann*
133 Tet Truss
134 Zomes
135 Zome *Steve Baer*
136 Wooden Dome
136 Hypar, Domes *Peter Calthorpe*
138 Red Rockers
140 Ivy Dome *Burton Wilson*

BUILDERS

145 Free Form
148 Traditional Yurt *Aron Faegre*
149 Wooden Yurt *Bill Coperthwaite*
150 Hogan and Tea House
150 Tipi-Snail Shell
151 Sod Iglu *Ole Wik*
152 Ken Kern
152 Will Wood
153 Long Hair, Masonic Lodges and the Seeds of Architecture
153 Designer-Builders
154 Val Agnoli
156 Doug Madsen
158 Robert Venable

ENERGY, WATER, FOOD, WASTE

160 Woodlands *Ken Kern*
161 Energy
162 Northern Plains Power
163 Interview from Montana
164 Sun and Wind in New Mexico *Jay Baldwin*
164 Solar Water Heaters
165 Harold Hay on Solar Energy
166 Windmills of Murcia *Paul Oliver*
167 Windworks *Hans Meyer*
168 Bathroom
169 Arctic Circle Insulation *Keith Jones*
169 Heating and Insulation
170 Community Water *Lewis MacAdams*
170 China *Bob Willmott*
171 A Small Garden
172 Bibliography
174 Credits

COLOR

33-36, 141-144 Color photographs are described and credited on page 175.

SHELTER

Library of Congress Catalog Card Number: 90-60125
ISBN 0-936070-11-0

8 9 10 11 12 06 05 04 03 02 01 00
Lowest digits indicate number and year of this printing.

Shelter was first published in 1973. It was then reprinted twice in the next three years and eventually sold 185,000 copies. Its purpose was to show a wide range of information on hand-built housing and the building crafts and to maintain a network of people interested in building and shelter, with subsequent publication of the best available information. This reprinted version is identical to the original. *Shelter* has been translated into French, German, and Spanish.

Distributed in the United States and Canada by Publishers Group West.

Additional copies of this book may be purchased for $19.95 per book plus $3.95 shipping and handling from:

Shelter Publications, Inc.,
P. O. Box 279
Bolinas, California 94924
1-800-307-0131

Visit Our Website
SHELTER ONLINE
www.shelterpub.com

Our thanks to publishers and authors of the following books for permission to use copyrighted material. All written material from books is credited at the point of use. Drawings and photos from books are credited on page 175. Brief reviews and access information on most of these books is in bibliography, pages 172-73.

The American Heritage History of Notable American Houses. Copyright (c) 1971 by American Heritage Publishing Co., Inc.

Architecture 2000: Predictions and Methods, by Charles Jencks, published in 1971 by Praeger Publishers, Inc., New York. Reprinted by permission.

At The Edge of History, by William Irwin Thompson. Copyright (c) 1971 by William Irwin Thompson. By permission of Harper & Row, Publishers, Inc.

Bamboo. Copyright (c) 1970 by Robert Austin, Koichiro Ueda, Dana Levy. John Weatherhill, Inc./Walker & Co.

The Barn, by Eric Arthur and Dudley Whitney. Copyright (c) 1972 by M.F. Feheley Arts Co. Reprinted by permission of New York Graphic Society, Ltd.

Black Elk Speaks, by John G. Neihardt; Simon and Schuster Pocketbooks. Copyright (c) by John G. Neihardt.

The California Indians, edited by R.F. Heizer and M.A. Whipple. Copyright (c) 1971, originally published by the Univ. of California Press. Reprinted by permission of the Regents of the Univ. of Calif.

Everyday Life in Prehistoric Times, by C.H.B. and M. Quennell. Copyright (c) 1959 by Marjorie Quennell. Reprinted by permission of B.T. Batsford, Ltd.

Experiencing Architecture, by Steen Eiler Rasmussen. Copyright (c) 1959 by Steen Eiler Rasmussen. By permission of the MIT Press, Cambridge, Mass.

Five California Architects, by Esther McCoy. Copyright (c) 1960, Reinhold Publishing Corp. Reprinted by permission of Van Nostrand Reinhold Co.

The Foxfire Book, by Eliot Wigginton. Copyright (c) 1972 by Brooks Eliot Wigginton. Reproduced by permission of Doubleday and Co., Inc.

In a Sacred Manner We Live, photographs by Edward S. Curtis, text by Dan D. Fowler. Copyright (c) 1972, Barre Publishers, Barre, Mass.

Indians in Overalls, by Jaime de Angulo. Copyright (c) 1950, The Hudson Review, Inc. Reprinted by permission.

Handmade Houses, by Art Boericke and Barry Shapiro. Copyright (c) 1973 by Arthur Boericke. Scrimshaw Press.

Harvey Wasserman's History of the United States. Copyright (c) 1972 by Harvey Wasserman. By permission of Harper & Row, Publishers, Inc.

House Form and Culture by Amos Rapoport. Copyright (c) 1969. Reprinted by permission of Prentice-Hall, Inc., Englewood Cliffs, New Jersey.

Houses and House Life of the American Aborigines by Lewis H. Morgan, published 1965. Reproduced by permission of The University of Chicago Press.

Living The Good Life, by Helen and Scott Nearing. Copyright (c) by Helen Nearing. By permission of Schocken Books Inc.

Master Builders of the Middle Ages, by David Jacobs. (c) 1969, American Heritage Publishing Company, Inc. Reprinted by permission.

Sketch on page 1 is reprinted with the permission of Farrar, Straus & Giroux from *Indian Tales* by Jaime de Angulo. Copyright (c) 1953 by Hill and Wang, Inc.

Memories, Dreams, Reflections by C.G. Jung, recorded and edited by Aniela Jaffe, translated by Richard and Clara Winston. Copyright (c) 1963 by Random House, Inc. Reprinted by permission of Pantheon Books, a Division of Random House, Inc.

Medieval Structure: The Gothic Vault, by James H. Acland. Copyright (c) University of Toronto Press, 1972.

A Museum of Early American Tools, by Eric Sloane. Copyright (c) 1964 by Wilfred Funk, Inc. with permission of Funk and Wagnalls Publishing Company, Inc.

The Pattern of English Building. Copyright (c) 1972 by Alec Clifton-Taylor. Reprinted by permission of Faber and Faber, Ltd.

"Let's Voyage into the New American House" is from the book, *The Pill Versus the Springhill Mine Disaster* by Richard Brautigan. Copyright (c) 1968 by Richard Brautigan. Seymour Lawrence Book/Delacorte Press. Reprinted by permission of the publisher.

Red Men Calling on the Great White Father, by Katherine C. Turner. Copyright (c) 1951 by the Univ. of Oklahoma Press.

Rolling Stone (*Decoding Arabic Design,* by David Saltman). Copyright (c) 1973 by Straight Arrow Publishers, Inc. All rights reserved. Reprinted by permission.

Shelter in Africa, edited by Paul Oliver. Copyright (c) 1971 by Barrie and Jenkins Publishers.

Stone Shelters, by Edward Allen. Copyright (c) 1969 by the Massachussets Institute of Technology. By permission of the MIT Press, Cambridge, Mass.

Touch the Earth: A Self-Portrait of Indian Existence, by T.C. McLuhan. Copyright (c) 1971 by T.C. McLuhan. Published by E.P. Dutton and Co., Inc. (Outerbridge and Lazard, Inc.) and used with their permission.

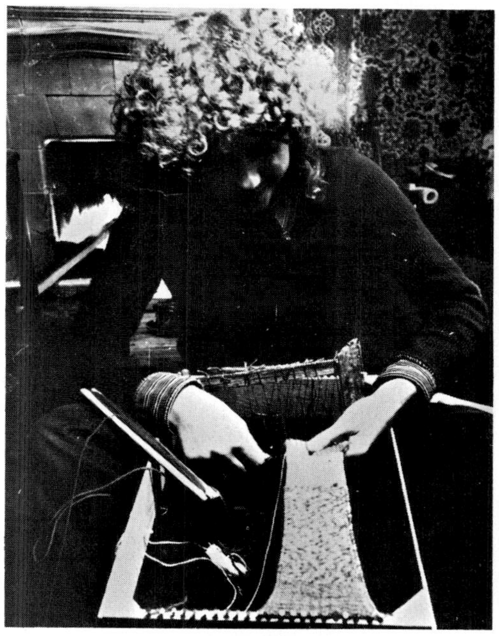

In times past, people built their own homes, grew their own food, made their own clothes. Knowledge of the building crafts and other skills of providing life's basic needs were generally passed along from father to son, mother to daughter, master to apprentice.

Then with industrialization and the population shift from country to cities, this knowledge was put aside and much of it has now been lost. We have seen an era of unprecedented prosperity in America based upon huge amounts of foreign and domestic resources and fueled by finite reserves of stored energy.

And as we have come to realize in recent years, we are running out. Materials are scarce, fuel is in short supply, and prices are escalating. To survive, one is going to have to be either rich or resourceful. Either more dependent upon, or freer from centralized production and controls. The choices are not clear cut, for these are complex times. But it is obvious that the more we can do for ourselves, the greater will our individual freedom and independence be.

This book is not about going off to live in a cave and growing all one's own food. It is not based on the idea that everyone can find an acre in the country, or upon a sentimental attachment to the past. It is rather about finding a new and necessary balance in our lives between what can be done by hand and what still must be done by machine.

For in times to come, we will have to find a responsive and sensitive balance between the still-usable skills and wisdom of the past and the sustainable products and inventions of the 20th century.

Of necessity or by choice, there may be a revival of hand work in America. We are certainly capable, and these inherent, dormant talents may prove to be some of our most valuable resources in the future.

This book is about simple homes, natural materials, and human resourcefulness. It is about discovery, hard work, the joys of self-sufficiency, and freedom. It is about *shelter,* which is more than a roof overhead.

SHELTER

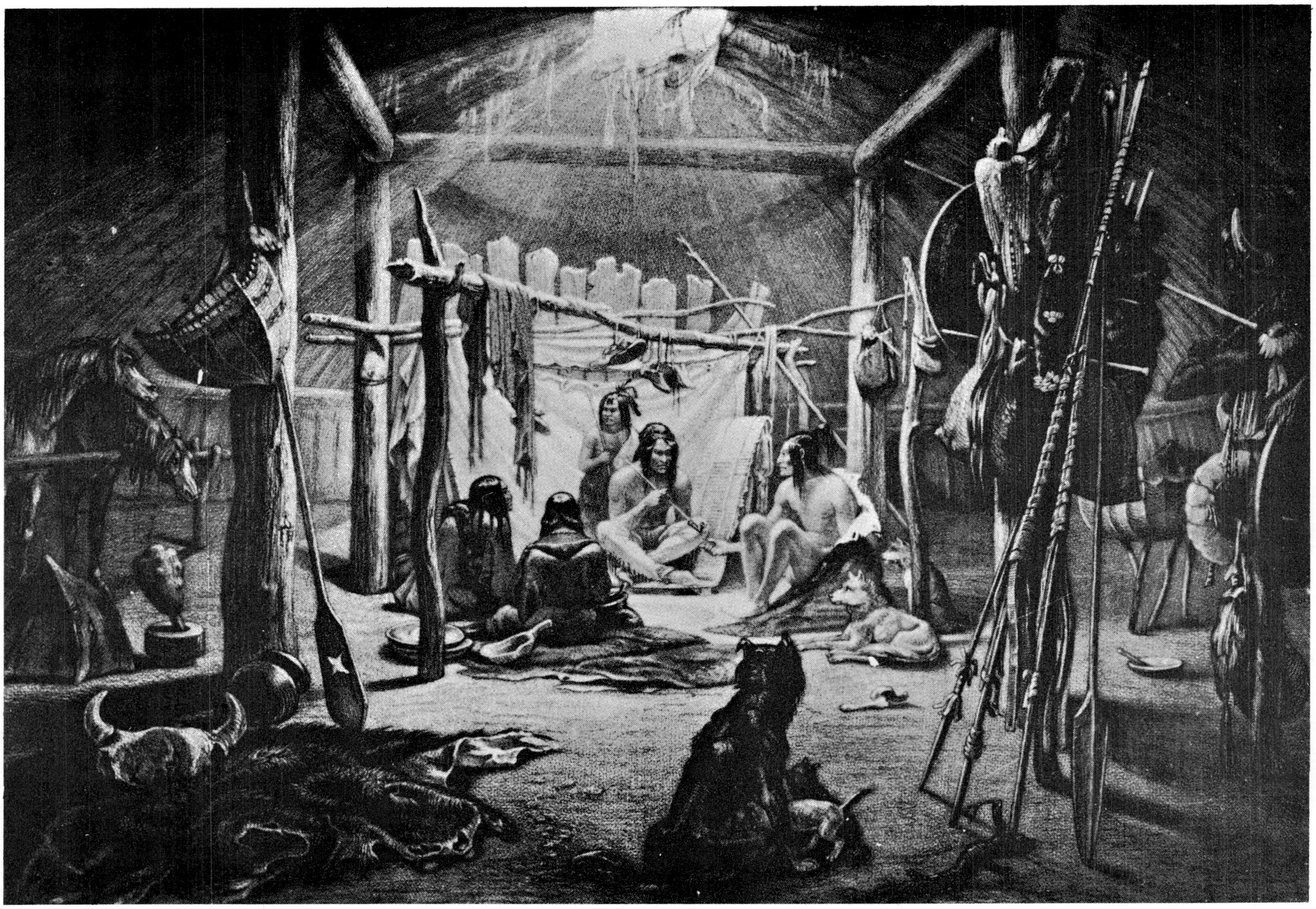

Interior of the lodge of a Mandan chief in the 1830's, a picture made on Prince Maximilian's expedition.

Early man lived under the trees and stars. At some time he found or improvised shelter.

...Hunters and fishermen in primeval times naturally sought shelter in rock caves and these were manifestly the earliest form of human dwellings; tillers of the soil took cover under arbours of trees, and from them fashioned huts of wattle and daub; while shepherds, who followed their flocks, would lie down under coverings of skins which only had to be raised on posts to form tents....

The first section of this book is our attempt to trace the evolution of these simple shelters — caves, huts, tents — to what we have now. By no means a complete history (we are not scholars or historians), it is rather an attempt to organize and understand cultural and structural concepts that appealed to our aesthetic senses.

Changing weather conditions, expanding agriculture and population, and the development of metal tools altered early man's shelter needs.Response to local materials, climate, and changing conditions has created an incredibly diverse number of structures: neolithic farmers who developed early rectangular wood frameworks from circular earth lodges; desert nomads with camel-packed, goat-skin tents; the Dogon tribes of Timbuktu whose mud cities reflect their view of the cosmos; American wood-stud palaces.

Sir Bannister Fletcher, quoted above and on pages 5, 6 and 11 was concerned with the history of architecture. What we are concerned with here is man and *shelter*.

Looking into courtyard of Tunisian troglodyte dwelling.

...Nature's caves, with their rough openings and walls and roof of rock, inevitably suggested the raising of stone walls to carry slabs of rock for roofs and old models of Egyptian houses show how rock caves influenced the plan, design and material for primitive structures....

"A simple cave could be enlarged, changed in shape, have another chamber added to it behind, to one side, above or below, linked by ramp, stair or doorway, and then another chamber beyond that, and yet another, perhaps, in a different direction, or branching off from one of the new chambers."

From *Stone Shelters*

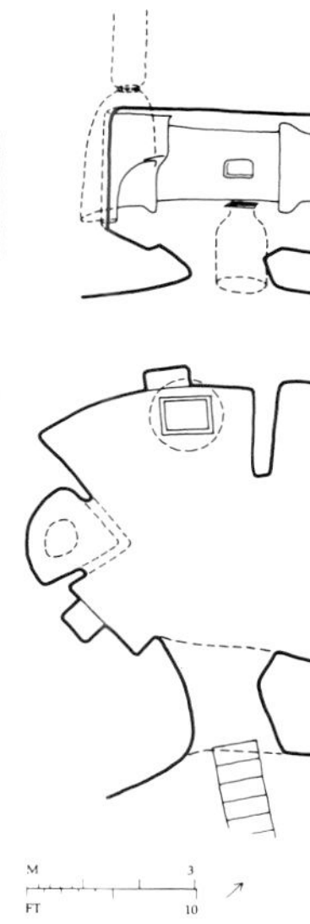

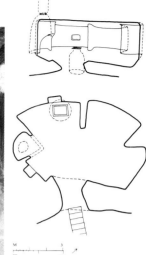

Drawings of Neolithic house plan and cave.

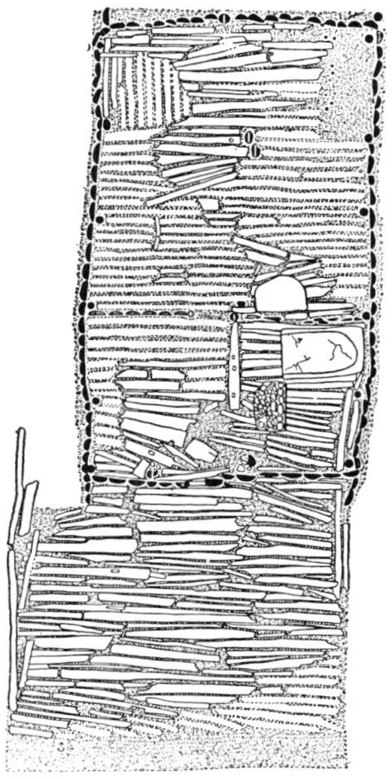

With suitable soil and climatic conditions, caves have provided shelter for man and other animals throughout history.

In addition to natural caves, there are numerous regions where people have carved their homes from the solid rock, often with spectacular results. In conventional construction space is enclosed by adding together pieces of material. These shelters were formed by the opposite process, that of subtraction.

In central Turkey there are the Cones of Cappadocia, where man and nature have combined to produce a spectacular landscape of natural cones and minarets sculpted by erosion, then hollowed into shelter. The soft volcanic stone of the area crumbles under the fingertips, and it was obviously easier to carve out a house than to construct one. Cappadocians carved entire cities, one to a depth of 265 feet, one an early 16 story skyscraper. Two cities, one containing 20,000 people, were connected by a 6 mile long tunnel. See pp. 98-99 for story and more photos.

Cones of Cappadocia

Caves at Massafra

Tule lodges of Yokuts

...Natural arbours, again, would suggest huts with tree trunks for walls, and closely laid branches, covered with turf for roof. Huts of this character are still in use amongst primitive peoples and the writer has seen them, as well as huts with two stories with external stairs, in the villages of old Jericho....

Early man had to build with readily available local materials -- canes and grasses, leaves and twigs. From these pliant and insubstantial materials he had to create a relatively permanent and structurally rigid unit. He did it by interweaving them to cover, vault or span a useful space. At first he may have simply tied together the tips of canes leaving the roots embedded in the ground. In this snug little arbour, a dome or roof springing directly from the ground he created a protected equable environment by using natural materials adapted with little change to his needs....

As early as 15,000 B. C. migratory hunters in Europe had discovered that turfs and earth were excellent insulators. They buried the simple round hut in a shallow excavation, building up a wall of turfs about the framework. The light saplings of the animal roof they covered with animal hides. Little more than cramped and smoky shelter, this primitive pit house became the source for the elaborate earth lodges of Europe, Asia, and North America.

The Gothic Vault

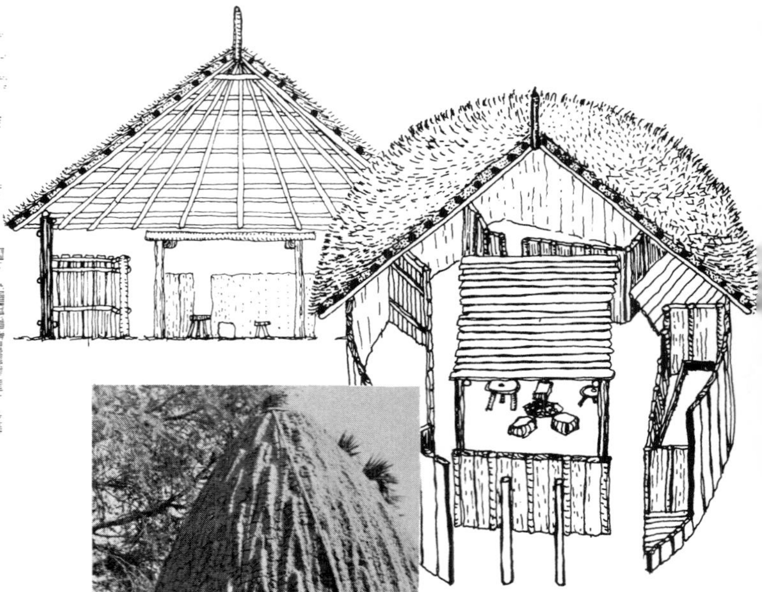

East Africa hut

Massa "bomb" houses, Northern Cameroon

Compound at Bongo, Nabdam

Traditionally when a man takes a new wife, the general distribution of new buildings is discussed by the family and the Tendaana, and marked out on the ground. The buildings are then erected by the men and rendered and decorated by the women who are to occupy. The wall paintings are bold, and within limits, excitingly varied from house to house, so that each woman's domain is physically and stylistically delineated.

When the compound head decides to build one or two huts, a sufficient quantity of earth is excavated from the nearest borrow pit and transported to the building site. Here the earth is put in heaps of approximately 80 to 90 cm high, water is added and the mud trampled until it has the consistency of mortar. This material is then molded into bricks of traditional type which are shaped like circular cones, about half the size of a rugby football, and then left to dry in the sun for at least two weeks. Meanwhile the foundation of the house is dug about 50 cm deep, to penetrate below the loose top soil. The dried bricks are then laid in courses, each of which is covered with a layer of mud mortar, consisting of mud and horse manure, or short cut grass which is mixed carefully before the mixture is used.

Shelter in Africa

Rural housing in Togo

It cannot have been so very different in Ur 5,000 years ago: the same laboriously fashioned bricks of sandy mud, then as now, the same sun weakly bonding and then harshly disintegrating them; the same spaces around a courtyard; the same enclosure; the same sudden transition from light into darkness; the same coolness after heat; the same starry nights; the same fears perhaps; the same sleep....

Meaning in Architecture

DOGON

Aldo Van Eyck — From *Meaning in Architecture*

The beautiful woven basket which the Dogon woman uses to carry grain and onions on her head, and as a unit of measure, has a square bottom and a round rim; the cosmos is represented by the basket inverted: the sun is round and the heaven above it is square....

Because of lack of grazing lands, they have become planters. They build on bare rock so the dwellings will have good foundations, and so that no arable land will go to waste; new fields will be established only where the bushland is cleared or soil laboriously transported in baskets and surrounded with rows of stones to keep it from being washed away....

Unlike people of other religious convictions, the Dogon expect of the hereafter no compensation through a better life. They find their life not bad, and learn in their well-ordered world that one is immediately rewarded for goodness, and that one pays for doing wrong and need not trail it with him after death.

The paradise of the Dogon, where the deceased reside, looks like Dogonland itself. The villages are like those in which the living dwell, the rich are rich, the poor are poor. All live with their families, planting millet and onions as they did on earth. In the dry brush the same trees stand, though the fruits they bear are more beautiful in colour, more lustrous, so that the dead can tell they are in paradise and no longer in the land of the Dogon....

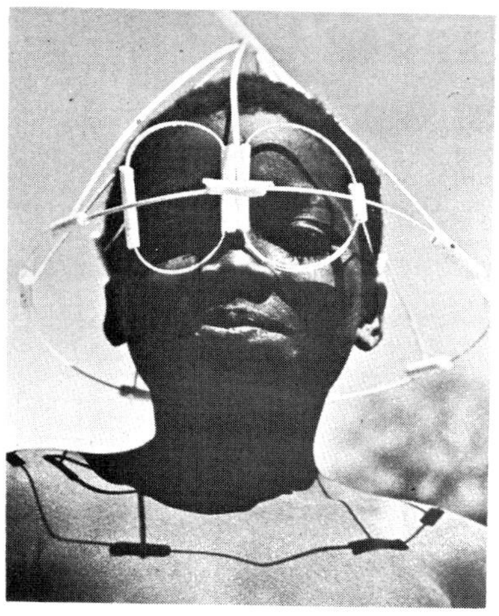

Among the Dogon and Bambara of Mali every object and social event has a symbolic quality, and the Dogon civilization, otherwise relatively poor, has several thousand symbolic elements. The farm plots and the whole landscape of the Dogon reflect this cosmic order. Their villages are built in pairs to represent heaven and earth, and fields are cleared in spirals because the world has been created spirally....

House Form and Culture

A Shrine of Ogol.

Yugo Doguru, granaries. Ogol.

In the hot and dry season of the year, from March to June, they patch their houses with a mixture of clay and water so they will be watertight at the coming of the first rains in June, when the entire population celebrates the sowing feast....

Banani type house.

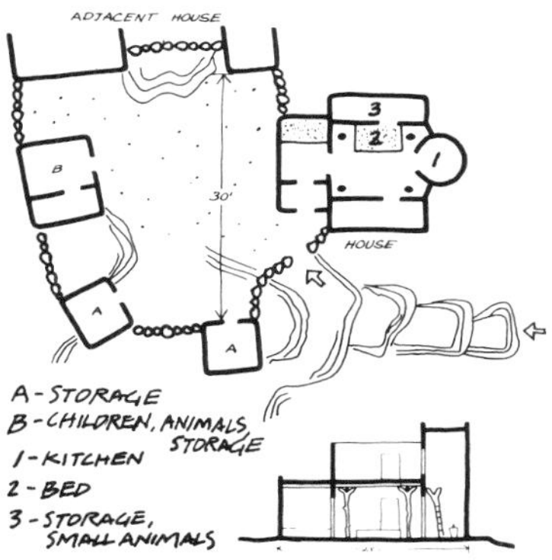

The Dogon people, numbering about a quarter million, live in the steppe region south-west of the bend in the Niger river at Timbuktu. There are some 700 villages along the 120 mile-long Bandiagara plateau, of two basic types, the plateau type built on flat rock outcroppings between arable fields, and the spectacular cliff debris type, built on rocks fallen from the high cliffs.

These collectively built cluster dwellings consist of main house, granaries, and smaller huts centered around a family yard, and interlinked with stone walls. Walls are built of mud bricks, then plastered with straw and mud. The kitchens are round with flat roofs that are supported by wooden posts and beams where needed. A notched ladder leads from the kitchen to the flat roof where the families will sleep during hot dry months. The granaries also have flat roofs, but are often covered additionally by conical straw roofs which are assembled on the ground, then lifted into place on the square based grain towers.

MASAI

Devoted cattlemen, the Masai of East Africa designed their homesteads for the best care and protection of their herds. The young warriors, "il morani" live apart from their families in loaf shaped mud barracks plastered with cow dung over a brushwood frame. Women build the barracks and provide food. Married men and their families live in cattle camps nearby. A thornbush fence surrounds the camp to keep out intruders, especially hyenas. Each woman has her own hut for sleeping and cooking which she shares with her children. Her hut, like the warriors', is easily dismantled and carried on donkey-back as the family moves to new grazing grounds.

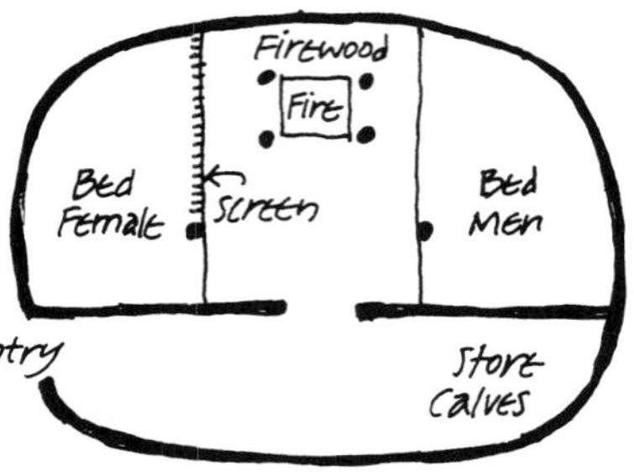

Huts are built for security — a low doorway and a dog-leg bend at the entrance slows down any intruding human or hyena and a sharp "panga" or machete is kept beside the bed of branches and sheepskins. Gourds full of milk or water are stored near the door and the fireplace where milk and cow's blood are prepared for the warriors, and cornmeal and millet for the married men. Each adult has his own hut; days are generally spent in the shade of an acacia tree drinking honey beer and talking about cattle and grazing.

Dennis Huckaby
Oakland, Ca.

Traditionally the Masai people of Kenya and Tanzania have been nomadic cattle herders, not game hunters. Because of this, Masai land has had an abundance of wild life. In recent years, with the growth of the tourist industry, much Masai land has been made into game preserves, causing competition for food between their cattle and wild game. The Masai have thus become less nomadic, and more dependent upon agriculture than in the past. Denny Grindall, who sent us these photos, has been helping Masai of the Rift Valley in East Africa construct earth fill dams, and is attempting to help them build structures that are more permanent than their temporary mud and dung nomadic houses.

Kikuya type Masai houses in village. Grass roofs on pole frame. Dung around walls.

Fresh cow dung seals sides. And later entire exterior.

Masai houses unsealed at base. More found in warmer areas.

ETHIOPIA

I managed to get my hands on *Domebook 2* in Athens on my way here. I thought you might be interested in the *tukul* of Ethiopia for your book, *Shelter.* The tukul could be easily built with resources available in the U.S. of A. Even as primitive as they may seem, if built right, a tukul will last many years. There is one near my house that was built in 1940. It is still a beautiful construction. One nice thing about the tukul is that if you live in the right area it can be built at the cost of your labor only. Not a bad price for a dwelling.

I have learned a few things here I would like to share. When King Menelik II moved the capital of Ethiopia to Addis Ababa the population grew rapidly and the wood resources declined. By using the trees for houses and firewood the mountains literally became bare. King Menelik searched for something to take the place of the native trees. He chose the Blue Eucalyptus of Austrailia. It grows fast, new trees grow from old stumps and it is an easy wood to work. To see the trees you would think they were native because of the great number of them. There is plenty of wood for everyone and the mountains are never bare.

I have seen towns where all the houses are made from empty tar barrells used in road building. Not exactly attractive but it is recycling something.

I am a college dropout and not a very good writer but I have tried my best to gather this information. I hope you can use it.

Chink Battle
HQ, Maag, Ethiopia

The tukul or sarbet (grass house) of the Ethiopian high plateau is a structure which utilizes simple building techniques and excellent use of natural resources.

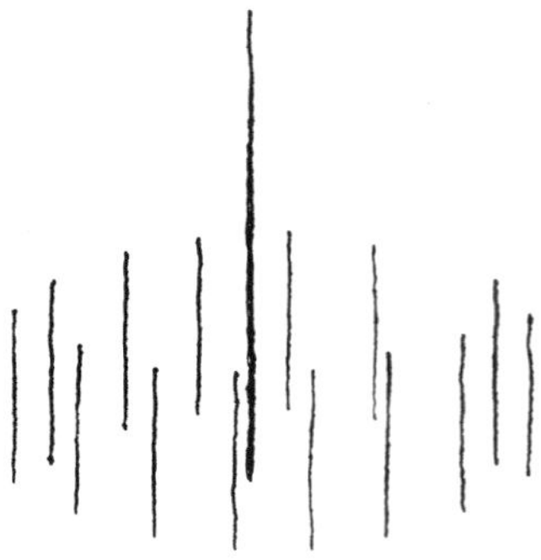

1.
First a circle with a diameter of 9 to 20 ft. is drawn at the building site. Eucalyptus poles are placed in the ground at one yard or so intervals along the circumference. The poles should be long enough so that at least 7 or 8 feet of the pole is above the ground. Next the center pole is set. It should be tall enough to give the roof an angle of at least 50°.

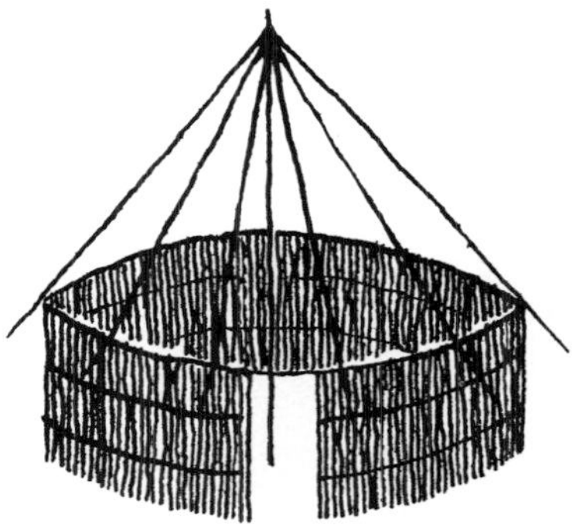

2.
Now the walls are filled with upright poles set close together and stuck in the ground. Rope is used to tie supports to the side of the wall. Green wood is used for ease in bending. Now the roof supports are attached 1 foot from the top of the center pole and extend about two feet past the top of the wall. This helps shed rain away from the wall. The supports can be extended even farther and used as a type of veranda.

3.
More supports are added to the roof. These again are of green wood and tied with rope. Now it is time to put on the roofing material. Sumbalit, a straw type grass is used. It is thatched or tied. This work is done from the top down, working carefully to insure a good roof. A pottery jar is added over the top of the center pole both as a decoration and to help shed water from the center of the roof.

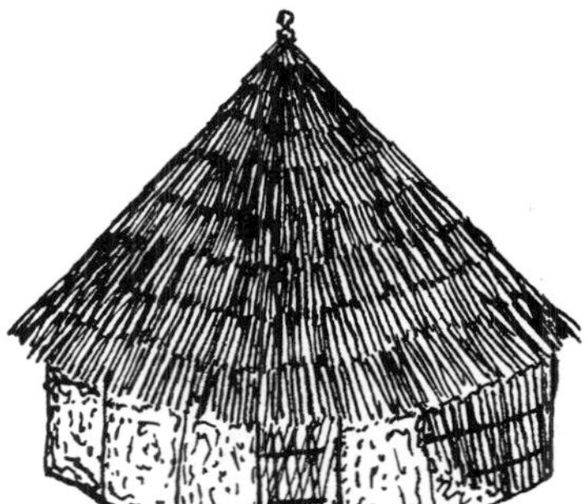

4.
Now an adobe plaster of straw and mud is put on the wall. After the plastering is finished, a door is built and installed and two small holes are put in the wall to allow sunlight to come in.

5.
Now with the outside finished it is time to work the inside. This is done according to the use. It can be used as living quarters, kitchen, stable, or storage. Many times it is used for all four and therefore must be partitioned accordingly.

KABRE

A Kabre compound near Lama Kara, Togo. The Kabre, renowned farmers, terraced hillsides, used compost, practiced crop rotation and planted several crops in association in order to maintain soil fertility.

Ann Maurice
Oakland, Ca.

IRON AGE HUTS

Charcoal burner's turf hut at Weald Open Air Museum, Sussex, England. In reconstruction a pole frame was covered with sacking, then branches with leaves, then turf stacked horizontally. This hut, used for sleeping, was constructed to specifications of Mrs. Arthur Langridge, whose family were traditional charcoal burners until the 1940's. Charcoal burning is an ancient craft, practised as early as 4,000 B.C. in Central Africa and unchanged until relatively recently with the introduction of metal kilns. Because charcoal gives off twice the heat of the equivalent weight in wood it was important in smelting iron.

Theoretical reconstruction of British Iron Age house at Avoncroft Museum of Buildings, Worcestershire, England. An excavation in 1959 revealed a circular wall over a meter thick and 10 meters in diameter, with no evidence of a center post; a pitch of 45° was chosen as the pitch necessary to shed water, and builders discovered two important facts during reconstruction: that the building has to have a wall plate so that the outward thrust of the roof rafters will not knock down the mortarless wall [this consists of notched and jointed poles set into the inside edge of the stone wall.] Secondly, it was necessary to include a ring beam in the roof apex lashing together all rafters with leather thongs. After frame erection, branches were interwoven among rafters and covered with hay before thatching....

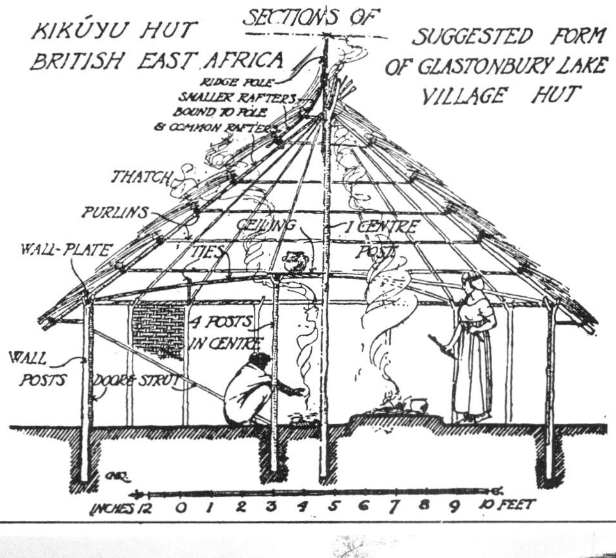

Replica of Iron Age hut at Avoncroft Museum.

Ancient Latvian summer kitchen "Slietenis." Latvian Ethnographic Museum, Riga, USSR.

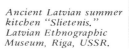

Theoretical reconstruction of Anglo-Saxon weavers hut, Weald Open Air Museum.

SOUTHERN CALIFORNIA

Made of mud, straw and clay mixture. Patted on by hand between lodge pole frame. Roof is bamboo thatch. All materials local Los Angeles county! The people who live in the apartments below complain that the hut is an eyesore.

Carey Smoot

Court Johnson saw this twig framework in a dream and built it on a Santa Barbara mountain side. He wired together bent green eucalyptus branches set in a circular concrete and rock footing. Since this photo the frame has been plastered with adobe mud. Court's dream revealed to him an archetypal structure as used by various primitive cultures — thin saplings tied to achieve the strength of compound curves. The same frameworks are found in Ethiopia.

TIN & THATCH IN TOGO

Here is a letter we received from Kelly Jon Morris, who spent 3½ years as a construction worker in Africa, then did some thinking which:

...led me to reflect on the meeting of "white-man technology" and indigenous architecture in Niamtougou, Togo, West Africa....Specifically, I was comparing the two houses I lived in where I was at Niamtougou — both of them marriages of African and European ways of building. One was a happy, comfortable union while the other was an unhappy one which eventually ended in divorce. Before comparing the two it is necessary to take a look at traditional building methods of the Lasso of Niamtougou:

THE BASIC ROUND HOUSE

The basic living unit is the circular mud-walled house with conical straw roof. This is built on the ground with little thought of a foundation as we know it, save an occasional shallow one of laterite stone and mud mortar. While some of the men dig the laterite soil, others mix it with their short-handled cultivation hoes (*konfirgou*) then form softball-sized balls of mud for the third group, who then mash the balls together in a circle, layer upon layer to build the walls, moulding and smoothing them like a sculptor his clay.

Door openings begin about a foot above ground level for various reasons — traditional ones relating to protection from enemies or animals, dangers which no longer exist, or ever-present ones, such as protection from rainwater, etc. Door and window openings seldom need lintels — when they are large enough to demand them, a branch from a tree will do.

Speaking of door and window openings: indigenous African homes are often criticized for their poor air circulation due to the small size of door and window openings and the specter is raised of tuberculosis spreading from one dark, dank hut to another. This dark picture is painted by people who live *in* their houses and who fail to realize that most Africans (most African peasants, that is, and they constitute 90% of the population) live *outside* of their houses. In the midday sun when temperatures hit 106-108° F, the shade of a mango tree is much preferable to being inside a house — any house, African or European style. And at night, mats are easily and often spread outside unless it is raining.

The conical roof is easily framed with three branches tied together at the top with locally-made cord, forming the basic unit. Additional branches complete the vertical piece and dried millet and sorghum stalks provide horizontal bracing, as well as something to tie the straw to.

While all this is going on, bundles of straw are brought to the site. The straw is cut at a height of about six feet. Each bundle is then spread out on the ground and the old men weave a single line of indigenous rope down the middle, forming a thick loose straw mat tied in the middle. It is then rolled up and stood up so that the thicker bottom edges will be even and passed up to the roof. There they are unrolled and tied onto this framework starting at the bottom and an inverted earthen jar caps the peak. The resulting cone is strong enough to span substantial distances without lower cords and braces characteristic of truss-systems and consequently relatively short-walled homes still have more than ample head room. The pitch allows rapid runoff of rainwater avoiding the two potential problems of the added weight of water-soaked straw and the eventual rotting that would occur. (The straw is changed every other year and lasts through two rainy seasons. Straw, needless to say, is in abundant supply there for the cutting.)

The round, conical-roofed house built by the Lasso is, in short, a simple, economical, and comfortable structure which is easily adapted to any terrain.

The first house I lived in was built in the European style — in terms of materials (as far as possible) and design (general architectural, that is, exterior style as well as room arrangement). It was rectangular in shape, had a spacious verandah in front, and a corrugated metal hip roof. The foundation and walls were made of rocks with mud mortar. The walls (inside and outside) and the outside of the foundation were plastered with a sand/cement mortar. The doors and windows were wood (mahogony was all that was available!) with the two upper panels louvered. The floor was a thin smooth layer of sand and cement mortar on top of packed dirt.

What was it like to live in? Miserable. One great source of misery in this tropical climate was the corrugated tin roof. In the dry season when temperatures shot to 115° F, tin-roofed houses are like ovens and this one was no exception. Even the addition of false ceilings of 4'x8' Masonite panels did little to help. In the rainy season, rain pounding down on the tin sounded like a thousand horses' hooves — impossible to carry on a conversation or to hear a radio playing full blast. And leaks? I never knew of a tin roof that didn't have at least one. Also, tin roofs have a nasty habit of being blown off by the high winds that accompany the arrival of the rainy season in April and May. The same fate always escaped nearby straw-roofed houses.

The second house I lived in proved to be a much more reasonable and comfortable, and, I might add, far less expensive, marriage of European style and materials with African ingenuity and locally available materials. This second house was in the compound of a man named Djonna — a tall, handsome, muscular man in his early 50's who held an important religious office in the village. The compound was the home for Djonna, two of his five wives and assorted children, grandchildren, and relatives of uncertain proximity. The houses were arranged in the following manner:

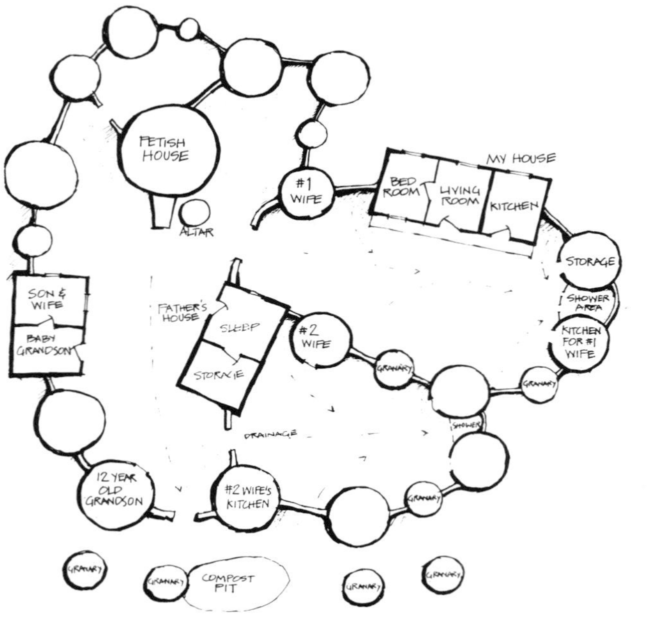

The house had three rooms, was rectangular in shape and had a straw roof. The foundation was, in the indigenous fashion, a shallow one of laterite rocks with laterite mud mortar. The walls were made of laterite mud bricks, made by the "paddle" method and sun baked. The walls were plastered with a laterite and fine sand mixture then coated with coal tar and finally white-washed (coal tar and white-wash are readily available and cheap). Window and door frames were made of *coquairs* — rough-hewn 3"x3" boards cut from a species of palm tree, the *romier* in French — I don't know what it is in English. (This amazing wood by the way, has great possibilities for low-cost housing — it is widely available, cheap, has great resistance to rotting and termites, and has great resilient strength. The only problem is that it's hard as hell to drive a nail into it, but that's not at all insurmountable.) The house was covered with a straw roof, built basically in the indigenous fashion but modified in the framing system by the addition of a ridge pole, making a hip roof, sort of.

The straw roof in itself was great — for keeping a place cool and dry, it can't be beat. (The straw of this region was round whereas that of the more humid south is flat like blades of grass. Don't know how they compare.) It's only disadvantage other than the danger of fire is that it must be changed every other year — the climate being one of a six-month dry and six-month rainy season, the straw is subjected to at least 2 months of heavy rain without much of a chance to dry out.

There were two features of this house that were particularly notable. The first was the false ceiling which served to keep the house much cooler in the daytime and much warmer at night — the latter being OK during the cool damp nights of the rainy season and during the hot nights of the dry season you sleep outside anyway.

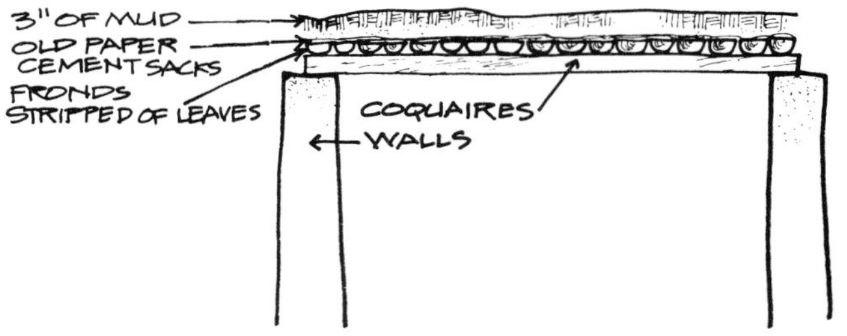

INDIGENOUS CEILING

First coquaires were laid about 6 feet apart from the top of one wall to the top of the other. Then, palm fronds stripped of their leaves were laid close together across and perpendicular to the coquaires. Old paper cement sacks were then ripped and spread out with overlapping edges on top of the fronds, and lastly a layer of about 3" of mud was packed on top and allowed to dry in the sun. And voila! Indigenous daytime air conditioning.

The second notable feature was the perfectly executed indigenous drainage system in the compound. First contrast it with the lack of adequate drainage in the first house I lived in — pools of rainwater always stayed on the verandah and frequently drained under the front door into the living room. The other house, however, was part of a compound perched on a gentle slope. The floor of the compound was covered by a hard, packed-mud surface. The coming of the dry season saw the forming of cooperatives by neighborhood women for the digging up and re-packing of compound floors on a rotating basis, just as the neighborhood men formed cooperatives for housebuilding. On one occasion as I was sleeping on the narrow porch in front of my house, I was awakened at 4 a.m. by the sound of a woman singing a Lasso work song to the accompaniment of fifteen pairs of hands slapping the ground beside me. After the floor had been dug up it was repacked and tamped by the women's slapping hands. The floor was smoothed down and the corners where the floor met the walls of a house or a dividing wall were rounded into the wall so that water draining off the walls and straw flows directly away from the wall and into the drainage flow. As the finishing touch, the women rub on a watery red oil made from the outer shell of the fruit of a tree the French call *merre* as a hardening agent. The result is a smooth hard surface on the floor of the compound. It is easy to keep clean and the floor of the compound is on a slope and is hand-contoured so that rainwater flows to the middle of the floor, down the slope, and out of the compound through holes in the base of the walls. As a result there is never any standing water in the compound after a rainstorm.

(As a sidelight, I might mention another indigenous surface hardening agent. The mainstay of the Lasso diet is a nutritious low alcohol millet beer they call *dam* which has a thick consistency. As one finishes drinking a gourd full of dam, there is a sediment of millet flour left in the bottom of the gourd. Quite often this pasty sediment is then thrown onto the side of a mud house where the sun soon dries it to a hard surface.)

I'm not sure what all this proves except that the introduction of "modern" materials (cement, corrugated tin sheets, nails, etc.) did not in this and many other cases make shelter. Shelter in this case was that which was natural, which made best use of locally available materials, and which in the end was the most life-supporting. More than that, shelter was the kind of life and the relationships which happened around that house, but that's another story...

Kelly Jon Morris
Durham, N.C.

P.S. I spent a lot of time building CINVA-RAM block buildings. Would be interested in hearing from people who already have or would like to build with it in the U.S.

Berber tents at tribal gathering "Fantasia" in Moroccan mid-plains area.

...Tents of sheepskins speak for themselves and are still as much in use among Bedouin Arabs and other nomadic tribes as they could have been in prehistoric times; and our thoughts turn naturally to the Tabernacle for the Ark of the Covenant, with its sheepskins and many woven hangings of silk and linen, which was carried by the Israelites through the desert, and was the apotheosis of the tent of the shepherds in the dawn of man's life on earth.

Such then were the first rough structures evolved from the three natural prototypes, when man began to build dwellings for himself and temples for his gods.

"Tent" symbol, motif used in Tunisian carpet weaving by nomads. (Symmetrical because "tents have two faces".)

Moroccan tent.

Interior of Moroccan tent.

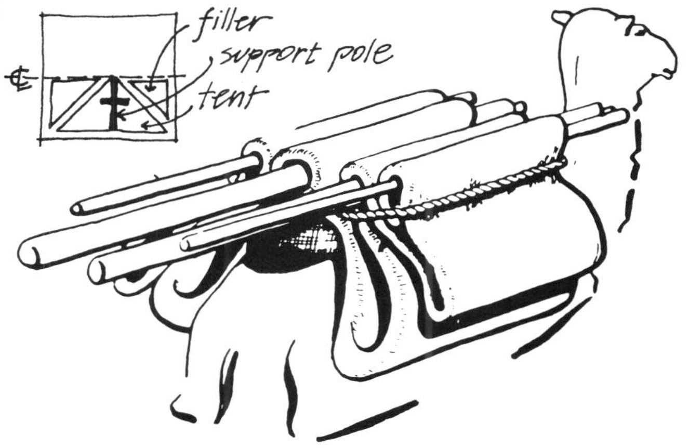

Camel loading procedure.

TUAREG

...On the whole the Tuareg of Ahaggar take a great interest in their tents. They like to have richly-decorated dwellings, and when a tent is made for a newly-married couple, the women do their utmost to make the sheet very beautiful. These aesthetic feelings seem to be unknown in regard to other types of dwelling used among the Tuareg, including the mat tent, and I think that they are responsible for some Tuareg using skin tents even when other dwelling types would be more suitable. I believe this to be true, the more so as the skin tent among the Northern Tuareg plays a great part in the wedding ceremony, which is called *ehen*, that is "a tent"....

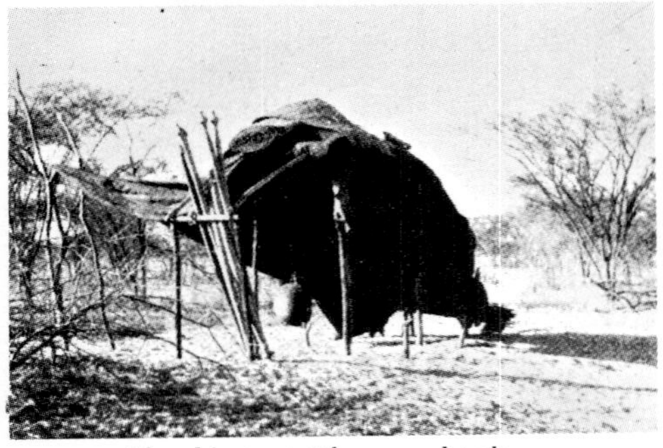

The skin tent. The carved poles leaning on the frame are main elements of a narrow wooden bed.

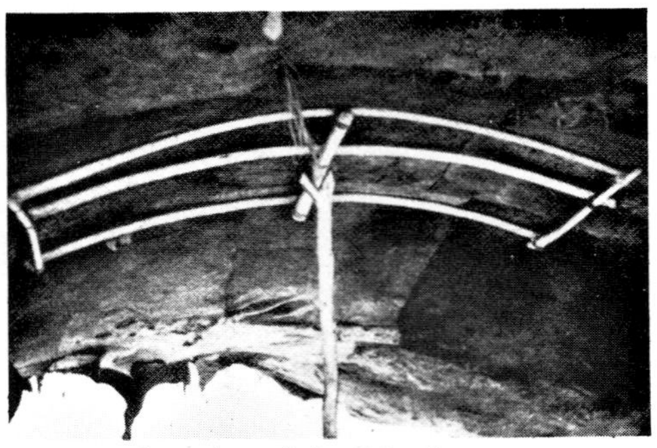

Cross bar of the B.2.g.7 tent. Ahaggar Tuareg of Tamesna, the Iregenaten tribe.

...There are advantages and disadvantages in mat tents as well as in skin tents. During the very hot season the mat tent is by far the best type of dwelling; it is comparatively high and the mat covering is not heated to any great extent by the sun. During the rainy season the mat tent is less practical than the skin tent. Rain may fall so heavily and abundantly that it penetrates the mat covering, and even if it does not it will be necessary to remove all mats after a heavy shower in order to dry them in the sun. The skin tent is normally erected quicker than the mat tent, it protects efficiently against rain, but is, on the other hand, most unpleasant during the hot season....

Riding camels in Ahaggar.

Leather bag for camel riders.

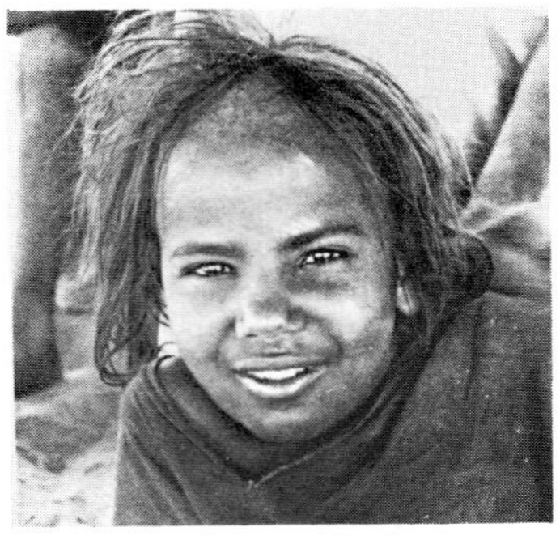

Boy of the noble class.

...The mat tent is normally erected as here described, but all women do not erect their tents exactly in the order mentioned. The work seems to be very easy to the women who measure no distances. They seem to carry out their work automatically. I have not been able to control the amount of time needed for one or two women to erect a mat tent of the type mentioned, for the women are frequently interrupted in their work by many other household duties, by their crying babies, or by visitors from other camps, but I believe that one woman can erect this elaborate tent structure in about half an hour....

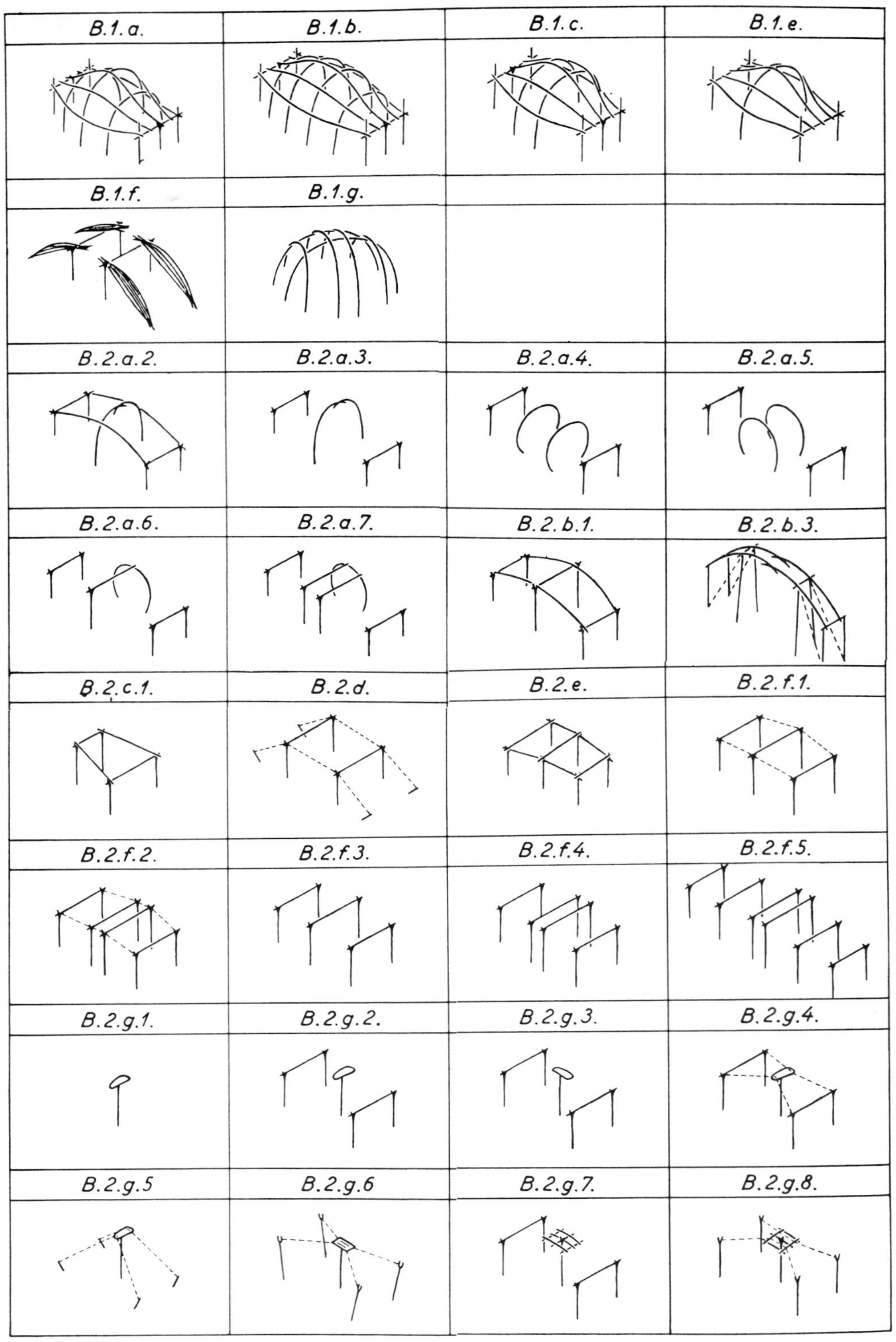

The main types of Tuareg mat tents (top 2 rows) and skin tents (bottom 6 rows).

Erection of mat tent in Ayr:

The wooden bed (tedabut) is placed upon the cleared ground. The pronged posts carrying riding cushions, skin bags, etc., and the curved wooden pieces constituting the three arches known as tekkekkewat *and* t'illisawin *are buried in the ground. The cushions and decorated skin bags resting in the heavy wooden prongs are temporarily covered with tent mats so that their colours may not be damaged by sunshine.*

*The arch pieces are lashed together and the two horizontal crossbars (*isgar*) are fixed to vertical sticks (*tigettewin*) buried in the ground at the narrow ends of the tent. A little discus-like thickening carved at the middle of each cross-bar is seen on this photograph.*

A long cord is wound around the high middle arch to form loops for the insertion of the slender rods forming semi-arches in the longitudinal direction. The thick ends of these rods are fastened to the crossbars but are not yet lashed together at their upper, narrow ends.

The narrow ends of the pieces of longitudinal semi-arches are lashed together. The tent structure is now ready to be covered with mats. The interior cover consisting of two rectangular mats made from plant stems and known as iwerweren *(sing.* ewerwer*) is seen in this photograph.*

*The dum palm mats of oval shape (*isfal*) and the long narrow dum palm mats (*asalemamas*) are placed over the* iwerweren *mats and are tied to the dome-shaped structure with cords. The narrow dum palm mat (*ereli*) surrounding the lower part of the tent is fixed to the structure.*

The mat tent is taken down for erection in a new place. The woman standing within the structure is loosening the looped cord which attaches the semi-arches to the high middle arch.
It is not the same tent as that depicted above, but the two tents belong to the same camp of the Imezzureg *tribe near Agadez in Southern Ayr.*

This page is from *Ecology and Culture of the Pastoral Tuareg.*

BEDOUIN

Text and drawings by Dale McLeod

THE BLACK TENT

The Bedouin people of the Arabian desert–the Rub' al-Khali or "Empty Quarter"–are primarily nomadic shepherds. This culture, like that of the Hopi Indians has evolved over thousands of years to meet harsh environmental conditions with minimal resources. Sheep and goats are the Bedouins' main source of livelihood, and it is the sparse grazing that causes them to move frequently from place to place.

The tent is a simple, adaptable system. The main structure is a roof rectangle made of woven goat's hair and/or sheep's wool. Together with poles and ropes it acts as a kind of space-frame on which are pinned two long narrow exterior curtains (*ruaq*) and one or more interior curtains (*qata*). Generally the exterior curtains enclose the sides and the interior curtain(s) divide the tent into men's and women's quarters; however in practice an infinite number of enclosures can be generated by hanging the curtains out onto the long tent ropes.

All of these pieces are made up of narrow woven strips–the roof of the tent is known by the name of the goat's hair strips (*filjan*). Generally the strips are purchased in town and sewn together; however the women of the household usually can and do spin and weave this material. The women are responsible for the ongoing maintenance of the tent.

The prosperity of the household is reflected in the length and corresponding number of poles in the roof, and in the decoration of the dividing curtains; a shiekh might have a four-pole tent with elaborate curtains.

METHOD OF TYING ROPES TO TENT ROOF

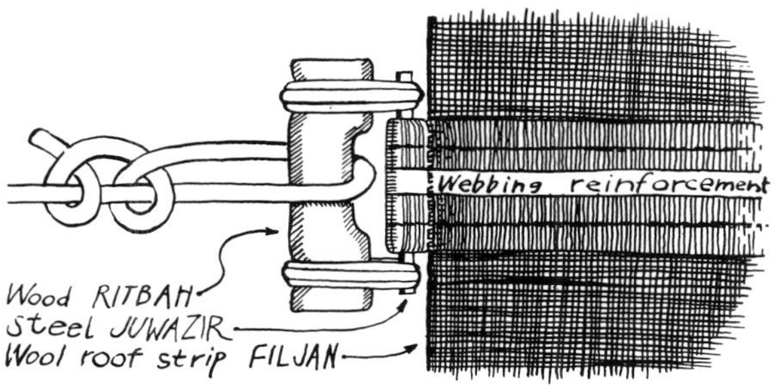

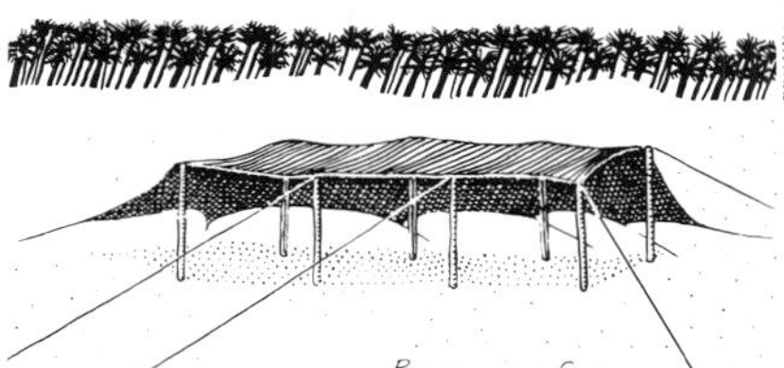

Basic roof structure, FILJAN

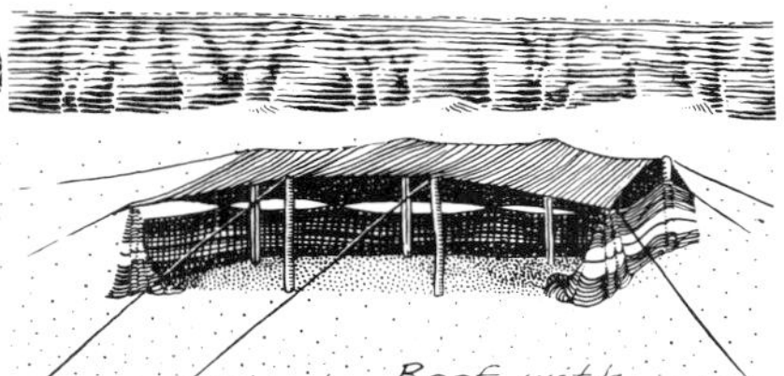

Roof with exteria curtains, RUAQ

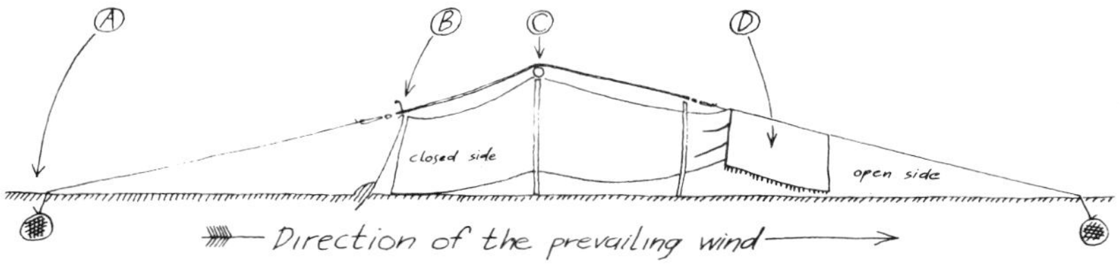

A. BURIED BUSH. method of anchoring tent in sand.
B. EXTERIOR CURTAIN. pinned to the tent roof & buried in the sand on the windward side.
C. CARVED PIECE OF WOOD. articulates tent pole with the tent fabric, sausage-shaped, it is fastened to the fabric.
D. FANCY END OF THE INTERIOR CURTAIN. identifies the front.

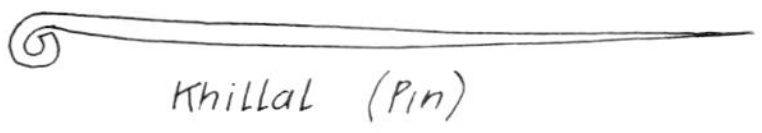

Khillal (Pin)

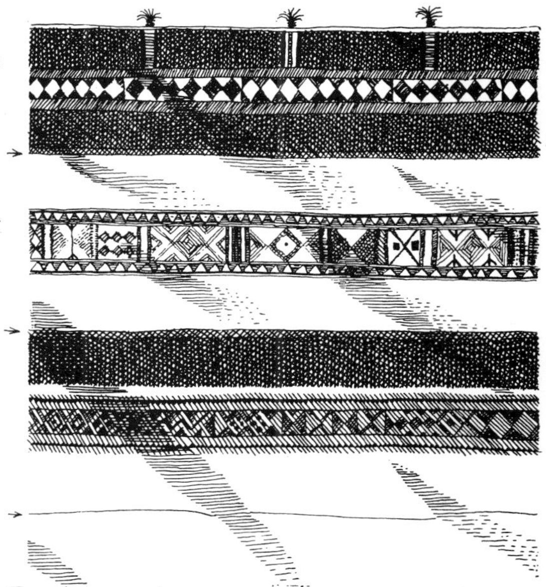

Traditional woven designs on the dividing curtain, QATA, of a well-to-do man's tent. note seams.

BLACK COFFEE

Bedouin society contained many examples of convention-turned-ritual. Traditionally, a traveller in the desert could always claim three days hospitality in a Bedouin tent, the only refuge there was in a thousand-mile wilderness.

When guests made their appearance it was the duty of the men of the household to prepare coffee. The implements and fuel for the fire were always set out in or near the men's quarters. Coffee making involved first roasting the beans in a long-handled pan and then allowing them to cool in a special box. After the beans were roasted they were pounded in a heavy mortar, then brewed with cardamom in three successive pots. All this took time–it was a getting-to-know-you period. When the taste was satisfactory, the man handed the pot to his eldest son. Holding the small cups in a stack on his wrist the boy poured, just so, a small amount and offered it to the most important guest first. When each guest had drunk as much as he wanted he inverted the cup in a certain way to signify, "enough".

The Bedouins are representative of a diffuse culture that has existed for thousands of years in an area including parts of Asia Minor, Eastern Europe and North Africa. Weaving, and especially rug weaving was brought to an extremely high level by these people before the familiar disintegration-through-contact-with-industrial-civilization set in. I read in a book about carpets that the emmissary from the Belgian court to a certain Persian prince remarked to his king that he had been the first man in history to have trod on the carpets of that palace in boots. When I was a kid in Arabia, during an era which I think of as the million-barrel a day period, the Americans were in the habit of referring to Arab workers in general as "the coolies", and were accustomed to being addressed as "sahib". Now the desert is criscrossed with roads and pipelines and the paraphernalia of American civilization. Soon the Bedouins will be gone....

...I hope you can use my drawings. I'm trying to do a few on the marsh Arabs of Iraq, also. The best book I know on the subject of the Bedouins is The Arab of the Desert, *by H.R.P. Dickson (London: Hodder & Stoughton, 1957).*

Dale McLeod

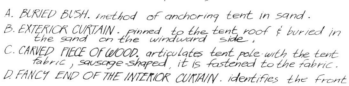

TEKNA

The Tekna tribes of southwest Morocco, as described in *Shelter in Africa*, wander with their herds of goats, sheep and camels in hot dry semi-desert country. Their tents are made of woven animal hair and carried by camel. Sketches by Peter Alford Andrews.

Rear view.

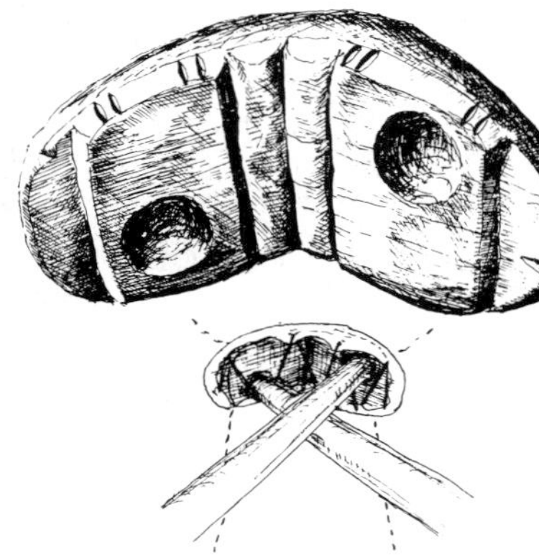

Ridge pieces.

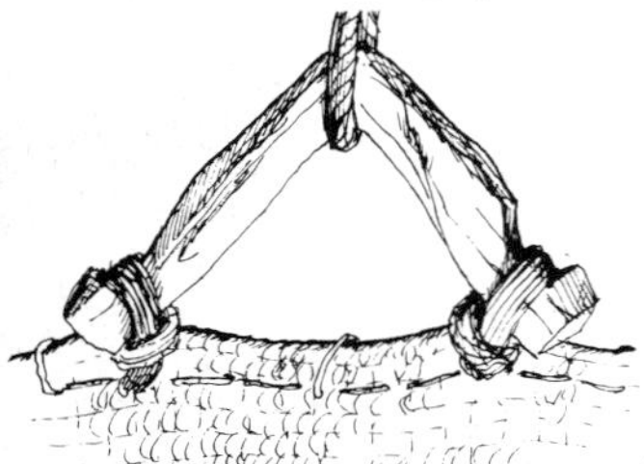

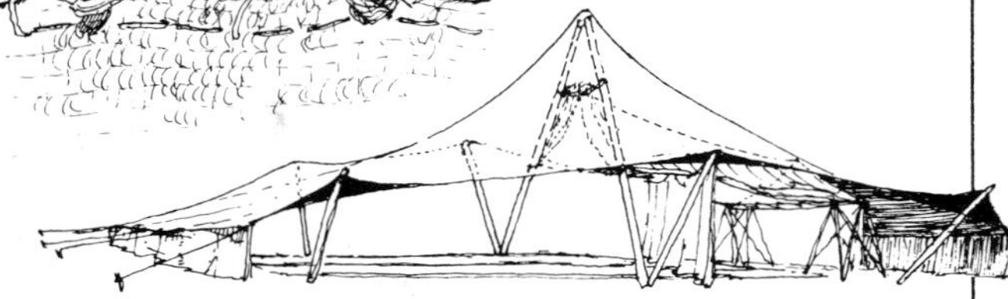

Front view of tent open for day use.

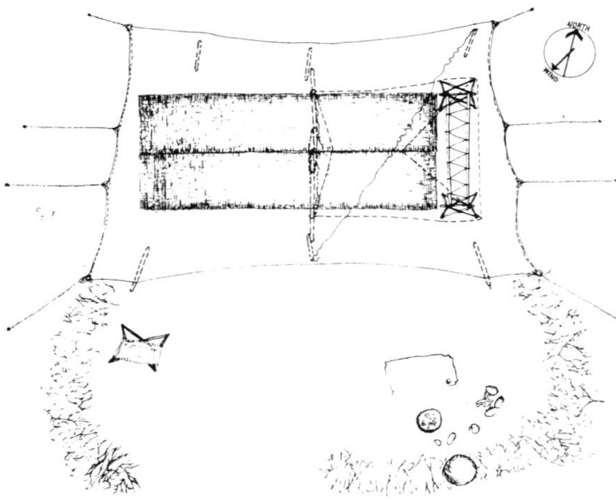

Plan of tent. Two floor mats are shades, with the womens' side on the right. The fireplace and oven pit are in the forecourt to the right, on the left is a water skin on its trestle. Brushwood surrounds the forecourt.

Side view.

NORTH AFRICA

Moroccan tents.

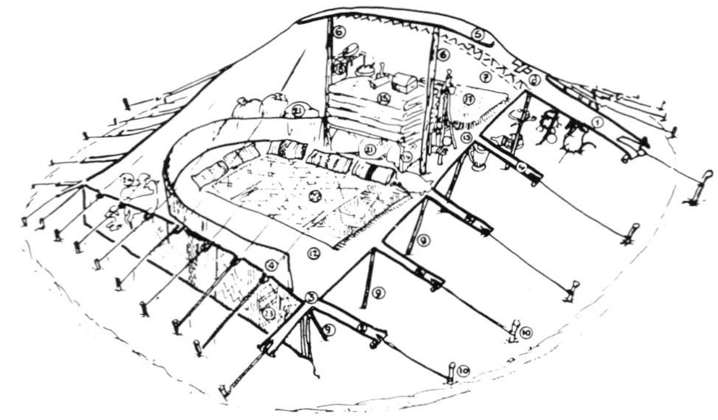

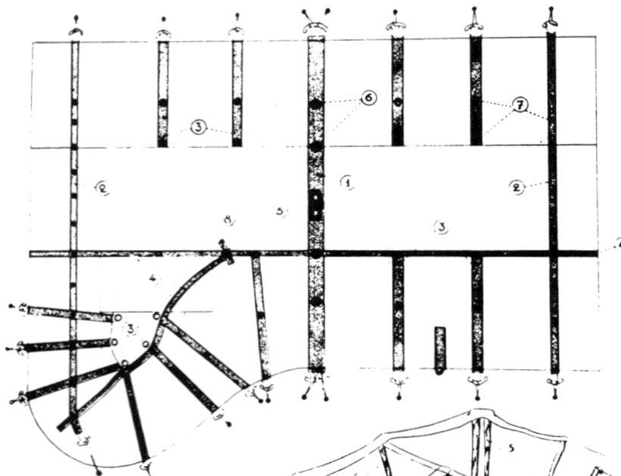

Plan.

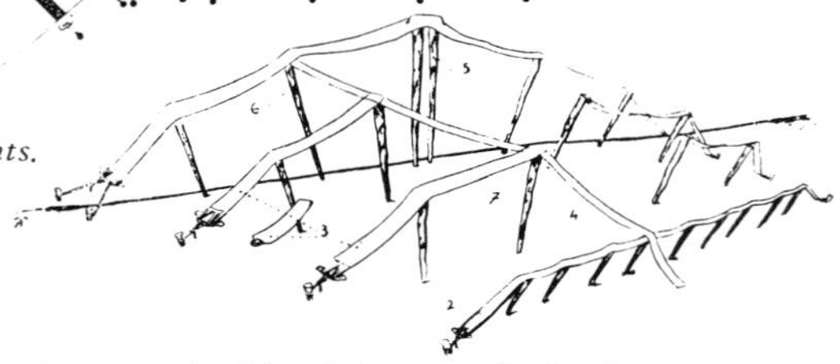

Berber tents.

The contrast between the life of the nomad who lives, not so much in a tent but in the desert, and the oasis dweller hardly needs emphasis. Shelter to the nomad might mean the shade of a rock or tree, and even when he visits the oasis his tents will be pitched at a respectable distance from it.

Shelter in Africa.

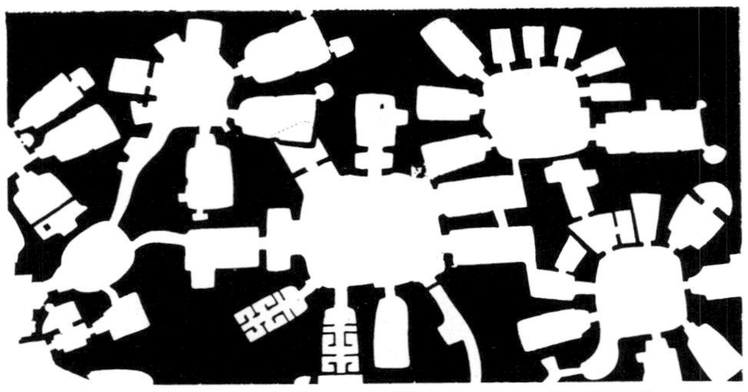

Plan of large troglodyte dwelling now used as hotel.

Central court of subterranean dwelling in Tunisia.

Berber and Arab tribes in North Africa have developed several traditional shelter types, each responding to distinct local living patterns. These types could be characterised as *stretched onto, dug into* or *built up.*

Stretched onto: The light mobile goat-and camel hair tents of the pastoral nomads.

Dug into: The troglodyte cave and hillside dwellings, prehistoric in origin, product of the subtractive, excavation process.

Built up: Sedentary dwellings built of earth, stone and masonry, from the isolated two-room-and-animal-enclosure plains homestead to massive, fortified hillside clusters to the blanket matrix of up to a few million connecting, adjoining and interwoven cells found in the dense urban "medinas."

These shelter types, beautiful to the eye and mind of the Westerner are being too rapidly abandoned for a new (Western) model: the detached, in-a-row, windowed "house," complete with garage and closet space, and dependent on expensive, often imported materials.

The traditional shelter types and their hundreds of variations have in common a tight dependence on local, cheap materials, a hand technology and a form derived from need for relief from exposure to the forces of sun, wind and endless openness.

The nomad tent, *stretched onto,* seen from Tibet to the African Atlantic among the tribes dependent on seasonal vegetation neatly fills the need for a quick-change, Camel-packed shelter. The low-lying manta-like tents seen in North Africa — the "beit kebir" or "Khyyma" (pl. "Khyyam," as in Omar Khyyam, great 11th Century mathematician, poet and tentmaker) are typically single-family shelters. They are modular in construction, limited in size to the carrying capacity of the family camel (which, with the family goats provides the hair from which the tent is woven), measuring around 20 by 40 feet, and weighing about 350 lbs. The modular unit of the tent is a woven band ("flij") about 2 feet wide (woven by the women), the length of the tent. Ten or so of these bands, laced together (by the men), reinforced by a skeleton of narrow tension bands, propped up by poles and staked to the ground make up the tent.

The *dug into* troglodyte burrows are the most (least) substantial form of shelter found in North Africa, particularly in southern Tunisia in the region around Gabes and inland.

As a concept, these subterranean shelters offer the perfect solution to the visual clutter of the landscape: They appear not to be there, even at short distance.

Tools and technology are rather simple, but the basic building material, even though free, is critical. The best is a form of spongy sandstone—easily worked and relatively free of moisture, though more clayey soils are sometimes used. Rainwater falling into the open courts is collected in cisterns. Surface water is carefully routed away from these light and air holes. The inside surfaces of the dwellings are of course virtually constant in temperature throughout the year. Only air temperature changes and air circulation can be somewhat controlled, even cooled, by the use of (moist) hanging screens. The troglodyte shelter is suited to bright, dry regions, whether hot or cold. They prôvide the ultimate in insulation and invisibility, the minimum of well-lighted and ventilated interior space.

Of the *built up*, "standard construction" type, two contrasting technologies have developed: the thin-shell brick vault and dome covering, and the massive rammed-earth dwellings.

The dome and tunnel-and-groin vault forms that you see particularly in Tunisia probably derive from the technical influence of the Roman Empire in that part of North Africa. The interior and exterior appearance of the buildings is livened by these roof forms which are often used when a flat roof might cost less.

The beauty of this technique lies in its simplicity. No formwork is used, and a team of three masons working with a piece of string, lightweight bricks and quick-setting mortar can put up a vault, say nine feet wide by fifteen feet long, or an eighteen foot diameter dome, in a couple of days.

The semicircular vault is laid up "feet first" (along its edges), sloping back to its crown. The mason setting the bricks is guided by another who holds a taut string against a scribed arc defining the vault section, or profile. He is served up bricks pre-spread with mortar by the third team member crouched at the feet of the mason. The quick-setting mortar has just enough adhesion to hold the weight of the brick. The vault though is barely holding its own until a layer of wire-reinforced cement covers it. Don't crawl on it or stand under it until it has been reinforced.

Constructing a dome is a similar, but simpler, process: The guide string is nailed to the focus of the dome (eliminating one of the team members) —dead center— and, like a leash, tied to the wrist of the mason. The string defines an accurate hemisphere as the mason lays the bricks, spiralling from the base upward. Like coil-building a pot in reverse.

Another, far less elegant "technology" has developed recently in the *bidonvilles,* literally, tincan cities that shelter the poorest and/or the most recent, and jobless, immigrant to the big city. These are largely clandestine, squatter dwellings, housing tens of thousands and growing fast, as in most traditionally agricultural lands experiencing an explosive rural-to-urban migration without sufficient industrial development to employ the hopeful immigrant.

The *bidonvilles* are put together with sticks, branches, mud and stone, flattened tins, wooden crates, canvas, paper and cardboard, string and wire, glass and plastic bottles, old tires — any kind of scrap (in an already waste-free economy). They are built without the security of land ownership, thus without the pride that accompanies investment. They are virtually without water and sewerage, not to mention electricity, paved roads, openness, green space and quiet. The use of materials is occasionally ingenious, but as shelter the *bidonville* is sordid, built and lived in in hardship; the environment demeans; humor, dignity and hope diminish.

In contrast, the old earthen dwellings built particularly in the southern regions of Morocco, the *Ksars,* developed originally as fortified, multi-family clusters by rural and mountain Berber tribes, are well constructed, secure and amenable environments: architecture. Their primary function is insulation — against extreme temperature (the wall mass provides an even, mean temperature through the diurnal cycle, relatively cool during the heat of the day and warm during the night chill), against light wind and noise (openings are small and few), against intruders (the *ksar* is a bastion, often several storeys in height, with few penetrations from the outside).

The common construction is of compacted, rammed earth built up from heavy stone foundations. Roof and floor structure are usually wooden, with an earthen covering, or sometimes arched masonry.

Rich Storek, Rabat, 13.5.73

السلام للولايات المتحدة

...My last few years have been in and around Africa and the Near East, working with volunteers in town planning, housing and building programs for various governments, concentrating on North Africa. It has been a priceless education for me and for my family, and good work, learning something about thin economies, foreign/local technologies, languages, belief, fresh food, diplomacy and America. Probably made a lot easier by my ignorance of most of these things elsewhere.

One of the most valuable aspects of being over here is seeing firsthand what a housing crisis in a nonindustrialized country feels and looks like. Along with teaching, town-and-building restoration and some research work, we have been plugging at an ever-worsening habitat program in several countries, seeing a continuing string of failures by deliberate-but-inadequate efforts to solve the shelter problem, but which make little more than a small dent in the whole mess. Living with the daily behind-the-scenes complexities that go to make up the dilemma gives Instant Wisdom: so you smile when some foreign flash comes saying, 'hey, ever think of a twoandahalf gainer, prestressed, sprayed-on fiber-reinforced dung units!?' Fortunately the notion seems to be gaining ground that the solution lies in other areas first. The technocrats, bureaucrats and politicians are devising real land ownership reforms and redesigning laws to hamper speculation, but that means eventually political reform which hits the powers where it hurts, and what kings and presidents want to avoid as long as possible. Hmmm.

Not without hope, on a trip to the US some months ago I invited the head of the Moroccan urbanism/habitat ministry to come along, to see some examples of planning, housing and building: some selfhelp, some subsidized, some experimental, some dreary, some successful, some un-. I wanted him especially to see some people's alternatives to the big operators' methods. We did manage to see some things that effectively proved my point: that some Americans are unplugging the machine and rediscovering their brains and hands (and looking for a Moroccan simplicity in their lives). He was impressed by that and was to be heard reporting to his people on return that "...There is a new revolution coming from the Americans...."

Well, more later. If you are anywhere near visiting this part of the world I would enjoy showing you some things and of course the carpet is down in our tent.

Best to you,

Rich

"Ghorfa" vaulted structures in Tunisia. Originally as grain storage lockers for nomad tribes, now being used for living. Making vaults.

Brick dome on square concrete base.

YURTS

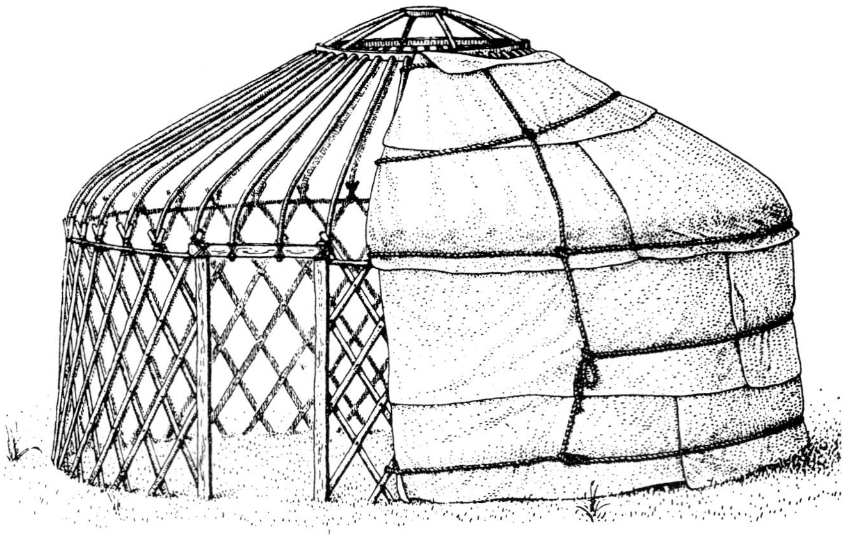

Several of the nomadic tribes of the Near East and Central Asia, from Iran to Mongolia, have for some several thousand years lived in a remarkable form of shelter: the yurt.

Yurts are particularly interesting shelters because they are so easily transported yet so solid in look and construction. The basic element is the expanding wall lattice: strips of wood are fastened together at intervals such that they can be expanded to form a larger wall section. Several of these sections are expanded and tied together with a door frame to form a circular wall. A compressive band/rope is then drawn around the top of the wall to help support the roof. Poles are then run from the top of the wall to a higher central compression ring. Sometimes there are two pillars helping hold up the central ring — and sometimes there are no pillars, the roof being self supporting like a truncated cone. The wood structure is then covered with various amounts of felt and canvas depending on the climate and weather.

The whole shelter is carried on one or two camels. It can be erected by several men in a half hour. After the outside covering is tied on and door shut it is astoundingly solid and sedentary looking.

Aron Faegre

The tent is rather like a ship without bulkheads or lockers, for although the partitions do not exist they are nonetheless real; everything is stowed in its traditional position. The only exception is that if a bridal yurt cannot be provided for a newly-wed couple a be-ribboned curtain screens off a small part of the tent to give them some privacy.

The yurt is always pitched facing south, so the pool of sunlight shining through the smoke hole in the roof acts as a clock. The area between the hearth and the door is the entrance hall; young animals may be tethered to the right of the door in winter. The rest is divided into thirds. To the west is the women's part; cooking-pots and maybe a samovar are kept on a shelf near the hearth; clothes storage, perhaps in chests but always also in large packs to protect it from the all-pervasive dust of the summer, is ranged along the wall; furthest in is the cradle, slung like a hammock. The east is basically the guests' side; sacks of dry goods like rice, barley and flour are also kept here. At the back of the tent two large bags hold the weaving equipment, and a gun may be hung between them. A mat is spread there where the men may sit and talk. At night the bedding is laid out on this mat.

On the Iranian side of the north-east border more and more Turkmen grazing is going under the plough. Now the yurts tend to be poor men's dwellings associated with villages, or they act as workshops for richer, settled tribesmen. Less than ten years ago it was a common enough sight to see groups of tents on the vast plain of the Gorgan; now it is a rare event.

Elisabeth Beazley
Country Life, 2/3/73

The interior plan of an Altai Tartar's yurt was firmly established by rules of etiquette. These rules were adhered to from Mongol to Tibet, by emperors in palaces to Tartars in their tents.

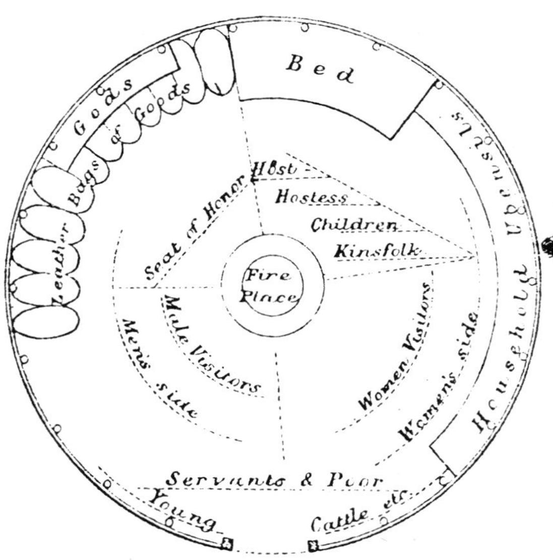

I can remember that winter of the hundred slain [1866] as a man may remember some bad dream he dreamed when he was little, but I can not tell just how much I heard when I was bigger and how much I understood when I was little. It is like some fearful thing in a fog, for it was a time when everything seemed troubled and afraid.

I had never seen a Wasichu [white man] then, and did not know what one looked like; but everyone was saying that the Wasichus were coming and that they were going to take our country and rub us all out and that we should all have to die fighting.

Once we were happy in our own country and we were seldom hungry, for then the two-leggeds and the four-leggeds lived together like relatives, and there was plenty for them and for us. But the Wasichus came, and they have made little islands for us and other little islands for the four-leggeds, and always these islands are becoming smaller, for around them surges the gnawing flood of the Wasichu; and it is dirty with lies and greed.

I was ten years old that winter, and that was the first time I ever saw a Wasichu. At first I thought they all looked sick, and I was afraid they might just begin to fight us any time, but I got used to them.

I can remember when the bison were so many that they could not be counted, but more and more Wasichus came to kill them until there were only heaps of bones scattered where they used to be. The Wasichus did not kill them to eat; they killed them for the metal that makes them crazy, and they took only the hides to sell. Sometimes they did not even take the hides, only the tongues; and I have heard that fire-boats came down the Missouri River loaded with dried bison tongues. You can see that the men who did this were crazy. Sometimes they did not even take the tongues; they just killed and killed because they liked to do that. When we hunted bison, we killed only what we needed.

Black Elk Speaks
from Touch the Earth

Jicarilla girl (Apache)

We always had plenty; our children never cried from hunger, neither were our people in want. The rapids of Rock River furnished us with an abundance of excellent fish, and the land being very fertile, never failed to produce good crops of corn, beans, pumpkins, and squashes...Here our village stood for more than a hundred years, during all of which time we were the undisputed possessors of the Mississippi Valley...Our village was healthy and there was no place in the country possessing such advantages, nor hunting grounds better than those we had in possession.
If a prophet had come to our village in those days and told us that the things were to take place which have since come to pass, none of our people would have believed him.

Ma-ka-tai-me-she-kia-kiak,
or Black Hawk, Chief of the
Sauk and Rox

Princess Angeline (Suguamish)

Sioux tipis on the Little Bighorn.

Once horses reached the northern Plains, several groups including the Chipewyan, Arapaho, Cree, Atsina, Piegan, Blackfeet, Blood, Cheyenne, the several Sioux or Dakota tribes, the Assiniboin and Crow abandoned their earth lodges and farm plots to become full time buffalo hunters, living in skin tipis and developing military societies and a highly structured system of warfare.

Some tribes, especially the Arikara, Mandan and Hidatsa, although adopting horses, remained as earth-lodge dwellers along the Middle Missouri. These tribes became middlemen between other Indians and fur traders in the early part of the 19th century.

In a Sacred Manner

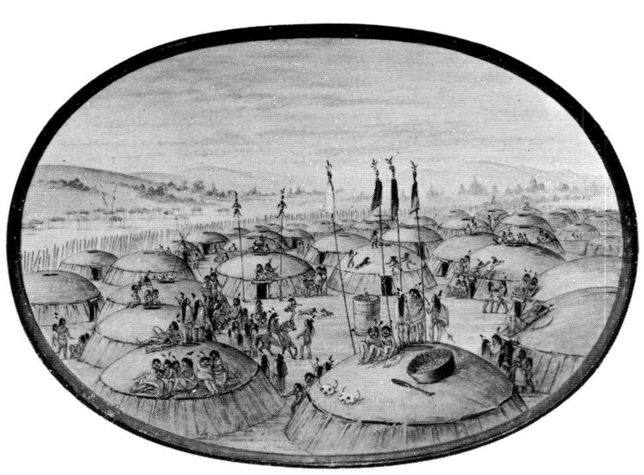

Mandan village. When there was heavy rain, the boats were used to cover the smoke holes.

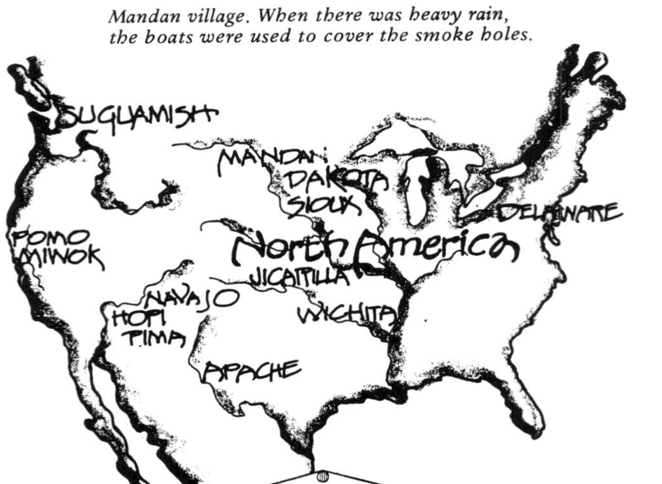

POMO DANCE HOUSE

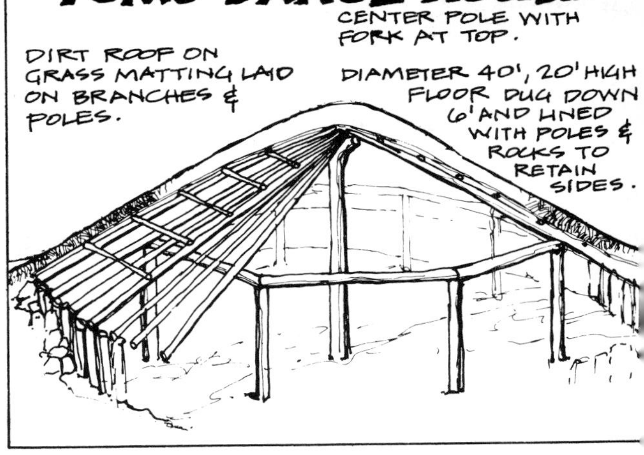

South of the Yurok, Klamath and others in the coast ranges and the central valley of California were numerous other small Indian groups, including the Wailaki, Yuki, Pomo, Wintun, Maidu, Miwok and Yokuts. These groups subsisted on acorns, fish, nuts, berries and deer. The Pomo, for example, occupied the Russian River valley, the adjacent coast and the Clear Lake basin between the Coast ranges. The people lived in plank or mat houses in settled villages. Unlike their northern neighbors most of these groups had definite political organizations with chiefs, often hereditary. Most groups had well-developed systems of myth and ceremony. Shamans were important but seemingly less so than to the north. Some groups, notably the Pomo, made excellent baskets, perhaps the finest in North America. Many of their feather-decorated baskets are now highly prized museum pieces.

In a Sacred Manner

MANDAN EARTH LODGE

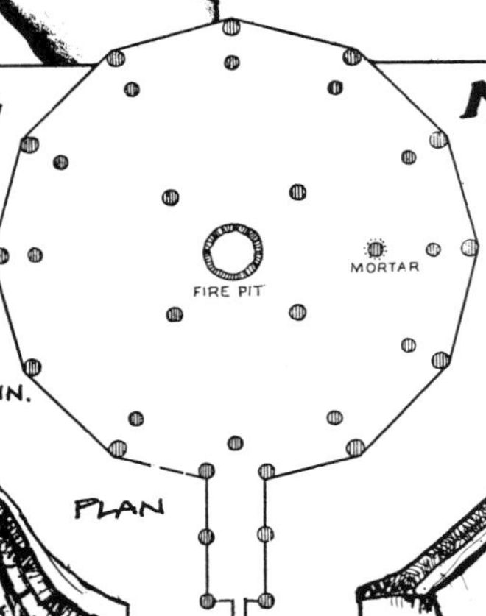

The Mandans and Minnetarees of the Upper Missouri constructed a timber-framed house... the houses were circular in external form, the walls being about five feet high, and sloping inward and upward from the ground, upon which rested an inclined roof, both the exterior wall and the roof being plastered over with earth a foot and a half thick.

These houses are about forty feet in diameter, with the floor sunk a foot or more below the surface of the ground, six feet high on the inside at the line of the wall, and from twelve to fifteen feet high at the center...An opening was left in the center, about four feet in diameter, for the exit of the smoke and for the admission of light. The interior was spacious and tolerably well lighted, although the opening in the roof and a single doorway were the only apertures through which light could penetrate. There was but one entrance, protected by what has been called the Eskimo doorway; that is, by a passage some five feet wide, ten or twelve feet long, and about six feet high, constructed with split timbers, roofed with poles, and covered with earth. Buffalo-robes suspended at the outer and inner entrances supplied the place of doors. Each house was comparted by screens of willow matting or unhaired skins suspended from the rafters, with spaces between for storage. These slightly-constructed apartments opened towards the central fire like stalls, thus defining an open central area around the fire-pit, which was the gathering place of the inmates of the lodge. This fire-pit was about five feet in diameter, a foot deep, and encircled with flat stones set up edgewise. A hard, smooth, earthen floor completed the interior. Such a lodge would accommodate five or six families, embracing thirty or forty persons.

From *Houses and House-Life*

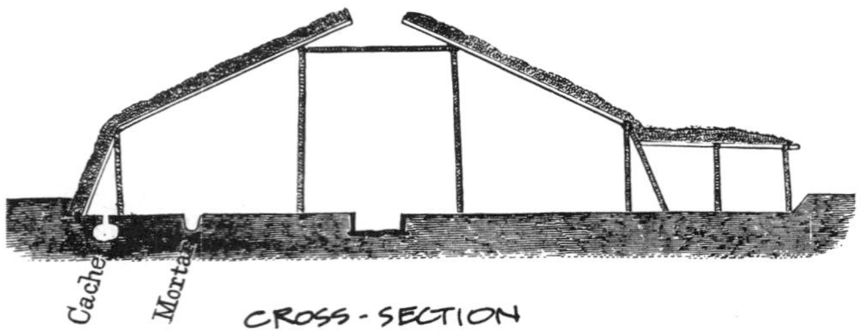

MIWOK ASSEMBLY HOUSE

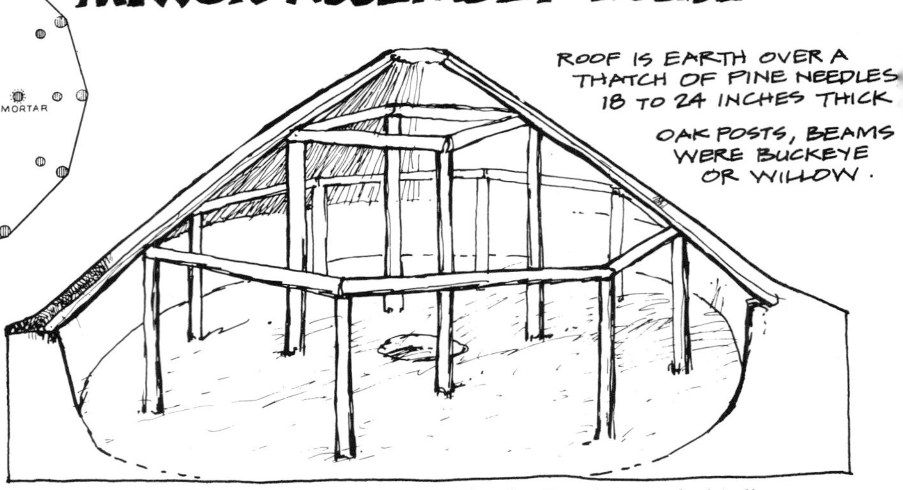

ROOF IS EARTH OVER A THATCH OF PINE NEEDLES 18 TO 24 INCHES THICK
OAK POSTS, BEAMS WERE BUCKEYE OR WILLOW.

After the timbers for the building had been gathered it took only four or five days to erect the building, everyone in the village helping. The wood used was oak, usually obtained by burning down the trees. If only two or three men were employed in obtaining the timbers, it took them two months.

The first step in the actual construction of the house was the excavation. The size of the area to be excavated was carefully measured. The measure of the radius was called *oyisa yana,* literally "four men." Four men actually stretched out on the ground, the head of one man touching the feet of the next man. If we consider the men as averaging 5½ feet, the diameter would be 44 feet. The excavating was done with digging sticks.

Next the four center posts which supported the roof were put in place, forming the four corners of a square, each side being the reach of a man in length. Four horizontal pieces were tied with withes to the tops of these posts. From these, radial beams were laid sloping to the sides of the pit, but supported midway by an octagon of stringers resting upon the side posts. The four center posts were each about a foot in diameter, the eight side posts smaller. The stringers were about 6 inches, the radial roof beams about 5 inches, and the numerous horizontal closely laid cross sticks upon which the roofing material was laid about 3 inches, in diameter. The posts were of oak, the stringers and roof beams of buckeye or willow. The four center posts were imbedded 2 feet, the others a foot. The two rear center posts were treated with "medicine" and only dancers could approach them closely. Posts were either notched or naturally forked at the top to hold the stringers.

A thatch of brush, topped with digger or western yellow pine needles, never sugar pine needles, was next put on. This was followed by the final covering of earth. Altogether the roof was 1½ or 2 feet thick. The opening in the top of the conical roof served as the smoke hole, the fire being built directly under it. The entrance was on any side.

Certain niceties appear in placing brush and earth on the roof. The first layer of brush, which was laid radially over the numerous horizontal roof timbers, was of willow. On this another layer at right angles was placed. The third layer was of a shrub with many close parallel twigs that kept the earth covering from leaking through and resisted rot. The proper depth of the earth layer was 4 or 5 inches and was measured by thrusting in the hand. The proper depth came to the base of the thumb.

The digging of the fireplace in the center of a new assembly house took place at the celebration following its completion. A digging stick was the tool; the depth to which dug was about a foot; its diameter between 2 and 3 feet.

At Chakachino, a post-Caucasian village near Jamestown, there have been four assembly houses within the memory of the informant, Tom Williams. When one became old and rotten it was torn down, the occasion being one for merrymaking. Also, the death of a chief was followed on one occasion by the burning of the assembly house as a mourning observance, as was the usual Miwok custom. Following the construction of each new assembly house at Chakachino, Miwok from various villages came to the opening ceremonies.

From *The California Indians*

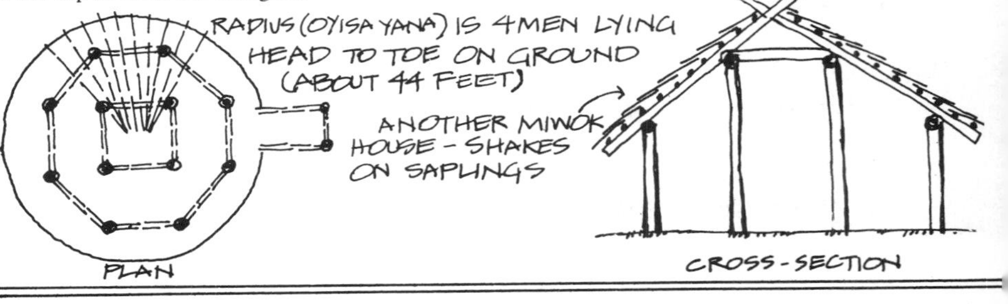

NAVAJO

The basic Navajo dwelling is the hogan. Originally it was a framework of heavy poles laid up like a tipi, with a projecting entrance, plastered with mud, suggesting a cross between an earth lodge and a tipi. Later the form changed to an eight-sided wall frame, with a domed, cribbed log roof—called *whirling logs* as shown above.

Curtis visited most of the Southwestern tribes, but he was most taken with the Hopi and the Navajo. Of the latter he wrote: "The Navajo is the American Bedouin, the chief human touch in the great plateau-desert region of our Southwest, acknowledging no superior, paying allegiance to no king in name of chief, a keeper of flocks and herds who asks nothing of the Government but to be unmolested in his pastoral life and in the religion of his forebearers."

In a Sacred Manner

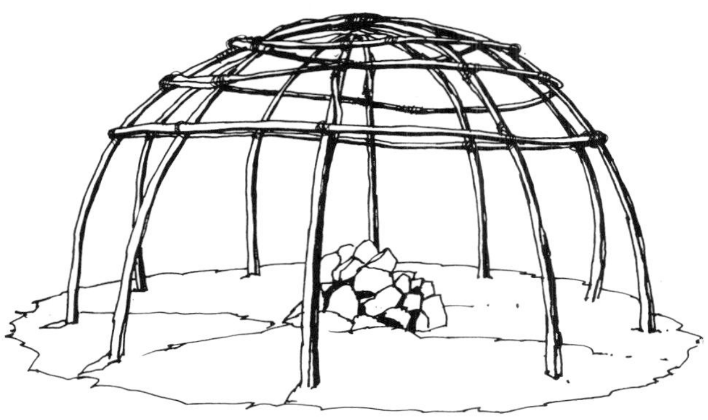

Frame of Peyote Sweat Lodge

HOPI

Wolpi, the "Place of the Notch" on the extreme end of the first Mesa (The Moqui Towns).

Ceremonial Kiva, dedicated to the earth spirits.

North kivas of Shumopavi

WICHITA GRASS HOUSE

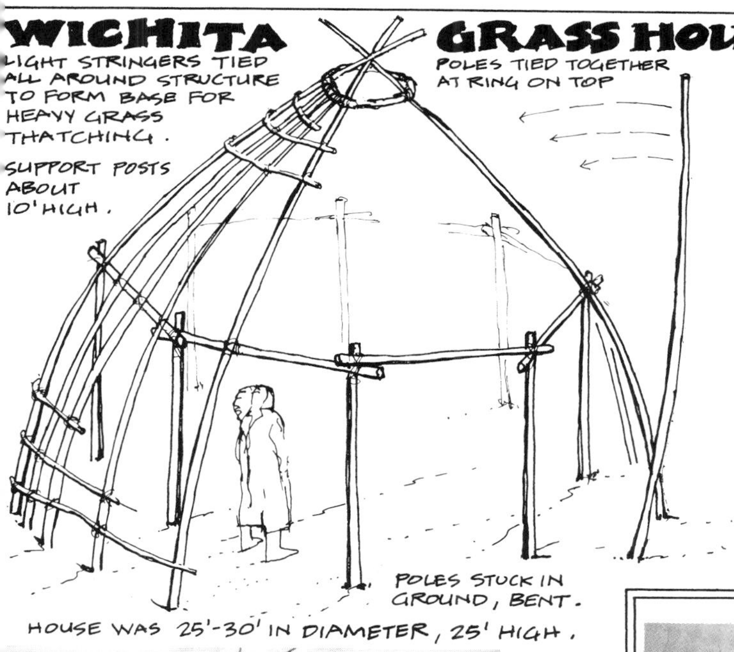

Apache wickeyup at Anadarko, Oklahoma

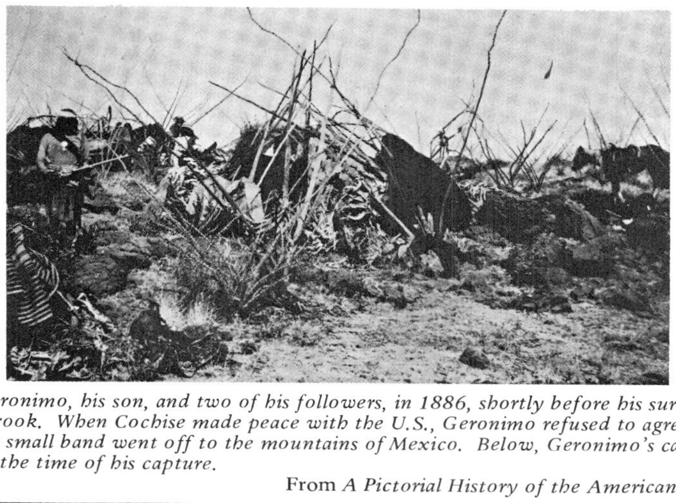

Above, Geronimo, his son, and two of his followers, in 1886, shortly before his surrender to General Crook. When Cochise made peace with the U.S., Geronimo refused to agree to it, and with a small band went off to the mountains of Mexico. Below, Geronimo's camp in Mexico at the time of his capture.

From *A Pictorial History of the American Indian*

PIMA ADOBE

PIMA BASKETS WERE WOVEN TIGHT ENOUGH TO HOLD WATER.

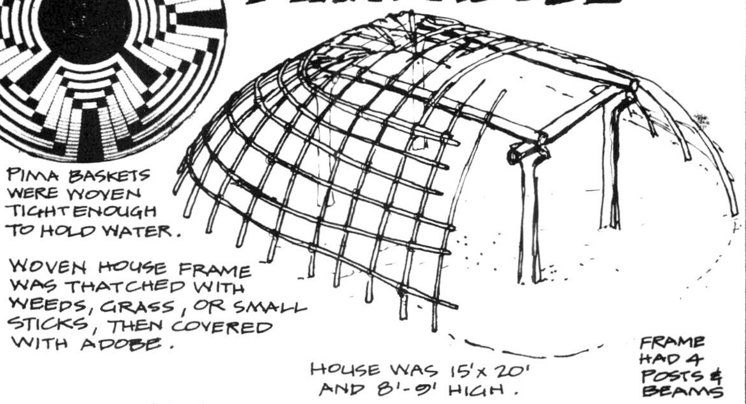

WOVEN HOUSE FRAME WAS THATCHED WITH WEEDS, GRASS, OR SMALL STICKS, THEN COVERED WITH ADOBE.
HOUSE WAS 15'x 20' AND 8'-9' HIGH.
FRAME HAD 4 POSTS & BEAMS

Living in country that is warm for most of the year, the Pimans wore little clothing. They built flat-roofed houses of poles—often digging out the ground inside—walled with cactus rods or brush and sometimes plastered with the clay called "adobe," which dries extremely hard...The Pimans, in their religion, paid a great deal of attention to visions which gave power, in which we see the old, shamanistic pattern. Their principal religious officials, however, worked their way up through a course of training and their major ceremonies required priestly control....

Pomo woman weaving latticed twine basket.

COYOTE & SILVER FOX

Jaime de Angulo

"Real Primitive People, not like those 'cultured' Indians of the Southwest..." wrote Jaime de Angulo of California's Pit River Indians. Born in Spain, reared in Paris, a doctor, cowboy and anthropologist, de Angulo spoke 17 different Indian languages and lived off and on with the Pit River Indians for over 40 years.

Of him, Bob Callahan writes:

Season after season he camped with Old Jack Folsom and Lena in the sage brush on the Plateau. His eye caught the detail of gambling games and healing rites, and his ear, always his ear, picked up the short, clipped cadences of the Pit River tongue. He never lost his fascination with that language. In de Angulo's later novels, David Olmsted writes, the Indians continue to speak "in sentences which are undeniably and perfectly grammatical Achumawi!"

In winter camp de Angulo began to translate the Dilasani qi, the old time stories of the Pit River People, the spirit history of the tribe. In the beginning was the Word...the stories, he felt, dated back into the furthest reaches of the stone age, were more ancient than myth. And the Word was with God...in these stories he felt he had found one of man's earliest attempts to make articulate the movement of the Spirit. And the Word was God..."The symbolism in these stories is so crude and so little disguised that it can't really be considered symbolism at all. In this early, primitive stage of civilization ideas are still immanent in objects, and have not yet been separated either through identification or projection. In these stories we find the Tinihowi — the primitive religious spirit — reflected throughout...and yet, the reader might ask, if the Pit River Indians have no religious ceremonies, no priesthood, no ritual of any kind, and not the slighest approach to any conception of Godhead, how can one speak of their having any spiritual or religious values? I grant that it may sound somewhat paradoxical, but I must answer on the contrary, the life of these Indians is nothing but a continuous religious experience...The spirit of wonder, the recognition of life as power, *as a mysterious, ubiquitous, concentrated form of nonmaterial energy, of something loose about the world and contained in a more or less condensed degree by every object — this is the credo of the Pit River Indians. Of course they would not put it precisely this way. The phraseology is mine, but it is not far from their own." Jaime de Angulo had rediscovered the Logos. Formed and transformed by a hundred Sierra mountain Homers, sung back and forth through these hills for thousands of years, the Dilasani qi were born that first morning. Dilasani qi. The Origin...*

From *Indians in Overalls:*

Wild Bill said he would stay here and wait for Jack Folsom and the rest of the party to come back from the *atsuge* country. That evening he told me a lot about Coyote and the Coyote saga. The Coyote stories form a regular cycle, a saga. This is true of all of California; and it extends eastward even as far as the Pueblos of Arizona and New Mexico. Coyote has a double personality. He is at once the Creator, and the Fool. This antinomy is very important. Unless you understand it you will miss the Indian psychology completely — at least you will miss the significance of their literature (because I call their tales, their "old-time stories," literature).

The wise man and the buffoon: the two aspects of Coyote, Coyote Old Man. Note that I don't call them the good and the evil, because that conception of morality does not seem to play much part in the Pit River attitude to life. Their mores are not much concerned with good and evil. You have a definite attitude toward moral right and moral wrong. I don't think the Pit River has. At least, if he has, he does not try to coerce. I have heard Indians say: "That's not right what he is doing, that fellow...." "What d'you mean it's not right?" "...Well...you ain't supposed to do things that way...it never was done that way...there'll be trouble." "Then why don't you stop him?" "Stop him? How can I stop him? It's his way."

The Pit Rivers (except the younger ones who have gone to the Government School at Fort Bidwell) don't ever seem to get a very clear conception of what you mean by the term God. This is true even of those who speak American fluently, like Wild Bill. He said to me: "What is this thing that the white people call God? They are always talking about it. It's goddam this and goddam that, and in the name of the god, and the god made the world. Who is that god, Doc? They say that Coyote is the Indian God, but if I say to them that God is Coyote, they get mad at me. Why?"

"Listen, Bill, tell me...Do the Indians think, really think that Coyote made the world? I mean, do they really think so? Do you really think so?"

"Why of course I do...Why not?...Anyway...that's what the old people always said...only they don't all tell the same story. Here is one way I heard it: it seems like there was nothing everywhere but a kind of fog. Fog and water mixed, they say, no land anywhere, and this here Silver Fox..."

"You mean Coyote?"

"No, no, I mean Silver Fox. Coyote comes later. You'll see, but right now, somewhere in the fog, they say, Silver Fox was wandering and feeling lonely. *Tsikuellaaduwi maandza tsikualaasa*. He was feeling lonely, the Silver Fox. I wish I would meet someone, he said to himself, the Silver Fox did. He was walking along in the fog. He met Coyote. 'I thought I was going to meet someone,' he said. The Coyote looked at him, but he didn't say anything. 'Where are you traveling?' says Fox. 'But where are YOU traveling? Why do you travel like that?' 'Because I am worried.'[1] 'I also am wandering,' said the Coyote, 'I also am worrying and traveling.' 'I thought I would meet someone, I thought I would meet someone. Let's you and I travel together. It's better for two people to be traveling together, that's what they always say....' "

"Wait a minute, Bill....Who said that?"

"The Fox said that. I don't know who he meant when he said: *that's what they always say.* It's funny, isn't it? How could he talk about *other* people since there had never been anybody before? I don't know....I wonder about that sometimes, myself. I have asked some of the old people and they say: That's what I have been wondering myself, but that's the way we have always heard it told. And then you hear the Paiutes tell it different! And our own people down the river, they also tell it a little bit different from us. Doc, maybe the whole thing just never happened....And maybe it did happen but everybody tells it different. People often do that, you know...."

"Well, go on with the story. You said that Fox had met Coyote...."

"Oh, yah....Well, this Coyote he says: 'What are we going to do now?' 'What do you think?' says Fox. 'I don't know,' says Coyote. 'Well then,' says Fox, 'I'll tell you: LET'S MAKE THE WORLD.' 'And how are we going to do that?' 'WE WILL SING ,' says the Fox.

"So, there they were singing up there in the sky. They were singing and stomping[2] and dancing around each other in a circle. Then the Fox he thought in his mind: CLUMP OF SOD, come!! That's the way he made it come: *by thinking.* Pretty soon he had it in his hands. And he was singing, all the while he had it in his hands. They were both singing and stomping. All of a sudden the Fox threw that clump of sod, that *tsapettia*, he threw it down into the clouds. 'Don't look down!' he said to the Coyote. 'Keep on singing! Shut your eyes, and keep them shut until I tell you.' So they kept on singing and stomping around each other in a circle for quite a while. Then the Fox said to the Coyote: 'Now, look down there. What do you see?' 'I see something...I see something... but I don't know what it is.' 'All right. Shut your eyes again!' Now they started singing and stomping again, and the Fox thought and wished: Stretch! Stretch! 'Now look down again. What do you see?' 'Oh! it's getting bigger!' 'Shut your eyes again and don't look down!' And they went on singing and stomping up there in the sky. 'Now look down again!' 'Oooh! Now it's big enough!' said the Coyote.

"That's the way they made the world, Doc. Then they both jumped down on it and they stretched it some more. Then they made mountains and valleys; they made trees and rock and everything. It took them a long time to do all that!"

"Didn't they make people, too?"

"No. Not people. Not Indians.[3] The Indians came much later after the world was spoiled by a crazy woman, Loon. But that's a long story....I'll tell you some day."

"All right, Bill, but tell me just one thing now: there was a world now; then there were a lot of animals living on it, but there were no people then...."

"Whad'you mean there were no people? Ain't animals people?"

"Yes, they are...but..."

"They are not Indians, but they are people, they are alive...Whad'you mean animal?"

"Well...how do you say 'animal' in Pit River?"

"...I dunno..."

"But suppose you wanted to say it?"

"Well...I guess I would say something like *teeqaade-wade toolol aakaadzi* (world-over, all living)...I guess that means animals, Doc."

"I don't see how, Bill. That means people, also. People are living, aren't they?"

"Sure they are! that's what I am telling you. Everything is living, even the rocks, even that bench you are sitting on. Somebody *made that bench for a purpose,* didn't he? Well then *it's alive,* isn't it? Everything is alive. That's what we Indians believe. White people think everything is dead...."

"Listen, Bill. How do you say 'people'?"

"I don't know...just *is*, I guess."

"I thought that meant 'Indian.'"

"Say...Ain't we *people*?!"

"So are the whites!"

"Like hell they are!! We call them *inillaaduwi*, 'tramps,' nothing but tramps. They don't believe anything is alive. They are dead themselves. I don't call that 'people.' They are smart, but they don't know anything....Say, it's getting late, Doc, I am getting sleepy. I guess I'll go out and sleep on top of the haystack...."

One of the last old style Pomo ceremonial lodges, used for Kuksu dances. Built in the late 1920's in Ione, southeast of Sacramento, Calif. Although the Pomos were mainly coastal, this was an offshoot tribe in the Sacramento area.

Pomo Shaman William Benson and wife. How the World was Made, *the Pomo Indian Creation Myth as told by Benson to de Angulo, will be published by Turtle Island Foundation.*

Indians in Overalls, *by Jaime de Angulo, is available in a $6 hard cover edition from Turtle Island Foundation, 2907 Bush St., San Francisco, Calif. 94115. Please write to publishers directly as the book is not available through local bookstores; enclose 15 cents postage and handling.*

Notes:

1. To be worried, *—insimallauw—* (conjugation II). When an Indian is worried, he goes wandering, *—inillaaduw—*. When he is "wandering" he goes around the mountains, cries, breaks pieces of wood, hurls stones. Some of his relatives may be watching him from afar, but they never come near.

2. Indian dancing is not like the European, by lifting the heels and balancing the body on the toes; on the contrary, one foot is raised *flat* from the ground while the other foot is pressed into the ground (by flexing the knee); then a very slight pause with one foot in the air; then the other foot is stamped flat into the ground while the first one is lifted. That is the fundamental idea; there are many variations; besides, the shoulders and head are made to synchronize or syncopate.

3. The word for "people" is *is*. Nowadays it is applied especially to Indians, in contradistinction to the term applied to the whites: *enellaaduwi*.

EARLY TIMBER

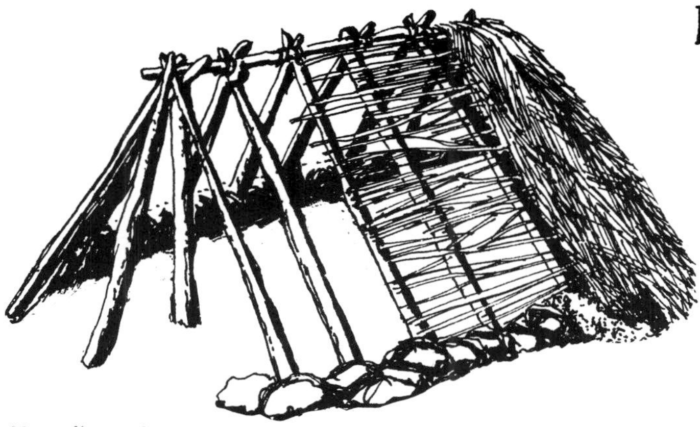

Nomadic Anglo-Saxon shelter.
After drawing from The
Family House in England.

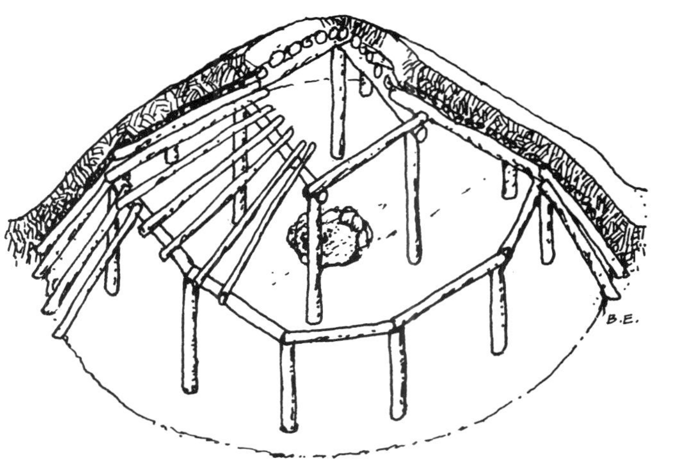

Circular earth lodge, with center supporting posts. With advent of agriculture, need to shelter crops and animals, it was expanded in neolithic times to...

Prehistoric man's wooden shelters were simple, crude, and suitable for his nomadic life. Generally round in shape and made of saplings lashed together, they did not require expert building skills. The transition from these early shelters to earth lodges, then elongated pole houses, then to mortise and tenon timber frames is covered in Medieval Structure: The Gothic Vault, *by James Acland. Following drawings (except as noted) and text are from this excellent book.*

When agricultural techniques and animal husbandry were introduced into Europe about 2500 BC, the early farmers soon found that they needed extra space to store grain and crops. Though the circular earth lodge sheltered animals as well as men, it was a form difficult to expand. Size was limited by the length of poles available for rafters. The Neolithic farmers met this new need by repeating the central square of supports used in the *earth lodge,* to create an elongated regular plan. The section through this long house of the north is identical to that through the earth lodge, but the repetitive bays allowed indefinite expansion for storage. The frame was, as in the past, made of lashed poles. Earth berms were banked up against the low walls of split logs or interwoven wattle. The builders set the butt ends of the long tapering poles into the berms and braced them over the two rows of posts. Over the rafters they placed a close mesh of light horizontal purlins to carry the thatch, turf or bark roof....

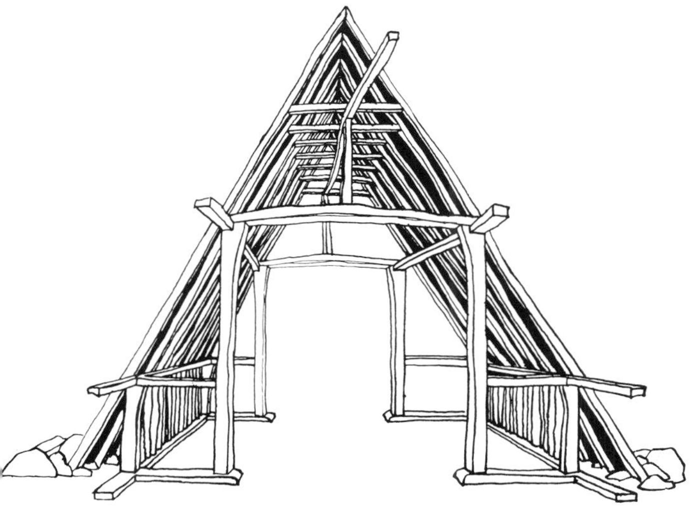

Early notched frame. From
The Family House in England.

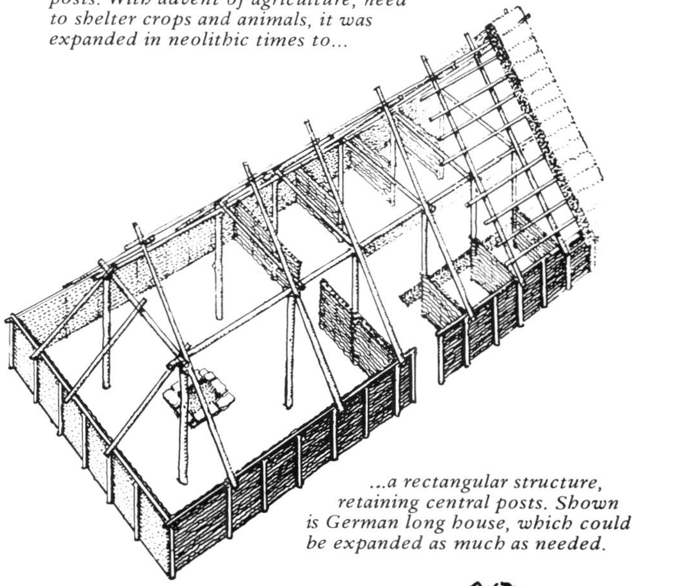

...a rectangular structure, retaining central posts. Shown is German long house, which could be expanded as much as needed.

The tightly lashed shell of poles carrying a protection of thatch or turf gave adequate insulation in the north but was subject to rapid decay. Moisture penetrated into the rafters, lashings rotted away, and the bark or sheathing became damp and mouldy in a short time. With only a central vent as a flue and one entry opening, the long house of the north tended to be dark, damp, smoky, and uncomfortable. The solution was to use the same basic plan but construct it of heavier timbers so that openings could be placed between the supports. To build a true frame of this type required better tools than the bronze axes and polished stone celts of the Neolithic. Only about 700 BC when iron tools became generally available in Europe, do we begin to see the emergence of carefully fitted framed structures in wood.

The peasant farmers exploited the new efficiency of iron axes and adzes to shape and model whole logs and develop a new form of structure. They set up vertical posts, cut with grooves into which were hammered tongued logs, and thus created a rigidly braced grid wall. No longer was it necessary to tie the building fabric together with complex and delicate lashings. Shaped timber connectors did the job better and were more lasting. Later builders learned to dispense with the heavy log or plank infills between the posts. They put up an open framed grid of timbers, mortised and tenoned at the joints to ensure rigidity. The apertures could then be left open as windows or doors or they could be filled with panels of woven wattle and daubed clay, effectively separating structural frame and sheathing skin....

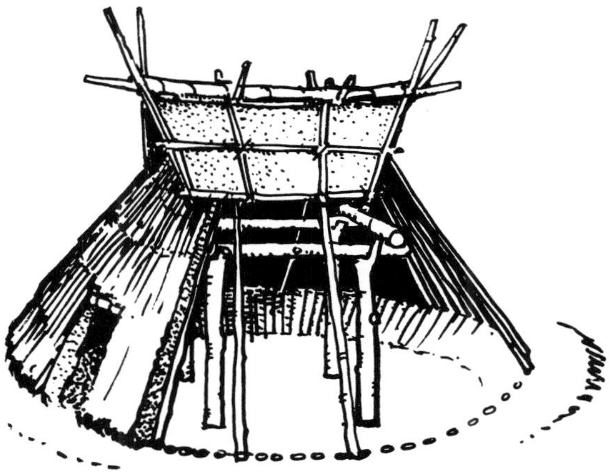

Neolithic northern Japanese shelter.

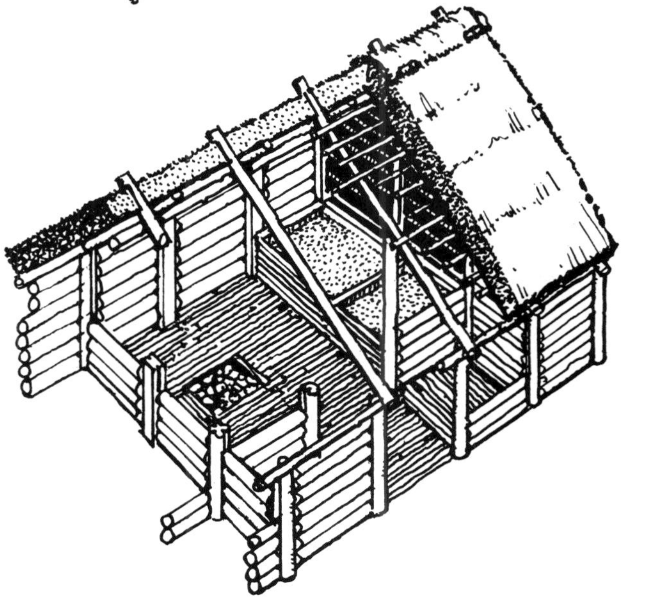

House in Poland, built in 700 B.C. with grooved posts to support tenoned logs.

During the long slow centuries of the early Middle Ages in northern Europe, wood remained the dominant building material. As the broadleaf and coniferous forests were cut to make way for agriculture, the bulk of timber, the pole, the plank, and the peg were combined and recombined with growing skill and sophistication to solve a host of technical problems.

The peremptory demands of incessant warfare led to a refinement of Roman siege machinery, with mangonel and catapult increasing in range and weight of projectile. The need for metals dug from ever deeper mines forced the invention of geared assemblies for hoisting and pumping. Animal, wind, and water power sources were pressed into service to crush, grind, and carry material. At the waterfront port, improved derricks and hoists were devised to load the ships, while shipwrights embarked upon that long series of refinements and adjustments which converted the long boat of the north to a serviceable oceangoing merchant craft, the Hansa cog. In this dawn of European technology the carpenter was in the forefront of technical invention, using cordage, wooden poles, and timber gearing to control and manipulate energy in response to the demands of peace and war.

Before AD 1000 court and castle, monastery and abbey, town and fortified wall were rough timber constructions. After AD 1000 this new technical skill began to be reflected in the increasing scale and technical artistry of building. The timber-framed structure became a carefully interlocked rigid frame suitable for monumental structures....

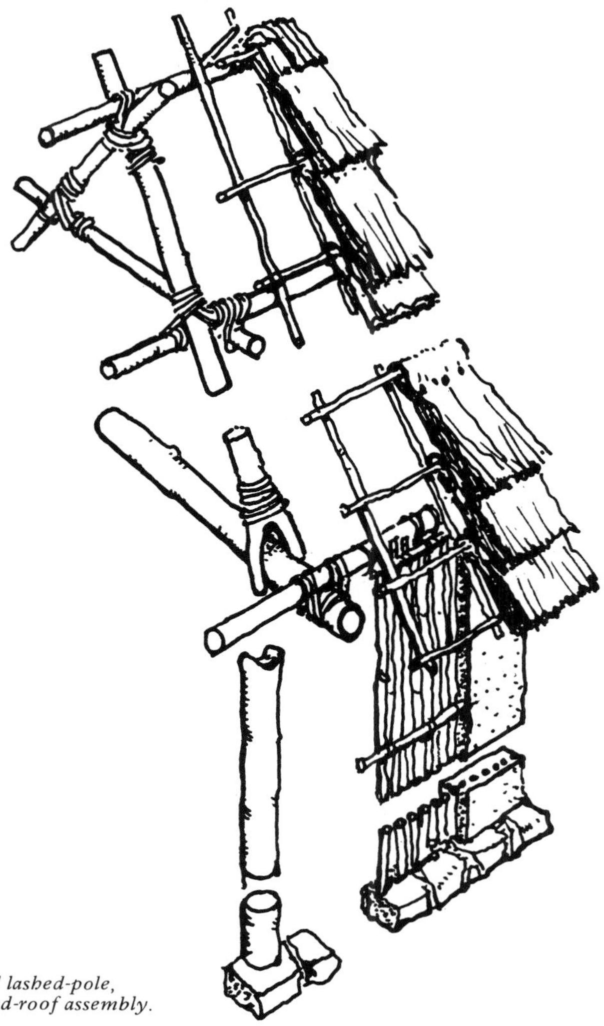

Typical lashed-pole,
thatched-roof assembly.

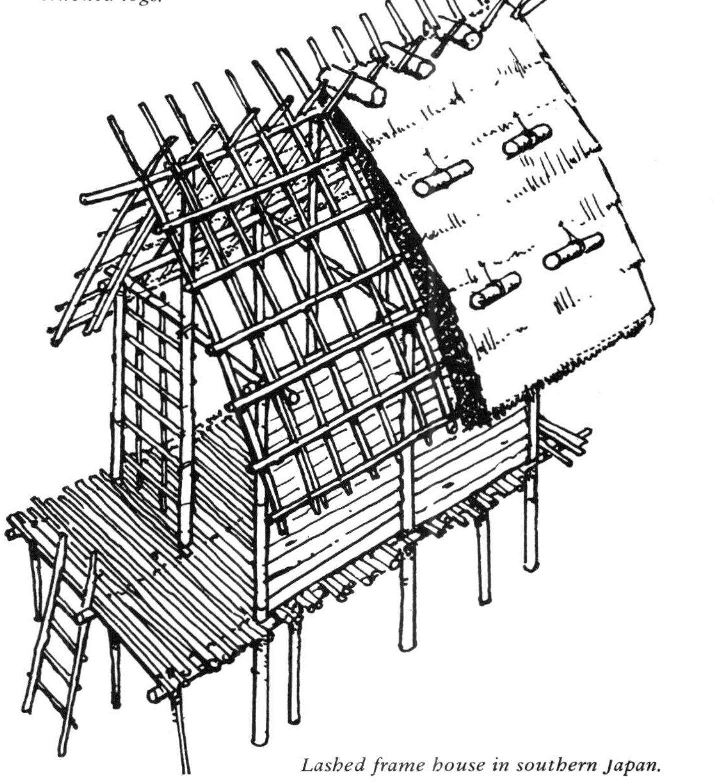

Lashed frame house in southern Japan.

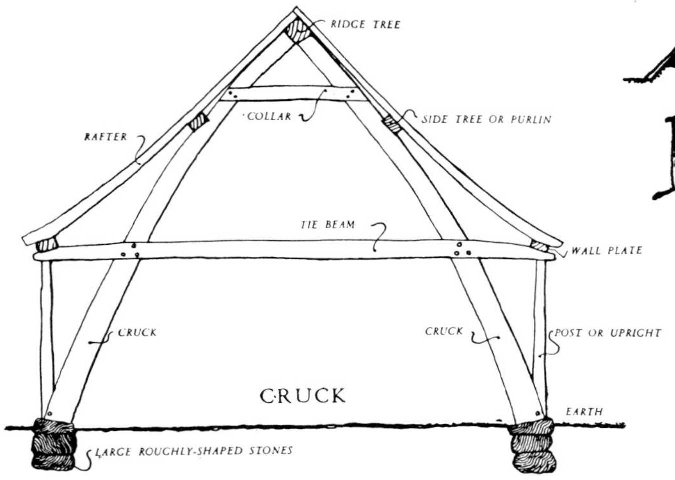

Development of the Cruck Frame

ENGLISH COTTAGE FRAME

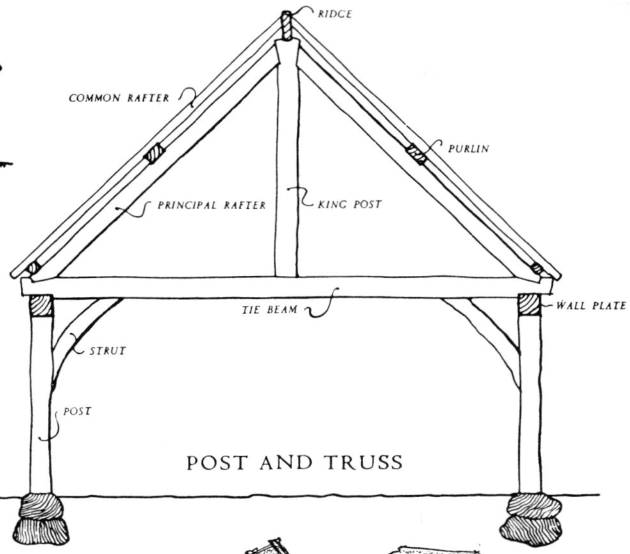

In early wooden structures there was no distinction between walls and roof. Frameworks consisted of tree trunks rammed a foot or so into the ground [their ends having been charred to prevent rot], lashed together at the top, supported at each end by forked uprights. A light ridge pole was lashed to the rafter poles at their apex for horizontal support. These structures were basically isoceles triangles.

From this was developed, in England, what is known as *cruck* construction. Crucks are pairs of naturally curved trees, split in half, used as opposing rafters. Resembling an upturned boat, these structures had sloping walls and little interior headroom. Each pair of crucks was usually braced by a cross beam or other type brace for rigidity.

As cruck construction developed, and houses became larger, these cross beams were extended to a point vertically above the base of the crucks, and connected to the base by posts. This gave the roof greater spread, and occupants more interior space.

Gradually cruck construction gave way to more developed timber building. The post and truss method, in which timbers of wall and roof were separate allowed for a much greater variety of form.

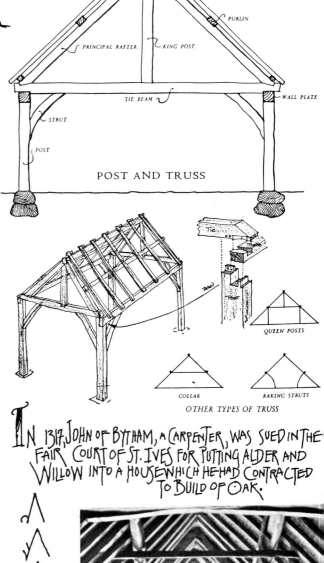

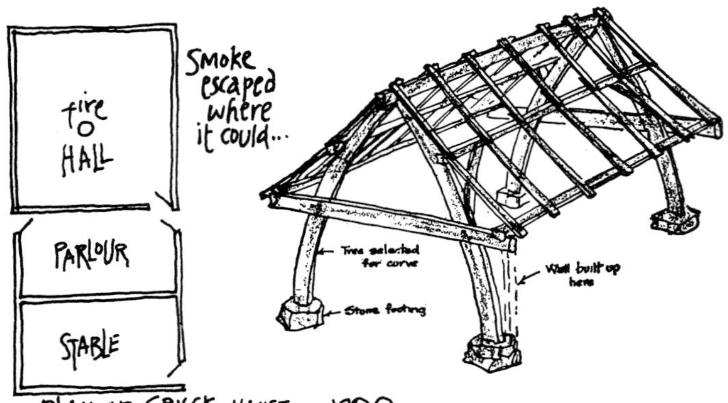

PLAN OF CRUCK HOUSE c.1200

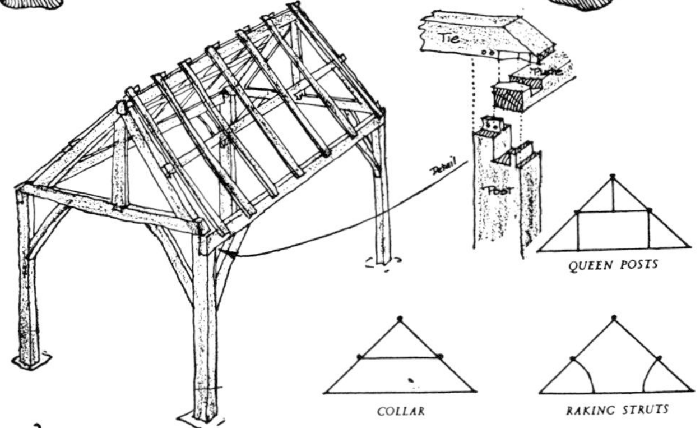

OTHER TYPES OF TRUSS

The Black Death of 1349 was the indirect cause of a revolution in domestic building at all levels. Almost overnight the heavy mortality rate caused an acute shortage of labour, thereby doubling its market value. The farmer, left with no one to work his land, changed from arable farming to pasturage for sheep and unwittingly laid the foundation of his own future wealth and the general prosperity of the country. Within a short time the wool trade was booming and sheep farmers, weavers, and merchants became rich, many of them becoming free men for the first time in their lives. Wealth spread rapidly in the fifteenth century and there was a great social upheaval in England, followed by many changes. The new middle classes emerged with a great desire to better themselves and to build houses which would bear comparison with that of the lord of the manor, so by the first decades of the sixteenth century there were innumerable owners of houses between the wealthy nobleman and the lowly cottager.

The Timber-frame House in England

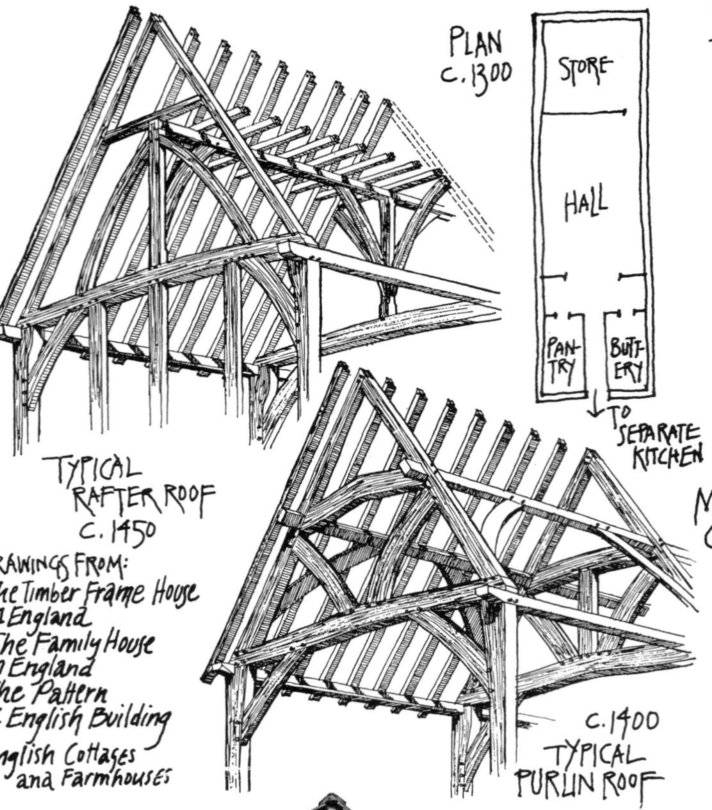

TYPICAL RAFTER ROOF c.1450

C.1400 TYPICAL PURLIN ROOF

DRAWINGS FROM:
- The Timber Frame House in England
- The Family House in England
- The Pattern of English Building
- English Cottages and Farmhouses

IN 1317, JOHN OF BYTHAM, A CARPENTER, WAS SUED IN THE FAIR COURT OF ST. IVES FOR PUTTING ALDER AND WILLOW INTO A HOUSE WHICH HE HAD CONTRACTED TO BUILD OF OAK.

MEDIÆVAL CARPENTERS MARKS

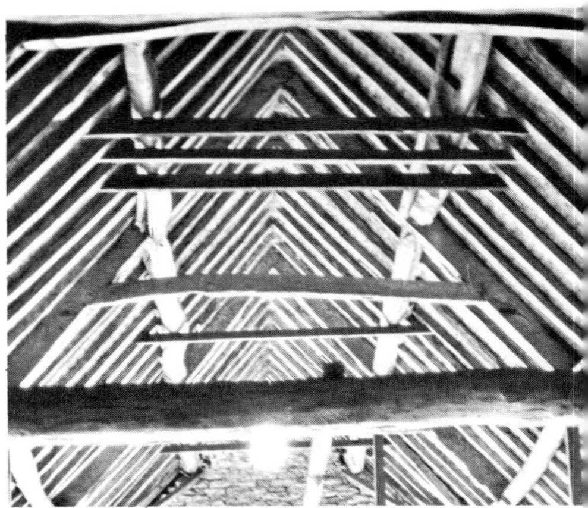

Stone barn in the Cotswalds, Interior.

Wellhouse, reconstructed framework c.1600. at Weald Open Air Museum, Sussex.

Bayleaf farmhouse Interior.

Bayleaf Farmhouse during reconstruction. Originally built c.1450. Weald Open Air Museum, Sussex.

Half Timber, wattle and daub house reconstructed at Avoncroft Museum of buildings, Worcestershire. Originally built before the 16th Century.

Interior of same.

Kitchen, Furlongs, Sussex. English Cottages and Farmhouses

THE SAD FATE OF MANY ENGLISH VILLAGES TODAY; CITY PEOPLE BUY COTTAGES FOR WEEKENDS AND HOLIDAYS, BRING FOOD AND GOODS WITH THEM FROM THE CITY, DON'T PATRONIZE THE LOCAL MERCHANTS. THE VILLAGE SUFFERS...

Driving down a small country lane in Norfolk, this building in the middle of a distant field. A peaceful *presence*, at rest with its surroundings.
After hundreds of years, abandoned, sinking gracefully back into the landscape its materials originally came from.
Vines climbing up the walls, in through the windows. Soon it will fold in the middle, kneel, in time become a mound in the field. Cycle completed.
Inside it's cheerful and light, unlike many other dank abandoned houses where death lingers.
Generations of shelter, births and deaths, sons and daughters. Countless fires built, meals cooked, needs tended.
What will the houses we are building now look like in 300 years?

Market Hall frame reconstructed. c1600. Weald, Sussex.

Inside every Englishman is a villager struggling to get out with such success that scarcely a dank hovel remains in the remotest hamlet that has not by now been turned into a gadget-happy country cottage, a Jaguar resting where the privy used to be. Even when the imprisoned cottager can't break out he shuts his eyes and, deep in the heart of Bermondsey or Brum, pretends to be a cottager....

Elspeth Huxley, The Sunday Times, 15 December 1968

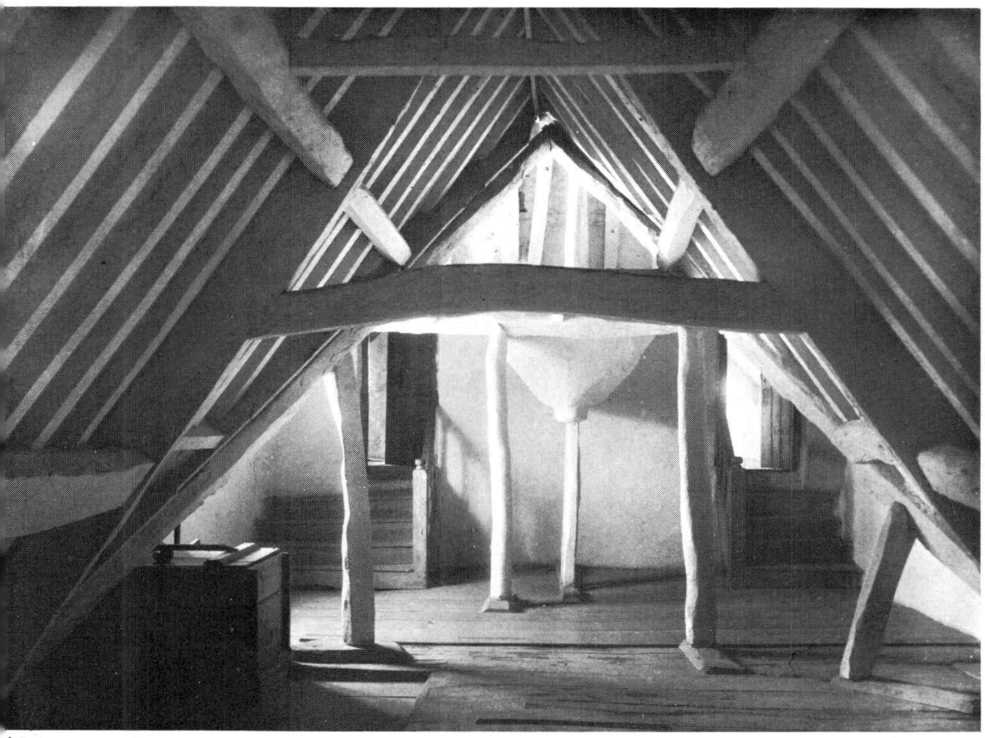

Kelmscott Manor: Attics, c.1897 Frederick H. Evans

Limetree Cottage, East Hagborn, Berkshire English Cottages and Farmhouses

YUGOSLAVIA

JAPAN

EASTERN EUROPE

NORWAY

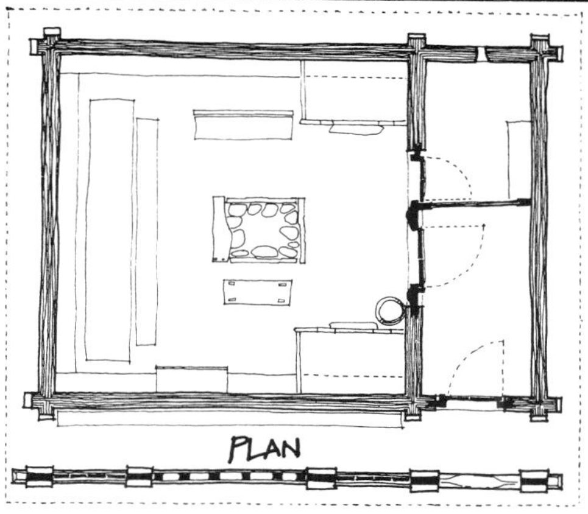

PLAN

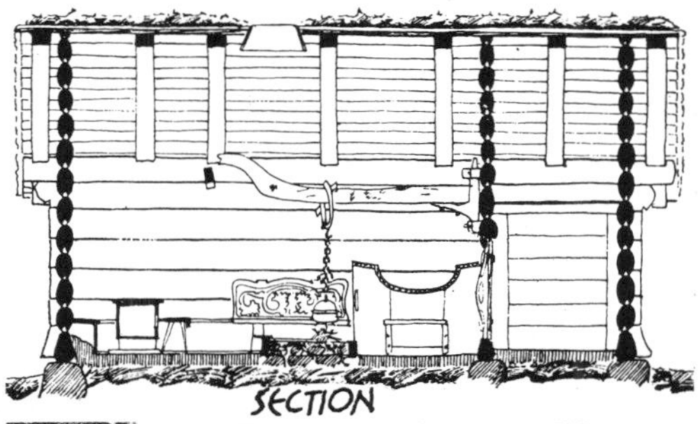

SECTION

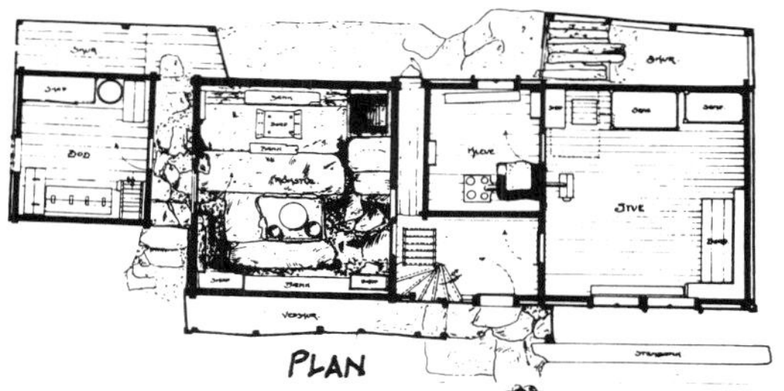

PLAN

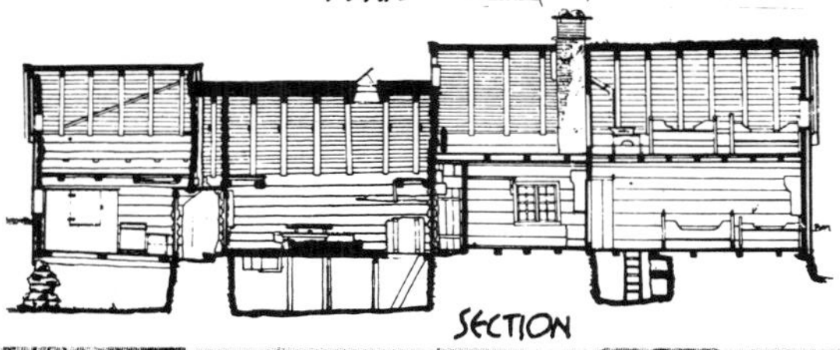

SECTION

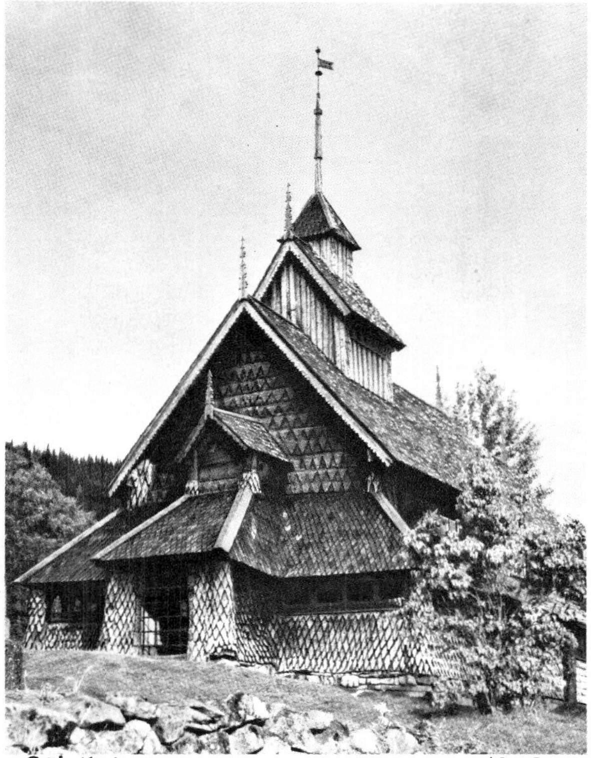

STAVE CHURCH. EIDSBORG, TELEMARK c. 1200

Along the fjords and in the hidden valleys of Norway stand the lofty, monumental wooden stave churches. The Norwegian term *stav* refers to the vertically placed structural timbers of these 11th and 12th century buildings, built by Viking boat builders when King Olaf spread Christianity. Yet some of the people did not always accept the new religion willingly, and the old myths — with dragons and epic heroes — persisted, and can be seen in the churches' rich carvings and decorations. Of the seven or eight hundred stave churches originally built, some 20 are still standing in Norway.

From travels into Russia the Norsemen learned log construction techniques, and since medieval times, folk building — homes and farm buildings — has consisted of combined log and stave construction, as shown in these small buildings.

The best material was fir wood rich in resin. The top of the tree was cut off, and it was left on its root for two years before being used in a building. The best trunks were reserved for the cottages and cow-stables. Log-construction demanded the highest skill, and the numerous terms employed in connection with the technique bear witness to a specialized and refined craftsmanship.

From *Stav og Laft*

URNES, SOGN c. 1150

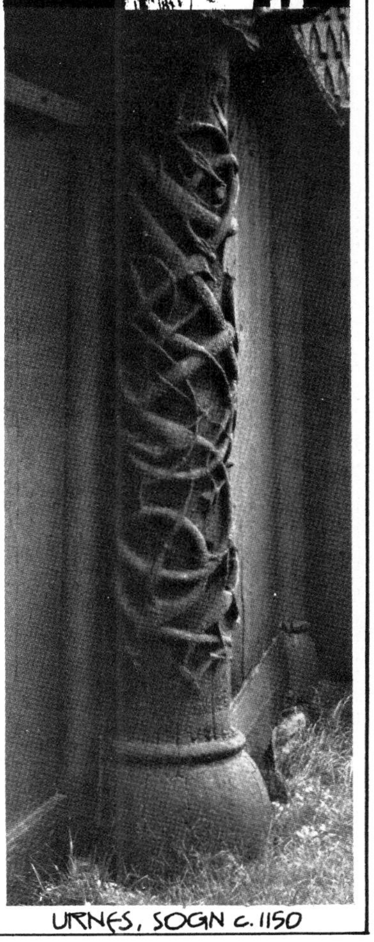

URNES, SOGN c. 1150

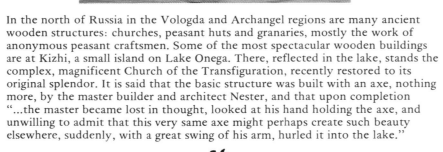

In the north of Russia in the Vologda and Archangel regions are many ancient wooden structures: churches, peasant huts and granaries, mostly the work of anonymous peasant craftsmen. Some of the most spectacular wooden buildings are at Kizhi, a small island on Lake Onega. There, reflected in the lake, stands the complex, magnificent Church of the Transfiguration, recently restored to its original splendor. It is said that the basic structure was built with an axe, nothing more, by the master builder and architect Nester, and that upon completion "...the master became lost in thought, looked at his hand holding the axe, and unwilling to admit that this very same axe might perhaps create such beauty elsewhere, suddenly, with a great swing of his arm, hurled it into the lake."

THE NEW WORLD

Pilgrim house.

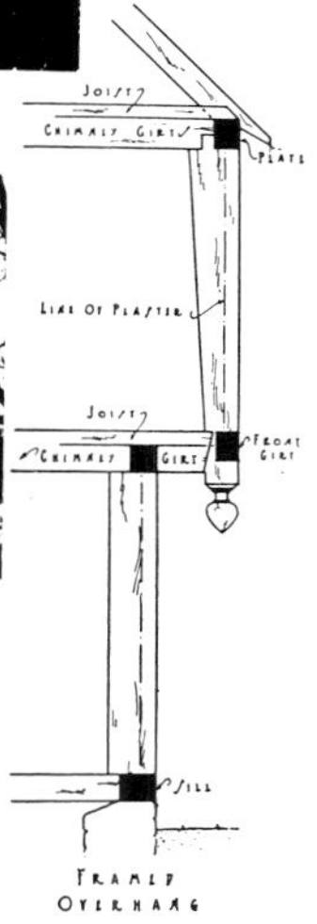

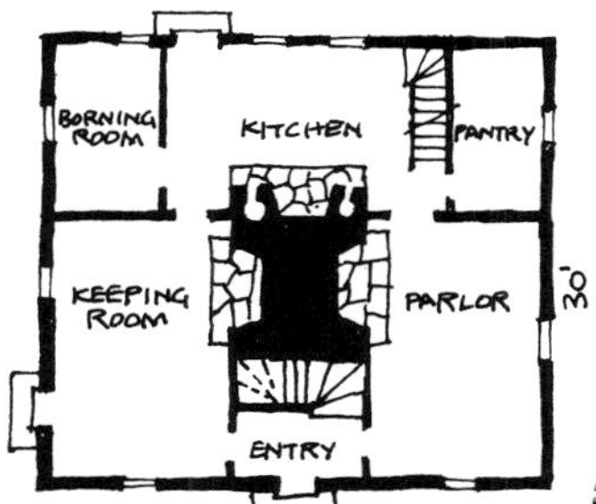

Saltbox floor plan.

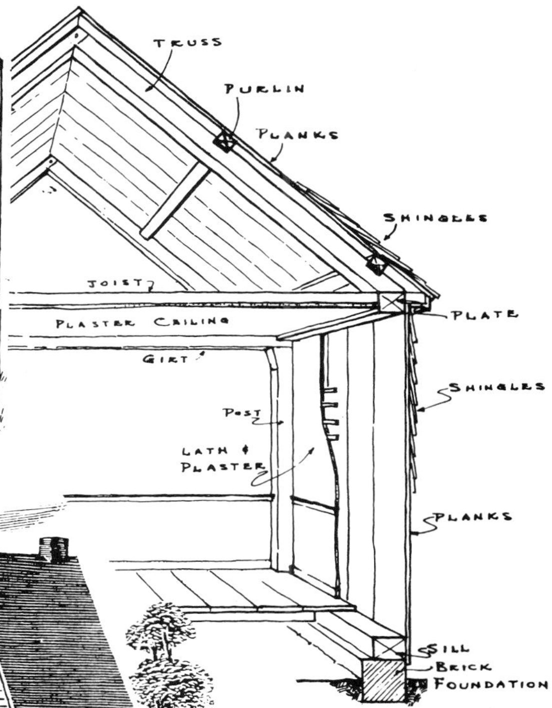

Typical plank construction observed in traditional houses of Lower Cape Cod. From Walter R. Nelson in *Pioneer America.*

Early New England houses.

The Peake house, c. 1700.

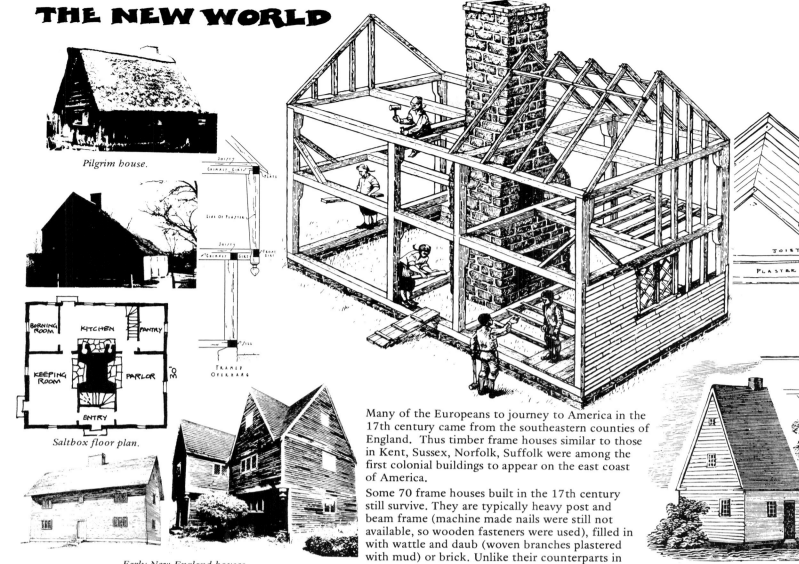

Many of the Europeans to journey to America in the 17th century came from the southeastern counties of England. Thus timber frame houses similar to those in Kent, Sussex, Norfolk, Suffolk were among the first colonial buildings to appear on the east coast of America.

Some 70 frame houses built in the 17th century still survive. They are typically heavy post and beam frame (machine made nails were still not available, so wooden fasteners were used), filled in with wattle and daub (woven branches plastered with mud) or brick. Unlike their counterparts in England, these structures were then covered with wooden clapboards as protection against severe weather. These early colonial buildings were simple, with little architectural ornament.

The pilgrims actually first landed in Provincetown, Cape Cod, before Plymouth. One of the first things they did, according to folks in Provincetown, was to steal the Indians' corn crop.

The great man wanted only a little, little land, on which to raise greens for his soup, just as much as a bullock's hide would cover. Here we first might have observed their deceitful spirit.

Delaware Indian view of the first arrival of the Dutch at Manhattan Island, about 1609.

...The era of the authentic Cape Cod cottage ended around 1850 when the advent of the stove eliminated the massive chimney block that had previously anchored the house to its site. At that time, home builders also had to import precut lumber from Maine. Thoreau explained the reason, "The old houses...are built of the timber of the Cape, but instead of the forests in the midst of which they originally stood, barren heaths...now stretch away on every side."

Notable American Houses.

LOG CABINS

Nels Wickstrom's log cabin, Florence County, Wisconsin, 1893.

...Simple, one-room dwellings of logs, notched together at the corners, were introduced to America around 1638 by Swedish settlers in Delaware. Subsequently, German and Scotch-Irish immigrants, as well as Russian explorers along the western coast and in Alaska, introduced their own forms of log construction. During the great westward expansion that began in the late 1700s, the log cabin was practically ubiquitous in timber-rich frontier areas; it could be built with only the aid of an axe, and required no costly nails. Intended to serve merely as way stations in the wilderness, cabins rarely became permanent homes. When families desired better housing with more amenities, they either abandoned their cabins — often to be occupied by new transients — incorporated them into larger dwellings, converted them into storage facilities, or in the South, used them as slave quarters. The myth of the log cabin as the sacrosanct birthplace of leaders, renowned for their honesty, humility, and other virtues, was inaugurated during the presidential campaign of 1840, when William Henry Harrison was touted throughout the country as a hard-cider swigging bumpkin who lived in a log cabin....

Notable American Houses

Hear ye, Dakotas! When the great father at Washington sent us his chief soldier [Major General William S. Harney] to ask for a path through our hunting grounds, a way for his iron road to the mountains and the western sea, we were told that they wished merely to pass through our country, not to tarry among us, but to seek for gold in the far west. Our old chiefs thought to show their friendship and good will, when they allowed this dangerous snake in our midst...

Yet before the ashes of the council fire are cold, the Great Father is building his forts among us. You have heard the sound of the white soldier's axe upon the Little Piney. His presence here is an insult and a threat. It is an insult to the spirits of our ancestors. Are we then to give up their sacred graves to be plowed for corn? Dakotas, I am for war!

Red Cloud, Oglala Sioux Cheif, 1866

J.C. Cram sod house, Loup County, Nebr., 1886

Inasmuch as there is nothing in bare earth to sell, no commercial group can be found to extol its merits.

Ken Kern

Adobe building near Truchas, New Mexico. Walls have been plastered.

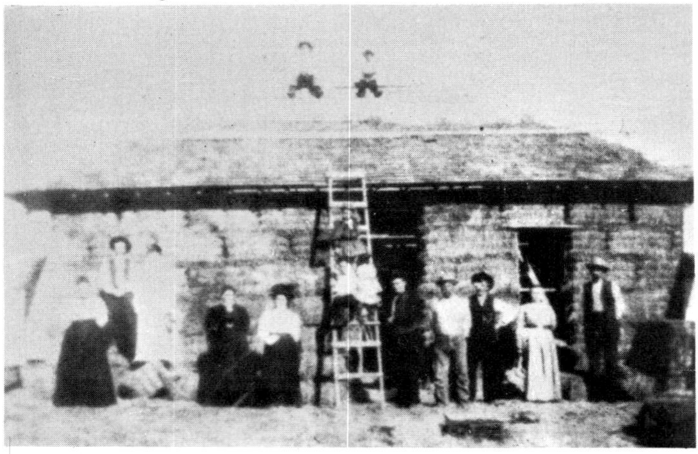

Simonton Purdum baled straw house, Nebraska sandhills.

Further information:
Adobe, pp. 43,66,67
Baled Hay and Sod, pp. 70-71

ADOBE

The most abundant building material known, the earth around us, has been used for house construction since early times. Adobe is perhaps the most popular and the oldest form of earth construction. Adobe blocks are shaped of wet mud and various additives, allowed to dry in the sun, then used to build a wall. Mortar used in holding the blocks together is often the same mud. Adobe houses have worked best in areas with little rainfall, as a wet climate is the worst enemy of an earth house.

The traditional adobe houses of New Mexico were built on well drained sites, laid on stone foundations to prevent moisture entering by capillary action. Walls were thick, both because adobe is structurally weak and for insulation. Roofs were slightly canted for drainage, and supported by pine pole rafters, called *vigas*. Wall heights were low, and wall spans short unless buttressed. Doors and windows were placed away from corners.

Indian pueblo builders allowed the vigas to project, never cutting them, so they could be used over and over. Some vigas in Hopi pueblos are 900 years old.

BALED HAY

In 1904 new lands in northwestern Nebraska were opened for homesteading under the provisions of a law authored by Moses Kinkaid, a Nebraska congressman. A substantial portion of the Kinkaid lands were the Nebraska Sandhills – a vast, desolate tract of grass-covered dunes. The Sandhills magnified the difficulties of the previous homestead lands: they were even more barren of trees and the weather was even more hostile. Furthermore, the sandy soil made poor construction sod, for if it did not disintegrate during cutting and handling, it would soon crumble after being laid up in walls. The oxen, which were preferred for sod cutting, had been traded for horses, preferred for general farmwork and transportation; the plows for cutting were gone; and, to some degree, the skills for sod construction had been forgotten. Again it was necessary to devise new techniques and materials for home building. Wild grasses and domestic hays were [and are] the most important crops of the Sandhills. They were cut and stacked for use on the same or neighboring farm, but if the hay was to be moved it was necessary to bale it. Balers were first introduced in the 1850's and by the 1890's hay presses were in general use. Indeed, by the '90's railroads refused to handle loose hay. It was then perhaps inevitable that some settler, desperate for a cheap, available building material, would eventually see the big, solid, hay blocks as a possibility. Soon baled hay was indeed a significant construction material – never reaching the importance, scope, and technology of sod, but widely known and used throughout the Kinkaid Sandhills.

Roger L. Welsch

SODDIES

European settlers who crossed the Missouri River came upon a barren, harsh land – what Lewis & Clark called the Great American Desert. Unlike their native countries where building materials – rock or wood – were abundant, there were few trees in the north central states, little rock, and though there was clay for making brick, there was no fuel to fire them.

...Since the acquisition of homestead lands required construction of a permanent dwelling within a limited time, would-be settlers had to develop some technique for home building: sod.

The Mormons, when they first crossed into Nebraska in 1846, found the Omaha and Pawnee Indians living in great earth lodges – heavy timber frames with a sod covering, and since the Mormons also built sod houses during their first winter here, it may well be that they picked up the idea from the Indians, while they were also bartering for food and animals. On the other hand, the Mormons and later settlers may have borrowed the concept from English or European antecedents, low turf and stone huts used especially for temporary field-housing.

Whatever the origin, the technique developed rapidly and spread throughout the region, where it dominated for fifty years. A special plow, the "grasshopper" or cutting plow, was developed to cut the sod cleanly and lift it gently from the earth; the turning plow had been designed after all to destroy sod, tumbling and breaking it. From crude dugouts the soddies developed into fine, permanent houses, many of which are still in use today. Indeed centuries of grass-roots grasping the sticky loess soil proved to be a superior construction material, supplanted by lumber only when it became prestigous to have a frame house – a sure sign of affluence because it cost so much to haul in the lumber and it took a fortune to heat the vastly inferior frame house...

Roger L. Welsch

Frame construction, in the meantime, had undergone a significant change. For hundreds – even thousands – of years men had framed their wooden buildings of heavy timbers, often more than a foot square, that were mortised, tenoned, and pegged together and then raised into position by group labor. In the middle years of the century a radical new method called "balloon framing" was evolved; a method of construction using light two-by-four studs nailed rather than joined together in close, basket-like manner, the studs rising continuously from foundation to rafters. Uninjured by mortise or tenon, with every strain coming in the direction of the fiber of some portion of the wood, the numerous, light sticks of the structure formed a fragile-looking skeleton that was actually exceptionally strong. As one contemporary observed, since no mysteries of carpentry were involved in such construction, houses could be thrown up by relatively inexperienced labor in quick time. The method had had to wait until cheap, machine-made nails were available in quantity, which they generally were as early as the 1830's. Balloon-frame houses were more than makeshifts; the method had been generally used throughout the country ever since it was first conceived. In passing, it should be remembered that without its time— and labor—saving advantages the mushroom cities of the West could never have risen as fast as they did.

Notable American Houses

Barn-raising of mortise and tenon frame at farm of Jacob Rober, Massillon, Ohio, 1888.

STUDS

The following is from The Construction of the Small House, *by H. Vandervoort Walsh, published in 1923 by Scribners. Mr. Walsh was Instructor of Construction, School of Architecture, Columbia University.*

It was only a matter of time when the thinning-down process began to make itself evident in the traditions of Colonial carpentry, and from its clumsy beginnings it evolved into the more or less standard form of construction which we call the brace-frame.

The difficulty of securing good labor in the West, and also the increasing use of the power sawmill, made it possible and necessary to standardize a quick and easy method of building which would meet the great demand for houses in rapidly growing communities.

Quoting from the New York *Tribune* of January 18, 1855, we have a very interesting account of the conditions which were then prevalent that brought about this later variation of the wooden-frame structure. The conditions there described seem almost like our modern difficulties with labor and materials.

"Mr. Robinson said: ... I would saw all my timbers for a frame house, or ordinary frame outbuilding, of the following dimensions: 2 x 8 inches; 2 x 4; 2 x 1. I have, however, built them, when I lived on the Grand Prairie of Indiana, many miles from sawmills, nearly all of split and hewed stuff, making use of rails or round poles, reduced to straight lines and even thickness on two sides, for studs and rafters. But sawed stuff is much the easiest, though in a timber country the other is far the cheapest. First, level your foundation, and lay down two of the 2x8 pieces, flatwise, for side-walls. Upon these set the floor-sleepers, on edge, 32 inches apart. Fasten one at each end, and perhaps one or two in the middle, if the building is large, with a wooden pin. These end-sleepers are the end-sills. Now lay the floor, unless you design to have one that would be likely to be injured by the weather before you get on the roof. It is a great saving, though, of labor to begin at the bottom of a house and build up. In laying the floor first, you have no studs to cut and fit around, and can let your boards run out over the ends, just as it happens, and afterward saw them off smooth by the sill. Now set up a corner-post, which is nothing but one of the 2x4 studs, fastening the bottom by four nails; make it plumb, and stay it each way. Set another at the other corner, and then mark off your door and window places and set up the side-studs and put in the frames. Fill up with studs between, 16 inches apart, supporting the top by a line or strip of board from corner to corner, or stayed studs between. Now cover that side with rough sheeting boards, unless you intend to side-up with clapboards on the studs, which I never would do, except for a small, common building....

...The rafters, if supported so as not to be over 10 feet long, will be strong enough of the 2x4 stuff. Bevel the ends and nail fast to the joist. Then there is no strain upon the sides by the weight of the roof, which may be covered with shingles or other materials – the cheapest being composition or cement roofs. To make one of this kind, take soft, spongy, thick paper, and tack it upon the boards in courses like shingles. Commence at the top with hot tar and saturate the paper, upon which sift evenly fine gravel, pressing it in while hot – that is, while tar and gravel are both hot. One coat will make a tight roof; two coats will make it more durable. Put up your partitions of stiff 1x4, unless where you want to support the upper joist – then use stiff 2x4, with strips nailed on top, for the joist to rest upon, fastening all together by nails, wherever timbers touch. Thus you will have a frame without a tenon or mortise, or brace, and yet it is far cheaper, and incalculably stronger when finished, than though it were composed of timbers 10 inches square, with a thousand auger holes and a hundred days' work with the chisel and adze, making holes and pins to fill them.

"To lay out and frame a building so that all its parts will come together requires the skill of a master mechanic, and a host of men and a deal of hard work to lift the great sticks of timber into position. To erect a balloon building requires about as much mechanical skill as it does to build a board fence. Any farmer who is handy with the saw, iron square, and hammer, with one of his boys or a common laborer to assist him, can go to work and put up a frame for an outbuilding, and finish it off with his own labor, just as well as to hire a carpenter to score and hew great oak sticks and fill them full of mortises, all by the science of the 'square rule.' It is a waste of labor that we should all lend our aid to put a stop to. Besides, it will enable many a farmer to improve his place with new buildings, who, though he has long needed them has shuddered at the thought of cutting down half of the best trees in his wood-lot, and then giving half a year's work to hauling it home and paying for what I do know is the wholly useless labor of framing. If it had not been for the knowledge of balloon frames, Chicago and San Francisco could never have arisen, as they did, from little villages to great cities in a single year. It is not alone city buildings, which are supported by one another, that may be thus erected, but those upon the open prairie, where the wind has a sweep from Mackinaw to the Mississippi, for there they are built, and stand as firm as any of the old frames of New England, with posts and beams 16 inches square."

The above address, which was delivered before the American Institute Farmers' Club, has been quoted in detail because of the interesting point of view of the days of 1855 which it reveals. When Mr. Robinson had finished there were other comments, especially one by Mr. Youmans, in which he described early conditions of building in San Francisco. He also said that he had adopted this plan of building on his farm in Saratoga County, where he found great difficulty in getting carpenters that would do as he wished. They could not give up tenons and mortises, and braces and big timbers, for the light ribs, 2 x 4 inches, of a balloon frame. Does this not remind the modern reader of comments he has heard upon all sides these days concerning labor which will not do what is wanted but insists on doing things in the old way?

Stud frame buildings built in Guthrie, after the 1889 noontime land run in Oklahoma.

San Francisco Victorians, stud frame.

More on studs: pp. 40-45

Great Coxwell barn.

GREAT TIMBER BARNS

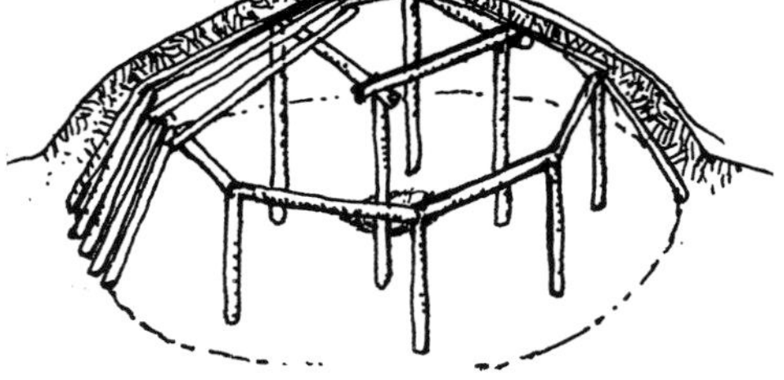

Circular earth lodge, with central square of posts was difficult to expand. Thus the round shape evolved into...

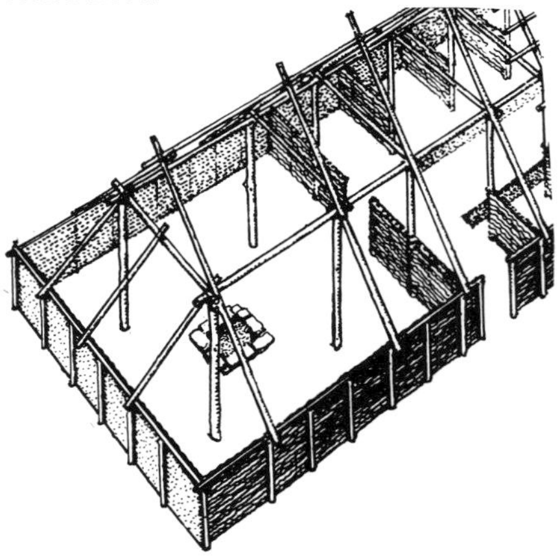

...rectangular farmhouse with the same interior square of posts and plates to support roof. This was the forerunner of the aisled and bay-divided *timber barn.*

Great Coxwell, Berkshire, England.

Tisbury, 15th century.

Great Coxwell.

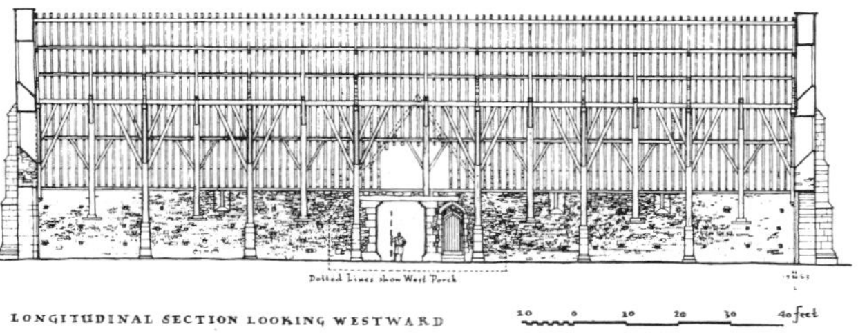

LONGITUDINAL SECTION LOOKING WESTWARD

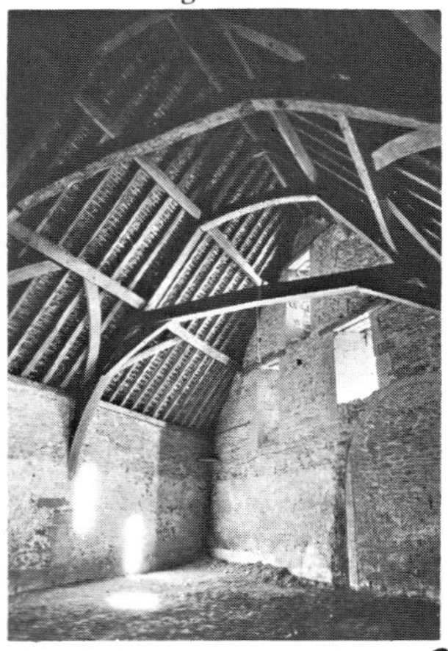

English barn.

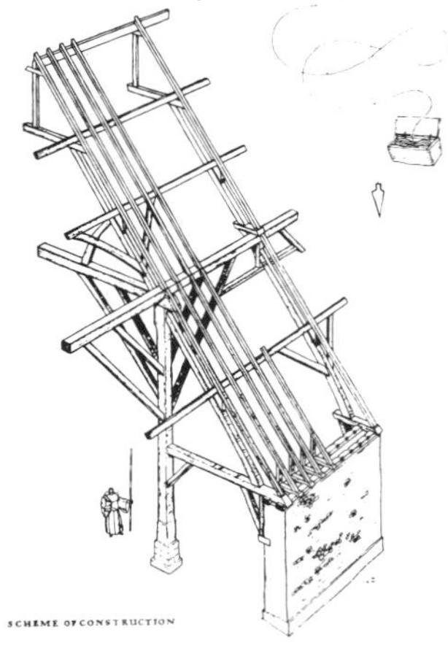

Beaulieu, St. Leonard.

SCHEME OF CONSTRUCTION

The cultural and economic significance of this building type has only recently begun to be understood. From prehistoric times through the Middle Ages this "all-purpose house" displayed an amazing functional versatility.

It was by origin the product of an agricultural economy and owes its incredible spread in space and time to its ability to respond to the needs of communal forms of living, to provide for large halls of assembly and for massive storage spaces. Before the invention of steel and ferroconcrete, it was the most logical and efficient way to create spacious shelters with a minimum outlay of materials. For this reason it was used from the time of its inception as an all-purpose structure wherever large protective roofs were needed: as dwelling for the chief and his clan, as shelter for the farmer and his cattle, or as place of storage for the harvest. It became in the medieval manor and palace hall, the principal seat of feudal administration. It served as church, as guest house, as hospital and as market hall....

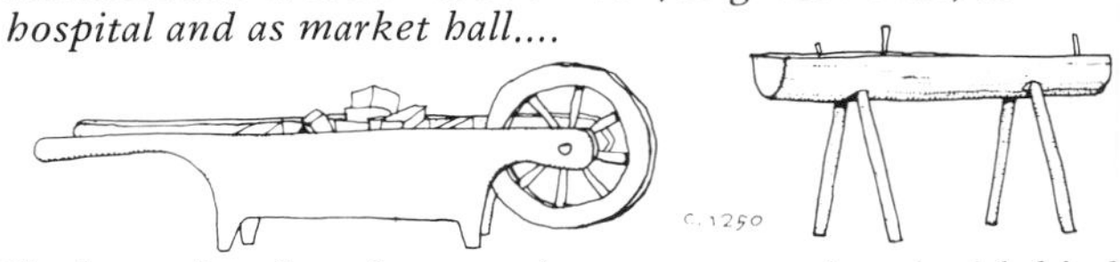

The large sketch and quote above are reproduced with kind permission of the authors and publishers of a unique, beautiful book, unfortunately out of print: *THE BARNS of The Abbey of Beaulieu at Its Granges of Great Coxwell & Beaulieu St. Leonards* by Walter Horn and Ernest Born. Published by University of California Press, 1965. It is a detailed study, with outstanding drawings of two gigantic 13th century English barns. Great Coxwell, 142' long, 38' wide, 48' high, is the finest surviving medieval barn in England. The other, Beaulieu-St. Leonards is the reconstruction, on a smaller scale, of England's largest medieval barn, which was originally 224' long, 67' wide, with 55' high gable walls. The authors are currently working on a three-volume set of books on aisled and bay-divided timber halls, which will take 3-5 more years to complete.

This basic structural framework, where roof rafters are supported by interior posts and plates [as opposed to clear spans] has been in use, with continued improvements, since the seventh century B.C. in Holland and Germany.

As Walter Horn points out, this structural type is still evolving, some 26 centuries after its early development. On page 32 are some more recent examples of this "all-purpose house."

NORTH AMERICAN BARNS

Prefabricated bent being raised. Riders on bent lower the pike poles as bent goes higher. Ontario, 1918.

Outstanding among barns for the excellence of their carpentry are the Pennsylvania barn, the English, and the Dutch barns of New York State. In one typical Ontario Pennsylvania example, all wood members were precut and numbered: posts and beams, purlins and rafters were adzed, but where carpentry approached the level of joinery, as it did in the shaping of mortise and tenons, the timber was sawn. Holes were bored in the pine logs and oak pegs locked the members in place. Nails, however admirably they might be forged, were forbidden. As someone said, prefabrication was elevated to a science before the term was even invented, and erection of the framework awaited only the arrival of friends and neighbours and the gift of a fine day.

THE BARN

If you're on property with a fine old barn, or if you're interested in the techniques and history of North America's finest (grandest) framing system, *The Barn* is worth maybe having ($25) and certainly viewing.

The book is excellent on the matter of taxonomy: Dutch barns, English barns, Pennsylvania barns, Connected barns, Polygonal barns, etc. Excellent also on history: barn design is traceable to upturned Iron Age ships and to ancient Basilican churches. The authors are "architectural" in their aesthetic appreciation; they pay due attention to the agricultural functioning and design innovations of various barns; they are less instructive about the dynamics of evolution in this splendidly vernacular and diverse medium. Happily, they cover the extraordinary barns of eastern Canada as well as northeastern U.S. The barns of the Deep South and Western States are not well explored.

The major resource in the book is scads of glorious photographs, many in color, many of fine structural detail.

Stewart Brand

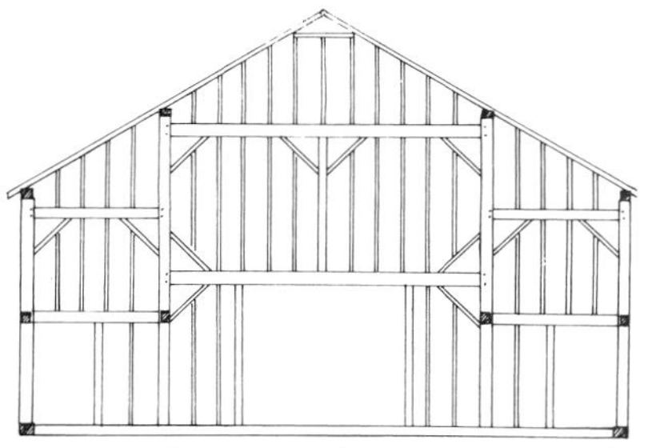

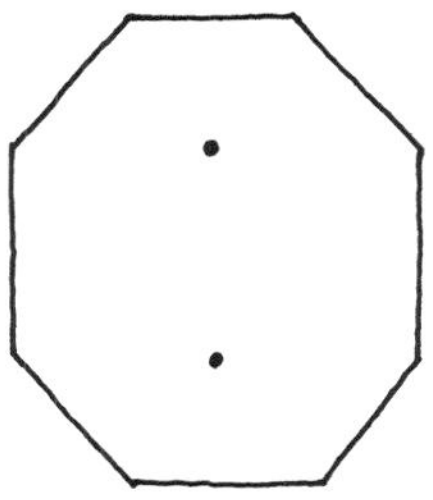

Polygonal barn in Quebec, an extended octagon with lean-tos added on.

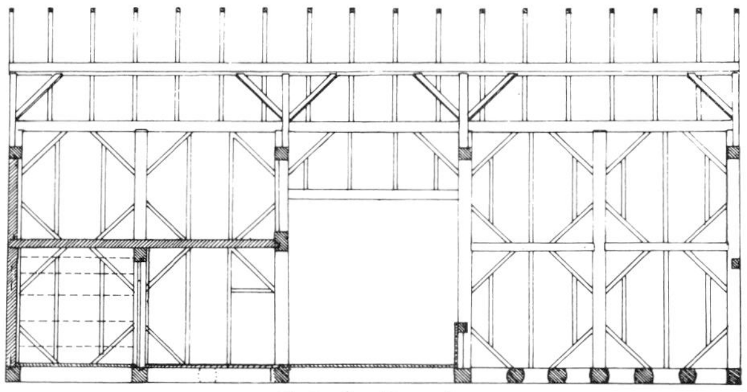

Ontario, Canada.

Mortise and tenon

Ontario

You put the barn good and far from the house in case it burned; like the night they woke up in a strange light and it was the glare of barn-flame in the bureau mirror: everything gone.

But at the home farm the three barns were still there, a square standing open for any child to wander in. Quiet and muffled in winter, the sound of the bull bashing impenitently against his stake and his wall; animal bodies bumping and squealing and rubbing; pock-pock-squawk from the henhouse, fierce indignant broody hens warming china eggs.

The air was floury at feed-time. You could sit (when you were alone) on the edge of the calves' pen letting the calves lick your bare knees as you felt their starred foreheads: cow-lick. You could sit in the buckwheat as if you were in hot-sand, pouring the smooth dark grain down your legs and wondering if there was a word for the rounded double-prism shape. Or watch old Molly the blind horse bringing load after load of straw in between the lofts after threshing; or was it hay after haying? Would a city child know?

The guineas stayed in the trees, but the banty hens played in the stable behind the horses' hooves, where no child went; and in the doorway, cats suckled whatever kittens came along. In the cowbarn, Ab put mangels through the mangle.

But the sheds fell down on the democrats and cutters; no stone-boats drew the milk-cans to the road; the families ran to girls and city-fellers; the hired men went away; and we were left with the stories.

Marian Engel

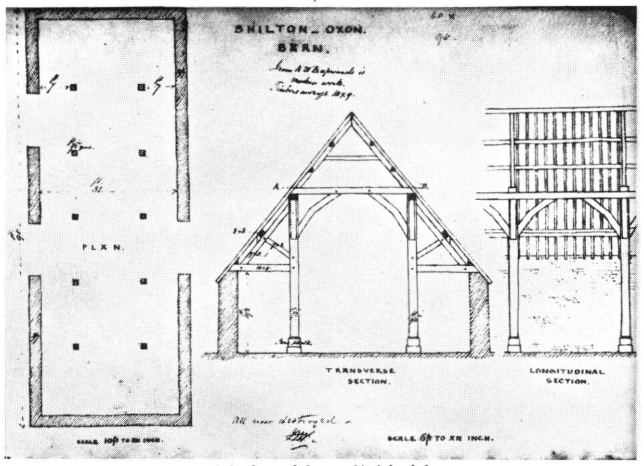

Aisled and bay-divided barn.

California

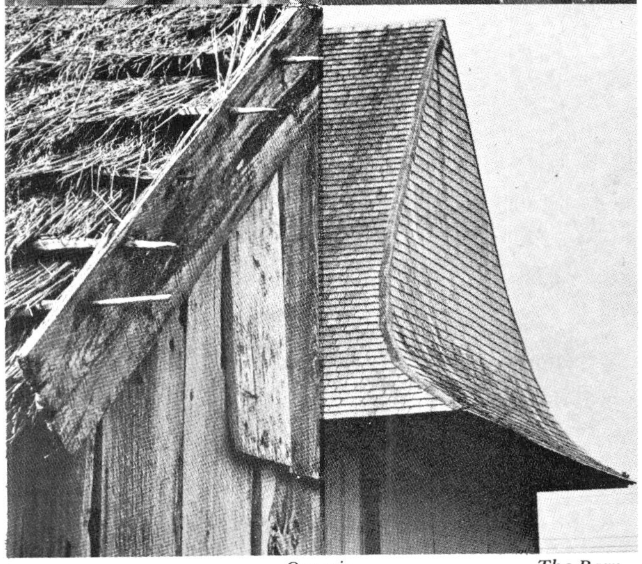

Ontario *The Barn*

The California barn, a further evolution of the great medieval English timber structures shown on page 30. Utilizing slender milled structural members, fastened with nails, these graceful buildings show the same "aisled and bay-divided" framework of their mortise and tenoned European predecessors. The American west coast barns, with lightweight tin or cedar shingle roofs, and no snow load requirements, are among the world's lightest weight large wooden structures.

A hundred years ago, over 80% of Americans lived on farms. Like the builders shown in the early pages of this book, American farmers built honest, practical, graceful structures that not only served their intended purposes, but somehow always seemed to harmonize with the surrounding countryside.

Alberta, Canada.

Log barn, Utah.

Simple, efficient framing system of small California barn. 4x4 posts, 8' apart. 2x4 plate and middle nailing piece. 1x12 vertical redwood boards nailed at bottom to sill, at top to plate help support roof load. Batts cover cracks between boards on outside. Y braces for diagonal stiffening. Compare with amount of wood used in a bldg. code stud wall. Note small amount of wood in roof frame. Lightweight roof minimizes earthquake force in the building.

Small California barn.

Kentucky

Cattle barn, Colorado.

A catalog of American and Canadian barns might make the best design manual for a novice builder. On these pages are some of the barns of North America, built by anonymous farmers and carpenters, designed with honesty and of necessity, and without an architect, contractor, or building inspector in sight. Look what we've lost.

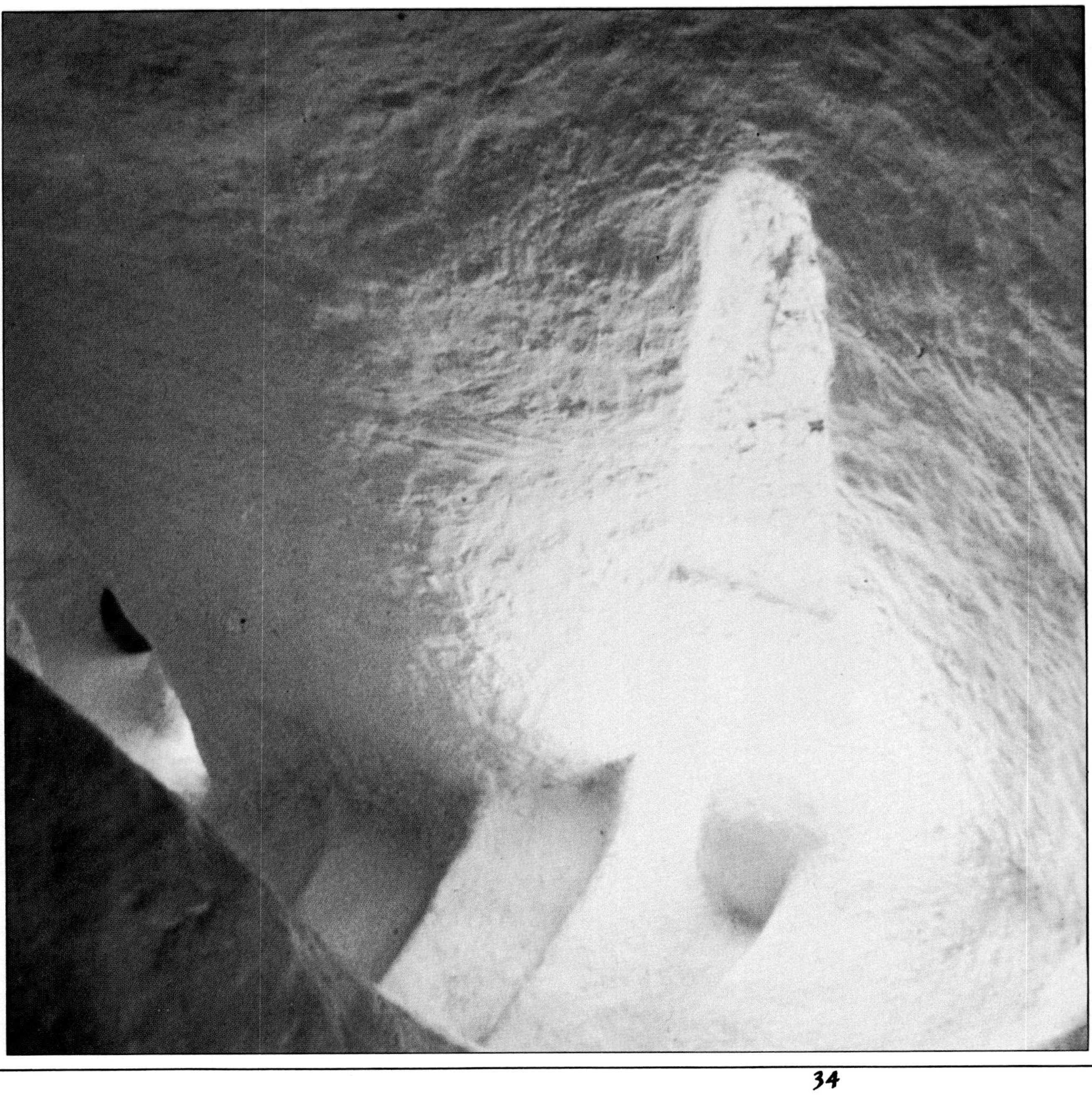

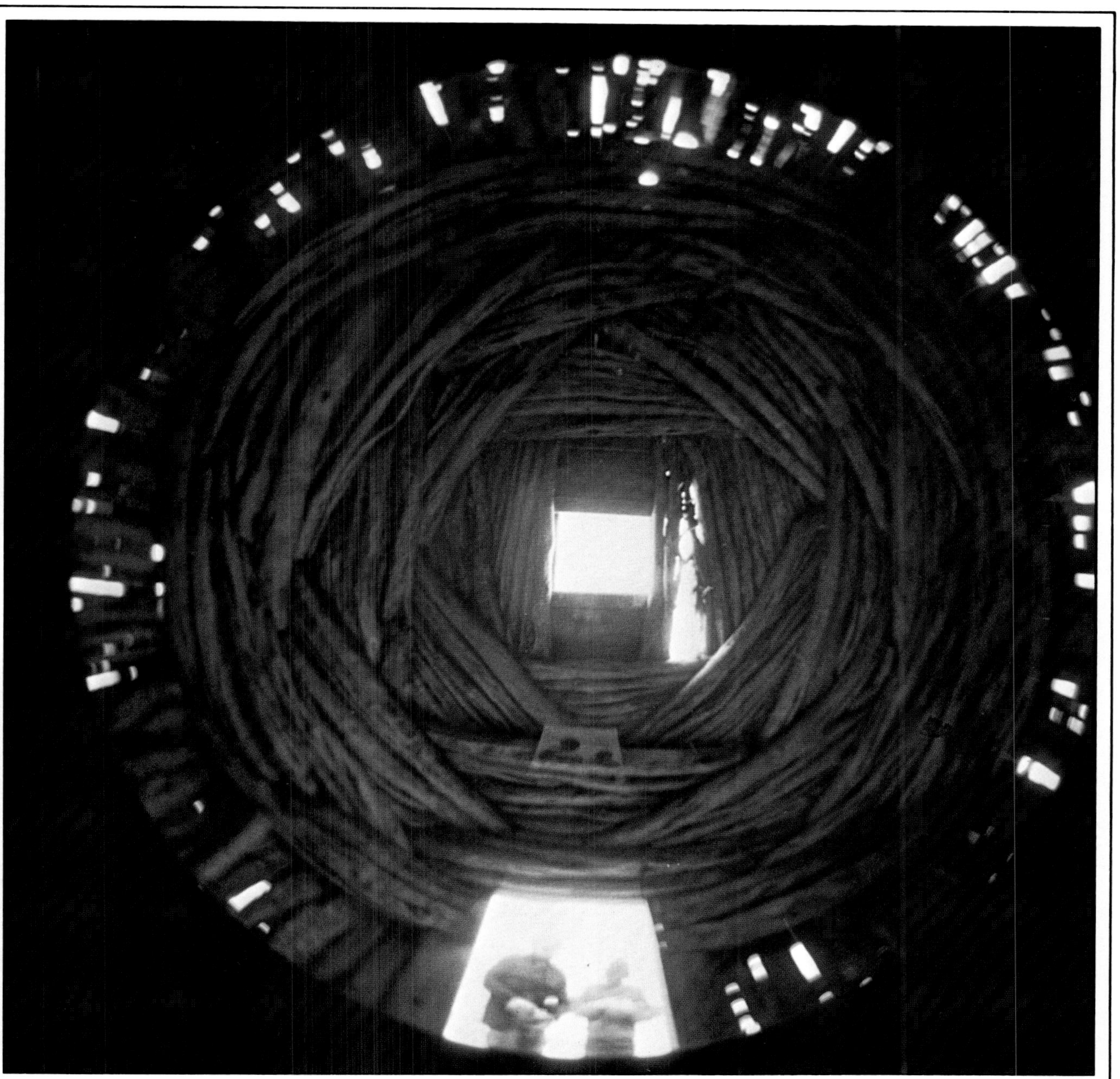

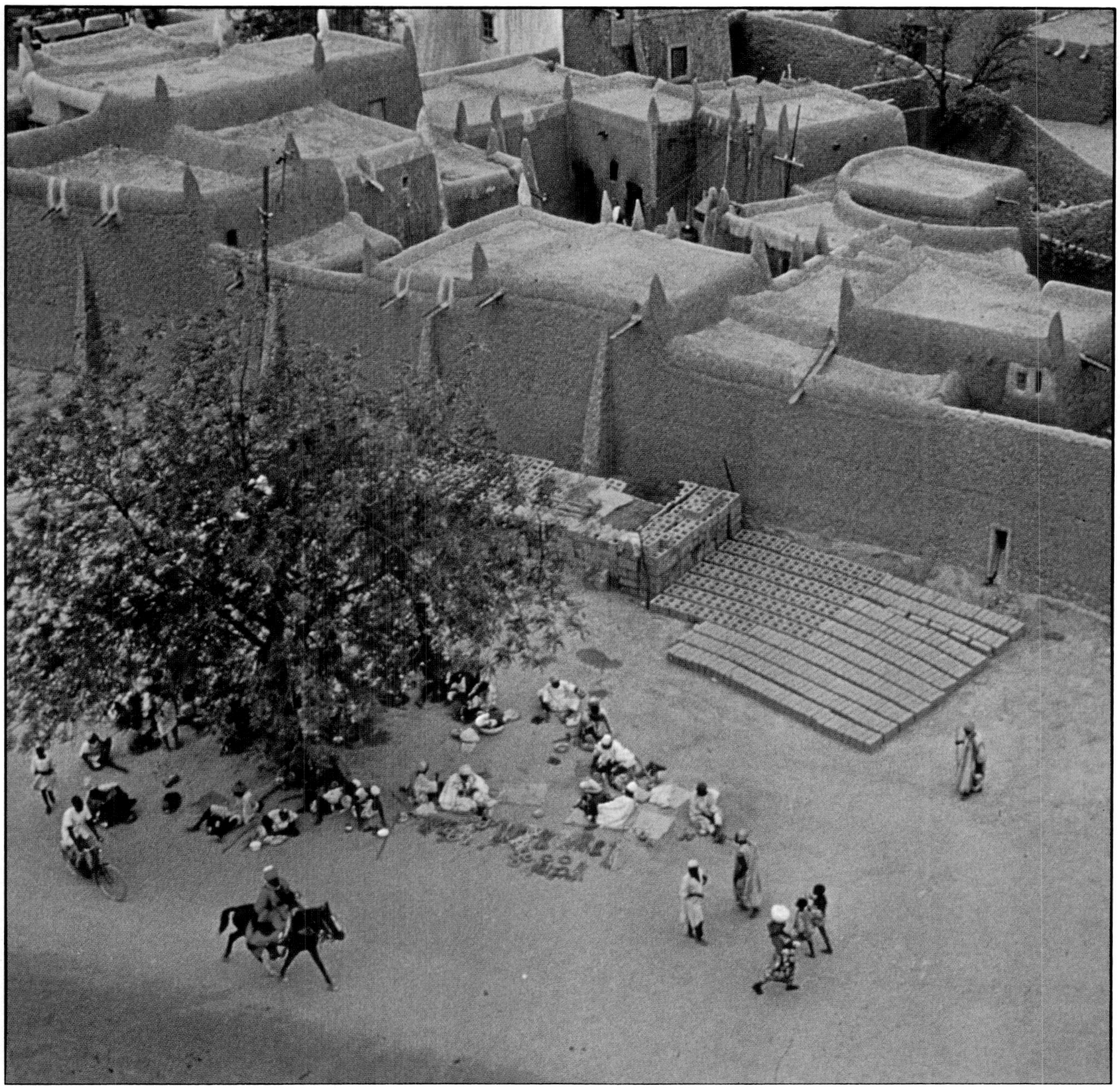

CHEVROLET
CO-OP SUPER

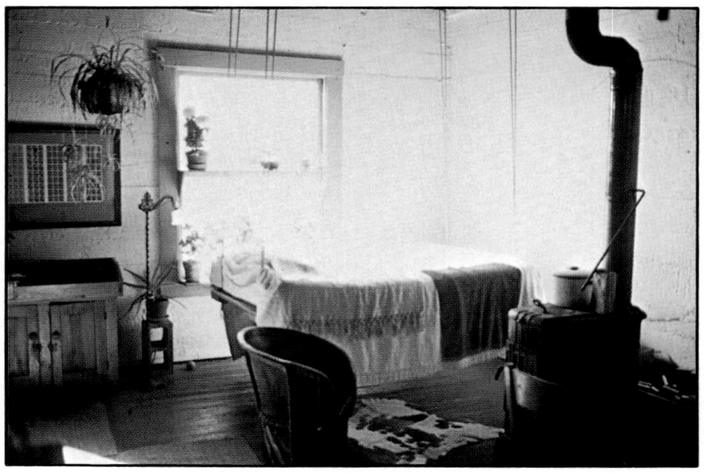

BUILDING

Finished tacking down the tar-paper roof and sealed it.
Stacked up
the hay and the feed mash, and the seed
together in a warm corner, piled the tools and the loose hay,
hung the door.
Beat the rains

Lewis MacAdams

Small buildings are quick to build, adaptable to used materials (cheap), easy to heat, simple for the inexperienced builder, and can later be added on to.

The next section of the book, pages 37-59, is on small buildings, including six small basic structures drawn by Bob Easton, and suggestions on floors, details, doors, and windows. Siting advice is on page 40. Plumbing and wiring books are listed on page 50, insulation and heating information on page 169.

Building small to start will give you basic experience. A small shed can be a place to live in or store things while you study the land and decide what to do next. You can watch the rising and setting of sun and moon, study outlook and orientation, learn about seasonal temperatures and wind direction, vegetation, rainfall: the many considerations that should help you decide what kind of house will suit your needs and fit the site.

The small building can then be expanded as needed. *You* will change during the building process, and building in increments gives you flexibility and adaptability as you go.

Building is hard work, costly and relatively permanent. Unlike a painter or potter, the builder cannot throw away an unsatisfactory result. There it stands, for all to see, for many years. Thus there is wisdom, especially for a new builder, in starting small, simply, and heeding local advice.

In Cape Cod we stopped at an old inn which had an amazing elliptical spiral staircase. I would have thought it impossible to build such a flowing shape of wood. The inn's owner told me that a traveling carpenter had built it and two others like it a hundred years ago on the Cape. My first thought was how fine for a builder to leave behind something so beautiful, useful and durable. My second thoughts were of the man himself, probably on horseback, saddlebag full of tools, his knowledge, skills and means of livelihood in his hands and head. No office to report in to, no bosses, the freedom of mobility. Then, as now, builders are needed. As materials and prices escalate, as unions, contractors and building inspectors tighten their grips, as mobile homes and prefabs proliferate in America, there will be an increasing need for builders with skills who don't charge $9 per hour, who will work with people.

If you'd like to become a builder but have no skills, a good way to begin is to get a pickup truck. Start by hauling, cleaning garages, gardening, odd jobs. If your work is good and wages fair, you'll get referrals. As you go, you can begin to pick up on skills. The lady who hires you to clean her attic needs some flashing put around the chimney, a new electrical receptacle installed. Learn what you can from books. Work with skilled craftsmen when possible for low wages in exchange for training. As time passes you'll learn carpentry, plumbing, wiring, painting. Each job is different. The better you get, the more freedom you'll have. Stay away from taxes, employees, becoming a contractor; maintain independence. Builders have always been in demand and if you can sleep in your truck (VW bus can be a home, carry large loads inside, and lumber on roof rack), you ought to be able to travel throughout the country. Even if people have no money you can always get room and board and appreciation.

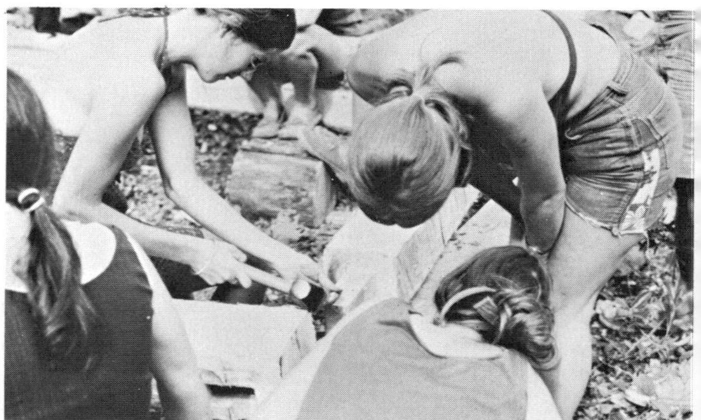

Each year over 150,000 families in America build their own homes. They either do the actual building, or act as general contractors in overseeing design and construction. Census figures in 1968 show that owner-builders are responsible for 20% of new single-family homes built in the United States. These houses are generally smaller, better built, and less expensive than developer-built housing. A significant statistic is that over 40% of them are paid for by cash — meaning no mortgage (and high) interest payments.

Further information in *Freedom to Build* by John C. Turner and Robert Fichter. Pub: Collier-Macmillan, 866 Third Ave., New York, N.Y. 10022.

s you work on a small building, let your experiences ide your progress. Build carefully and don't get apped by an idea that will dictate all your work. se shapes and materials that allow you freedom to provise, feature a piece of wood with interesting ain, or put in a window for a view you didn't know as there.

et involved with every detail, scavenge salvage yards r building materials, doorknobs, window latches. ake the experience enjoyable by allowing enough ne to appreciate your work as you build, let your ouse reflect your experiences during that time.

rawing and planning before you start can be helpful, you don't take it too seriously, and have had building perience. Draw mainly to figure out materials, room es and structure. Don't draw timbers you haven't cked up yourself, know how materials look, feel, and ork. Drawing equipment: board with soft surface, T-uare (30" long), adjustable triangle, scale (draw floor ans and elevations at ¼" equals one foot, details of ructure 1" equals one foot). Use tracing paper sold in lls at drafting supply stores, develop your ideas by awing on the tracing paper over your last sketch, eserving the best parts and redrawing the rest. If you ed help in planning, talk to local carpenters, builders d small contractors. (Architects are expensive and are ually interested in large buildings.)

art out with a small basic shape, and only alter the sign in response to the nature around your site: a ylight or high windows for extra light, or different oor levels on a sloping site. Unusual design is only od when its reason for being there is discovered rough its contemplation.

Bob Easton

When our family moved from the Bay Area to Big Sur in 1966 we cleaned up an abandoned chicken coop and lived in it for a year. Looking back, it was in many ways the best home we've had. Low ceiling, tin roof, humble architecture, built mindful of local weather conditions and with the economy and grace of necessity. It was easy to put in extra doors and windows because there was no grand scheme to interrupt. It was small and easy to heat, and we weathered one of the worst coastal winters there.

After about 10 years of building I've realized how simple and unpretentious a home can be. Building a house needn't (usually shouldn't) be a trip. Time and again I've been led by an abstract concept into a lengthy or impractical building project when a chicken house or small barn design would have been far better. So now when people ask for advice, I tell them to study the farm buildings near their site. They can serve as models for a house, with minor adaptations. One story, 2x4 framing, vertical walls, lightweight, roll roofing. Well insulated, with a fireplace to sit by on cold nights. Used wooden doors and windows. Kitchen that opens out into the garden. Quick to build so you can get on with life.

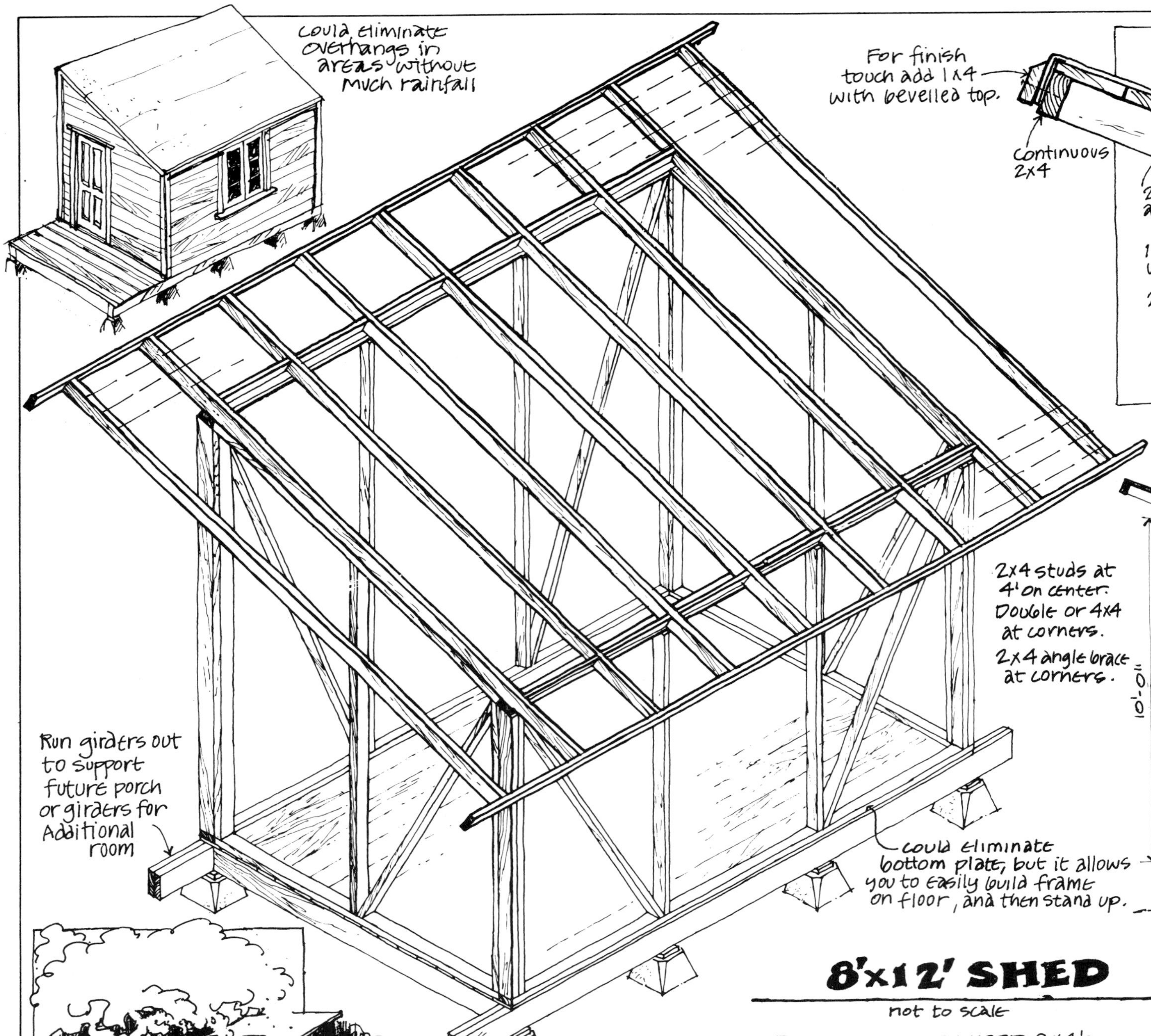

Could eliminate overhangs in areas without much rainfall
Run girders out to support future porch or girders for Additional room
Could eliminate bottom plate, but it allows you to easily build frame on floor, and then stand up.
2x4 studs at 4' on center. Double or 4x4 at corners.
2x4 angle brace at corners.

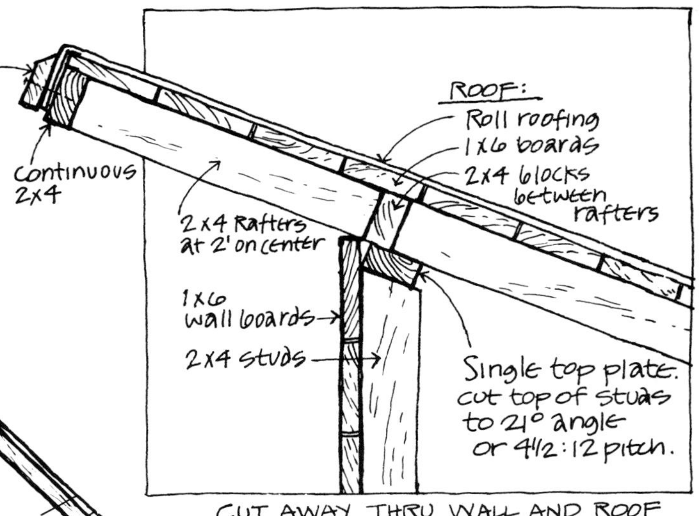

For finish touch add 1x4 with bevelled top.
Continuous 2x4
ROOF:
Roll roofing
1x6 boards
2x4 blocks between rafters
2x4 Rafters at 2' on center
1x6 wall boards
2x4 studs
Single top plate. cut top of studs to 21° angle or 4½:12 pitch.
CUT AWAY THRU WALL AND ROOF

All studs and rafters nailed with 2-16d common nails. Used wood is very hard, use american made nails.

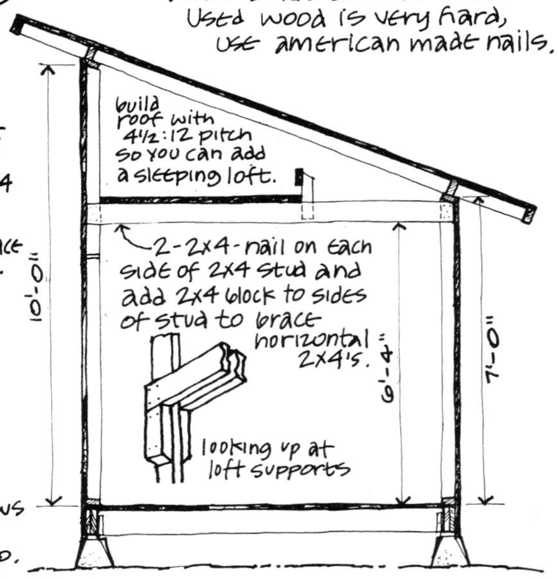

build roof with 4½:12 pitch so you can add a sleeping loft.
2-2x4-nail on each side of 2x4 stud and add 2x4 block to sides of stud to brace horizontal 2x4's.
looking up at loft supports
10'-0"
6'-4"
7'-0"

CUT AWAY THRU SHED

8'x12' SHED

not to scale

- BUILD FRAME WITH USED 2X4'S.
- ROOF AND WALLS COVERED WITH SALVAGED 1X6 BOARDS.
- SEE PAGE 58 FOR FLOOR FRAMING, PAGES 60-62 FOR WALLS, WINDOWS, DOORS, ROOFING.

add another room to shed

porch added to shed

Siting...

If you plan to build a dwelling, before you decide on a design or materials, you should consider the site: how you will be affected by sun, wind, rain, summer and winter climate, roads, outlook, trees, neighbors, cars, birds...Ideas about structure will appear to you from these considerations. It would be ideal to camp out on your site for a year and watch the changes before building anything.

An alternative would be to build a small shed to the side of your site and live there for a year before deciding what to do next. You could watch the angle of the sun change throughout the year, learn where the winter storms come from, and figure how to have the morning sun at your breakfast table. You could see how to catch cool breezes in the summer and see the stars at night. Also, you'd have time to meet the neighbors, study their houses, talk to the old folks about elements peculiar to the area: special winds, drainage problems, sources of cheap materials: local wisdom about local problems.

Point is: don't approach your site with preconceived notions. The site, and what surrounds it, should have the strongest influence on whatever is placed upon it. In the ancient Chinese art of *feng-shui*, "...building location and design is based on a belief that at every place there are special topographical features, either natural or artificial, which indicate or modify the cosmic energies present there. The forms and arrangements of hills, the nature and directions of watercourses, the heights and forms of buildings, the locations of forests, roads and bridges are all important factors. The influence of the sun, the moon, the planets and the stars are also considered important...."

Feng-Shui, Ernest J. Eitel, Hong Kong, 1873.

Some siting tips:

- If your buildable area is limited, first determine what and where your waste disposal (septic system or outhouse, compost pile) must be located.
- Sun angle: in most areas, a roof overhang can let in low winter sun, and block the hot (and higher) summer sun. Skylights that face south and/or west will let in a lot of summer heat.
- Winds: you can site your building to block the prevailing winds so you can have a protected outdoor area. Winds follow the canyons, in hot areas, there is usually a breeze in ravines. Storms batter buildings built on hilltops, exposed sites.
- Study old buildings that are still in good shape in your area, check their siting and materials.

Build your shed to the side of where you would imagine your main room. Later, you could easily attach a larger building to a simple shed, or leave it separate as a guest room or shop. This shed, and the buildings shown on the next pages are simple prototypes for new builders to start from, and are designed to be built from used and salvaged wood. On page 50 are listed other books that have practical building information.

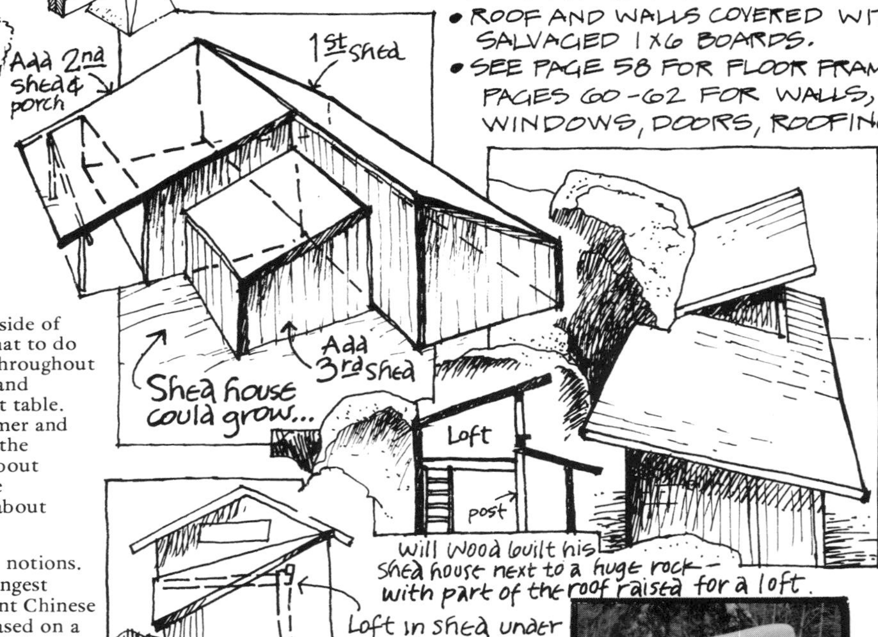

Add 2nd shed & porch
1st shed
Add 3rd shed
Shed house could grow...
Loft
post

Will Wood built his shed house next to a huge rock with part of the roof raised for a loft.

Loft in shed under raised roof. Built by Kelly Hart.

Kelly's shed, center left. Will Wood's workshop.

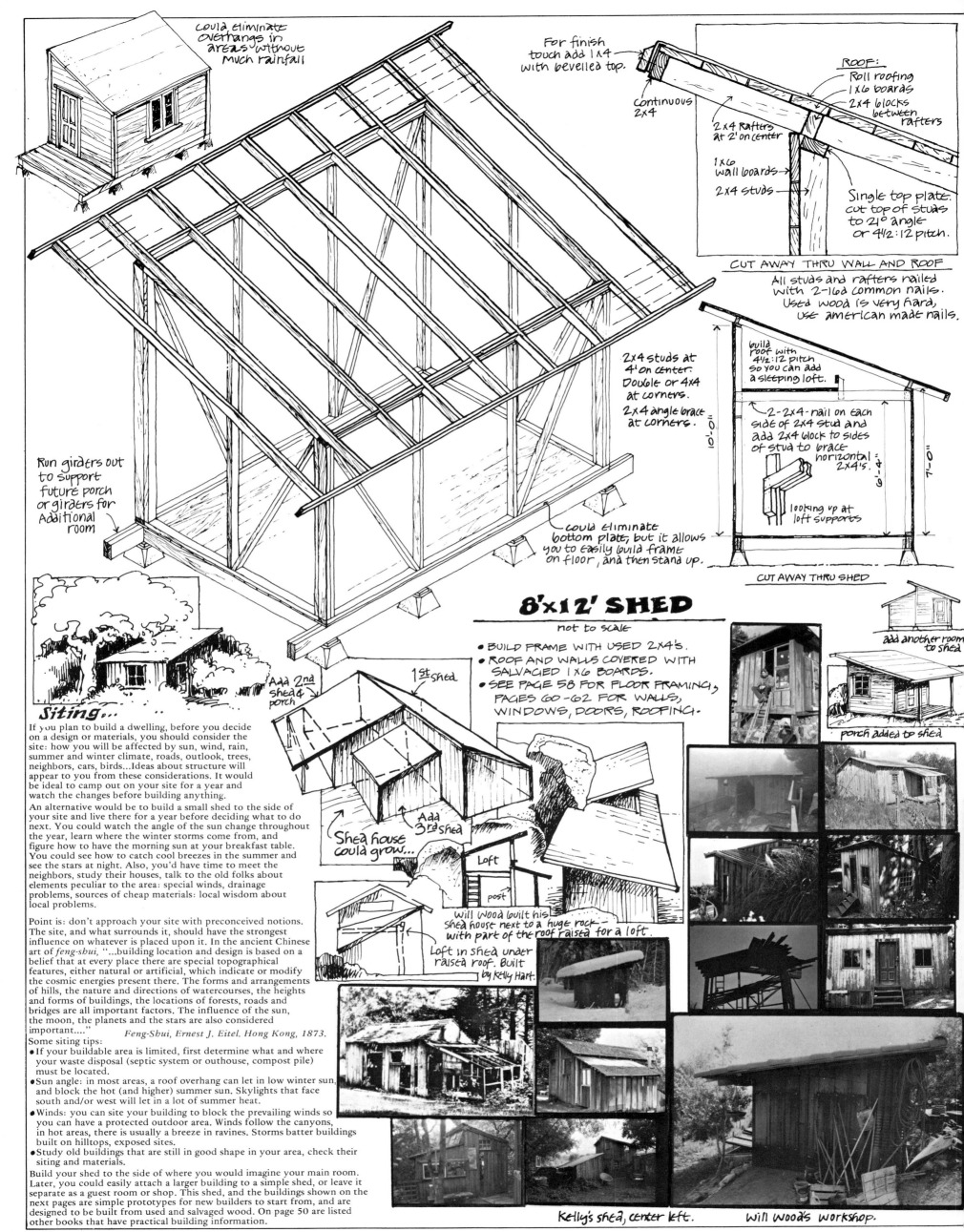

Skylights

Loft- use a natural pole to support it.

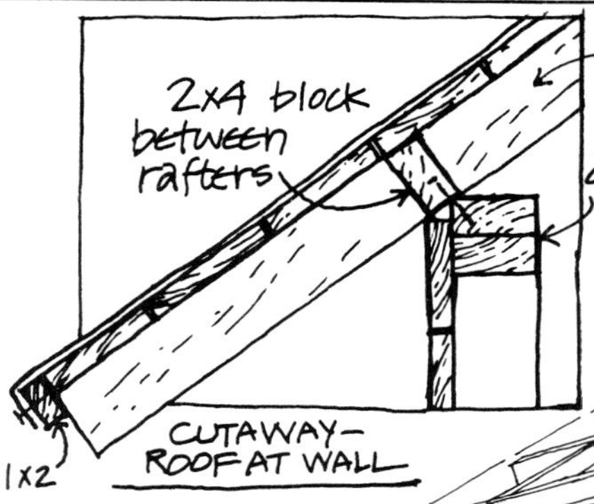
2x4 block between rafters
2x4 rafters at 2' on centers nail to plate at point of contact.
double 2x4 continuous plate
1x2
CUTAWAY- ROOF AT WALL

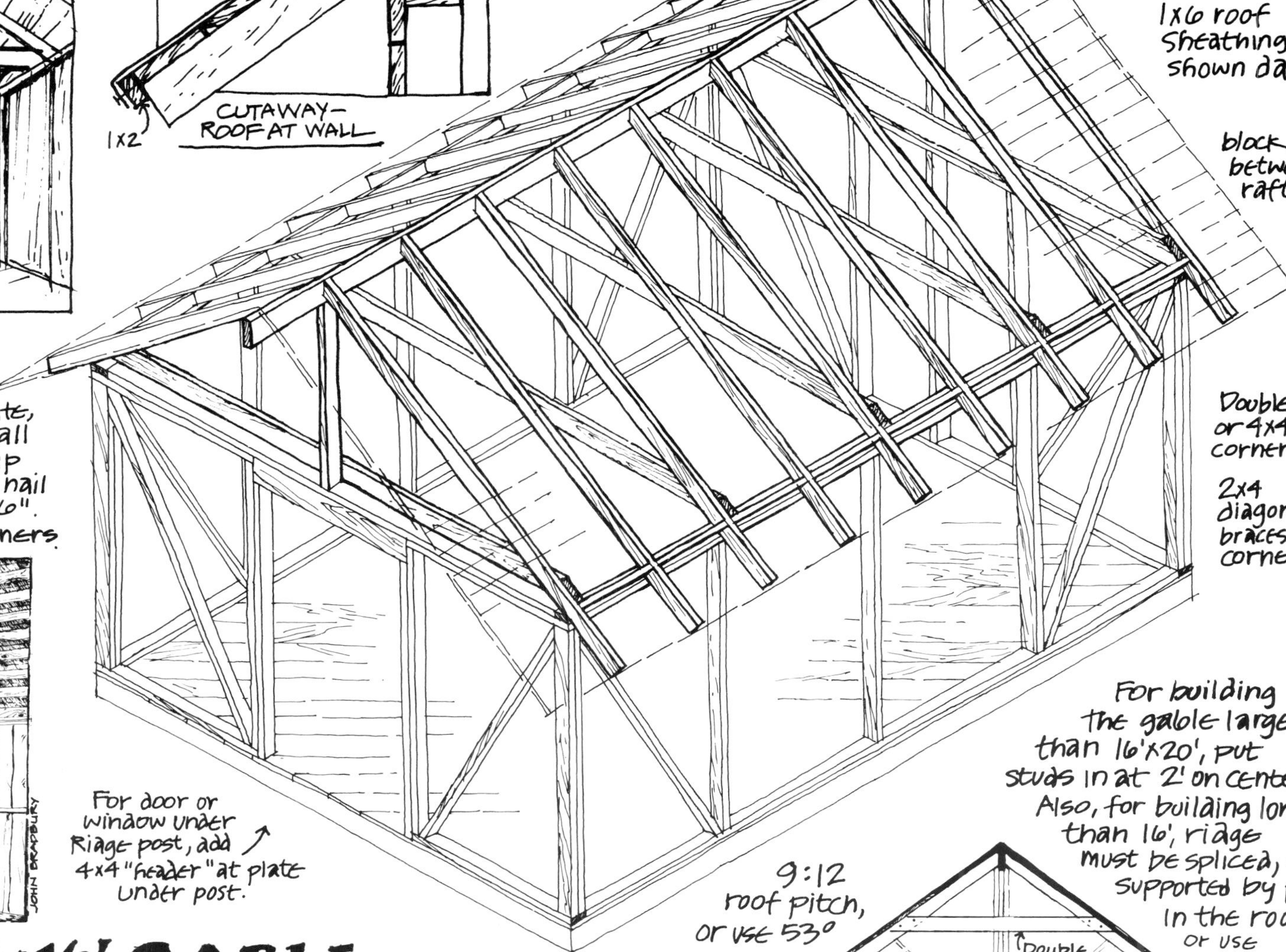
Put up 2x6 ridge board, then butt rafters to it. Brace post under ridge until rafters are on.
1x6 roof sheathing is shown dashed
block between rafters
Double 2x4 or 4x4 corner posts.
2x4 diagonal braces at corners.
For door or window under Ridge post, add 4x4 "header" at plate under post.

Continuous 2x4 plate, build level around all 4 sides, always lap joints by 4', and nail together every 16". Lap at corners.

JOHN BRADBURY

For building the gable larger than 16'x20', put studs in at 2' on centers, Also, for building longer than 16', ridge must be spliced, or supported by post in the room, or use trusses.

12'x16' GABLE

not to scale

- ALL USED 2X4'S EXCEPT FOR 2X6"X20' LONG RIDGE BOARD.

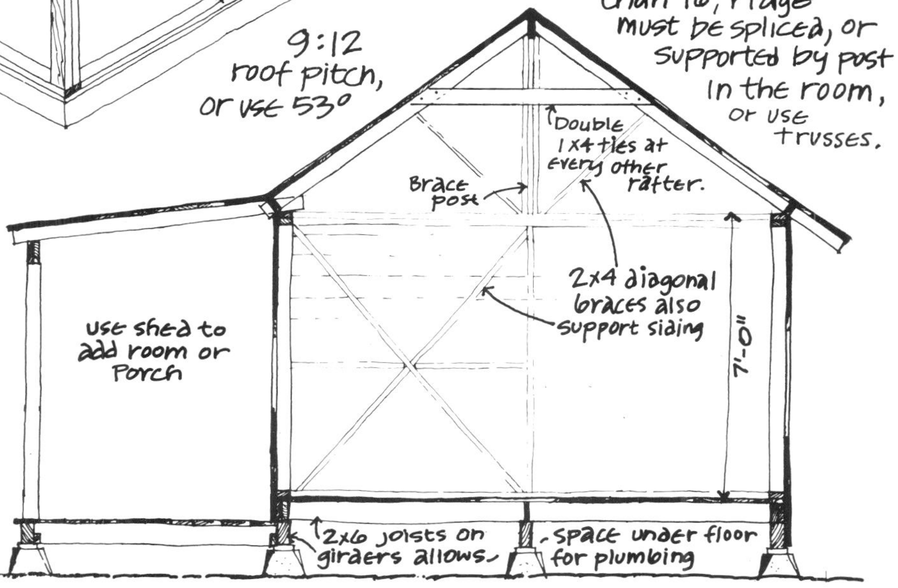
9:12 roof pitch, or use 53°
Double 1x4 ties at every other rafter.
Brace post
2x4 diagonal braces also support siding
7'-0"
Use shed to add room or Porch
2x6 joists on girders allows
space under floor for plumbing

CUTAWAY THRU GABLE AND SHED

Whereas the shed roof is the simplest 2x4 structure, its size is limited. Still using 2x4's the small gable roof building is the easiest way to build a larger structure. It has the structural strength of the triangle and the tent shaped roof gives a sense of shelter. The pitch can be made steep for a shake roof or interior loft space, or shallow to conserve on materials and for cozy interior feeling. The gable and shed shapes fit together well in various combinations, as shown in sketches on these two pages.

Roof at Dutch Open Air Museum

Gabled room built by Bill Beckman

A 8'x13' Gabled building built with all used materials.

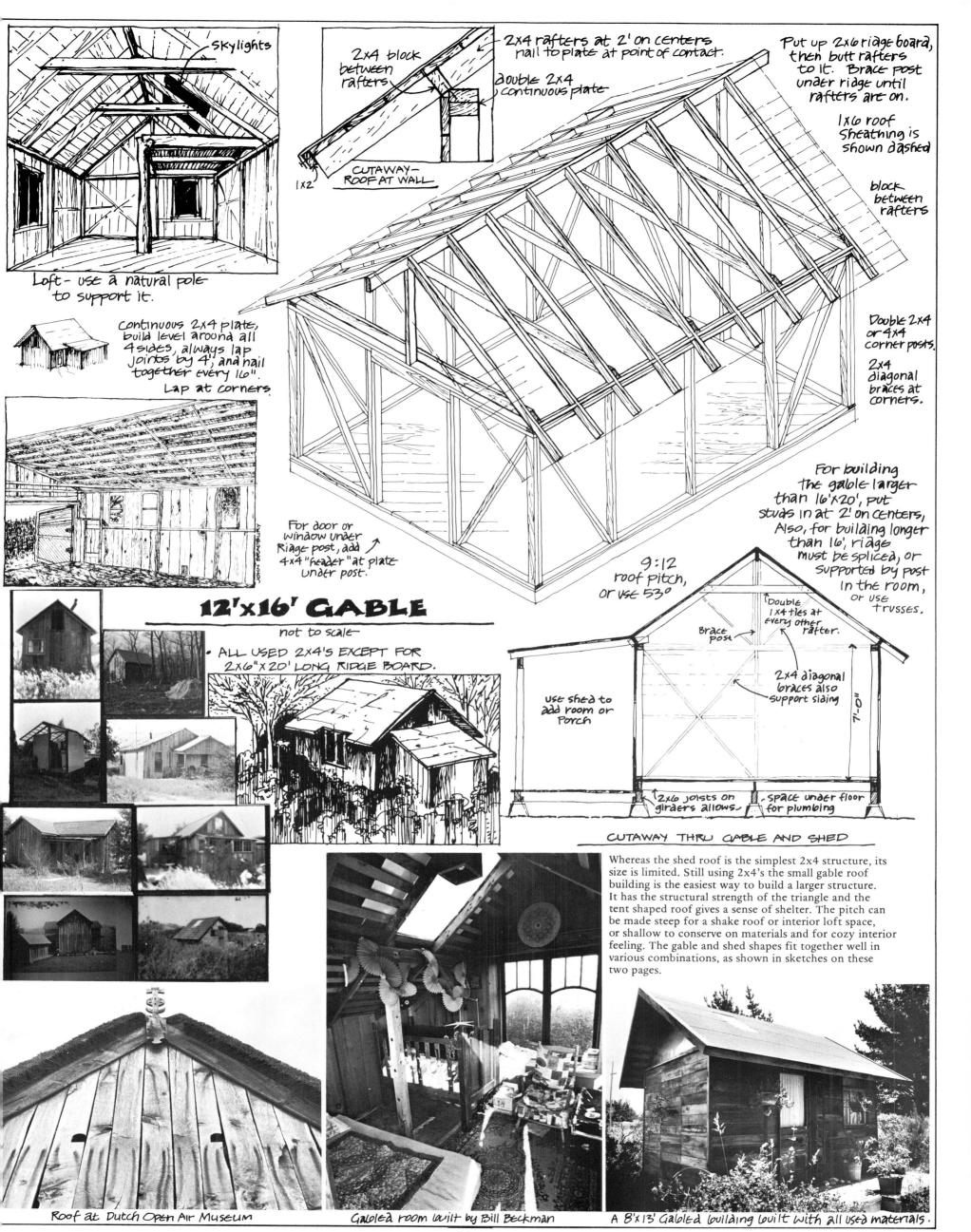

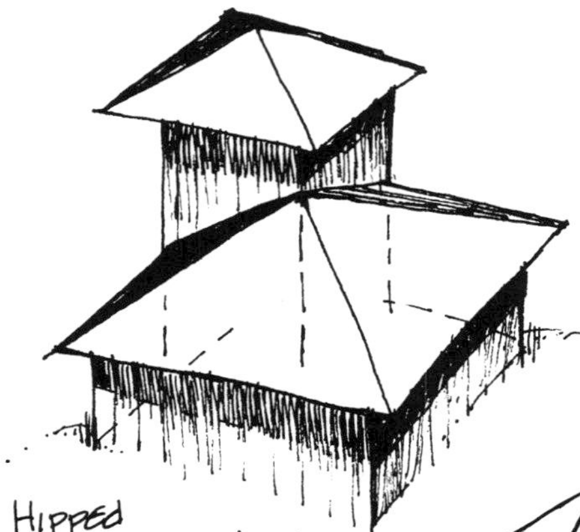

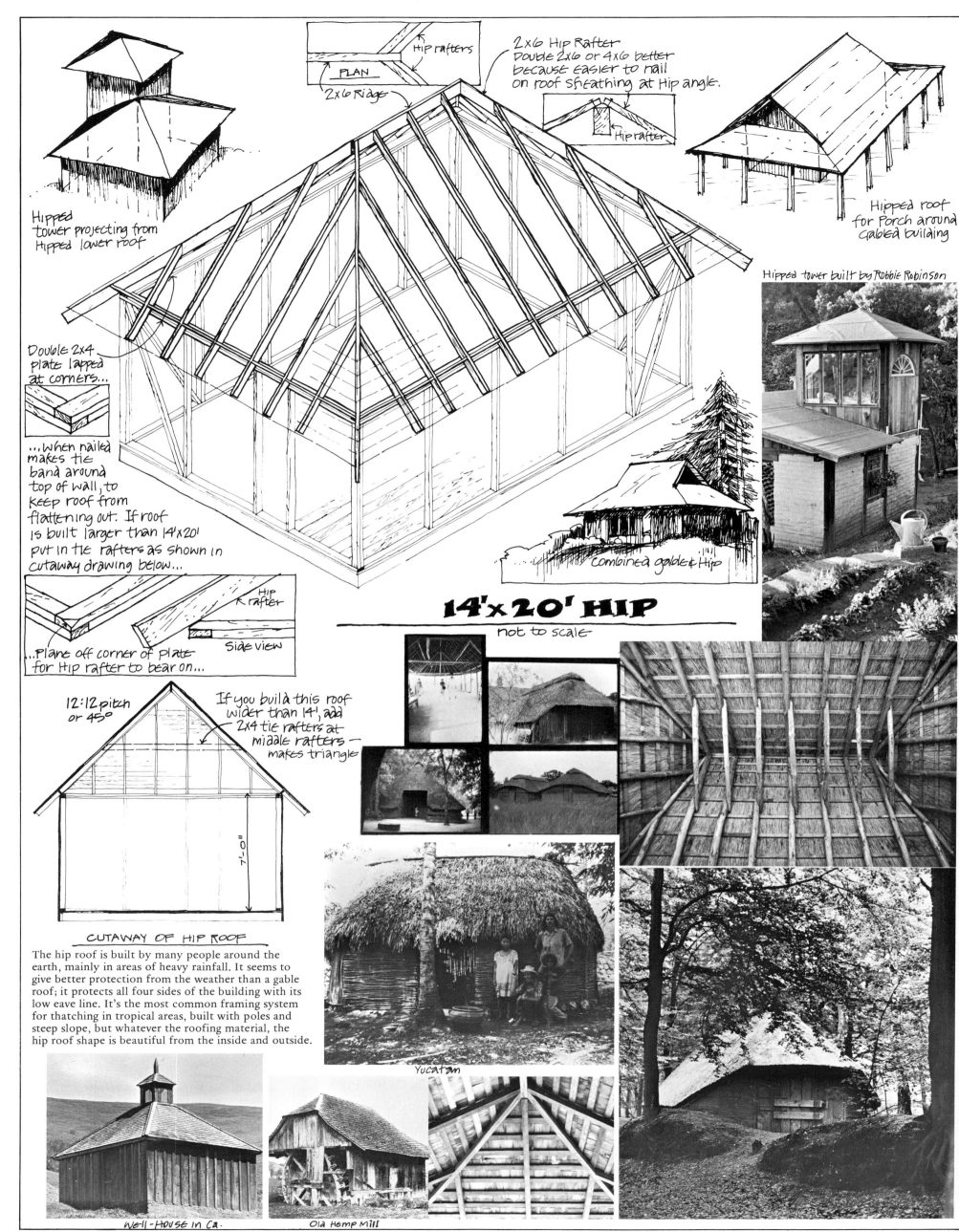

Hipped tower projecting from Hipped lower roof

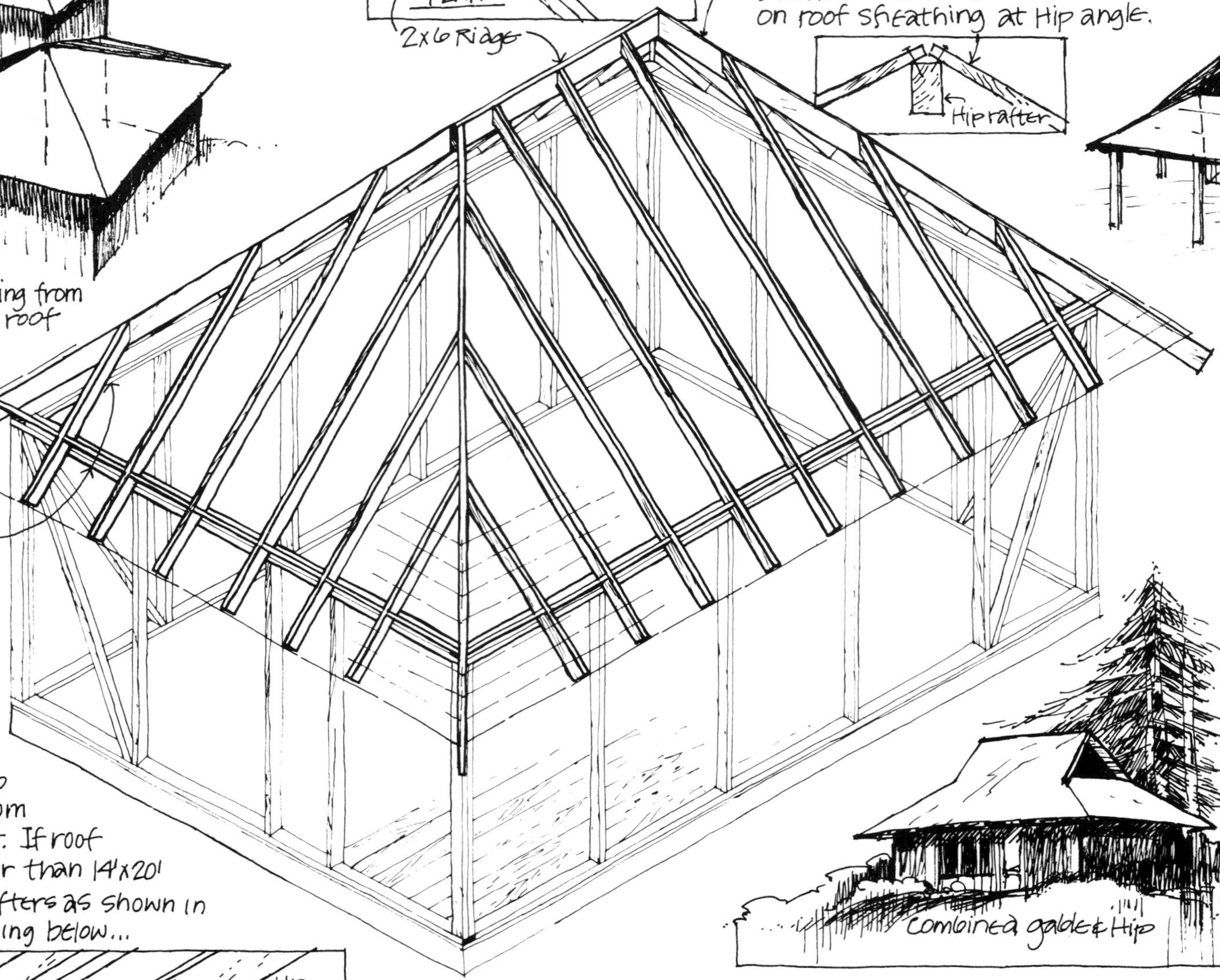

Hip rafters
PLAN
2x6 Ridge
2x6 Hip Rafter
Double 2x6 or 4x6 better because easier to nail on roof sheathing at Hip angle.
Hip rafter

Hipped roof for Porch around Gabled building

Hipped tower built by Robbie Robinson

Double 2x4 plate lapped at corners...

...when nailed makes tie band around top of wall, to keep roof from flattening out. If roof is built larger than 14'x20' put in tie rafters as shown in cutaway drawing below...

Combined gable & Hip

Hip rafter
Side view

...Plane off corner of plate for Hip rafter to bear on...

14'x20' HIP

not to scale

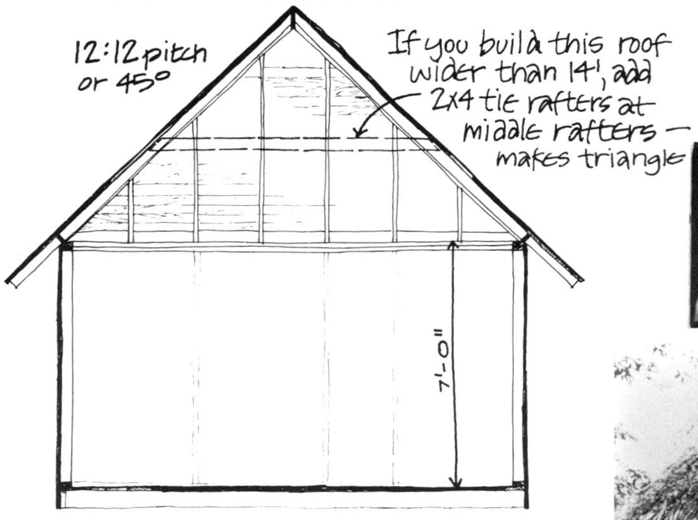

12:12 pitch or 45°
If you build this roof wider than 14', add 2x4 tie rafters at middle rafters — makes triangle
7'-0"

CUTAWAY OF HIP ROOF

The hip roof is built by many people around the earth, mainly in areas of heavy rainfall. It seems to give better protection from the weather than a gable roof; it protects all four sides of the building with its low eave line. It's the most common framing system for thatching in tropical areas, built with poles and steep slope, but whatever the roofing material, the hip roof shape is beautiful from the inside and outside.

Yucatan

Well-House in Ca.

Old Hemp Mill

Round wall makes adobe very strong

Roof:

Slope Vigas for roof slope So flat roof drains.
Used 2x6's or 2x4's flat can span 4' or 5'... 1x6's can span only 2'.
Roofing paper over roof boards.

Traditional New Mexico Adobe wall and Roof:

1. Small peeled poles laid in zig-zag pattern on vigas.
2. Then Roofing paper and Tar.
3. 8" of soil or pumice for insulation.
4. Then more Roofing paper and tar.
5. Final paper goes up and laps adobe wall.
6. 4" metal roof drain thru parapet.

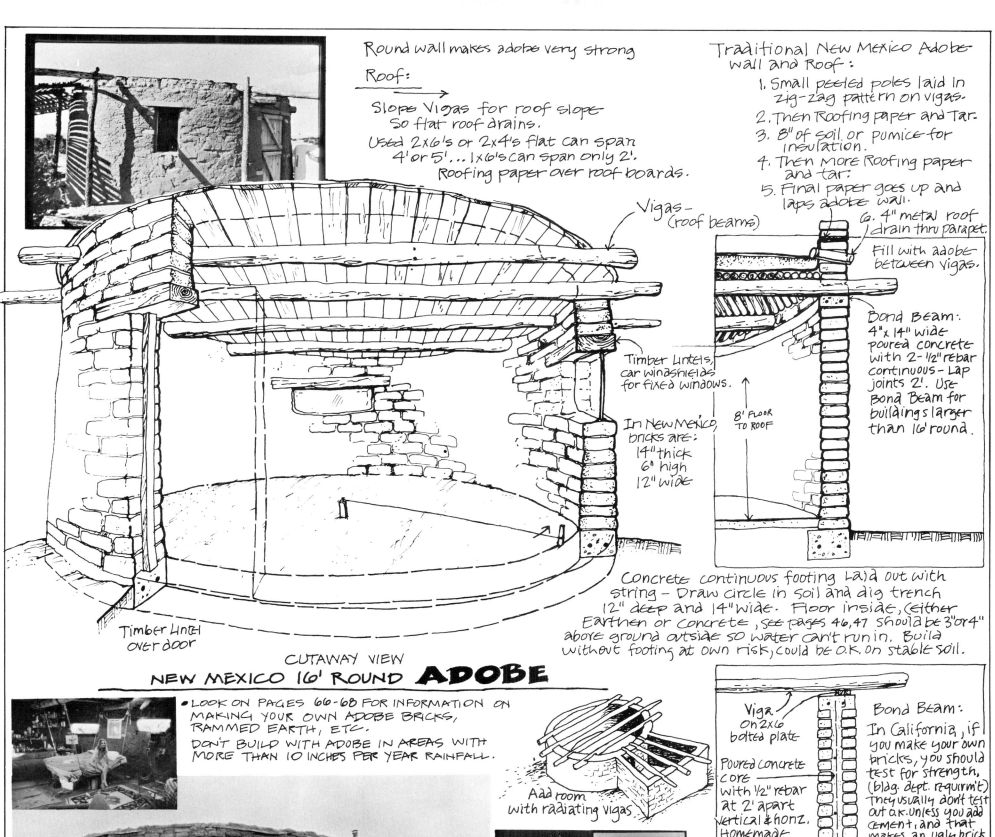

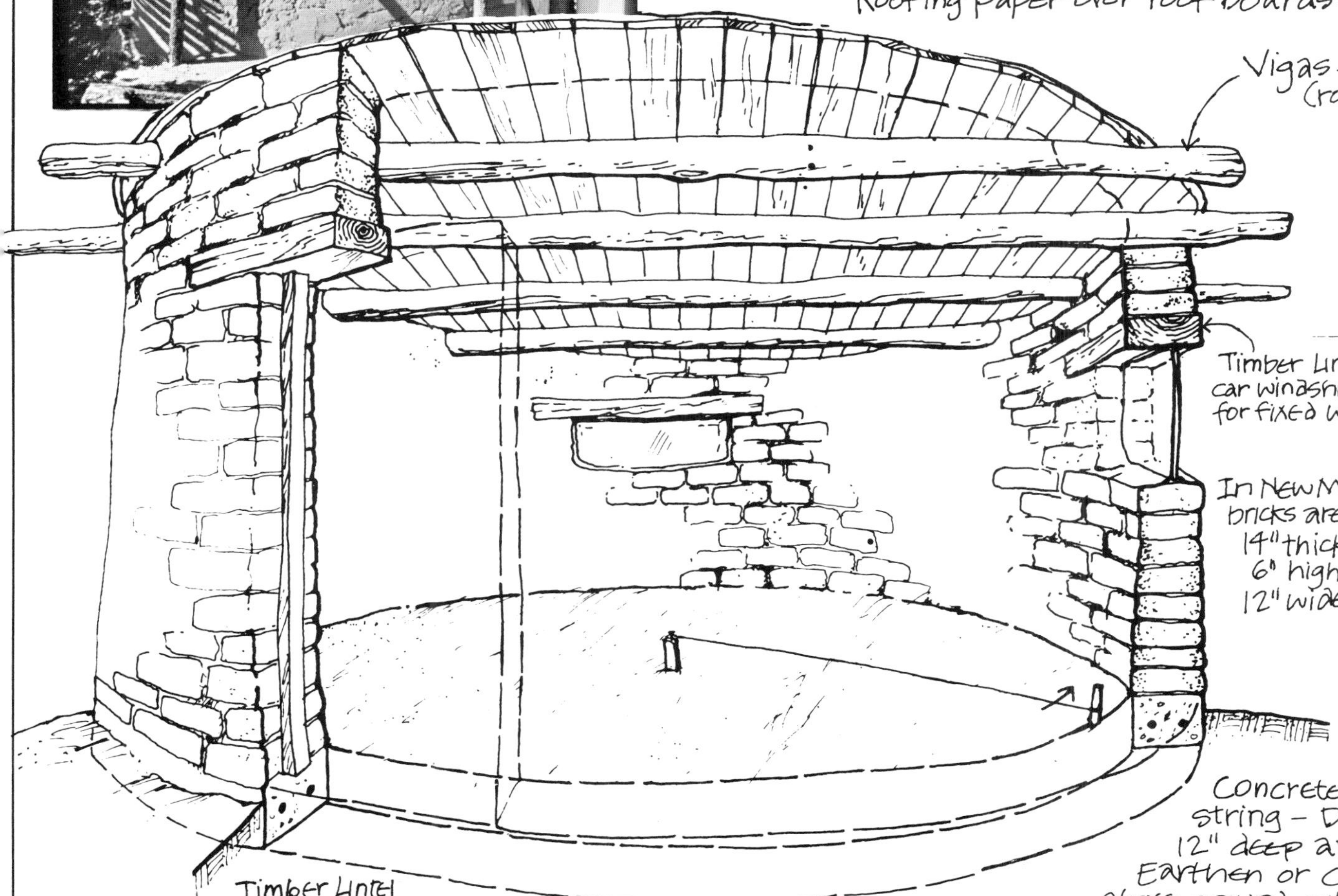

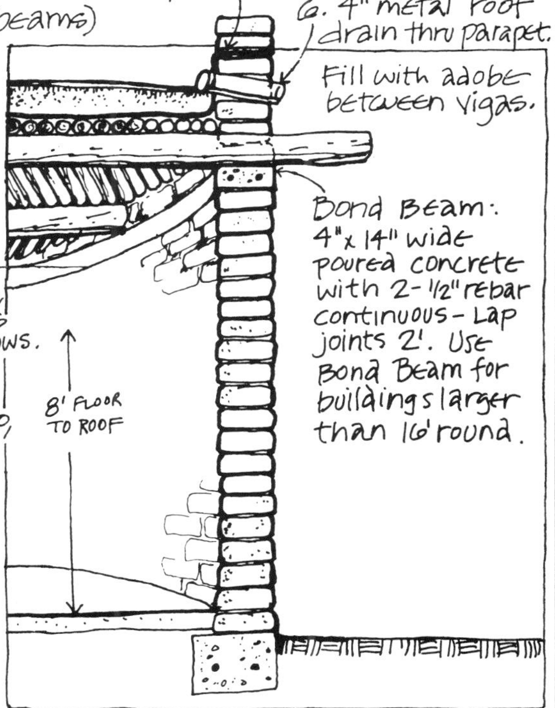

Bond Beam: 4"x 14" wide poured concrete with 2 - 1/2" rebar continuous - Lap joints 2'. Use Bond Beam for buildings larger than 16' round.

Concrete continuous footing laid out with string - Draw circle in soil and dig trench 12" deep and 14" wide. Floor inside, (either Earthen or Concrete, see pages 46, 47) should be 3" or 4" above ground outside so water can't run in. Build without footing at own risk, could be O.K. on stable soil.

CUTAWAY VIEW

NEW MEXICO 16' ROUND **ADOBE**

- LOOK ON PAGES 66-68 FOR INFORMATION ON MAKING YOUR OWN ADOBE BRICKS, RAMMED EARTH, ETC.
  DON'T BUILD WITH ADOBE IN AREAS WITH MORE THAN 10 INCHES PER YEAR RAINFALL.

Add room with radiating vigas

Interior, Exterior, Round Adobe

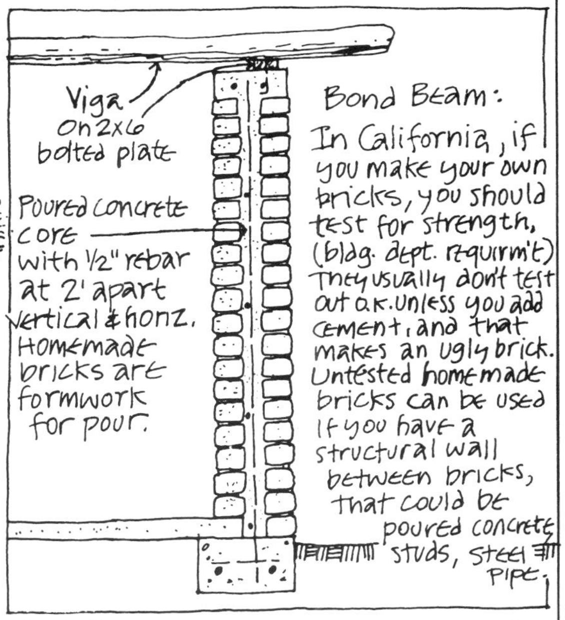

Viga on 2x6 bolted plate
Poured concrete core with 1/2" rebar at 2' apart vertical & horiz. Homemade bricks are formwork for pour.

Bond Beam: In California, if you make your own bricks, you should test for strength, (bldg. dept. requirm't) They usually don't test out o.k. unless you add cement, and that makes an ugly brick. Untested homemade bricks can be used if you have a structural wall between bricks, that could be poured concrete studs, steel pipe.

California (Earthquake country)

California Adobe with Sod Roof Built by Bill Greenough

California Post Adobe

- Support roof on 6x6 posts and beams (posts 4' or 6' apart).
- Fill between posts with bricks.
- For buildings larger than 16' square or round, posts must be diagonally braced with 2x6 bolted at posts.

Jock Favour Adobe, New Mexico.

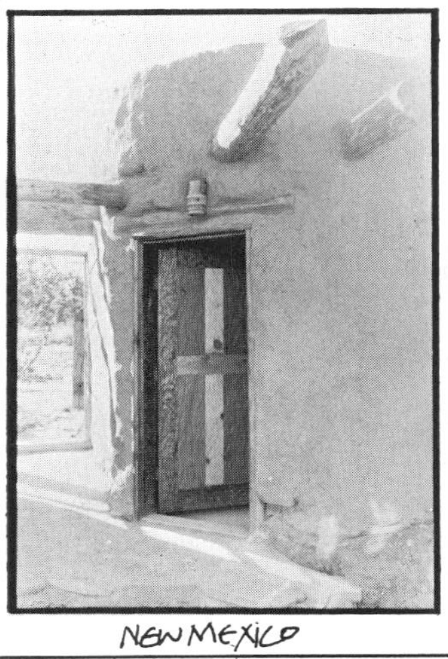

New Mexico

2 story Adobe, Yucatan

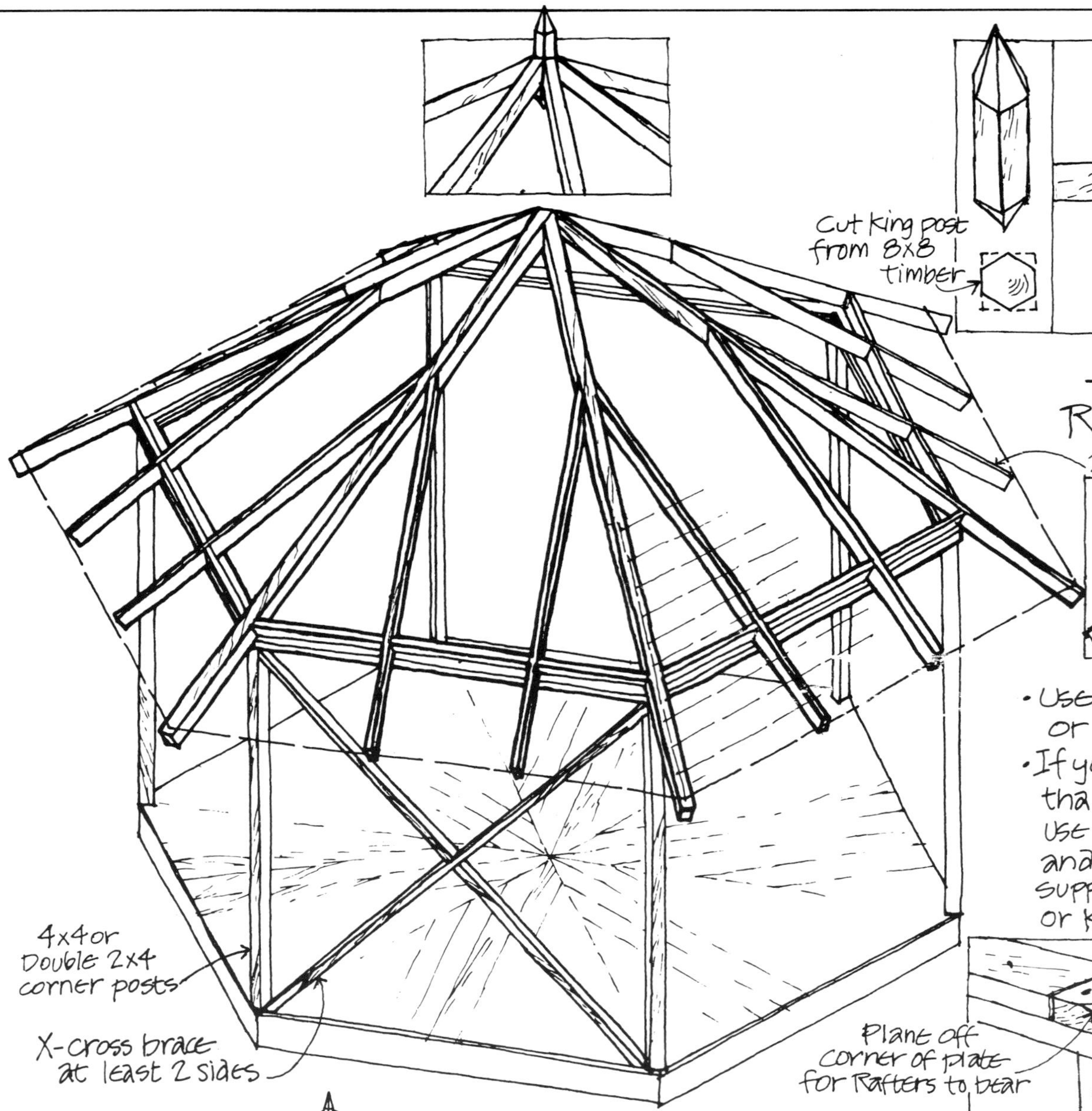

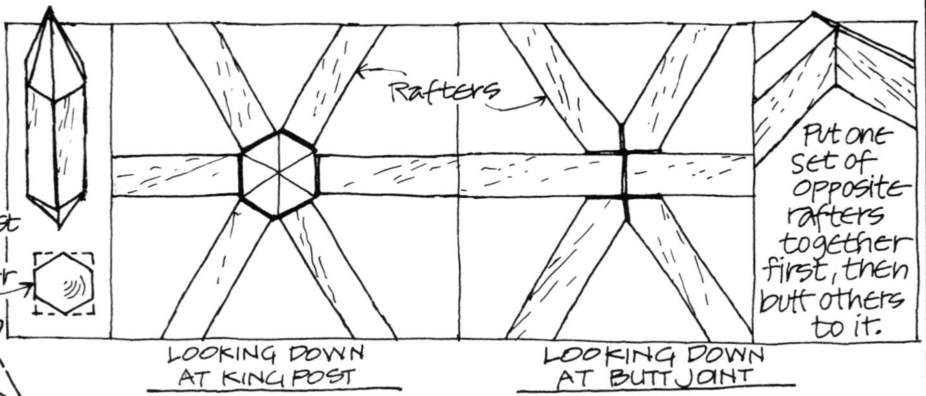

Rafters at Ridge point - 2 ways to frame.

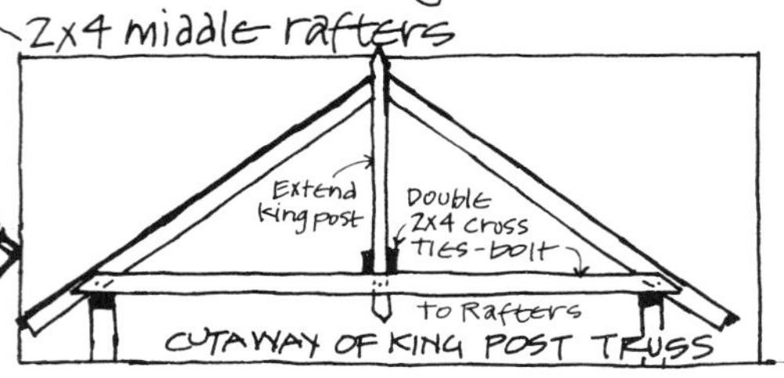

- Use 4x4 Rafters or double 2x4's.
- If you build larger than 12' diameter, use 4x6" or 4x8" Rafters, and either use center support post to floor or or king post truss.

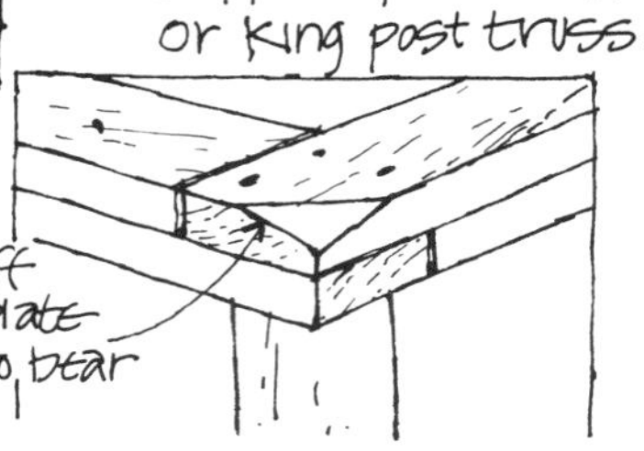

Pentagonal building at Ananda

HEXAGON 12' DIAMETER

- HEX IS EASIEST POLYGONAL BUILDING
- SEE PAGE 131 FOR MORE MANY-SIDED STRUCTURES.
- EVEN-SIDED POLYGONS EASIEST.

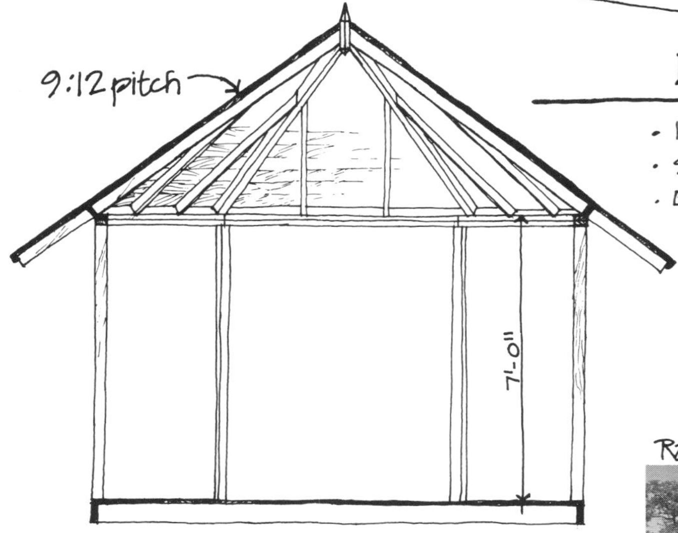

CUTAWAY THRU HEXAGON

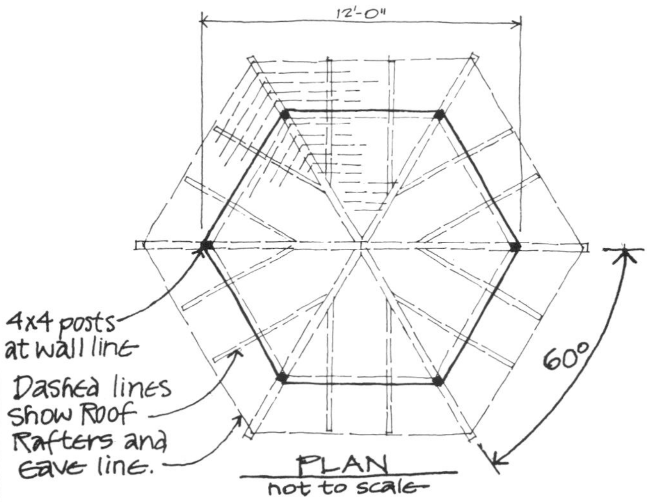

PLAN
not to scale

Random polygon at Black Mountain, Ca.

1/2 Hex Room on house, Texas

Gazebo with carved King Post

12-sided Haberhedron built by Jock Angus

Another way to frame a small Hex roof Needs all 4x6 Lumber

Small Hex built in California by Scott Sheldon

POLYGONAL BUILDINGS...

Polygonal, or many-sided buildings, make a more interesting interior space than rectilinear buildings, but are harder to build and fit furniture into. The angle cuts you've got to make can mean that it will take you twice as long to build. Compared to geodesic domes, however, polygonal buildings are easier to build, more adaptable to used materials, easier to enlarge, and are more durable. It's also easier to put in used doors and windows in the vertical walls.

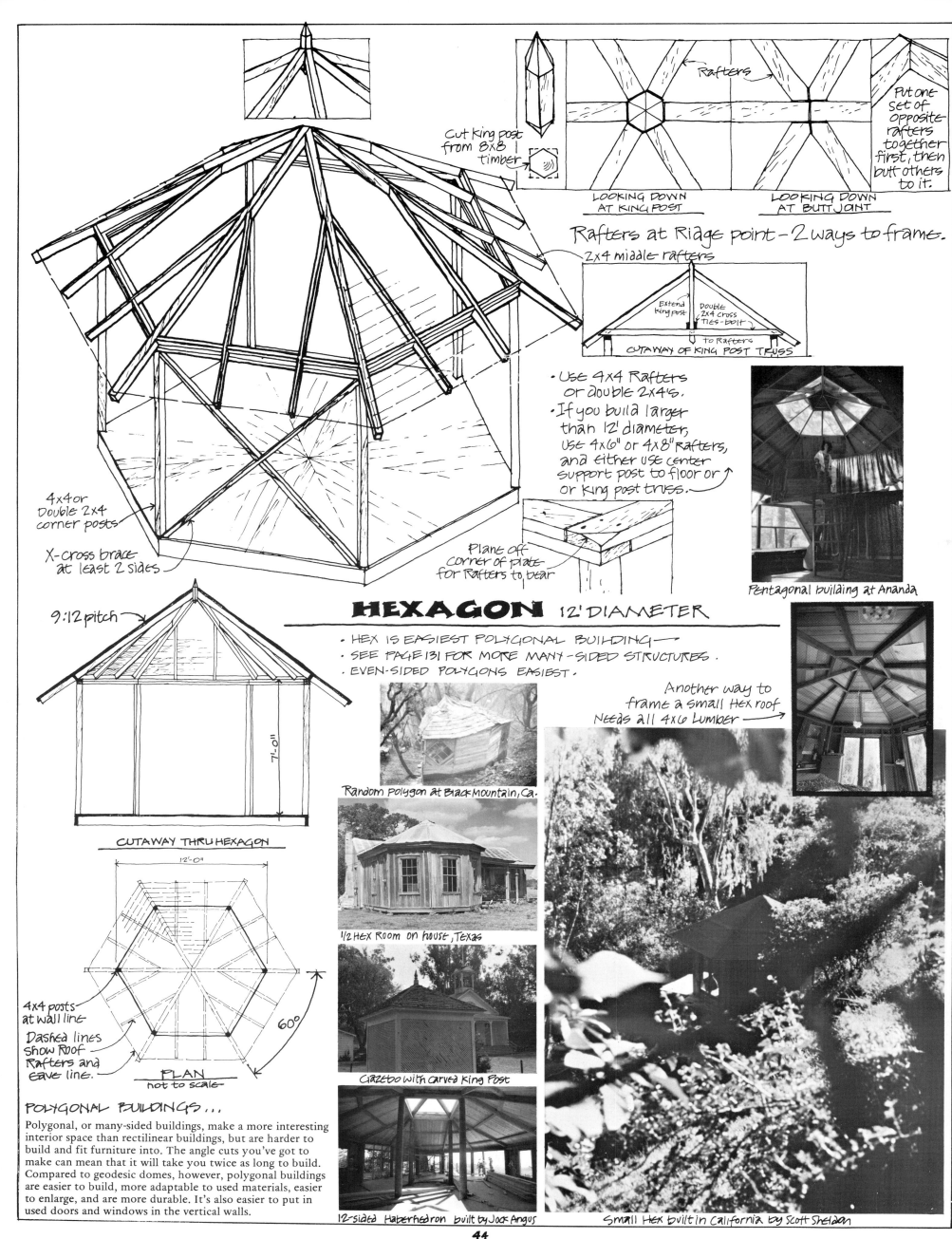

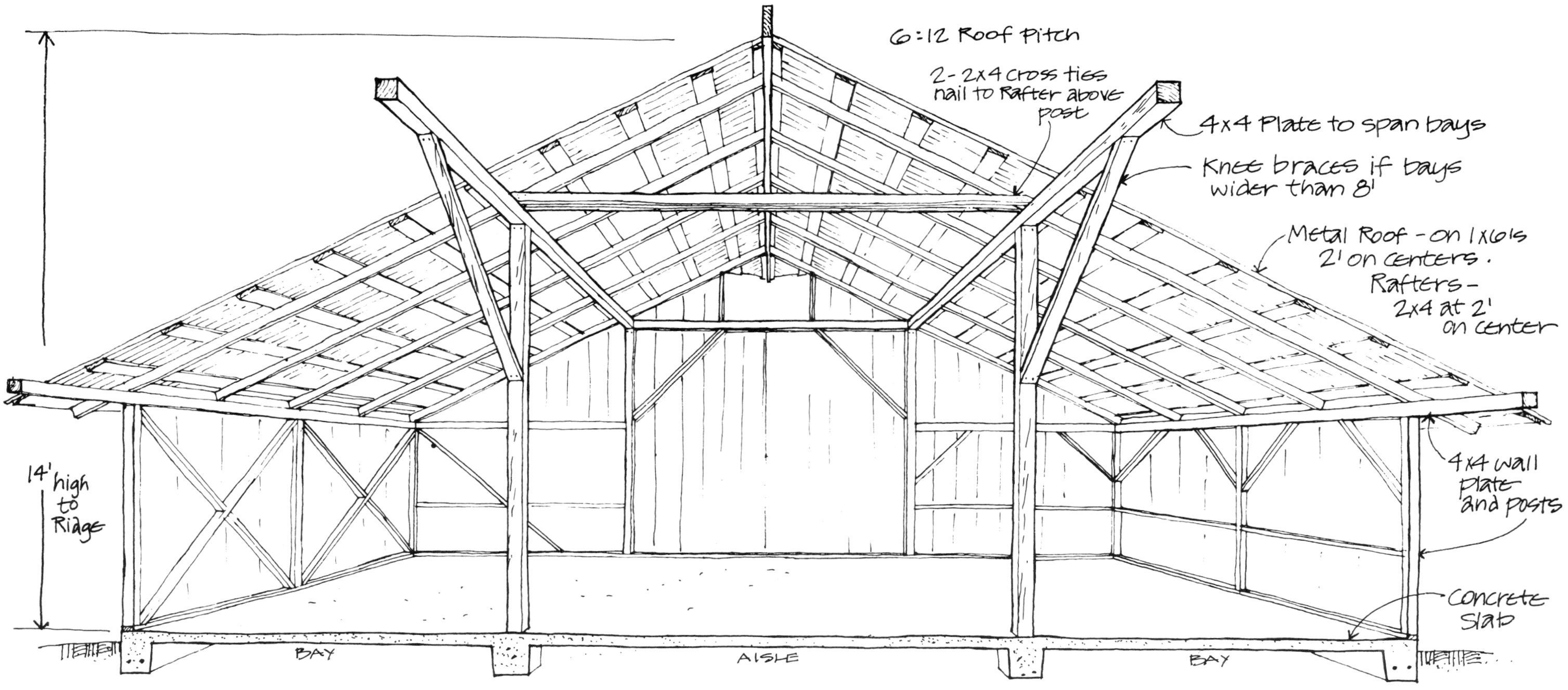

CUTAWAY VIEW

AISLED & BAY-DIVIDED CALIFORNIA **BARN** 30' WIDE — BUILD TO ANY LENGTH.

not to scale

- A WAY TO FRAME A LARGE SPACE WITH LIGHTWEIGHT STRUCTURE - USES: 1x6, 2x4, 2x6, 4x4. USE HEAVIER MEMBERS IF YOU BUILD IN AREAS WITH SNOWFALL. CHECK OUT LOCAL BARNS - ALWAYS - BEFORE BUILDING.

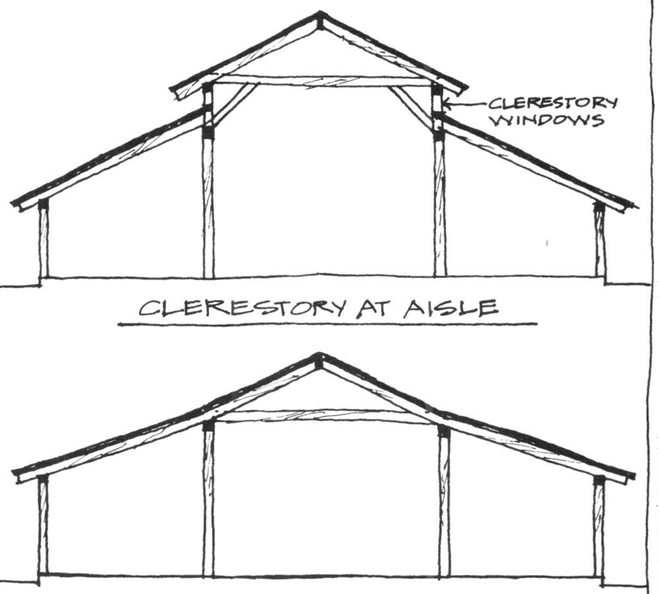

CLERESTORY AT AISLE

CHANGE ROOF PITCH AT AISLE MAKES MORE HEADROOM IN BAYS.

·2 VARIATIONS·

California

The aisled and bay-divided California barn, like a miniature of the great English barns on page 30, uses the 2 rows of posts and plates down the middle to support a lightweight rafter frame. The middle posts also make interior divisions easy, and could support a loft. This frame is easy to expand or add to, and looks best when the eaves are very low on the sides. The basic frame is seen throughout the world.

Small Barn built with 4" poles - California

Large Barn

Cape Cod

WOOD FLOORS for SHEDS

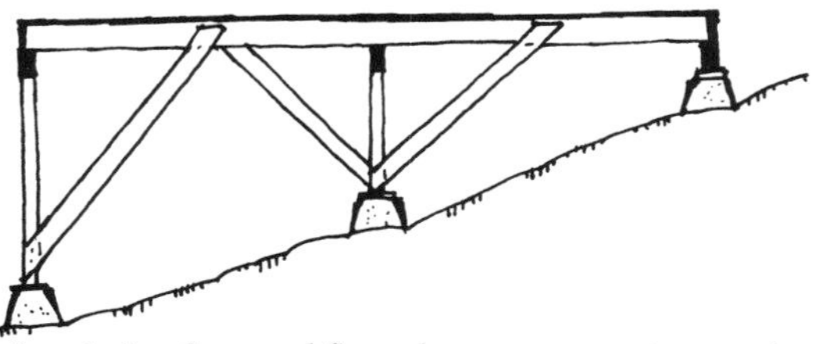

Simplest foundation for wood floors is to use pre-cast concrete piers. You can buy them in a lumber yard for about $1.25 each, or make them yourself in this manner:

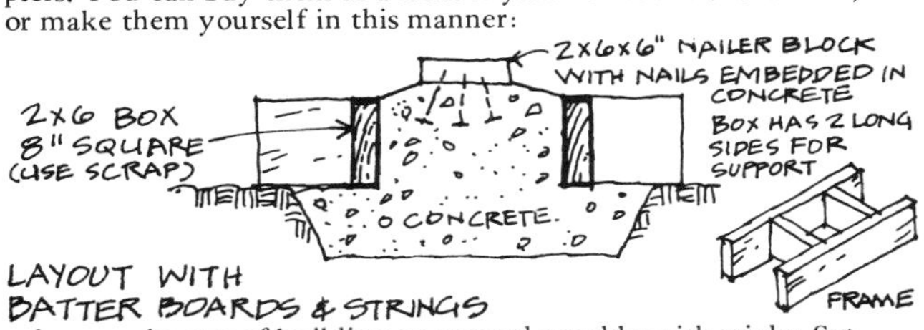

LAYOUT WITH BATTER BOARDS & STRINGS

—draw perimeter of building on ground roughly with sticks. Set stakes at corners.

—set up batter boards about 3' back from corner stakes as shown. Strings that stretch between batter boards allow you to determine then recheck pier locations. Batter boards should be at same height all around.

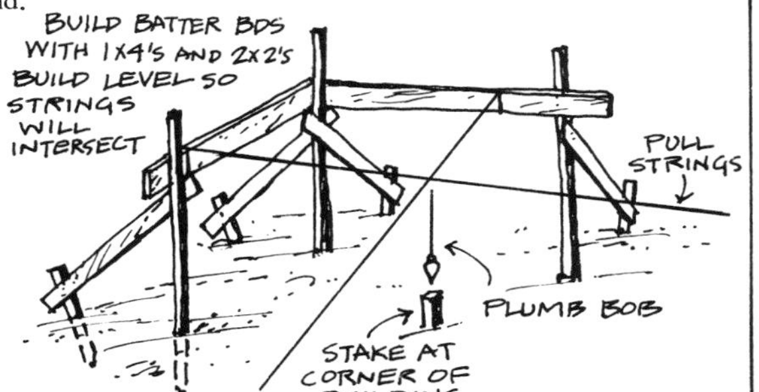

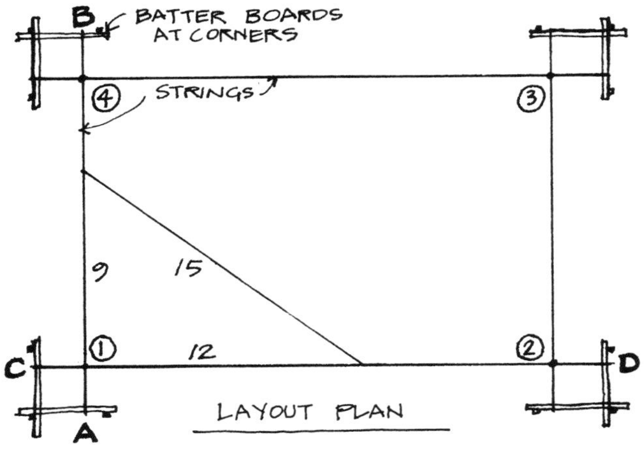

—drive small nails on top of boards at A & B to set one line of building. Make loops both ends piece of string, stretch between A & B.

—Now you want to establish line C – D at right angles to A – B. This is done by the *3-4-5 triangle.* Above corner 1, mark a point on line A – B with felt-tipped pen or piece of string. Now from that point, mark a point 9' out towards B from corner 1. Next stretch out string C – D temporarily, making sure with plumb bob that it crosses corner 1. Then measure 12' along C – D from corner 1, put a mark on the string. Now with tape measure establish third leg of the *3-4-5* (or 9-12-15) *triangle:* hold tape to line up a diagonal of 15' with two points on strings. Move line C – D (keeping string above corner 1) until everything intersects. Put in nails for permanent line C – D.

—corners 2 & 4 are established by measuring along strings from corner 1. Get corner 3 by putting in the other strings and measuring along them.

—Now take diagonals (corners 1-3, 2-4); they have to be equal for building to be square. Move nails and strings around until diagonals are equal. Also check corner measurements from intersecting strings.

PIER LOCATION

—Now you're ready to locate piers. Drop a plumb bob at each of four corners, and along lines for other piers. When bob is dropped on ground, take shovel, scribe 24" square around it. Remove bob, dig down to undisturbed earth, making hole about twice as big as pier base – this gives you room for moving it around in final alignment.

—In cold winter areas, you must dig down to the frost line or rest piers on well-drained gravel beds.

—For buildings over 100 sq. ft. you should put a sack of ready mix (or equivalent) under each pier.

SETTING AND ALIGNING PIERS

—set in piers. Line up with strings and plumb bobs. If no concrete underneath, tilt back pier, scrape high spots until level with straight-claw hammer. Use hand level.

—When all piers are in, determine the highest, then measure how much lower is each other pier so you can cut posts for joists. Transit is best if you can borrow one. A straight 2 x 4 with level on top works, but not suitable for long measurements. The old water level method, shown below is a good way to determine post heights.

—measure height of each pier, subtract from height of highest, cut 4 x 4 posts, toenail to pier block and you're ready to nail on girders and/or joists.

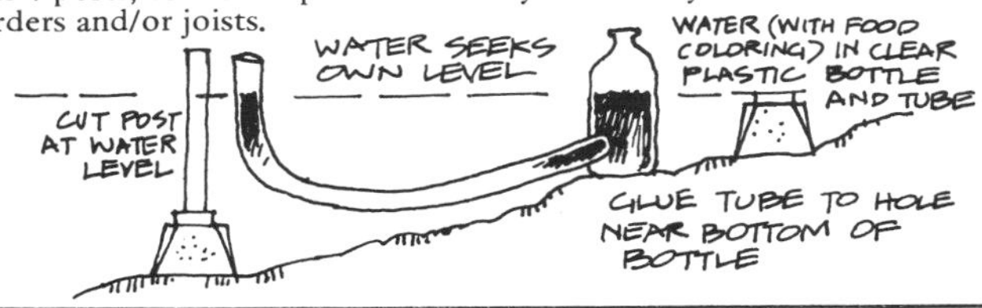

WOOD FLOOR FRAMING

There are 2 simple wood floor systems which you can use on the pier foundation. A joist and girder frame with ¾" flooring, or a girder frame with 2x6 T&G.

Joist and Girder

—2x6 joists spaced at 16" on center can span 6 feet. 2x8 can span 10'. 2x12 up to 16'.

—4x6 (or double 2x6's nailed together) girders spaced 6 feet apart to support 2x6's. This girder can span 6 feet between piers. 4x8 can span 10', 4x12 up to 16'.

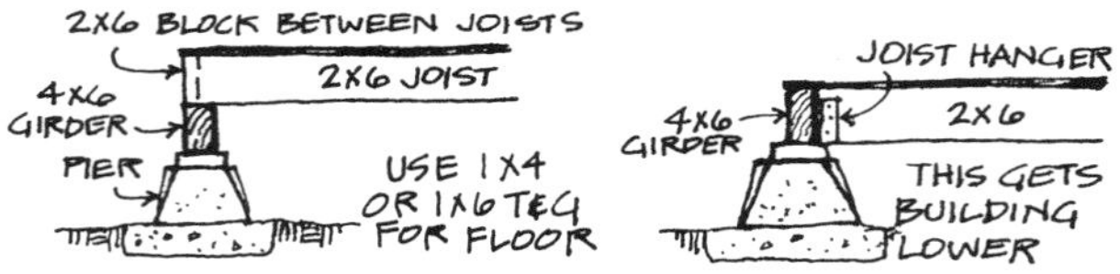

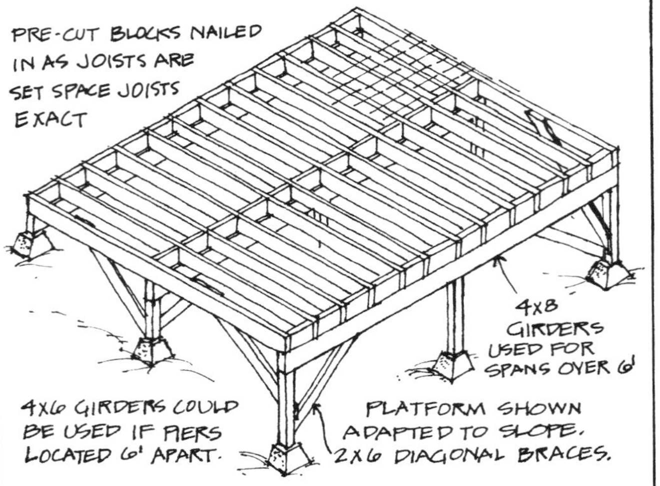

12'X16' PLATFORM FRAMED WITH 2X6'S AT 16" ON CENTER ON 4X8 GIRDERS.
- BUILDING ON SLOPE IS EASIER THE FEWER PIERS YOU HAVE TO LOCATE.
- JOIST HANGERS USE LESS LUMBER.

Girder and 2x6 T&G

—4x6 girders spaced at 4 feet on center can span 6 feet. 4x8 can span 12 feet, 4x12 up to 18 feet. Block between girders every 10 feet.

—2x6 T&G can span 4 feet.

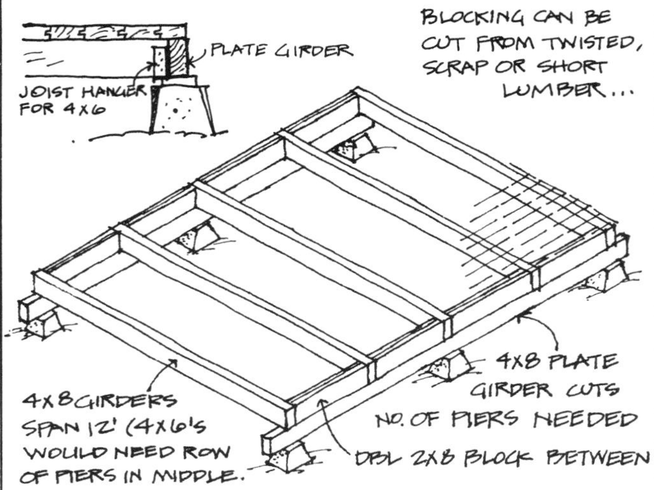

12'X16' PLATFORM FRAMED WITH 4X8 GIRDERS ON 4X8 PLATE GIRDERS.

Building codes near cities usually will allow sheds up to 400 square feet of floor area to use the pier foundation, but require a standard continuous foundation, pole or other type for larger buildings. Check books listed on the bottom of this page.

TUBE FORM

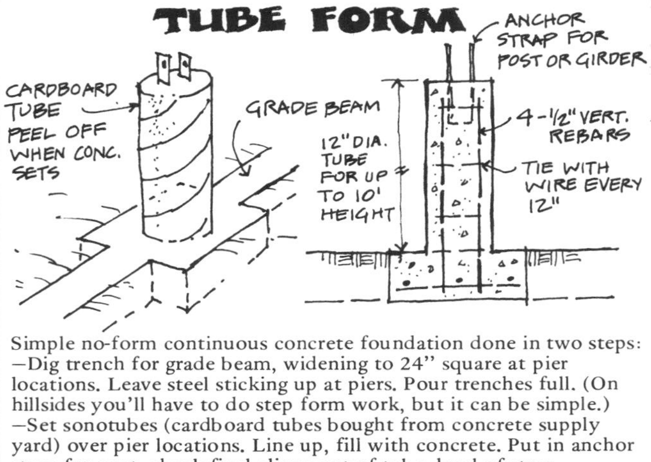

Simple no-form continuous concrete foundation done in two steps:

—Dig trench for grade beam, widening to 24" square at pier locations. Leave steel sticking up at piers. Pour trenches full. (On hillsides you'll have to do step form work, but it can be simple.)

—Set sonotubes (cardboard tubes bought from concrete supply yard) over pier locations. Line up, fill with concrete. Put in anchor strap for post, check final alignment of tube, level of strap.

EARTHEN FLOORS

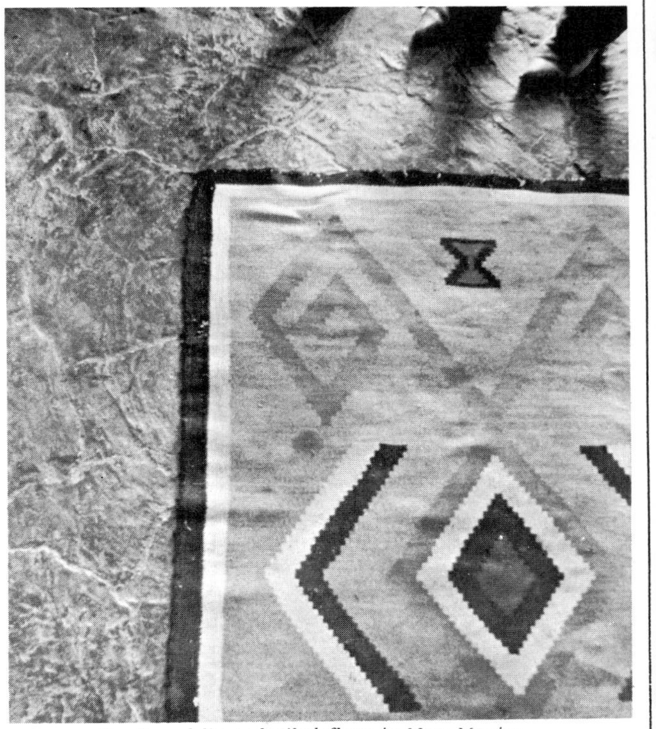

Earth and linseed oiled floor in New Mexico.

Today in the Taos area and in the vicinity of Santa Cruz one occasionally sees earthern floors of a different sort. A deep burnished brown in color, they appear to be constructed of carefully fitted flagstone, but they are actually made of mud. They are completely durable and satisfactory under ordinary conditions of wear and tear and are extraordinarily attractive. Thin high heels or hobnailed boots might mar them, but floors of this kind have withstood everyday traffic for twenty years protected only by an occasional waxing.

Anyone who has made mud pies can handle the simple though laborious process by which they are constructed. This is the opinion of Miss Augustine Stoll, who had these handsome floors in her Santa Cruz ranch house and more recently in her home at Placitas. The method of construction may vary a little, but this is Miss Stoll's recipe.

Any soil which can be used for making adobe bricks may be used for this floor. Screen the soil to remove lumps and rocks. Mix it thoroughly with water to the same consistency as the mud used to make adobes. "About like fudge," is the way Miss Stoll describes it.

Pour a slab about four inches thick. The area upon which the floor is to be poured must be smooth and level and dampened. All of the floor for one room must be poured in one day if it is to appear uniform and so it will not show a joint where the pouring began again. Level the slab with hand towels and by eye rather than with a screed which may leave marks on the surface.

Allow it to dry approximately ten days. At the end of this period it may be safely walked on. A number of cracks will have developed during drying. These cracks are now filled with adobe slip, mud thin enough to pour from a container with a spout such as a watering can.

After the slip has dried (usually three or four days), paint the floor liberally with boiled linseed oil. In about a week when the floor has cured and is dry again, apply a second coat of boiled linseed oil which has been thinned with one quart of turpentine to the gallon of linseed oil to hasten the drying time. Finish with several thin coats of a good floor wax.

If further cracks develop with time, the floor is easily repaired by filling the crack with slip and treating the area with linseed oil. If a section becomes damaged, you can cut away the damaged area, fill it with mud, and go through the steps outlined above....

From *Build with Adobe,*
Marcia Southwick,
The Swallow Press, 1965

POLE FOOTINGS & HOUSES

Pressure treated poles embedded in the ground can be used for a footing like the tube form concrete pier (shown at left), or used as footing and posts for a house frame (at right). Concrete or sand can be used to backfill the hole, depending on soil conditions and the weight of the structure.

Further information: *FHA Pole House Construction,* 50 cents. American Wood Preservers Institute, 1651 Old Meadow Road, McLean, Va. 22101.

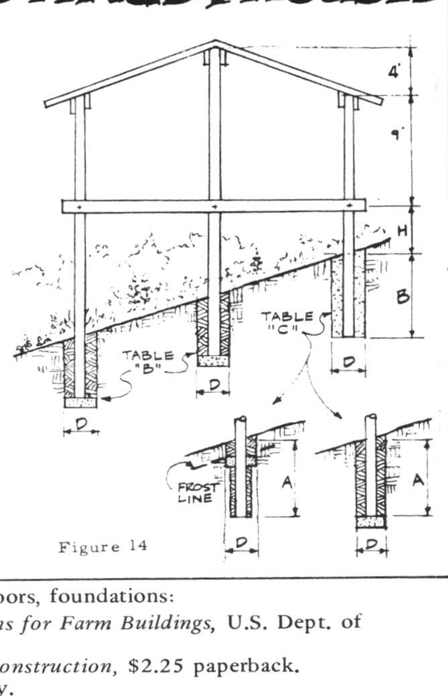

Figure 14

Further reading on floors, foundations:

Pamphlet: *Foundations for Farm Buildings,* U.S. Dept. of Agriculture.

Wood-Frame House Construction, $2.25 paperback. Details in bibliography.

CONCRETE FLOORS

Concrete is a practical floor material. It's cheaper than wood, and with proper precautions doesn't have to be icy cold. You need relatively level ground for concrete; if you're on a hillside, it's generally easier to build a wood floor than to cut into the hillside with a bulldozer.

TYPICAL CONCRETE FLOOR

–typical concrete floor is 3"-4" thick, reinforced with 6" X 6" X ¼" steel mesh. You can make a floor without the mesh, but it should be used in earthquake areas. The mesh stops the floor from drifting apart if the slab cracks. Thickness can vary. I built a 1½" thick concrete floor that seemed to work and Ken Kern describes a 1" concrete floor in his book *The Owner Built Home.*
–perimeter of floor is deeper, as some weight is there.
–½" reinforcing steel goes around footing; it should be suspended above ground (tie it with wire to rocks).
–gravel underneath for drainage, and to provide solid, level place for concrete slab to rest on.
–plastic sheet important as vapor barrier. Put it over the gravel and under everything else.
–hooked end of anchor bolt is embedded in the concrete.

The ground must be level. This is done with a long straight 2 X 4, that can rotate from a stake in the center. Use the 2 X 4 on edge, with a 4' level on top. Find the low spot by moving the board around and reading the level. You can tell how much you have to dig out at the high spots by lifting the low end of the level (which is on top of the 2 X 4) until the bubble reads true. If you have to lift it up 1", the grade gains 1" in 4', 2" in 8', etc.

If you have low spots that must be filled in (say from digging out a stone) never use loose fill, rather fill with gravel or decomposed granite, then tamp firmly. Loose dirt will settle and leave hollows under the slab which may later crack. Final level can be obtained after the forms and screed guides are set by using the screed board and reading under it for level.

FORMS

You can rent steel stakes for forms – they're very useful, much easier than making stakes of wood. If form starts to give when you're pouring, shovel dirt in behind it.

Position the form so it's level. The top of the wood is floor level and concrete is poured up to that level. Set one form board level, butt the next one to it, level it and so on. Check with the long 2 X 3 to avoid accumulating error. Don't let any stakes or nails stick up above the form boards, as they'll hang up the screed and cause grief when things are hectic. Once the form is level, no tightrope acts or anything that will change the level or push it out of line.

Setting the rebar is very important. The reinforced foundation supports the weight of the building, and anchors it. There should be no steel in contact with the ground. For example, it's fairly common to drive stakes of rebar into the ground to wire horizontal rebar to; this provides a path for rust to travel through all the steel in the slab, which it will do. Rather than this, wire the steel together at overlaps and prop it up by wiring it to rocks or broken bricks.

SCREEDING

Once you start pouring concrete you need a quick way of achieving an approximately uniform level; the method is *screeding.*

HOME MIXING

If you mix yourself, convert to tons and use the following table to determine how many sacks of cement to order, depending on how rich a mix you use. (1:2:4 is common for floors – that means 1 part cement/2 parts sand/4 parts gravel)

Mix Proportions

Cement	Sand	Gravel	Total Mix of These	Constant
1	2	3	1.5	4
1	2	4	1.6	3½
1	3	4	1.7	3
1	3	5	1.8	2½

Rent a big mixer, ½ sack or larger. Full sack is best. Don't mess with small underpowered mixers, as the first part of the pour will be setting up before you've poured the last. Get a contractor's wheelbarrow with pneumatic tire for carrying concrete from mixer to floor. It's important not to let the materials in the mixer get dry so that they pack. When water is added to a pack like this it takes much time to penetrate and slows things down. Therefore, keep careful track of how much water you use for the mix; then begin adding water first. Try to get a 55 gallon drum, with hose slowly filling, and two buckets to dip water. Or have a hose filling two buckets while shovelling.

First water, then gravel, then sand. Cement last.

READYMIX

The larger the job, the more you should consider ready mix. A standard truck can hold 7 yards, six in the hills. You should avoid getting a "hot mix" – a load that will set rapidly. One of the features of concrete is that it will not set while mixing, but if set time has passed while in the truck, you can end up with several tons of unwanted and very hard concrete. To avoid this, make sure the driver comes to the site beforehand so he doesn't get lost. Make sure he has washed out his last load, so it doesn't catalyze your load.

Get all the advice you can from the driver, in fact you can let him help run the pour. Most of what I know about concrete came from Bud Golden, a truck-driver from Monterey who delivered 12 truck loads of concrete to us while we were building a huge house. Each time he would arrive with a load we'd let him run the job – he'd tell us what to do.

Pumping: It's possible to pump concrete up hill, or in to inaccessable sites – as much as 250 ft. from the road. It may cost $50 or more and be well worth it. Ask the ready mix company.

With ready mix you must be prepared to work fast, especially if the weather is hot. A truck has about 20 ft. of chutes; if that's not enough to reach, you may have to build additional chutes, or wheelbarrow it from the truck, in which case you'll need ramps. With heavy wheelbarrows-full have a helper running backwards, holding the front. The main thing is to be ready before the truck arrives. You usually get 45 minutes per truckload before they charge overtime.

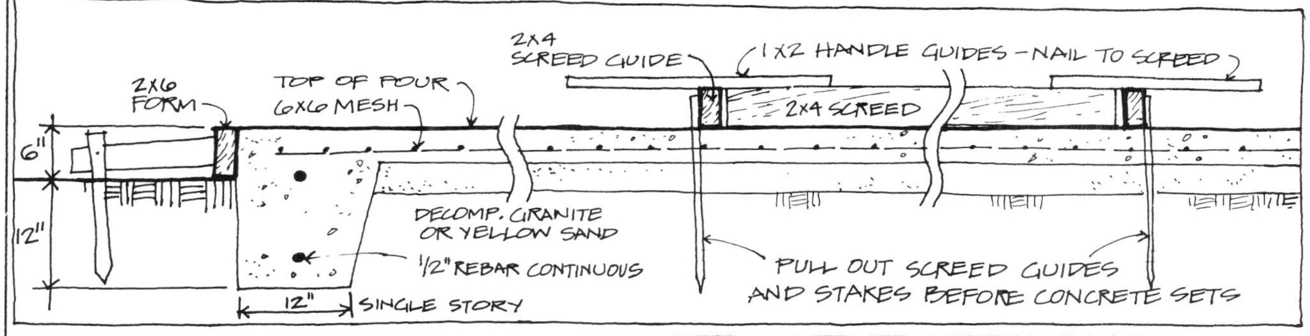

ABOVE:
CROSS SECTION VIEW THRU FOOTING & SLAB

The bottom of the screed board is at finished floor level. The 1 X 2 handle guides slide on top of the screed guides.

Note: when screeding between the screed guide and the *form,* one guide bar runs on the screed guide, and the bottom of the screed board runs along the top of the form. You'll see this if you mentally pick up the 2 X 4 screed board in the above sketch and move it over to the left, to screed the section next to the form.

As the concrete is poured, two men move the 2 X 4 back and forth – wading through the concrete with rubber boots on – and it levels the concrete, ready for final troweling. The push-pull action of the screed is very important, as it won't work to just pull the screed along.

Keep a flat-nosed shovel and rake handy to remove concrete if too much is dumped in front of the screed.

Once you have a level to pour concrete against, anchor bolts in place, gravel down, plastic on top, steel mesh on top, reinforcing steel in trench, screed guide installed, etc. you're ready to pour.

You can either mix yourself or order a ready-mix truck. The sand-gravel mix, or the concrete comes either by the cubic yard or by the ton.
–to change cubic feet to cubic yard, divide by 27
–to change cubic feet to tons, divide by 20 (someone said this should be 13)
–to find cubic feet of floor volume: multiply .7854 times diameter squared (.7854 d^2) and multiply by the depth in feet (4/12 if 4" deep floor)
–to figure content of footings (ditch around perimeter) multiply width of trench in feet (as 8/12 for form 8" wide) by depth of trench in feet (as 16/12 for a form 16" deep) by the length of the trench (circumference of dome)
–perimeter of circle: 2 r (2 x 3.14 x radius)
Add total for floor to total for footings to get total cubic feet of cement. Convert to tons or yards.

Salvaged chunks of broken concrete slab used for floor and retaining wall. Built by Jon Lazell.

POURING

Three is a good size crew, more is better. One on mixer or guiding truck chute, two working with concrete, alternating on wheelbarrow and leveling. Have floats and finishing trowels ready. Use gloves, lemon juice, or some kind of oil on your hands before the pour for protection. Concrete will crack your skin. If you haven't used a plastic ground sheet, wet earth before pouring. This gives you more time to work. Start early in the morning and keep moving. If concrete starts to set before you're ready you have problems.

Dump concrete in place, screed off level.

If you use steel mesh, pull it off the ground as pouring progresses. Do this with 4' hook made of rebar – reach through the concrete and pull it into the middle of the slab. Keep checking anchor bolts so they're upright and at proper height to grab ground struts. If you stop for lunch clean tools off.

Take a stick and puddle the footings – poke concrete up and down to fill any voids in the trench.

TAMPING

A tamp is an angle iron frame about 10" X 45" with a 3/4" mesh expanded metal screen welded to it, and waist level handles. Tamping pushes the gravel down from the surface so it won't get in the way of troweling. It helps to level the slab and brings water, sand and cement to the surface. Tamping is done right after screeding. You can rent a tamp.

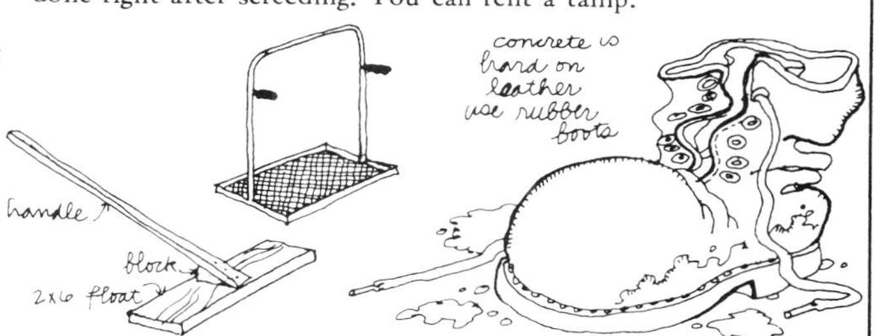

BULLFLOAT

The bullfloat is used after tamping to get an even flow of the mixture of sand, cement, water. Work it over large areas, while the concrete is still wet. Lift the leading edge slightly so it doesn't dig in. You can make one of 1" or 2" by 6" board about 12" long with a 1" X 2" handle long enough to reach the center of the slab from the outer edge.

WOODFLOAT

When the water raised by tamping and bullfloating begins to absorb and the surface looks sugared or frosted, use the woodfloat. It's a slab of wood with a handle – cheap to buy, easy to make. Use two trowels, one to lean on while reaching, the other to trowel. The leaning trowel can be the steel trowel to be used later. You should have two kneelpads of ¾" plywood a ft. sq. You are ready to wood float when the pads don't sink more than ¼" with your weight on them. Work in sweeps to obtain excess. A good woodfloat finish, or a spongefloat finish is an OK bonding surface for tile or brick. Keep in mind that the first part of the pour may be ready for woodfloating first, even though the rest of the slab is still wet.

STEEL TROWEL

The steel trowel is used to polish, seal and waterproof the slab. The slab should be very level by this time as only very small defects can now be corrected. The slab is ready for the steel trowel when the trowel raises a polish and the kneelboards make little or no mark on the surface. Again, use a trowel to lean on. Lift an edge of the trowel so it doesn't dig in.

If an area is too dry, position the steel as flat as possible and move in a fast 1 ft. diameter circle, pressing hard. This will draw water. If not, sprinkle water on, with a little cement.

You can get a variety of finishes, including a floor that is trowled too smoothly, which will invite later spills. You can get a slight texture by working up some moisture by the circular flat trowel motion, letting the suction under the blade draw up slight ridges.

TIPS

–if kneelboard sticks to slab, pry up one corner with the steel trowel to break vacuum.
–floats and trowels must be perfectly clean. Old cement or rust makes them useless.
–an edging tool can be used to go around the slab at the edge of the form boards to make a nicely rounded transition. This tool is used when the slab is still quite wet.
–if you suspect you'll run out of daylight get floodlights.
–put up a barricade to keep stray people and pets off the curing slab.
–don't walk on the slab for three days.
–try to cover the slab with plastic or fabric and keep it wet for a few days, especially in hot weather, as it is stronger if it cures slowly.
–you can insert wood in the slab for later nailing-to. Do this by driving 16d nails in 2 X 4's. Then put wood in concrete, nails down.

CONCRETE FLOOR

This is a lightweight concrete floor described in *The Owner Built Home* used in India. It is 1" thick, rests on plunger piles, which are formed by driving a crowbar into the ground 3' deep. They are spaced 3' on centers, and then filled with fine concrete. There are two layers of concrete, each ½" thick, spread over Hessian, a type of burlap. After a few weeks the loose earth filling settles, leaving an air space under the slab, which is held up by the piles. The floor reportedly took a load of 450 lbs. per sq. ft. (most building codes require 30-50 lbs. per sq. ft. minimum).

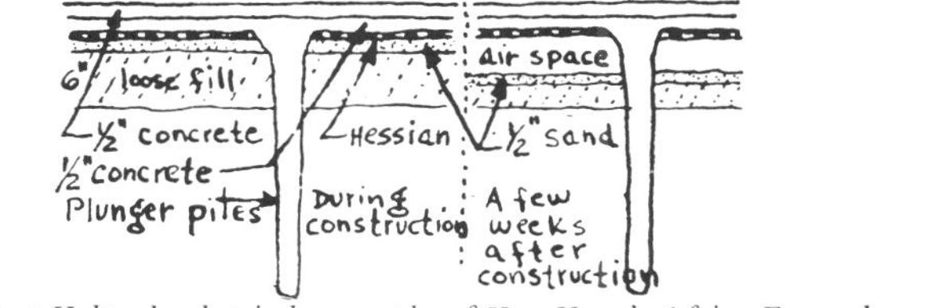

Bart Kaltenbach tried a sample of Ken Kern's Africa Formula floor: 8 parts sawdust, 6 sand, 4 cement, 1 lime, 10 days damp cure. The sample looked good, Bart says it would make a good floor. The sawdust should make the floor warmer, softer on bare feet. See Kern's Owner Built Home *for other ideas on concrete floors.*

WINDOWS & DOORS

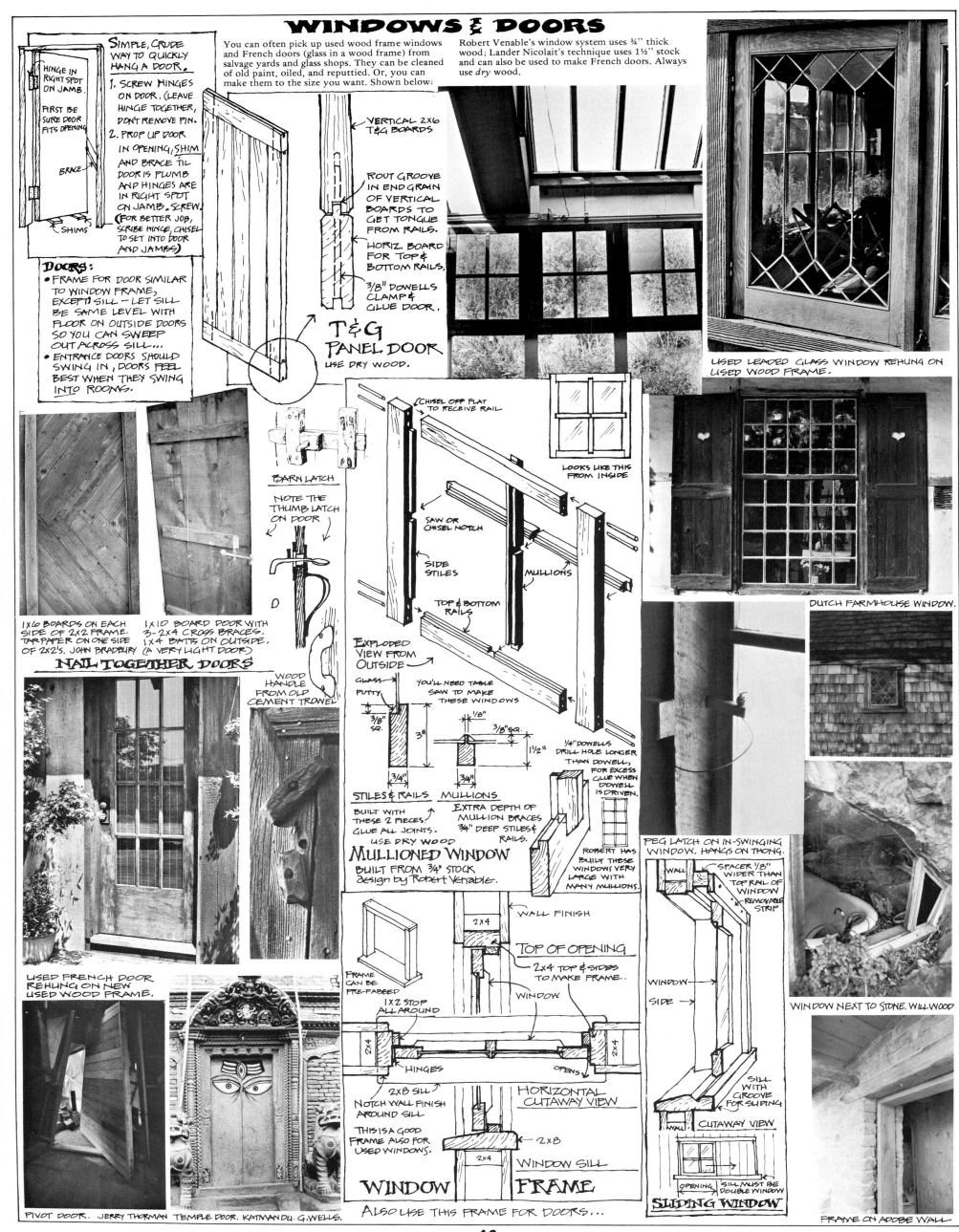

You can often pick up used wood frame windows and French doors (glass in a wood frame) from salvage yards and glass shops. They can be cleaned of old paint, oiled, and reputtied. Or, you can make them to the size you want. Shown below: Robert Venable's window system uses ¾" thick wood; Lander Nicolait's technique uses 1½" stock and can also be used to make French doors. Always use *dry* wood.

SIMPLE, CRUDE WAY TO QUICKLY HANG A DOOR.

1. SCREW HINGES ON DOOR. (LEAVE HINGE TOGETHER, DON'T REMOVE PIN.
2. PROP UP DOOR IN OPENING, SHIM AND BRACE TIL DOOR IS PLUMB AND HINGES ARE IN RIGHT SPOT ON JAMB. SCREW. (FOR BETTER JOB, SCRIBE HINGE, CHISEL TO SET INTO DOOR AND JAMBS)

DOORS:
- FRAME FOR DOOR SIMILAR TO WINDOW FRAME, EXCEPT SILL – LET SILL BE SAME LEVEL WITH FLOOR ON OUTSIDE DOORS SO YOU CAN SWEEP OUT ACROSS SILL...
- ENTRANCE DOORS SHOULD SWING IN, DOORS FEEL BEST WHEN THEY SWING INTO ROOMS.

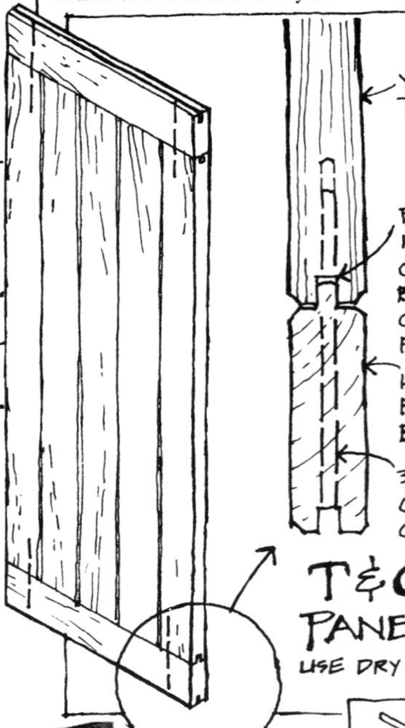

T&G PANEL DOOR
USE DRY WOOD.

USED LEADED GLASS WINDOW REHUNG ON USED WOOD FRAME.

1x6 BOARDS ON EACH SIDE OF 2x2 FRAME. TARPAPER ON ONE SIDE OF 2x2'S. JOHN BRADBURY

1x10 BOARD DOOR WITH 3-2x4 CROSS BRACES. 1x4 BATTS ON OUTSIDE. (A VERY LIGHT DOOR)

NAIL TOGETHER DOORS

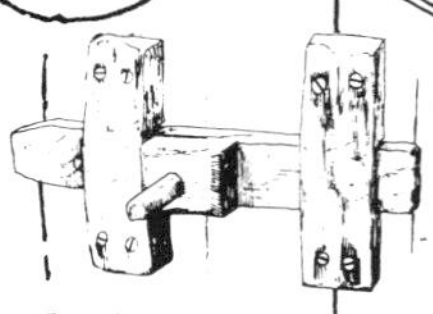

BARN LATCH

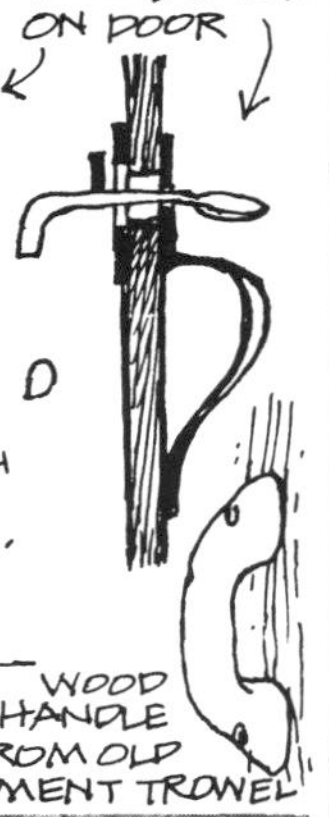

NOTE THE THUMB LATCH ON DOOR

WOOD HANDLE FROM OLD CEMENT TROWEL

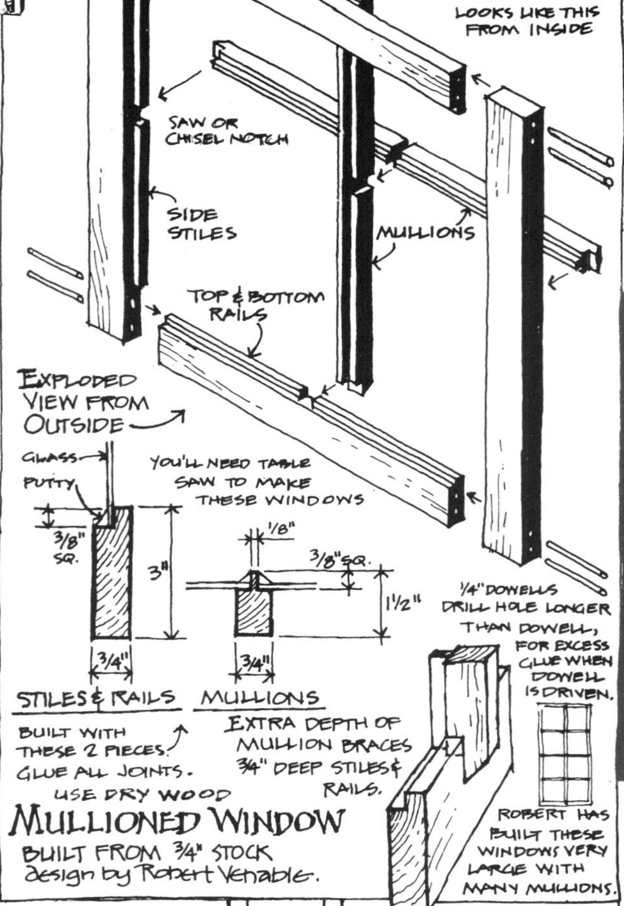

MULLIONED WINDOW
BUILT FROM ¾" STOCK
design by Robert Venable.

DUTCH FARMHOUSE WINDOW.

PEG LATCH ON IN-SWINGING WINDOW. HANGS ON THONG.

WINDOW NEXT TO STONE. WILL WOOD

USED FRENCH DOOR REHUNG ON NEW USED WOOD FRAME.

PIVOT DOOR. JERRY THORMAN

TEMPLE DOOR. KATMANDU. G. WELLS.

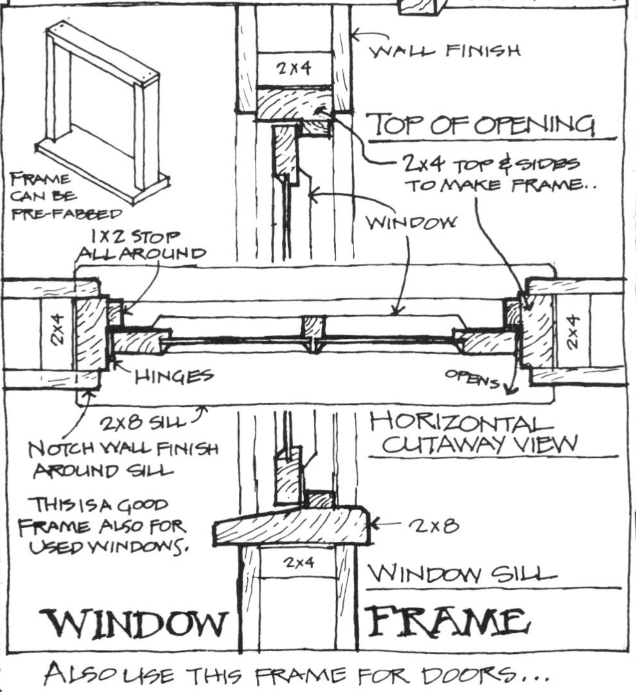

WINDOW FRAME

ALSO USE THIS FRAME FOR DOORS...

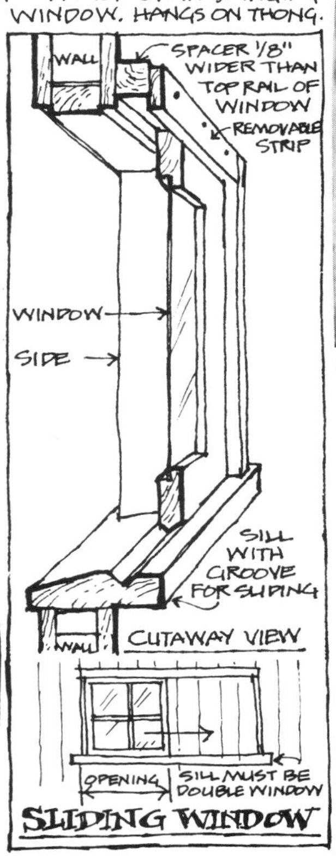

SLIDING WINDOW

FRAME ON ADOBE WALL

ROOFING & SKYLIGHTS

Roll roofing and asphalt shingles are easy to install (instructions come with each package), are fire resistant and by far the cheapest roofing material. Summer 1973 California prices of roofing products per 100 sq. ft.: 90 lb. roll roofing: $7.00; corrugated steel roofing: about $15.00; seal-down tab shingles: $17.00; corrugated fiberglass: $19.00; no. 2 cedar shingles: $52.50; medium cedar shakes: $67.50.

Translucent fiberglass (comes in rolls) is a good skylight material: about 45 cents per sq. ft.

JET AIRPLANE COCKPIT DOME SKYLIGHT

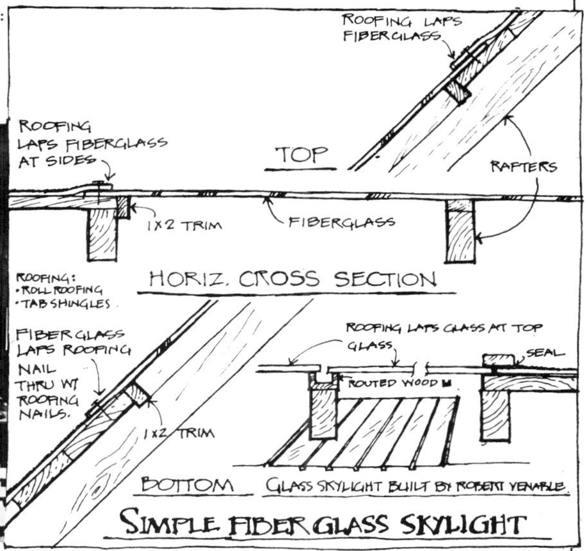

SIMPLE FIBERGLASS SKYLIGHT

STAIRS ARE BEAUTIFUL WHEN BUILT VERY SIMPLY.

FLAT ROLL TRANSLUCENT FIBERGLASS SKYLIGHT

ROLL ROOFING EASY TO USE ON POLYGONAL BUILDINGS

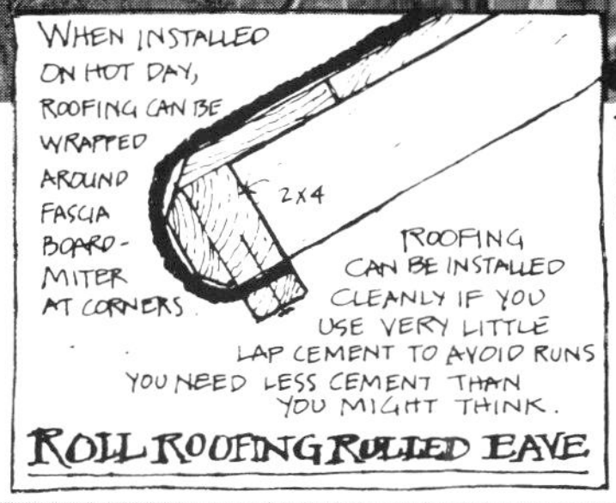

ROLL ROOFING ROLLED EAVE

MORE DOORS & WINDOWS...

TAB SHINGLES IN SNAKESKIN PATTERN (WITH VARIED COLORS) BY OLAF PALM

SIMPLE WINDOW FRAME USING 2X4S BY LANDER NICOLAIT

EASTERN EUROPE

CAR WINDSHIELD WINDOW IN ROUND ADOBE

DOOR AT VAL AGNOLI'S

TOOLS & TIPS

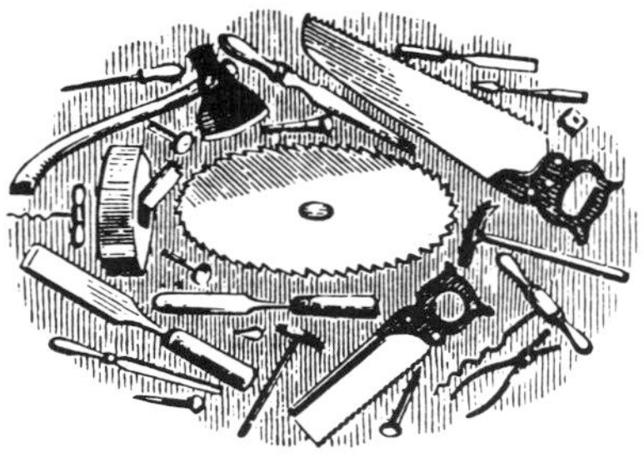

Hammers...wood handle most comfortable but won't last if you're wrecking, pulling lots of nails. Fiberglass shank 16 oz. Plumb has springy feel. Solid steel shank hammers are probably the toughest. Keep head dry and rough. Tubular steel shank 20 oz., good for wrecking and general heavy work and framing. Straight claw is good for digging and prying.

Saws...we've heard that hand saw steel has dropped in quality since 1963 (Jock Angus, who plays the saw, says you can't get high notes on new saws)...best new saw made is expensive Sandvik with black plastic handle, but try to find old saw...and keep it sharp...best all purpose saw is 8 pt. cross cut...have teeth of handsaws set extra heavily. This way you can sharpen with triangular file several times in between professional sharpenings.

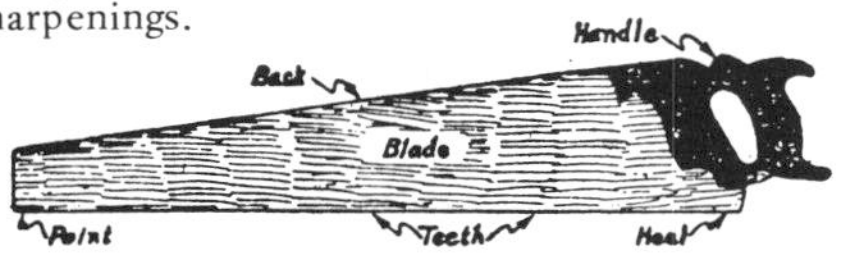

Circular saws...the worm-drive Skilsaw shown below (sells for about $100) is recommended by many carpenters, but we like the smaller, lighter Porter-Cable, it's easier to move around and use with one hand.

Framing square...get the Stanley, their instruction booklet shows how to do many cuts. Picture below shows how to determine a roof pitch, this one is 8 rise/12 run. Also use to cut notches in stair stringers.

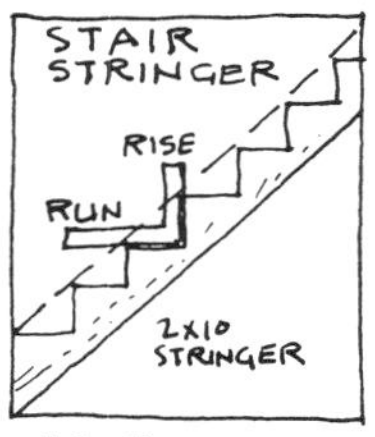

Chalk line...used for marking straight lines, stretch between 2 points and snap on surface.

Small plane...or Stanley surform for shaving off edges of exposed wood — called champfering.

THE TOOL BELT

For general small building, leather apron, costs $15, can last for 10 years if cleaned and oiled regularly. Right side: hammer, ¼"x8" screwdriver, pliers, flat carpenters pencil; left side: combination square, 1½"x7" chisel, nail set. Middle, Stanley 12' power lock tape, 20' tape if you're doing a lot of building. Also, a good sharp pocket knife.

THE TOOL BOX

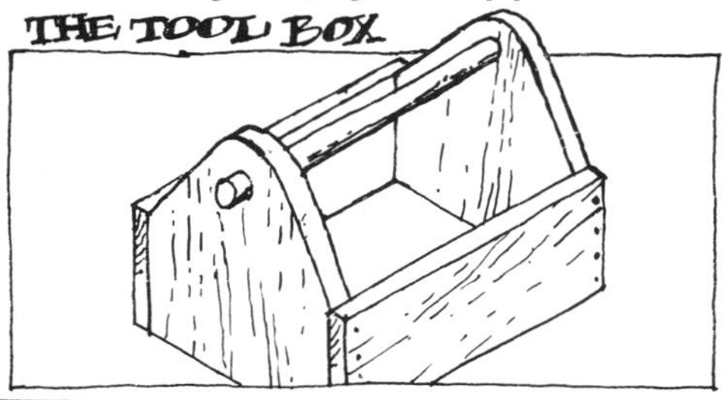

Sawing boards in Kashmir.

To avoid nails splitting wood: blunt nail tips. Some carpenters drill holes in bottom of wood handle hammers, fill with wax, stick nails in. Waxed nails drive easy.

To put a nail in up high, start it one-handed like this...

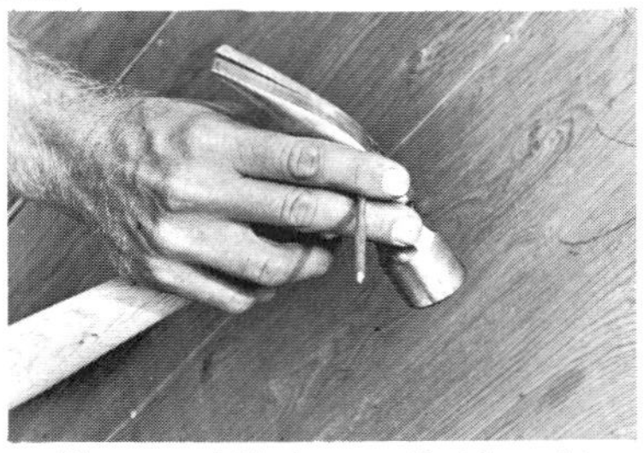

When nailing wood likely to split (dry, thin, near end) stagger nails, don't align on same grain.

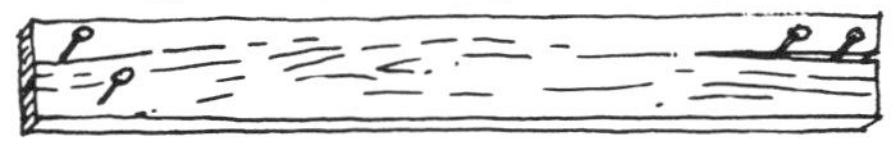

In laying out studs, cut and put top and bottom plates together on floor, mark stud locations with square across both plates. Mark all on one pull of tape so errors won't accumulate.

Diagonal braces...building codes require for single story buildings one diagonal brace every 25' of wall (at corners). Can be done with 2x4 as shown on building pages, or as 1x6 let-in brace...tack 1x6 on studs, mark, take off, cut to ¾" depth with circular saw, chisel out. Another way...put 1x6 or 1x8 wall boards on diagonally.

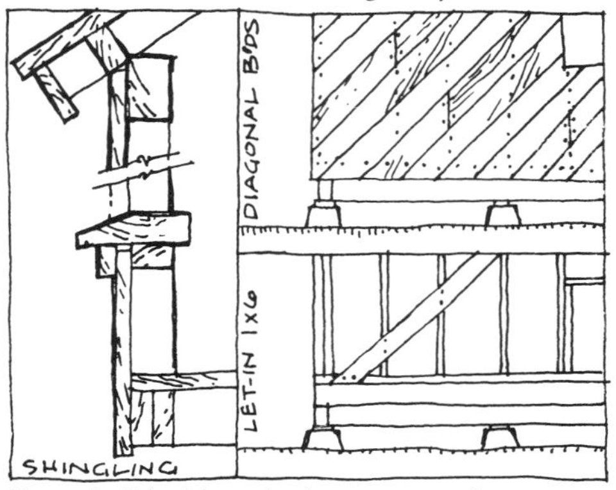

Try to build without having to use caulks, plastics. Overlap outside wood joints to get shingling effect.

Carved corbel block by Will Wood.

Select best pieces from your lumber pile to use on building where you can appreciate them, on doors, around windows, a special wall.

When building with timbers, dish out ends to insure tight joint around edges.

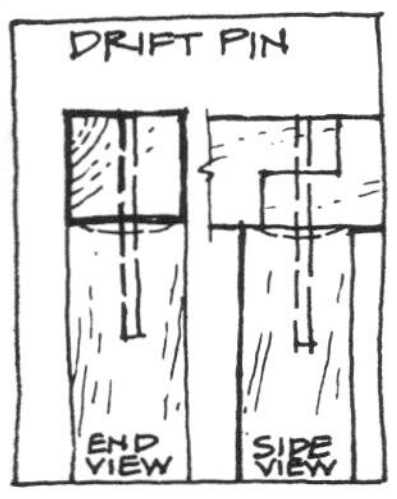

Drift pins...fast, easy way to build with timbers...set beam on post, drill with ½" electricians bit thru beam into post at least 6", then drive ½" dia. rebar pin with sledge.

Nailing on joists...precut blocks and use to space joists as you progressively work across girders.

2x6 T & G floor...if you lay boards radially, the joint can be routed for a cover strip...then sand smooth.

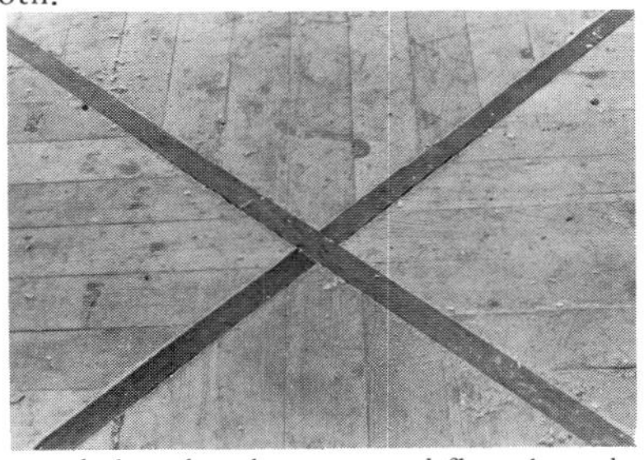

Zen temple burglar alarm...wood floor boards planed to fit so tightly together that they squeaked when walked on...but the Zendo always had earth floor so squeaks wouldn't distract meditation.

Curved footings...Ed Schertz built curved forms for a free form adobe house slab by bending and staking expanded metal mesh to the shape, then made it rigid by plastering it.

Curving bond beam on adobe walls can be formed with sheet iron strips held in place with 2x4 braces.

Laminated 1" wood frame (nailed and bolted) of Nepenthe Restaurant in Big Sur, designed by Rowan Maiden.

Old doors and windows...use extra long screws when rehanging to get bite into new part of frame.

A quick hinge...finishing nail in frame, screweye in window, bend nail up.

Door handle...find a branch...also good for drawer pulls, coat hooks. You don't have to go to the hardware store for hardware.

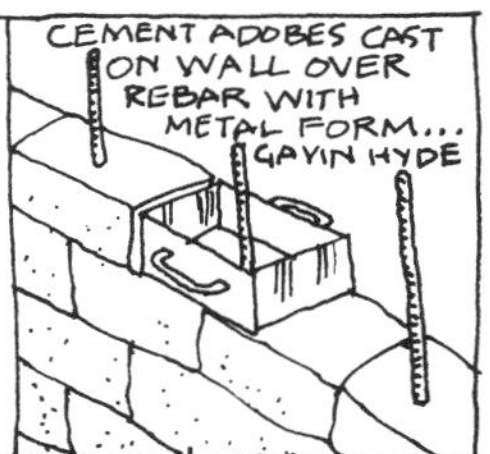

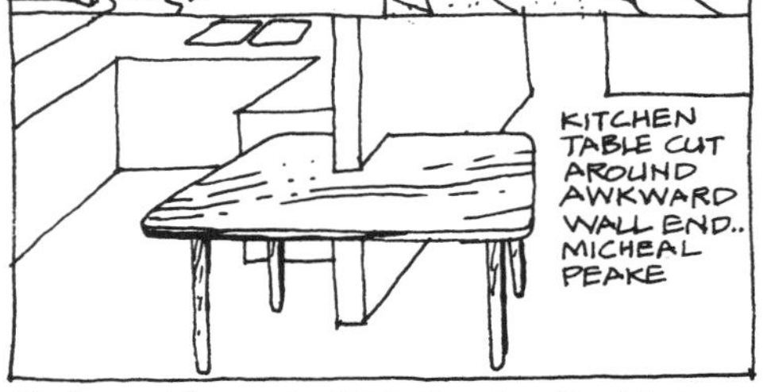

Kitchen cabinets...2x6 T & G makes fine kitchen counter...for drawers, get an old desk or dresser at a yard sale, strip or paint, cut down if necessary to fit under counter.

MITER BOX

FOR CLOSE CABINET WORK

Conrad's scaffold, Cape Cod. Block and tackle lifts heavy pieces of driftwood for loading on jeep.

Further reading on simple building, wiring and plumbing (Details in bibliography):

Wood-frame House Construction
Hand Wood-working Tools
The Practical Handbook of Plumbing and Heating
Wiring Simplified
Sears books on Electrical and Plumbing

used to be... when a rafter was first put in place, folks would say a prayer of thanksgiving and lash a small tree to it...
Rob Thallon

Edward S. Morse travelled to Japan in 1877 to study various species of brachiopods. He began to also make notes on Japanese houses, and during his final visit in 1882 he was admonished by a friend to stop "frittering away your valuable time on lower forms of animal life," and concentrate on documenting traditional Japanese life, predicting it would soon disappear.

Morse published *Japanese Homes and Their Surroundings* in 1885 and it remains today the best source of detailed information on the traditional Japanese residence and its construction.

JAPANESE HOMES

AND

THEIR SURROUNDINGS

EDWARD S. MORSE

LATE DIRECTOR OF THE PEABODY ACADEMY OF SCIENCE;
PROFESSOR OF ZOOLOGY, UNIVERSITY OF TOKYO, JAPAN;
MEMBER OF THE NATIONAL ACADEMY OF SCIENCE;
FELLOW OF THE AMERICAN ACADEMY OF ARTS AND SCIENCES

WITH ILLUSTRATIONS

BY THE AUTHOR

The translation of the terms applied to many parts of the house are quite curious and interesting. The word mune, *signifying the ridge of the house, has the same meaning as with us; the same word is applied to the back of a sword and to the ridge of a mountain. In Korea the ridge of the thatched roof is braided, or at least the thatch seems to be knotted or braided at this point; and the Korean word for the ridge means literally* back-bone, *from its resemblance to the back-bone of a fish.*

In Japan the roof of a house is called yane. *Now,* yane *literally means* house-root; *but how such a term could be applied to the roof is a mystery. I have questioned many intelligent Japanese in regard to this word, and have never received any satisfactory answer as to the reason of its application to the roof of a house. A Korean friend has suggested that the name might have been applied through association: a tree without a root dies, and a house without a roof decays. He also told me that the Chinese character* ne *meant origin.*

In Korea the foundation of a house is called the foot of the house, and the foundation stones are called shoe-stones.

The Japanese word for ceiling is ten-jo, *– literally, "heaven's well." It is an interesting fact that the root of both words,* ceiling *and* ten-jo, *means "heaven."*

It may be interesting, in this connection, to mention a few of the principal tools one commonly sees in use among the Japanese carpenters. After having seen the good and serviceable carpentry, the perfect joints and complex mortises, done by good Japanese workmen, one is astonished to find that they do their work without the aid of certain appliances considered indispensable by similar craftsmen in our country. They have no bench, no vise, no spirit-level, and no bit-stock; and as for labor-saving machinery, they have absolutely nothing. With many places which could be utilized for water-power, the old country saw-mill has not occurred to them. Their tools appear to be roughly made, and of primitive design, though evidently of the best-tempered steel.

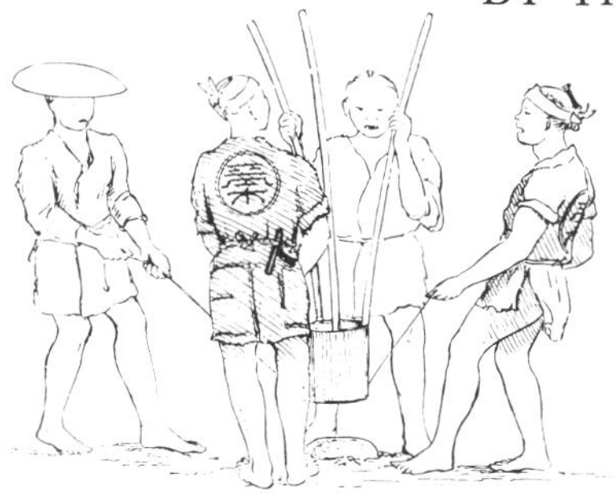

Fig. 5. — Founding down Foundation Stones.

Fig. 4. — Side Framing.

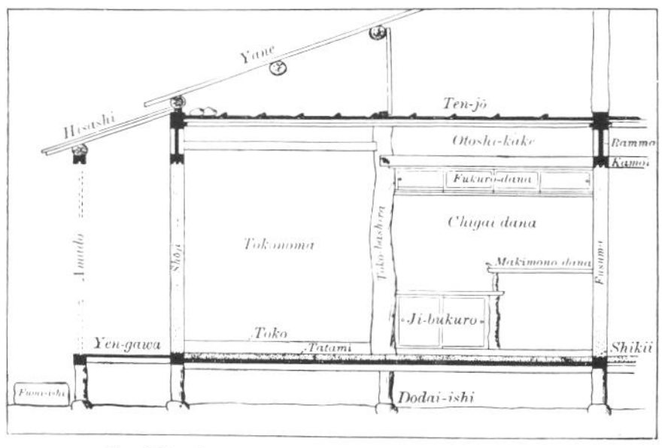

Fig. 103. — Section through Verandah and Guest-room.

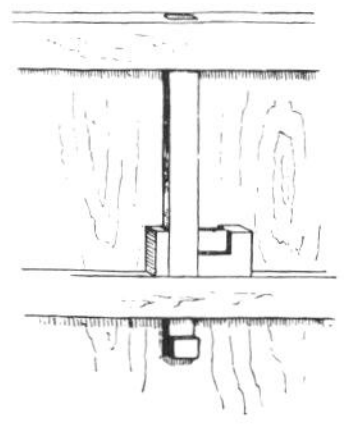

Fig. 233. — Rain-door Lock unbolted.

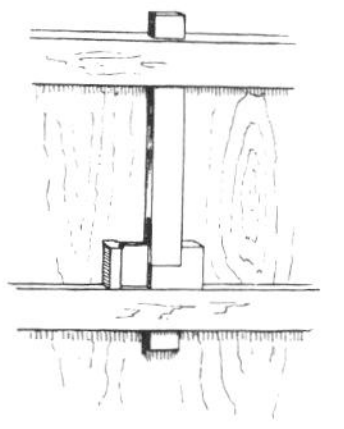

Fig. 234. — Rain-door Lock bolted.

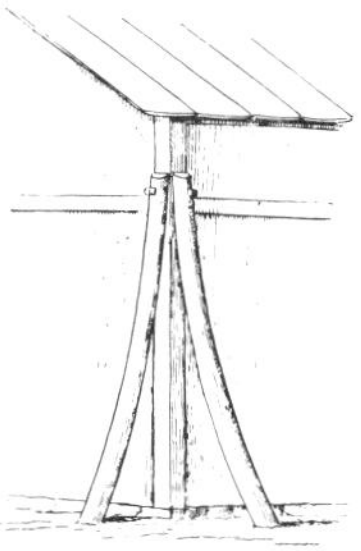

Fig. 13. — Outside Braces.

Fig. 127. — Guest-room of Dwelling in Tokio.

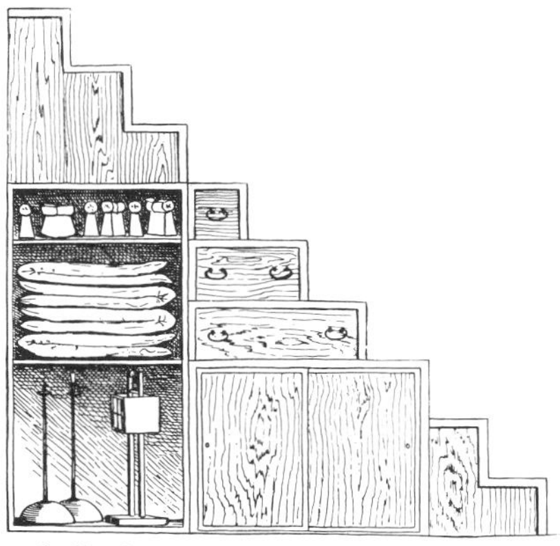

Fig. 177. — Kitchen Closet, Drawers, Cupboard, and Stairs combined.

Fig. 271. — Summer-house in Private Garden, Tokio.

Fig. 36. — Street View of Dwelling in Tokio.

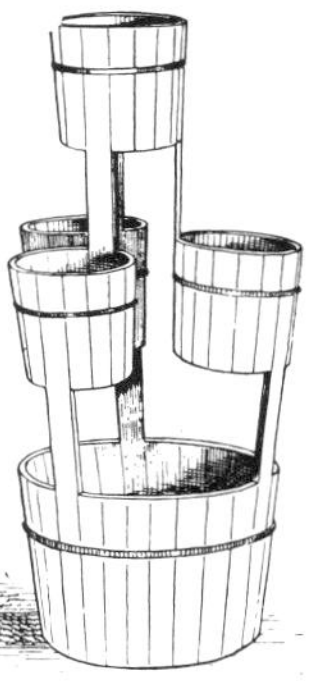

Fig. 297. — Curious Combination of Buckets for Flowers.

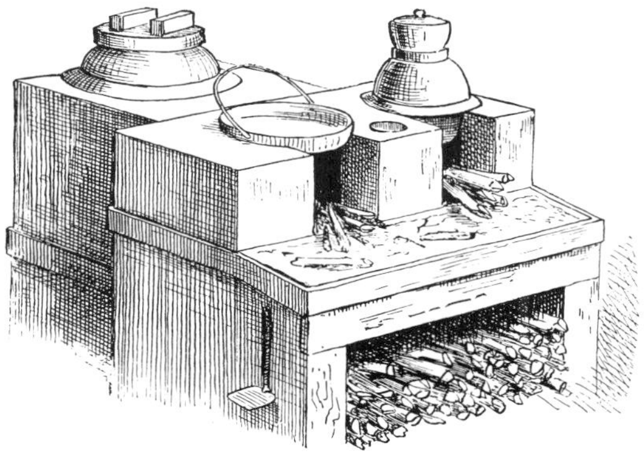

Fig. 168. — Kitchen Range.

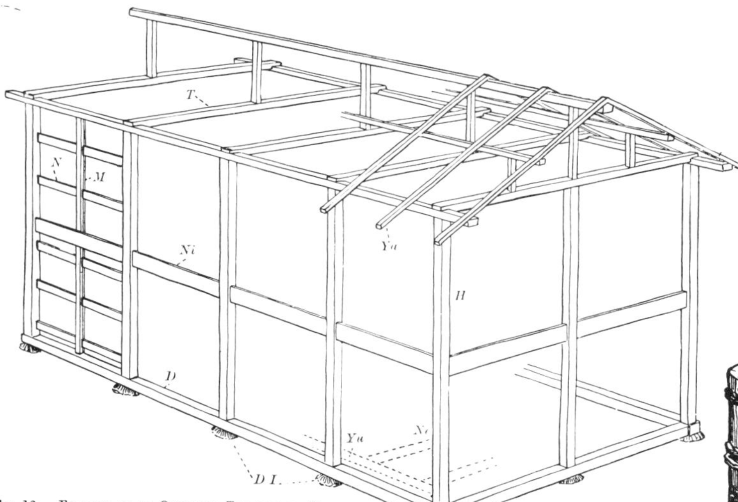

Fig. 12 — Framing of an Ordinary Two-storied House.

Fig. 130. — Tea-room in Nan-en-ji Temple, Kioto.

Fig. 176. — An Adjustable Device for supporting a Kettle.

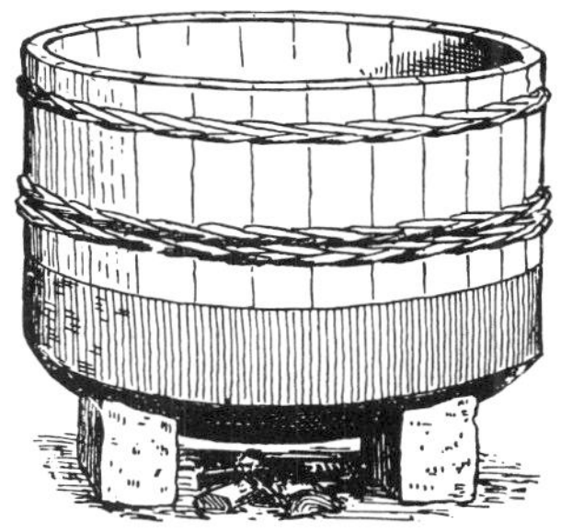

Fig. 184. — Bath-tub, with Iron Base.

A more minute description of the mats may be given at this point. A brief allusion has already been made to them in the remarks on house-construction. These mats, or tatami, are made very carefully of straw, matted and bound together with stout string to the thickness of two inches or more, – the upper surface being covered with a straw-matting precisely like the Canton matting we are familiar with, though in the better class of mats of a little finer quality. The edges are trimmed true and square, and the two longer sides are bordered on the upper surface and edge with a strip of black linen an inch or more in width.

The making of mats is quite a separate trade from that of making the straw-matting with which they are covered. The mat-maker may often be seen at work in front of his door, crouching down to a low frame upon which the mat rests.

As we have before remarked, the architect invariably plans his rooms to accommodate a certain number of mats; and since these mats have a definite size, any indication on the plan of the number of mats a room is to contain gives at once its dimensions also. The mats are laid in the following numbers, – two, three, four and one-half, six, eight, ten, twelve, fourteen, sixteen, and so on. In the two-mat room the mats are laid side by side. In the three-mat room the mats may be laid side by side, or two mats in one way and the third mat crosswise at the end. In the four and one-half mat room the mats are laid with the half-mat in one corner. The six and eight mat rooms are the most common-sized rooms; and this gives some indication of the small size of the ordinary Japanese room and house, – the six-mat room being about nine feet by twelve; the eight-mat room being twelve by twelve; and the ten-mat room being twelve by fifteen. The accompanying sketch shows the usual arrangements for these mats.

Fig. 179. — Steps to Verandah.

A very common form of bath in the country consists of a large and shallow iron kettle, upon the top of which is secured a wooden extension, so as to give sufficient depth to the water within. The fire is built beneath the kettle, – the bather having a rack of wood which he sinks beneath him, and upon which he stands to protect his feet from burning.

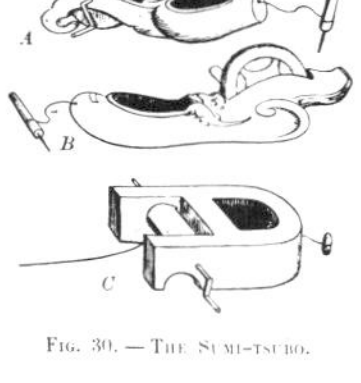

Fig. 30. — The Sumi-tsubo.

Fig. 245. — Bolt for little Sliding Door in Gateway.

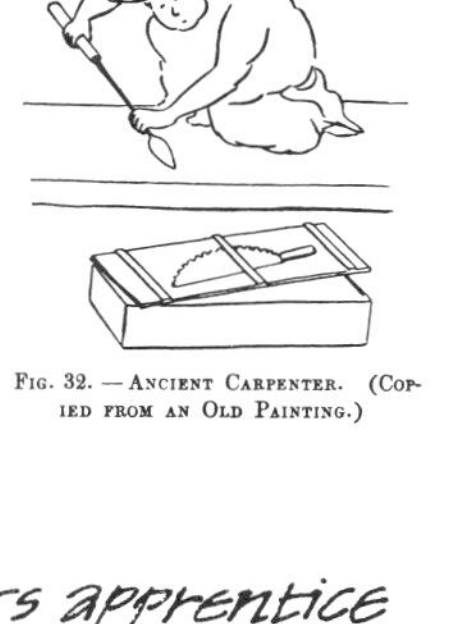

Fig. 32. — Ancient Carpenter. (Copied from an Old Painting.)

In my remarks on house-construction I made mention of the plaster walls, and of the various colored sands used in the plaster. There are many ways of treating this surface, by which curious effects are obtained. Little gray and white pebbles are sometimes mixed with the plaster. The shells of a little fresh-water bivalve (Corbicula) are pounded into fragments and mixed with the plaster. In the province of Mikawa I saw an iron-gray plaster, in which had been mixed the short fibres of finely-chopped hemp, the fibres glistening in the plaster; the effect was odd and striking. In the province of Omi it was not unusual to see white plastered surfaces smoothly finished, in which iron-dust had been blown evenly upon the surface while the plaster was yet moist, and, oxidizing, had given a warm brownish-yellow tint to the whole.

In Japan, traditional carpenters apprentice for 15 years. Certain skilled people are named as National Treasures, among them carpenters.

Eric & Debby Park bought an old house in Springfield, Oregon for $200 down, $75 per month, fixed it up while living in it and sold it a year later for $8700 cash. Eric has written a small book on their experiences, available for $2.00 from Exchange Magazine, 454 Willamette, Eugene, Oregon 97401. *Here we reprint portions of the book. One word of caution: not all remodel jobs work out this well; you've got to choose a house with care. But if you are interested in such a venture, here are extracts from Eric's treatise: good advice on the American economy, house remodeling, and neighbourly relations: "...Nobody heckles a hippy when he's stoned on work...."*

RESIDENCE RENAISSANCE

RECYCLING A SMALL CITY HOUSE

by Eric Park

PREFACE

It's about hard work, it's about craft, it's about recycling low-cost housing. Debby and I have just recycled our first home in Springfield, Oregon, over a period of thirteen months, from August through August. We bought it for $3600 [at two hundred down and seventy-five a month], an abandoned house, and sold it for $8700 cash. But it was more than just a living, and more than just a place: it was a job, a craft that we learned as we went along.

We've read William Nickerson's memoirs, and we're convinced that any dunderhead with sufficient singleness of purpose, by concentrating exclusively upon Mammon and Nickerson's Way, can make a million in real estate. But we're not into finance juggling, contract labor, rapid turnover, rentals, or even, strictly speaking, profit. We're into making obsolete the American obsolescence economy. We're into being willing to "rough it" for a few weeks or months in a substandard dwelling, in order to make it attractive and livable. We're into saving the millions of trees cut down each year by builders, when there are homes that can be saved through a little honest labor and elbow grease. And we're finally, at three removes, into making a living by patching the roof over our heads.

The following pages offer case history and general advice on buying a home, city vs. country, basic plumbing, windows and roof, exterior painting, finish plumbing, wiring, flooring, ceilings and walls, garage, grounds, and selling. And probably I couldn't avoid putting in a lot of nonsense about the way I look at the world.

Buying a home is eliminating the middleman: if you rent, you're but giving the landlord sufficient cash to buy the place himself. If you cut out the landlord, you're free to improve the place, resell, get back your initial investment, and pay yourself for the intervening labor — *if* you avoid biting off more than you can chew.

BUYING A HOUSE

1. "I'd rather rent." A renter, as renter, invests nothing in his own future. He might as well be throwing the rent money away. Inflation murders him. The buyer, on the other hand, eventually gets back all, or nearly all, or even more than his original investment, while inflation merely multiplies his holdings.

2. "Nobody should own the land." You buy, or you rent [unless you live in a truck or bus, paying dues to Standard Oil, or unless you're *really* a nomad]. True, if you buy, you're pretending you "own" the land; but if you rent, you're merely buying the property for your landlord, who pays his house payments, taxes, and insurance out of your rent money, and slices some off the top for himself. So who ought to buy the land with your money, you or the landlord?

3. "I don't have that kind of money." Yes you do. If you have a $500 car, and can find a deal like we did, you could sell the car for the down payment *plus* the first four monthly payments. Even doctors, even lawyers, even mayors and presidents buy their homes on time. Time payments, and not having a car, are a part of our scheme.

4. "I don't have that kind of time." In his book on real estate for profit, Nickerson recommends a turnover rate of two years maximum. But though we took our time, we needed just thirteen months. During nine of those months I went to the U. or Oregon as a full-time grad student, putting in just nineteen actual weeks on the house. Had we been content just to double our money and get out, we might have sold after redoing only the plumbing, cleaning, painting, windows, and temporary roofing — ten weeks. If you haven't got ten weeks, then you're in need of therapy.

5. "I don't want to be a capitalist." Every musician passing the hat, every poet selling a book, every artist with his paintings, every teacher on salary, yea, every guru or minister collecting an offering, is, in effect, into capitalism. In short, then, the way we see it, unless you're dead, you're probably a capitalist. But does capitalism control you? Or do you curb its evils [it *is* evil] and properly distribute your wealth? As Gurney Norman puts it, "Buying land is saving the earth — an acre or so at a time."

Three No's [any one of these disqualifies the property]:

1. No concrete foundation. Build your house upon a rock. At this point, the cost of a new foundation is prohibitive to you. And you can't get a cash sale without one.
2. Rotten or sagging structure. If roof sags, walls are out of plumb, or if major supporting members crumble easily or show evidence of heavy insect damage, the house is probably beyond repair. The trick is to know what's critical. Don't overestimate your energies or your strength.
3. Hopeless neighborhood. Nobody wants a palace next door to the garbage dump. Avoid jet-bomber flight patterns, pollutant industry. What you're after is the crumbiest house in an otherwise decent neighborhood.

Six Yes's [low-cost, high-return repairs involving mostly energy and labor]:

1. Paint in bad shape, boards weather-beaten and unsightly but structurally intact [will look like new with sanding and repainting].
2. Plumbing old-fashioned or in disrepair [if you're not afraid to learn. There's nothing cosmically difficult about plumbing].
3. Interior walls, ceilings, floors chipped and disfigured with plaster falling out [merely patch and refinish].
4. Broken windows, doors falling off, trash all over [ideal! Simply clean, air, replace windowglass and hinges].
5. Roof solid, but roofing in disrepair. [Any roof can be covered and made waterproof for about $50, leaving you an indefinite period to raise money for shingles].
6. Grounds gone to seed [gardening is an art, and one of man's great joys, and makes a spectacular difference in appearance].

A final note: your chances of scoring a substandard, sub — $5000 home without going through a realtor are practically nil. Such homes have lost owners by death or default, have lain vacant and fallen into disrepair, been inherited by disinterested relatives, or simply abandoned. No one wants to waste money advertising it — so someone, at long last, turns it over to a realtor. What has he got to lose? The realtor takes his cut on time, depending upon a sale. That's where you come in.

CITY vs. COUNTRY

So you'd rather move to the country? Fine. You can just as easily recycle a country home, using many of the same techniques as those detailed in these pages, if you can mount these persistent hurdles: time, tools, materials, transportation, and market. [I'm dismissing for now the question of the urban cultural advantages, concerts, FM, theatres, crafts centers, and coffee houses.]

Time: when you get right down to it, the country is about farming, and farming takes time. You've got to harvest and sow, truck and haul, keep the earth producing. You've got to feed, milk, nurse, herd, round up and probably slaughter the livestock, and pay the vet. How much energy can you spare from that for the house?

And if you're not into farming, are you really more than halfway into the country?

Tools and Materials: in order to get them, you've got to have *transportation*. On work days, I found myself making as many as five or six trips to the hardware store for nails, screens, plumbing and wiring parts, or just advice. I'm sure I averaged at least two. I was able to make the trip in about a minute each way, by bicycle, since I lived in the city — and what I couldn't carry home, merchants were usually glad to deliver. But in the country, you can't make it without a car, or more probably, a truck. You won't have the time to bicycle in and out, or hitch, dodging the shit the rednecks throw at you, twice a day or more. Nor will you have the money to support a farm, a house recycling project, *and* a truck. Can you add gas, oil, tires, and repairs to your other bills — all the bouncing down country roads — and make it?

What remains, then, is the work. Here are some general tips on how to keep your operations running smoothly [there will be more as we go along].

1. Keep a log of projects undertaken and accomplished.
2. Keep a running total of expenses, with receipts. If possible, keep track of your labor hours.
3. Try to estimate the value of your repairs, to compare with cost. This helps you estimate resale value. Painting, for example, may cost only $60 yet add $1500 to the value of your home.
4. Take "before" and "after" photos. Amaze your friends.
5. Open charge accounts once you've established yourself, particularly at the hardware store. AND PAY YOUR BILLS.
6. Have fire insurance [usually required].
7. Establish priorities. If you buy during the summer, prepare for the winter. Make roof watertight, repair windows and doors, paint weather-beaten exteriors to withstand extremes of climate. *Emergencies first, aesthetics second.*
8. Build yourself a tool bench as soon as you can, to pound on, hang tools on, lay out problems on. I made one out of an old oil tank stand. Brace it diagonally, and every which way. Make it massive, and daintiness be damned.

BASIC PLUMBING

We had no water. When the water company arrived to turn it on, all they found was a disconnected, shallowly buried pipe leading to the neighbor's hose faucet. Apparently the previous tenants had been content to drain water from next door. No record at the courthouse of any previous service. But just to be sure, Jan the water man, a soft-spoken fellow from Holland, began sweeping his metal-detector over the lawn, like a spaceman, or a Kentucky water-diviner, looking for pipe. Nothing. No water.

This meant that I had to pay $1.50 for a permit from the city to hook onto the water main. The city would, for free, dig up the street and make the connection — but I would have to dig an 85-foot trench from the existing hose connection out to the street, dodging around the apple tree and under five feet of sidewalk. There was, Jan assured me, no other way. Hiring it done was, of course, out of the question.

I borrowed a shovel and set to work. My back ached. The ground was so hard that I had to set the shovel, leap into the air, and drive the blade into the ground pogo-stick fashion. I was able to make no more than five feet an hour, but by God, every hour, five feet of ditch materialized, bearing toward the street. The street was Nirvana. These inauspicious beginnings were so tangible, so soothing even, after all the metaphysics of real-estate negotiation!

Just short of the sidewalk, the blade gave out a metallic ring. I'd been hoping for buried treasure — I was pretty far gone, saving old china fragments — that's how China got its name, it's at the other end of this mountain of old china fragments — so I stopped to inspect. It was — surprise, surprise — the old water main.

Thus began my long and fruitful association with the water company.

I stormed the city offices, black with rage, demanding my $1.50 back — and got it. Even I was a little surprised. Jan seemed to think he was in some kind of trouble. From me? Well, I let them think what they wanted, and got back to work on my trench.

Eighty feet out, I ran into the sidewalk. Nothing for it but to start tunnelling. I had made one foot in about three hours when I noticed that I had an audience, a curious neighbor. I explained about the trench and the tunnel. He said, Why didn't I bore it out, and disappeared. Five minutes later he reappeared with a length of cylindrical rain-guttering and, waxing enthusiastic, set to work opposite me. Twisting in the pipe, pulling it back out, and emptying six inches of earth at a time, he managed to join his tunnel to mine, beneath the sidewalk, in about fifteen minutes. Thus began my program of neighbor relations through property improvement.

The water company arrived promptly to turn on the water, but when pressure was applied, something gave way beneath the house. Fortunately, we had a big hole in one of our floors, near the old water heater [I wouldn't have known to look for a crawl space], so down I went to inspect. Did I need a flashlight? The water company's flashlight located a leaking elbow joint just inside the foundation. Did I need a new, adjustable elbow joint, ordinarily available only through a plumber's supply? The water company's joint sealed the leak in a matter of minutes. The water was turned back on, and shortly there came a sputter and snort, a jet of something brown, and then a steady stream of pure, clear water. A simple thing, easily taken for granted, but to us it meant that we had carted in the last load of water from the neighbor's hose-tap. No more bucket brigade.

continued

ıe water company, by the way, is one of the best
•urces of those cable-spools so readily convertible into
ble-tops. When Jan heard I was interested [we had no
rniture], he personally delivered a five-footer good
•r a coffee table and, later, an outdoor picnic table. Jan
so tried to give us his refrigerator, as he'd just bought
mself a new one, but by that time we'd scored our own
an auction.

word on appliances — don't throw away your savings
new ones. We found an old four-burner stove in
quisite condition with oven, night light, and functioning
ock [!] attached for $10 at a used appliance store. To
t it home, I traded the neighbor some labor — buried a
ment step he had just torn out, and was too tired to
ul away — for a ride in his pickup. Debby's winning
d on the refrigerator was $25. Auctions are a gas, but
ey move too fast for me. The auctioneer was trying
r $30, and I was still trying to decide if it was worth
5, when Debby wrapped it up. ["Sold to the
wlyweds!" the man said.] Debby was delighted
th her bargain until the freezer door fell off, the
ilk froze, and the front door-latch fell off. We rigged
me catches for the doors, but had to learn to live
th semi-solid milk.

ɔw we had running water, free drains, stove, icebox, and
ffee table, two friendly neighbors, and the water company
xious to please. The hot water could wait [just heat up
ld], as could the finish-plumbing [we took hip and
onge baths, flushed the toilet with a five-gallon bucket].
e had to weather-proof our exterior before the winter
orms [in Oregon, the rains] set in.

EXTERIOR PAINT

I had the most previous experience with house painting,
ıich brings the highest return per invested dollar of any
•usehold repair, so the temptation to philosophize is
rong — but I'll sit on my prose, and merely pass along
ese nine Paint Commandments:

DON'T SCRIMP ON PAINT (part one). Beg or borrow the best going, which in my opinion is Dutch Boy. Cheap paints spatter, cheap paints run, cheap paints force you to use two coats when all you need is one. One coat will hardly need to be supplemented 50% sooner than two coats.

PREPARE THE SURFACE: the most important step in painting of any kind. Nail in loose boards, replace the irreparably damaged; hammer in loose nails, remove useless nails, hooks, wires; putty cracks and holes; prune back obstructive shrubbery; knock down hornets' nests; and *clean*. Spend at least two days sanding and scraping for each anticipated day of painting. In extreme cases, with very old, peeling paint, rent a disc sander and strip each board down to the blond wood. Use coarse paper, as paint quickly clogs fine grades. *Don't* rent a belt sander, as you'll have to hold it sideways, you won't be able to keep the belt on straight, the paper wears unevenly and, anyway, disc sanders are made for this job. The disc follows the curve of a weathered board. Cost: three to five dollars a day.

PAINT TO PROTECT THE WOOD. This is the primary function of exterior paint, which adds its own life to that of the board it covers. Paint lasts 5-6 years if cared for, cleaned occasionally and patched in spots. Don't leave unprotected surfaces, or skip sections simply because they're hidden from street view.

AVOID LOUD COLORS. Don't go on a virility trip painting your house. The best tones come to life at a *second* glance. Exterior paint serves best as a backdrop to the architecture, the trim, the gardening and the interior finishing. Bright orange, green or purple houses inevitably lead me to question the sanity of their owners.

DON'T SCRIMP ON PAINT (part two). When in doubt, use more. Your brush has a good load if it glides smoothly over the board. If it drags or "scumbles," reload. Give brush a quick tap against ferrule, cover a small section completely, and move on.

BOILED LINSEED OIL. Use it — which means using oil-based paint — unless you've got metal or asbestos siding. It replaces natural oils soaked out by rain and winds, makes wood more waterproof.

START AT THE TOP. Work downwards and cover drips as you proceed. This may mean that you must start with the trim, since the undersides of eaves, which you paint upsidedown, are notoriously messy. Good painters inevitably drip, but use quality (thicker) paint, start at the top, and spread tarp where necessary, so that no one is the wiser.

TRIM window and door frames, eaves, "corner boards," and decorative architecture. A sub-trim (black) may be advisable for doors, window edging, etc. Both trims together are likely to require twice as much time as the primary painting, as did surface preparation: but it's worth it, since trim "makes" the visual impact of the paint job.

PAINT DISCOUNT: get one by explaining to the paint supply store of your choice (Dutch Boy in my case) that you recycle houses professionally. You roof them, plumb them, *and paint them*...all unimpeachably true. Professional painters' discounts amount to 50% or more, or about five dollars a gallon.

WIRING & INTERIOR FINISH

house, like some peaceful, domestic animal, has windows
r eyes, heating as a kind of breath, plumbing not unlike
our own, frame and siding as ribs and skin, and wiring for
ains. You may put its brains in order without fear if you
member to TURN OFF ALL THE ELECTRICITY.
omplete plumbing and wiring, which involve knocking
oles in walls, before finishing interior surfaces. You will
uestion the sanity of the electricians who have gone before
ɔu. Our hot-water heater line ran ten feet to the chimney

after ascending into the attic, circled the chimney, dropped into the spare bedroom and ran through a hole in the wall, over the doorsill and into the utility room, before connecting with the heater — a yard from where it had begun.

Simplest is always best. Replace old switches, such as the circular spring-type that came with our house, and reposition switches or plugs located near water, such as the switch beside our new shower. Set fixtures into leaded boxes, using wiring caps. Wiring is most often organized in the attic. Locate in and out wires [which is which doesn't matter, since they alternate] and organize circuits accordingly. For every light, a switch and four wires, two for the light and two for the switch cutting into the circuit. Simply stare at the circuit until you understand it. In my case, this sometimes involved hours.

Start at the top. Patching ceilings first will save on heating bills, as heat tends to rise through cracks into the attic, there to warm the spiders. DON'T BUY SPACKLING COMPOUND. Buy plaster at a builder's supply for about $2.50 per hundred pounds. This grainy plaster is ideal for base coats. Now sift out the grains, and you have spackling compound at one-tenth the price. Soak surfaces thoroughly before applying plaster; but above all, and this is the cardinal rule of all plastering, KEEP TROWEL SOAKING WET. If you don't, your ruin is assured.

Now, should you require a texturized paint, simply add back into the paint the grains you removed from the plaster, and save *that* markup. For kitchens and bathrooms, where water resistance is needed, or where any wood surface is involved, use an interior, semi-gloss enamel. Otherwise, latex will do.

For any interior decoration, unity is the watchword. Harmonize colors, or the *tones* of colors. Our coved, or rounded, ceilings suggested warmth, which we tried to emphasize by leaning heavily toward olives, creams.

A lady we know employed a single trim, white, and then wallpapered each room in her house differently. Old wallpaper may be steamed off with a soggy towel and household iron. For hopeless walls, consider panelling. Panelling is less un-ecological than you might think: It's a very thin veneer of actual wood, over a plywood or pressed-board backing — a kind of lumber baloney sandwich.

Flooring: our leaky, departed water heater, plus three years of rains beating through a broken window, had rotted out a part of the back-room floor. Here's what to do in such a case: first, *tear out* rotten flooring and throw it away, following the rule that what you *can* tear out, you probably *should* tear out. Exterminate any bugs mercilessly — if they eat your house, a forest will be ravaged to replace it. If joists, the heavy [probably 2 x 6] beams supporting the visible floor, have been affected, then gouge away the unhealthy sections of joist, and bolt a shorter section of 2 x 6 to the healthy joist, using heavy-duty, quarter-inch bolts. *Paint* the completed and repaired, and carefully levelled, joist. Now lay the subflooring, usually in diagonal strips of overlapping 1 x 6 planks. Nail to the joists. As flooring supports walls, you may be forced to tear out parts of walls in order to re-lay flooring

Joist lumber is expensive, up to 50 cents a foot, so look around for it. While jogging one morning I located an enormous pile of 2 x 12's behind a neighbor's garage, healthy lumber he'd salvaged from an old schoolhouse that was being torn down. He had vague plans of using them for a carport, but under pressure from his wife [who wanted them out of her yard] he let me have them, even helped me saw them to fit my flooring needs: each beam sawn lengthwise made two 2 x 6's of equal length. I had a great deal more than I needed. Joist repair, then, cost me nothing but a dollar for the bolts.

Now the flooring proper: the advantage of tongue-in-groove is that it interlocks the entire floor. No single board may sag of its own accord. The disadvantage is that's expensive. A very small patch of floor cost me almost $30, so it pays to measure very carefully, and to use odd lengths.

Finished floors last longer, and make a dramatic visual difference, but it's a lost art. You'll have to decide between linoleum, the old cover-up, and a genuinely refinished wood floor. *If your floor planks are scarcely two inches wide, and you can't nick them with a fingernail, then you've got hardwood floors, usually oak or maple. These floors are treasures. Don't cover them up in any but the most desperate circumstances. Ordinarily they're better left unstained, but do apply a protective coat of urethane.* Otherwise, remember that the floor sander, which may be run only with the grain, and which requires several feet in which to maneuver, is a refinishing necessity.

The refinishing formula? — *sand, stain,* and *urethane.* When sanding, control the momentum of the machine by pressing down on the handle, thus lifting the drive belt off the floor. For corners and edges, use a handsander, or rental floor-edger made for the purpose. Don't run the machine sideways, as it gouges into the grain. Paper should be coarse enough to strip old varnish or deck paint in one movement, and tight enough not to slip: a one-dollar belt lasts about 15 seconds if not properly tightened, and at $240 an hour even Rockefeller would tighten his belt. Results may later be fine-sanded smooth.

Use a "penetrating wood stain," not a "varnish stain." The darker the better, particularly on aging or much-abused surfaces. Avoid redwood finishes on floors, including so-called "maple." Ordinary paint thinner lightens stain. *Urethane* brings out gloss and grain, and resists scuffs, corrosion and water better than varnish. Use a "clear gloss," not a "satin" or "semi-gloss" variety. Three coats [add paint thinner to the first, or "sealer"], six hours between each coat.

TILES

Acting on a neighbor's hot tip, I rode the Springfield-Eugene bike path, beside the Willamette River, to Floors Unlimited. This proved to be a retail flooring house featuring low prices and remnants of linoleum carpeting. I emerged in an hour, ready to invest in a flashy "Congoleum" remnant, which the salesman was willing to "sacrifice" to me for exactly the price of a new set of ordinary "block" tiles. Debby came with me the following day to inspect my choice, and immediately fell in love with the "block" tiles instead. Years before, she and her entire family had helped to install a similar flooring in a converted carport, and so, either out of nostalgia or good sense, she wanted me to pass up my "Congoleum." I grumbled a bit — we men have no authority anymore — but gave in, loaded up the tiles, and immediately had to admit that they were certainly a good deal more portable than the linoleum carpeting. In our bike baskets, we carried back home 115 square feet of color-coordinated tiles, two quarts of tile cement, and for good measure a half-gallon of paint that I needed, from a nearby paint yard.

We let the tiles gather dust for a month while I worked on other projects; but when we finally got around to them, we were surprised and pleased by their appearance and ease of application.

A word about your choice of hardware stores, since I'm advising you to spend so much time in them: I've had my best luck with locally-owned, personally-operated stores. Owners care, managers don't. Good hardware men know everything the plumbers and electricians know, and they'll spend fifteen minutes explaining to you how to install a fifteen-cent part. They want you back, again and again, getting nickle-and-dimed to death. If it's a good store, you'll endure this gladly.

Avoid department stores, auto supply stores. They'll carry auto tires, tape decks, televisions and vibrating chairs, but they'll be short on hardware.

Avoid slickly packaged items. They're disguised, they're dishonest, and they cost more. Why wrap a roll of electrical tape? A strip of aluminum molding? Screws? Hinges?]

WINDOWS & DOORS

We spent our first hours merely sweeping the debris from our abandoned house. Kids at target practice had demolished thirteen windows plus the door-pane; we fitted plywood into the door-frame, and replaced the windows for $28 plus the price of putty. Glass companies will deliver: they're used to it, as customers don't like to handle glass. Installation procedure: chip out old putty and broken glass with a wood chisel. Place new pane in, "set" with tacks or tiny nails, then putty all round. Mold putty into corner for a tight seal. While on the job, chip away and replace all loose, cracked putty from remaining windows.

Doors can make a difference. Front and rear doors should be lockable and weather-tight: pre-assembled lock and latch works cost us $3.50 per set. Using a beautiful "Hindu" plane, like a straight razor with two wooden handles borrowed from a neighbor, we planed all interior doors to close within their casings [natural house settling causes doors to stick]. To recycle screen doors, pry off the wooden stripping over the old screens, then cut new screens to fit, and nail stripping back down over new screens. Diagonal 19 cent bracing rods prevent screen door sag.

THE ROOF

Ten months and two loans later, our borrowing power enhanced by the growing value of the house, we continued roofing operations with new asphalt seal-down shingles and new rain-gutters. Materials costs for both, with nails and tar: $243. *Shop around for shingles:* Copeland's, the most expensive lumberyard in the Willamette Valley, happens to have the best prices on roofing. I saved some forty dollars by checking. Get *seal-down* shingles because they're only a dollar more per square, and hurricanes can't tear them loose, once the sun melts them together. Wooden shingles are three times as expensive, and a waste of lumber.

continued

Build shingles up in diagonal rows, but be careful not to lose control of horizontal lines. Diagonal method eliminates lateral movement. Cut shingles with tin snips, which may be cleaned afterwards with ordinary paint thinner. You'll need far less tar this time, as shingles self-seal, and nail-heads are unexposed. However, apply tar liberally around chimneys, vent stacks, etc. [the "flashings"]. You may wish to trim the finished product with inexpensive aluminum stripping, just for the aesthetics of it.

Shingles will cost you $15-30 per hundred square feet. Having no such money, we settled for "roofing felt", or tarpaper. $50 worth of medium ("thirty-pound") roofing felt bought us a year's time in which to raise money for shingles. I removed the old rain-gutters, which were in flitters, and tore off the top layer of asphalt roofing, to reveal a still-usable undercoating of wooden shingles. I pounded down the old nails, swept clean the resulting surface, and installed wooden patches in a few spots, caulking the seams with "cold-process" tar, which solidifies slowly on application. The tarpaper simply rolled on: start at the bottom, and secure lower edge of each layer with four galvanized roofing nails per yard, upper edge with one. Tar should be applied over each nail-head, and a strip of tar used to seal down each layer of felt to the layer below.

THE GARDEN

We were fortunate in having a great deal of shrubbery already in place: box hedges, ornamentals, hydrangea, laurel, rose, the blackberry and grape, and lemon balm. In the Willamette Valley, the problem is to keep what you *don't* want from growing. We had, besides the apple and cherry trees, two ten-foot maples growing out from under the foundation, and a yard full of weeds and dog shit.

The latter we gathered into a pile. We cut the weeds, stacked the hay onto the pile, and voila! our compost heap was established. Every day, then, for the next nine months, the german shepherd next door left us one or two more presents, which we dutifully spaded into the heap, which found its way eventually into the garden. Debby's garden was a constant joy to both of us. Fed by Willie [the german shepherd] and by our organic kitchen scraps from August through May, and seeded during the last month of the spring rains, the garden produced, all in a ten-by-ten space, beautiful crops of miniature corn, peas, green beans, carrots, beets, lettuce [partially ripped off by the neighbor's escaped pet rabbit] and squash aplenty — plus a lovely border of marigolds, both dwarf and large-sized. All but the marigolds, which kept the bugs away naturally, were delicious.

Debby trimmed the shrubbery down to a manageable size and shape, careful not to impose geometrical shapes on these living things. Trimming in the fall encourages blooms in the spring. On a sadder note, I had to take an axe to the two maples growing out of the foundation, which they would eventually have cracked open. Debby planted flowers along the back wall, where one of the maples had left a gap, and scattered more seeds at any breaks in the bushes. I filled one of these breaks myself by transplanting a three-foot strip of lemon balm [easily mistaken for mint...it makes a very subtle, mellow tea] from its niche in the neighbor's driveway to the rear corner of our house. To my infinite surprise [my previous gardening experience had come when, as a kid in Kansas, I had planted a watermelon seed in an old Franco-American spaghetti can, in vain], the lemon balm flourished, disdaining even the normal "transplant shock," delighted to have a little soil to call its own.

One of the cherries had to come down. I refuse to operate chain saws, so we had to contract the job. The contracting was an education. Our first contractor, Mulkey, big-time, radio-dispatched, took one look at me [long hair], figured I was an idiot, and bid $75. I must have turned pale. He came down right away to $60. I shook his hand and said good-bye, and ran out to the street to call his chief rival, Engleman. Engleman, a more easy-going fellow, was told only that he had to beat Mulkey. [Loggers' rivalries are intense in these parts.] Engleman looked at the tree, not me, and bid $25....

But surely my finest gardening hour, and my last of any kind to be spent on house improvements, came with my attack on the concrete slab which had lain, unused, for decades, between the alley and the fruit bushes. It was 20 x 10, and six inches deep. Electing single combat without benefit of power equipment, I set to work with an old pickaxe, sledge and shovel. The object was to crack the slab with a few swings of the sledge, then to loosen blocks of concrete from the main slab with the pick. This was by far the hardest labor, but at the same time the most satisfying, that I had done. Here was our personal philosophy in action: to remove a blight — concrete — by simple hard work, restoring the land to its natural, green state. Using the manual method I spent two days busting up the slab and piling it to one side.

Any man with an axe, pick, or sledge is immediately the attention focus for any kid within blocks. I had as many as a dozen crowded about the demolition site, begging for a turn with the tools. One kid, Greg, actually proved strong enough to use the sledge with results, so that — shades of Twain and the whitewashed fence — I managed to show the kids a good time by letting them do part of the work. I paid them off in orange juice.

Singing John Henry songs to myself as I swung the pick, I realized why it is that prisoners are set to work busting rock [I presume they still are] — because the last thing on your mind, at the end of the day, is raising any further hell.

Disposing of the concrete presented a problem, until I learned that a neighbor down on the corner, who was constructing himself a concrete porch, was in need of fill. So working together, he and I transferred the broken-up concrete to the site of his porch, each of us considering the exchange a favor. Here, indeed, was the principle of neighbor relations through property improvement in action. I had little in common with this neighbor, a weight-lifting Teamster with seven kids, and yet we had cemented our neighborly ties, if you'll pardon the pun, by my concentration on the improvement of my property. Nobody heckles a hippie when he's stoned on work — and no neighbor can help appreciating that, as your property value goes up, so does that of the entire neighborhood.

Now I needed some topsoil. Fortunately, Springfield is a working man's town: the fellow whose back yard had overlooked the old bramble-covered slab drove a gravel truck. He was happy to provide me with a free load of loam. This I mixed with the sawdust from the recently sanded floors, and spread it evenly over the new plot of ground, finally seeding it with a couple dollars' worth of rye. It was up in two or three days, and, aided by an unseasonable August rain, developed within a week into a genuine stand of lawn.

But if I had been repaid in no other way for my labor on the garage and the back-yard plot, my moment would have come when one neighbor, who had once instructed his kids to stay out of our dilapidated garage, wandered by with his wife in the early evening. I emerged into the alley on some errand, to find them stopped behind the garage, shaking their heads and smiling.

"Just admiring your work," they said.

SELLING

There are two ways to sell, by owner and through a realtor. Sale by owner means that you are your own mouthpiece, paying nobody a fee except in "closing" the sale. To close the sale, look up the nearest "title" or "escrow" company, which for about $50, plus a dollar a thousand [$59 on a $9000 sale], will handle everything: title insurance, assignment of contract, deeds, everything. Buyer and seller ordinarily split these "escrow" fees. Sale through a realtor means that you hire a real-estate agent, who handles the hype and promotion, footwork, negotiations, and takes six to ten percent of your money for his trouble. It's your choice: the business is largely a matter of nerve.

There are two kinds of payment, in cash and on time. Occasionally a buyer is rich, or has just sold his own place — in which case he simply hands you a check.

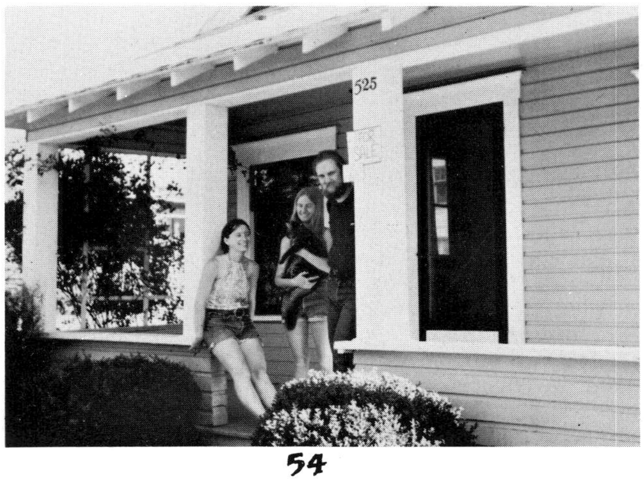

Usually, though, in a cash sale, the buyer takes out a loan, which means that the bank buys the place from you, and the buyer pays off the bank on time. *Bank closings are more expensive.* You need no credit whatsoever to prearrange a bank loan: they'll appraise your place for free, and tell you in a day or two how much they'll loan on it, usually 60-80%. The balance is your buyer's down payment. Be sure to advertise "bank terms available."

Time payments, regulated by "land sales contracts," entitle you to collect interest, usually about 7% these days. If you're looking for a little income, have no pressing debts, and plenty of time, then this is the sale for you. And if you've got credit [ha!] you can take out the home-loan yourself, and offer any kind of terms to the buyer [have him pay off the bank].

These, then, are the cards. How will you play them? Here's what we did: we determined to sell for cash, set $8500 as the minimum we would settle for, and put the house up for sale "by owner":

> $9750 — completely restored 2½ bedrooms, new roof, new paint, newly refinished floors. Concrete foundation, two-car garage. See and inquire at [address]. By owner till Sept. 1.

Notice, first, the price. We settled for $8700 cash after only six days, narrowly missing a $9000 sale in three. [The buyer backed out at the last minute, only to change his mind and show up shortly after we had signed the earnest money receipt.]

Notice we listed no phone number. If a buyer wasn't willing to come and look, then the last thing we wanted was to talk with him on the phone. [But then again, we didn't have a telephone.]

Notice the time limit. Assuming that on September 1, the sale is to be transferred into the hands of a realtor, whose asking price will be higher, house-hunters are conditioned, like the famous salivating dog, to hurry. And, true, we would have turned the sale over to Roy on September 1, but the price tag probably would have stuck.

Further tips on selling from the world's worst salesman:

1. Keep your mouth shut; let the house sell itself. The right buyer must simply happen along.
2. Don't hurry. Don't let anyone hurry you. Always sleep on any but the most spectacular offers.
3. Blitz the newspapers. Plan to spend $25-30, or more, on advertising. Hit the dailies, the weeklies, the "classified flea markets," the bulletin boards, and spread the word around.
4. Be sure the house is clean. Don't start selling until all remodelling projects have been completed. If you must engage in some yoga of the hands, tinker with furniture out in the garage.
5. *Be at home.* Don't go out until you are ready to have a nervous breakdown, and then leave a note ["having nervous breakdown," or the like]. On the subject of nerves, avoid coffee as the plague. This is no time to be wired. Meditate, recite calmative sutras, be prepared to *wait*.
6. Follow up all contacts. Keep people informed when loans come through, etc. Don't let them rest. We triggered our sale by chasing down a fellow who had said he would be back, then wasn't. He returned, and made an offer [a little low] right under the nose of the eventual buyer, who had just arrived [at this point my buddy Lee came around, taking pictures and looking important]. Bidding spreads like wildfire. A third party got into the act, making an offer and heading home for some earnest money. Number two, thinking fast, whipped out his checkbook and settled with me on the spot [I had told everybody I would settle with the first acceptable bidder to come up with the earnest money]. We might have waited a few days longer and settled for a little more — we have sworn to be demons on our next sale — but this time, buyer and seller were no doubt equally pleased.

I've said that people are getting their heads bent in the city; and during our year at "Old Lilac," a lot of people got their heads bent on the subject of hippies, or whatever the hell we are. We had moved into an eyesore, a firetrap that had been called hopeless; and we had moved out of a neat, freshly painted little house and garden, brimming with flowers, fruit and vegetables. We had surely raised the values of every house on the block. We'd gotten off food stamps [yes, you can qualify for food stamps while buying a home], and we had acquired a lifelong craft.

The neighbors paid us off in smiles and little gifts that never failed to delight us: a brass bed, an upholstered arm-chair, a shaky but serviceable kitchen table, greeting cards, vegetables and fruit, and chocolate cake [we don't eat sugar, but we ate the cake under the circumstances]. We received a lot of advice [much of it good] and even, when we most needed it, some home-grown smoke. Only one neighbor refused to like us, the fellow across the street who hadn't liked loaning me his pipe wrench [Chapter VI]. This fellow had a modern tract-style house, a truck, two cars, and a terrifying American eagle decal on his front door. You lose a few.

We've moved now, into Eugene, there to fix up another place, though starting this time from a little higher spot on the ladder. We're not getting ambitious, nor do we intend to move as fast as Nickerson did — we're into enjoying our craft, and remaining in tune with the earth. We're not going to be landlords. We'll tackle one place at a time, the one we're living in — leaving it up to you to get out and get started, to spread the work and the word — Residence Renaissance.

end

• CAVEAT VENDOR •

George Hoffman has inspected over 500 houses in the last six years, advising prospective buyers of the physical conditions of the structure. Now, he has written a small manual, Don't Go Buy Appearances, *which tells you how to check drainage, foundation, construction, plumbing, wiring, heating, tile, roof, paint, dry rot, etc. He explains that the real estate dealer won't do this for you, as he works for the seller, and although the seller is supposed to disclose any faults under the* full disclosure *law, he often doesn't. Or, he may not know that the wiring is bad, or the foundation about to sag. Tools needed for inspection: a ¾" diameter steel ball, a level, screwdriver, flashlight and a good eye. $3.00 from George Hoffman, Box 80, Corte Madera, Calif. 94925.*

Someone had told me about a house made of abalone shells and junk south of Big Sur, so on a trip down the coast I found my way early one Sunday morning to the unique, hillside house built by Art Beal. Art invited me up for breakfast, and as we sat in the crow's nest over toast, peanut butter and coffee, I listened to him talk about his life, his outlook, the changes he's seen. There's now a freeway visible & audible from his home, tract houses above have caused mudslides on his property, and many locals consider his eccentric house & behaviour bad for land values. Art is half American Indian, an ex-champion long-distance swimmer, a vegetarian, ornery, loud, and charming. There aren't many like him left.

Lloyd: Did you do most of this by yourself?

Art: What?

This building here...

Everything! I ask no help from nobody. I didn't want any. Didn't want 'em fucking around.

I created and built this junk heap. Back through my trail of life, I gathered all this shit. This was all open land. Just a wilderness. Now all these assholes, the big shots, they want to change this and change that. And I don't like it.

They (*meaning individuals*) can't build because it's against the law. It doesn't conform with your own employees' viewpoint and opinion. It's graft. You're forced to pay taxes, pay these assholes, and I mean assholes, you can't do this, you got to get a permit, a license, to do this. Now you got to be an engineer. Licensed engineer. You got to get a license to fart.

The white man, wherever he goes, he destroys every fucking thing.

These houses didn't used to be in here?

There was nothing. It was all open range. There was nothing here. One home, an old hacienda. That's all.

Were there a lot of trees up in here?

See 'em now? And up above 7 homes going in there, they're denuding the old gal. They've all got to come out. More of this equipment. More homes going in. Then they put that suicide avenue in. (*Freeway.*)

It's hard. It's difficult to explain to people who haven't *been, done or seen.* It's very difficult. When I say I built the road in here with idiot sticks, they don't know what the hell I'm talking about.

With the what?

Idiot sticks!

What's that?

A pick and a shovel. Only an idiot would use them.

Some students from the S.F. Art Institute had just spent a week with Art, filming him and his house.

They were here with you for about a week, you say?

Eight days, there were six of them stayed 8 days. Thanksgiving Dinner I had 15, they come and went.

You cooked the dinner?

No, fire did it. I prepared it and put on the fire and let the fire do it. I don't get that hot....

You built the abalone steps and the arches....

Everything.

I had junk here, on this road and up above. And this was all junk, woven right in this mountain side. Every time I'd find junk, I'd have a place for it.

I don't mince words or stutter. This is supposed to be a free nation. Was when the Americans owned it. White man come, off we go. All gone. No more. Then they say this is a free nation? They force you, like in the military, they force a man, draft his horse, teach him how to legally slaughter and murder. And then they force him to go to some other nation and kill those people there.

That don't kind of make sense to me.

What's going to happen, I don't know. And I don't give a hoot.

During the war this was all a blackout and I went back to San Francisco and realized two of my ambitions as a kid, to be a captain in Greyhound and be a fireman. And I done a lot of crap.

They say I'm a revolutionist. You're goddamned right. And I'm a good one.

I'm half American and half white.

If the American continent is the American nation, what in the hell have they done with the other 20 American nations? What a fraud.

There's so goddamned much shit that goes on here.

...I'm a big rebel. I'm the biggest revolutionist that ever put on a pair of shoes. I revolt against anything, everything, and even that. Whatever it is, I revolt against it...

DR. TINKERPAW

Nobody likes the truth. In fact, they don't know how to handle the truth.

This is real. All of this junk found. The majority of this lumber came right out of the ocean, also. I had to hunt to find this stuff. All these rocks come from this side of the lighthouse. All sand and stuff come right from the ocean.

How long have you been at it?

I don't know. If I keep breathing 3 more months, I'll be 77. And there's three of us left that have seen this all open range.

Is it alright to take some pictures around the place?

I don't give a shit. You won't be the first one.

All this is junk. Come out of other houses remodeled. The main part of the lumber, the studding, flooring, sheeting, and rafters and underneath, it's all right from the ocean. Yes. There's so much junk. A lot of times, I'd say I'll do it this way, and do it that way. *If. If. If* I'd done it. If the dog hadn't stopped to shit, he'd a caught the rabbit. So a lot of the set things, I just threw in any old way. I've made so many changes here. So many changes. This is my room. Right here is over a year's work for me. Lugging this out. See, everything was lugged in buckets or piggyback. And this is all solid rock. Each level is hewed right out of solid rock. Down below I used to chute that.

Are you still doing much work every day?

Oh, I just fuck around. Biding my time. Lot of stuff I don't like, I change around.

Have you done that all along?

Oh, yes. So many things I've changed. What I do now is in there for keeps. But before, I didn't know what the hell I was going to do, now I know what's got to be done. And what I've got to do. So this junk here, I ran out of junk and happened to see this stuff up above there, and I said, hell, that's the answer. That's what I want. So I bought it down and put it in here. And I say I had so much junk here that ... Yeah, I don't try to keep up myself anymore. If I see a little thing to do, maybe I'll work 2 or 3 hours, maybe I don't work period. There are so many damned little things I'd like to do. But you see, now, I don't go out and work no more. I wore four trucks out here in the last 25 years. I just quit.

If I last another year I'll be happy.

I've worked so goddamned hard I can only do a little now, a little here.

Superstition & religion is what kept this world back.

Do you think you'll stay here?

Well, I've been here all these years. Why let the assholes run me off now. I been here before they. And like these animals, I tell 'em those animals were here before you. Leave 'em alone. This is their home. They don't do you no bad. None of these animals do anybody bad. But they can't bear to see anything move. Run for the flit gun. Exterminator.

I'm a vegetarian. That's the way I get along.

Do you eat fish?

Once in a while, but I'm cutting that because all this shit they're putting out in the ocean now, it's all contaminated. I get along very nicely without eating any flesh, but that's the most flesh I do eat.

Have you been that way long?

Ever since I was a youngster. These long haired Methodists, they lied to me. I took a pet pig, I raised it up, they tooken it away and they slaughtered it. Then they lied, oh it's around somewhere.

What do you eat for protein, in place of meat?

I eat anything I want. I don't pay that shit no mind. I'm like any other animal, there's something out there that I like, I go eat it. Your system will tell you what you want. I get hungry, the food is there, I'm going to eat it.

This is the last. These iron men (*bulldozers that caused mud slides*) as I say, and all this stuff, I can't keep up with it, otherwise I always had a garden, stuff grows all the year here, all the year.

There was nothing here. All wild country. It wasn't wild, it was tame compared to now, you get out, open the door and you don't know if somebody's going to konk you in the brain or what they're going to do to you. The human animal is the only animal upon this world that manufactures arms and amunitions and go out and legally destruct one another. You see them birds go out and manufacture amunition to kill one another? Because they disagree?

I get along on my $98 a month. You don't see any wrinkles in that gut.

Carmen and Richie Quinones' Living Room.

Twig crib made by James Surls.

Colorado cabin.

Joanne's kitchen.

Tarks desk.

Sarah's collage on refrigerator.

Joanne and Peter's house.

Sarah Kahn's studio built by John Bradbury.

September

The grasses are light brown
and the ocean comes in
long shimmering lines
under the fleet from last night
which dozes now in the early morning

Here and there horses graze
on somebody's acreage

strangely, it was not my desire

that bade me speak in church to be released
but memory of the way it used to be in
careless and exotic play

when characters were promises
then recognitions. The world of transformation
is real and not real but trusting.

Enough of these lessons? I mean
didactic phrases to take you in and out of
love's mysterious bonds?

Well I myself am not myself.

and which power of survival I speak
for is not made of houses.

It is inner luxury, of golden figures
that breathe like mountains do
and whose skin is made dusky by stars.

Joanne Kyger

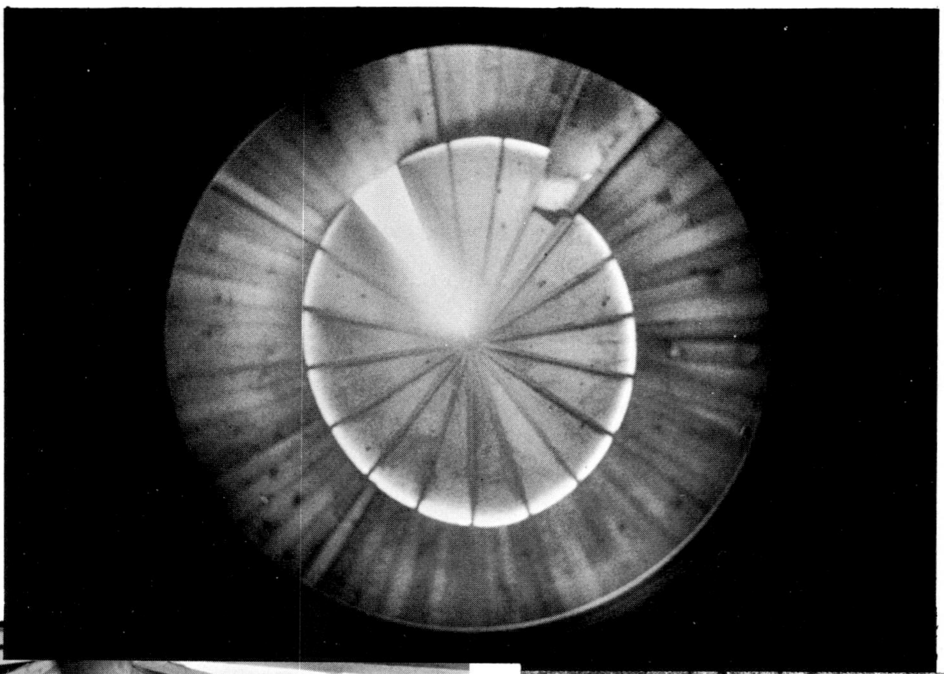

This spiral building evolved its shape as it was being built. Starting with the floor of a water tower, the builders decided to extend it to an oval, then egg shape. When they thought about fitting in a door, the nautilus shell spiral came about. It was built as a zendo, by Bob Anderson.

Hyperbolic paraboloid shelter with eucalyptus poles and plywood. There are no walls. By Barry Smith.

LET'S VOYAGE INTO THE

THERE ARE DOORS
THAT WANT TO BE FREE
FROM THEIR HINGES TO
FLY WITH PERFECT CLOUDS.

THERE ARE WINDOWS
THAT WANT TO BE
RELEASED FROM THEIR
FRAMES TO RUN WITH
THE DEER THROUGH
BACK COUNTRY MEADOWS.

THERE ARE WALLS
THAT WANT TO PROWL
WITH THE MOUNTAINS
THROUGH THE EARLY
MORNING DUSK.

DRIFTWOOD HOUSE

...the design, and the house itself in fact, were inspired by a guy named John, a surfer from the L.A. area who worked his way up the coast building some 10 houses at various locations convenient for supply of driftwood water and reasonable access to a town....

I spent four beautiful uncommonly sunny but windy summer months slowly working on the house — collecting wood (all driftwood from the beach) windows from various sources and nails. I had to buy a few spikes but most of the nails I used were bent or rusted nails salvaged from other jobs....

'69-'70 was a pretty wet winter — with almost 3 solid weeks of rain in January — culminating in a rare coastal thunder storm. John Hardcastle (a different John) was looking out the window digging the lightning when he started hearing rumbles and saw water coming thru the back wall. He got out the door just as the river pushed his house & him into the ocean....

Dick Keigwin

P.S. Months later I found some of the wood on the town beach which was salvaged and became part of Renee's house.

RENEE'S DRIFTWOOD KITCHEN

NEW AMERICAN HOUSE

THERE ARE FLOORS
THAT WANT TO DIGEST
THEIR FURNITURE INTO
FLOWERS AND TREES.

THERE ARE ROOFS
THAT WANT TO TRAVEL
GRACEFULLY WITH
THE STARS THROUGH
CIRCLES OF DARKNESS.

RICHARD BRAUTIGAN

MATERIALS METHODS

When men built in pre-industrial times they had to use materials and knowledge that were at hand. Their dwellings were as ingenious and varied as their cultures: cottages built of stone gathered from cleared fields in Ireland, animal hair tents of the nomadic Bedouin herders, mud and adobe buildings of dry desert areas.

The most important part of building is the materials you use. Today, unlike our ancestors, we have a choice of building materials from the entire earth; oil from the Mid East makes polyurethane foam, aluminum from Venezuela clads mobile homes. But we now know that these resources are limited, they are becoming too expensive for the individual builder and soon may be available only to the corporate interests.

Highly processed materials of the industrial society — plywood, metals, plastics — are expensive because of the high costs in extracting, processing and transporting. It may be that there is a correlation between the cost of a material and the exploitive damage done to the earth in its creation. Highly processed materials also appear to have a short useful life span, especially plastics.

We suggest you try to build with materials you find around you, close to your site. Use materials that are found on the earth, those that require a minimum of processing and transportation. Build with wood where it grows; wood is the only building material we can regenerate (part of the cost of lumber should go to reforestation). Use adobe or stabilized earth in dry climates, it is the earth's most abundant building material. Rock, bamboo, plaster, brick....

Industrial society creates wastes. A challenge to the builder is to utilize this fallout, as has been done by making bricks from newspapers, from garbage, and from the sulphur refined out of crude oil. Another challenge is salvaging used materials, building with scavenged junk and scrap. The *Zarch* structure on P. 132 is shingled with scrap aluminum offset plates. Inventive builders have veneered plaster surfaces with pieces of broken plates, built walls of hard coal, constructed curving shapes with broken pieces of concrete and short scraps of 2x4.

Using natural materials means a lot of work. Buildings that are pieced together with wood, dirt, rock, scrap, take time, but as the craftsman does it, he can choose to do it well, and shape the building with his imagination and changing perceptions. Use of local natural materials and hand labor results in a harmony of house and landscape. The character and warmth of less processed materials seem to make them easiest to live with over a period of time...a building is your skin extended.

Nature builds with what is available, and local, and can be studied, not copied. A nautilus secretes its shell. A swallow weaves its nest from mud, straw, horsehair and feathers.

You have leather?
You have thread and nails and dye and tools?
Then why don't you make yourself a pair of shoes?

Mulla Nasrudin

Zarch

Simon Rodia

Cutaway view of nautilus in its shell.

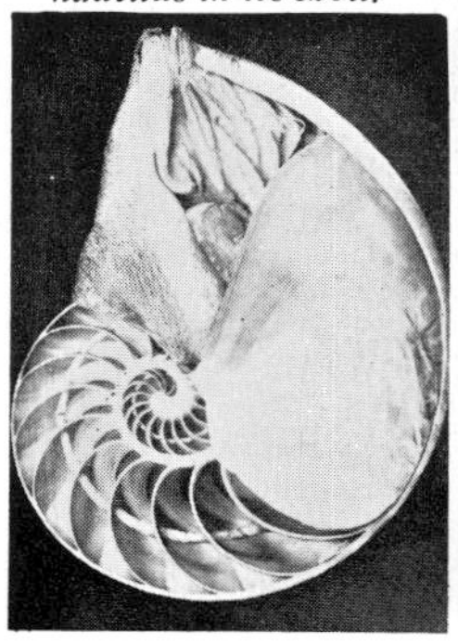

Bedouins

Latvian summer kitchen made from old boat.

scrap 2x4

MATERIALS & ANIMAL ENERGY

The three elaborations of shelter: Location, Transport & Processing

by Peter Warshall

The materials for non-human shelters are always elegantly simple and economical: the materials are easily available, require a minimum amount of transport from the source to the site of the shelter, and, except for the male bower birds of New Guinea who gorgeously advertise their boudoir, the materials are structurally crucial. Non-human animals are not interested in the baroque or fancy moldings.

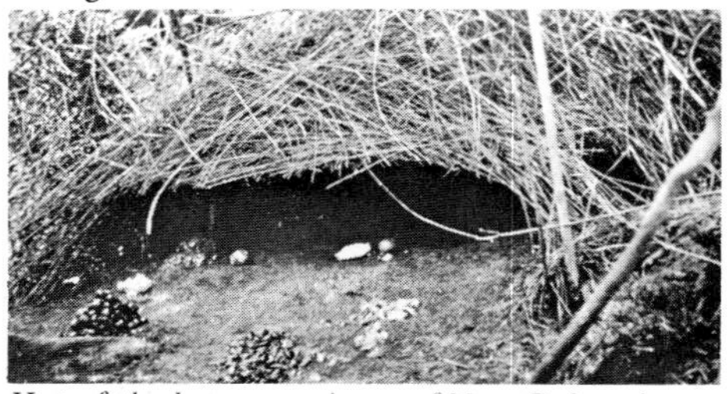

Hut of the brown gardener of New Guinea is built with central pole. The decorations on the display-ground are principally fruit and flowers.

Location

Most non-human animals are highly conservative about spending energy to construct shelters. Many never build shelters. Among mammals, all the ungulates (deer, zebra, gnu) never build any shelter. Neither do whales or dolphins. Like most living creatures, they rely on their senses and just use what's there. The Shelter Trip is simply finding the appropriate LOCATION. This is especially true for temporary shelters used for sleeping: the sea otter swims among algae and kelp, wraps itself among the seaweed to provide an anchor and floats asleep. More elaborately, certain tropical bats eat diagonal slits in palm leaves. The leaf droops and the bat sleeps in the "tent" formed by the droop. Many spiders, crickets, and other insects go to the extravagant limits of wrapping a leaf around themselves and tying the leaf into a cone with their body's silk. But most insects simply hide among the grasses and in the furrows of bark.

Excavate

Most rodents simply EXCAVATE their shelters and line them with soft grasses. Nothing is built. Insects, using the architecture of plant stems, tunnel into the stem's tube to rest or sleep or reproduce. Larger mammals, avoiding any strenuous labor, move in on already-made excavations. In Africa, I have seen warthogs use the abandoned burrows of aardvarks as sleeping quarters, retreats from danger and raising their young. Jackals, bat-eared fox, and spotted hyenas also use the burrows to raise their young. In North America, the interminably lazy coyote uses the burrows of badgers, skunks, or foxes.

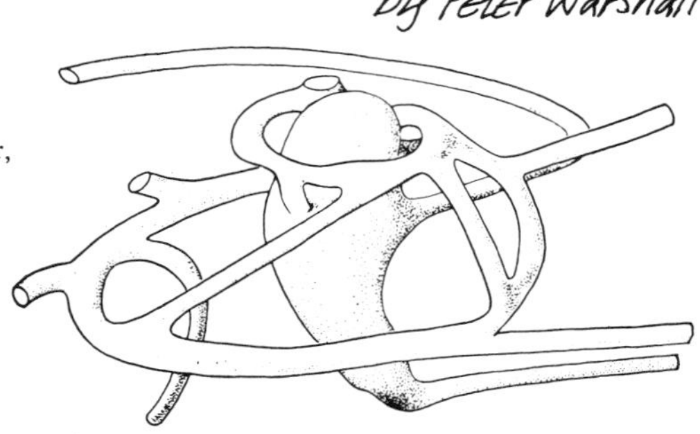

The nest of the European Mole is located carefully under brambles or a tree trunk. The "dome" is about a foot below the surface in silica-clayey soil. The bottom of the "well" serves as an emergency escape route and a drainage outlet for excess water. The walls of the nest are compacted clay, smoothed, and sometimes plastered. Only the permanent galleries *are shown. These permanent galleries extend out to the* hunting galleries *which are less permanent and altered daily.* From *The Natural History of Mammals.*

Coyotes limit themselves to enlarging, cleaning, and sometimes furnishing these burrows with an air hole. In short, most mammals use the self-support properties of the Earth itself. The only energy expended is finding the right spot and digging it out. The crustacean analog to using another's burrow is the hermit crab which uses another's shell. These crabs are totally dependent on the size and numbers of empty shells along the beach — something to be remembered by shell collectors. The avian analog is the owl who uses the woodpecker's excavations. All such re-use of another's burrow represents elementary re-cycling of shelters and energy conservation.

Some animals are partly passive agents in another animal's shelter trip. The malaria-causing protozoa is a fine example. It spends half its life in the belly of the malaria mosquito — subject only to the same problems as the mosquito. The other half of its life, the protozoa is injected into human blood. There, inside the red blood cells, the protozoa has the protection of the cell membrane and an ample food and oxygen supply to reproduce. When too many protozoa are living in one red blood cell, the cell bursts (causing fever in the human) and each protozoa enters another cell. Meanwhile, they receive all the sheltering benefits of being inside us. This protozoa and many other animals live in a *living shelter* and without even looking for a host, reap the benefits of an honored guest.

The extreme economy of non-human animal shelters is the chrysalis. Here, the very skin of the caterpillar turns into the shelter for metamorphosis. Only the thread from the caterpillar's mouth to the branch is added. The energy is daily eating.

Transport

In terms of energy and energy conservation, animals who use the immediate materials around them prosper. But, some animals have come to TRANSPORT materials from one location to another — requiring a new step in the transformation of materials to shelter. Transportation of materials usually occurs among animals when a *double-safety* situation arises: no adequate shelter is provided directly by the habitat and the young need protection *without the parents' presence.* The young go unattended because the parent may die before they are born or because the parents must forage for food at long distances from the shelter. A mud builder, the potter wasp, makes a little vase (with a cover). She fills the bottom of the vase with meat, then suspends the eggs from the cover so they dangle over the meat. She puts the cover on, then splits.

Potter wasp.

Unattended young are common among birds. Again, *location* is most important. But, additionally, carrying special materials to the location helps the young. Weaver finches (see *Thatching*) make shelters that will be somewhat safe from snakes and other predators. Carefully chosen long grasses are woven into a nest and then thatched. The extreme case is the mallee fowl whose shelter is a self-destructing compost heap. The eggs are large with lots of yolk and are buried under grasses covered by sand. The male returns to the compost heap-egg incubator and checks the temperature with his beak. He adds sand or takes it away to keep the eggs just warm enough. Finally, the young hatch and push their way out of the compost heap — never getting to know their parents.

Hornet nest made from pulp.

Processing

The third elaboration of shelter, after *location* and *transport* of materials, is PROCESSING raw material to make building materials. *Silk* is the most common bodily secretion used for homes and shelters. *Paper* is made by hornets by finding sawdust, rotted wood and discarded human litter and chewing it with their own saliva to form a liquid pulp. The pulp is spread to form the *carton* internal structure of the nest (brooding chambers, storage chambers, pillars between tiers) and spread out in sheets for the insulating layers around the "rooms" and "galleries". Bees process *resin* from wounds in pine trees into a substance called propolis. Propolis is one of the best sealers known on Earth. Bees use it to patch cracks in tree trunks, to cut down on drafts, to make doorways the right diameter, etc. Any rough object like unsanded wood that might harm a bee is covered with propolis. Any space that is too small to make a bee corridor or a bee honey comb (usually less than a quarter of an inch) is filled with propolis. This removes the shelters for harmful bacteria, molds and other insects that could ruin the bee hive. *Saliva* and *shit* and other excretions are used by termites in Africa in making their huge mounds. Almost all the clams, lobsters, and cephalopods (like the nautilus use secretions to form their shell-shelters.

SHAKES SHINGLING

Leon Henry

Mrs. Henry

Random pattern stone roof in Norway. Stone drilled and wired to wood frame.

Froe is tool to split shakes- steel blade is hit with wood mallet

The principle of shingling, or overlapping, as with a bird's feathers, is perhaps the best water-shedding device discovered by man, and is used in a variety of ways on roofs: shingles, shakes, tiles, tin roofing, mineral paper, composition shingles, slate, etc. All overlap and pass water along until it leaves the building.

Whereas shingles are cut by saws, shakes are handsplit. Best shakes come from old trees, with tight growth rings. A good shake tree will be at least two feet in diameter at a point two feet above ground, straight, and with no limbs in the first 15 feet. A sign of good grain is bark that runs straight from the roots to the first fork. (The best book on splitting and shakes is *Old Ways of Working Wood.* See bibliography.) Some of the best woods are oak, cedar, and redwood, all becoming increasingly scarce. But if you look around, on beaches or in the woods, you may find trees left behind by loggers, and shakes can often be made from short lengths of wood not good for much else. The best thing about shakes is that with a chain saw and a few hand tools, you can make your own building materials.

I've covered the walls and/or roofs of three houses now with home-split redwood shakes. For one, I found a *windfall* tree, knocked down years earlier by wind, in the Mendocino woods, and the owner let me split shakes from it in return for leaving some for him. The second tree, for the second house I found up a steep trail in a canyon abandoned by loggers. The last bunch of shakes came from driftwood redwood logs: the logs were about ½ mile from truck access so I went down to the beach in a wet suit, with a kayak paddle, levered a log into the water at high tide, sat on it, and rowed up to the beach road. I floated in as far as I could, left the log, came back at low tide, cut it up, and hauled sections home in my truck, to be split later into shakes. Another way to move logs along a beach is get them into the water, then tow them with a rope. Let the water carry the weight.

There's a good section on splitting oak shakes in *The Foxfire Book*. Here is what I've learned working with redwood:

—First thing, before cutting up an entire log: cut out a section and try splitting test shakes. Many trees aren't suitable — knots, bad grain or too young.

—You cut 24" sections out of the log for standard shakes. If it's a big tree and your chain saw won't cut through all the way, cut as deep as possible, then with wedges split sections out.

—Here are photos from *The Foxfire Book* of splitting a log into bolts.

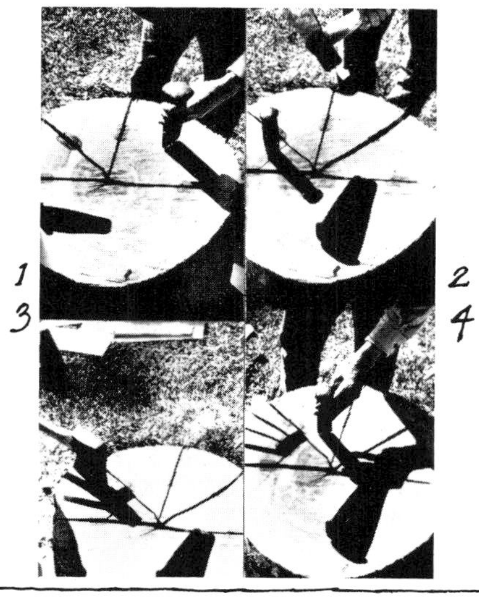

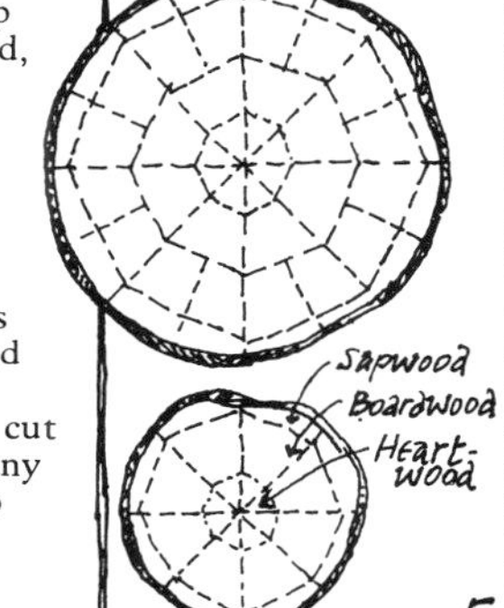

This procedure works if you are going to split standard *against-the-grain*-shakes. For these you must have a tree with tight rings.

—If the wood doesn't split well against the grain, as shown by the bolts in photo 5, it may still be possible to split *with* the grain, to make *bastard shakes.* In splitting bastard shakes, you do not split from the ends. Rather, and this is important, you split the bolt down the middle, and continue to halve the sections until you get as thin as you can. These are more likely to warp than regular shakes, better for walls than roofs.

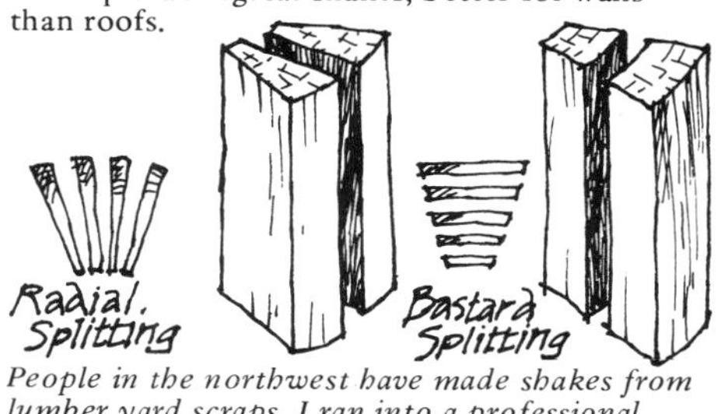

People in the northwest have made shakes from lumber yard scraps. I ran into a professional shake splitter once, he told me that commercial cedar shakes are made from wood after logging operations, and that there's much wastage; they only take the prime wood, and burn the rest.

Most cedar shakes are split into about 1" thick sections, then sawed in two with a taper at one end. They're sawn by eye on a band saw, the men doing it are paid by the number of shakes they turn out, and most of them have less than ten fingers.

You can try making shakes out of any type wood. Shakes to be used on walls don't have to be as good as roofing shakes. I've made them from old redwood highway markers, railroad ties, douglas fir, and even eucalyptus.

Roof pitch and exposure: Handsplit shakes should be used on roofs where the slope or pitch is sufficient to insure good drainage. Minimum recommended pitch is 1/6th or 4-in-12 (4" vertical rise for each 12" horizontal run). Maximum recommended weather exposure is 10" for 24" shakes.

Roof application: Along the eave line, a 36" wide strip of 30 lb. roofing felt is laid over the sheathing. The beginning or starter course at the eave line should be doubled. After each course of shakes is applied, an 18" wide strip of 30 lb. roofing felt is laid over the top portion of the shakes and extending onto the sheathing, with the bottom edge of the felt positioned at a distance above the butt equal to twice the weather exposure.

Nailing: Use two hot-dipped zinc-coated nails for each shake, placing them approximately one inch from each edge, and high enough to be covered an inch or two by the succeeding course. Nails should be long enough to penetrate at least ½" into sheathing.

Individual shakes should be spaced apart about 1/4 to 3/8 inches, to allow for possible expansion. These joints or "spaces-between-shakes" should be broken or offset at least 1½ inches in adjacent courses.

Literature: Complete details of handsplit shake application methods and grades are given in the 36-page Certi-Split manual. A free copy can be obtained by writing — Red Cedar Shingle & Handsplit Shake Bureau, 5510 White Bldg., Seattle, Wn. 98101.

THE FROE: you can make one yourself from old car springs, or see if local hardware store can order, or get one by mail from Foxfire: *made by blacksmith John Conley, 12" handle, 12-14" hand shaped blade, send $15.00 to* Foxfire, *Rabun Gap, Georgia 30568.*

Some tips from Leon Henry, shake splitter from the Russian River area, California, on redwood shakes:

How would you frame a roof for 24" shakes?

On top of the rafters, nail a 1 x 3 or 1 x 4" strip every 5-6". Nail here, two nails per shake:

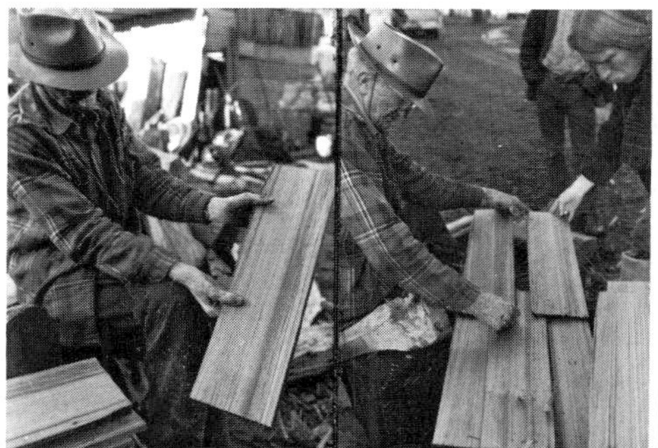

At right, Leon is laying out shakes as you could put them on a wall.

Leave about ½" between shakes. 10" exposure. Nails under above shake.

Put 'em on right, get the right pitch, you don't need tar paper. I've slept in a place, you could see the stars through the shakes, but when it rained, no water came in.

A lot of people make the mistake of putting shakes on a roof that's too flat. You shouldn't go shallower than 5" in 12" — that's pretty steep. If it's too shallow the wind can get underneath and pull shakes off, and water gets driven up inside them.

Building codes today require tar paper in between shakes. Leon thinks this isn't a bad idea, but not necessary with proper pitch, and that this ruling probably came from people trying to put shakes on shallow-pitched roofs, and roofing companies trying to sell paper. He says that what's good about nailing shakes onto strips is air circulation, drying out, less chance of rot.

THE NATURE OF WOOD

Structure of Wood. Wood, like all plant material, is made up of cells, or fibers, which when magnified have an appearance similar to but less regular than honeycomb.

The walls of the honeycomb correspond to the walls of the fibers, and the cavities in the honeycomb correspond to the hollow or open spaces of the fibers.

Softwoods and Hardwoods. All lumber is divided as a matter of convenience into two great groups, softwoods and hardwoods. The softwoods in general are the coniferous or cone-bearing trees, such as the various pines, spruces, hemlocks, firs, and cedar. The hardwoods are the non-conebearing trees, such as the maple, oak, poplar, and the like.

Moisture Content. While the tree is living, both the cells and cell walls are more or less filled with water. As soon as the tree is cut, the water within the cells, *free water,* begins to evaporate. When practically all of the free water has evaporated, the wood is said to be at the fiber-saturation point; i.e. what water remains is mainly in the cell or fiber walls.

Except in a few species, there is no change in size during this preliminary drying process, and therefore no shrinkage during the evaporation of the free water. Shrinkage begins only when water begins to leave the cell walls themselves. What causes shrinkage and other changes in wood is not fully understood; but it is thought that as water leaves the cell walls they contract, becoming harder and denser, thereby causing a general reduction in size of the piece of wood. If the specimen is placed in an oven which is maintained at 212° F., the water will evaporate and the specimen will continue to lose weight for a time. Finally a point is reached at which the weight remains substantially constant, meaning that all of the water in the cells and cell walls has been driven off. The piece is then said to be "kiln dried."

If it is now taken out of the oven and allowed to remain in the open air, it will gradually take on weight, due to the absorption of moisture from the air. As when placed in the oven, a point is reached at which the weight of the wood in contact with the air remains more or less constant. Careful tests, however, show that it does not remain exactly constant, for it will take on and give off water as moisture in the atmosphere increases or decreases. Thus, a piece of wood will contain more water during the humid, moist summer months than in the colder, drier winter months. When the piece is in this condition it is in *equilibrium with the air* and is said to be *air-dry.*

Felling a redwood.

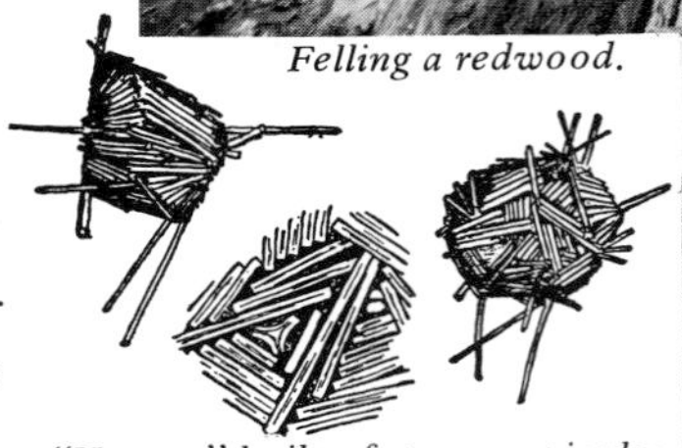

"Houses" built of sponge spicules by Psamnosphaera rustica.

Clear cutting in Washington.

Logs on way to mill.

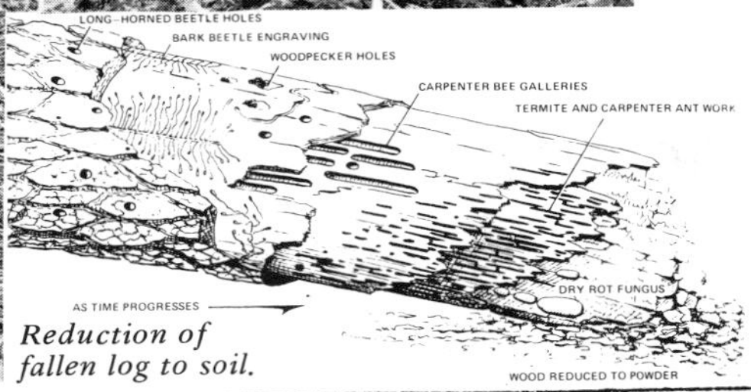

Reduction of fallen log to soil.

A piece of lumber cut from a green tree and left in the atmosphere in such a way that the air may circulate freely about it will gradually arrive at this air-dry condition. This ordinarily takes from one to three months, and the process is termed *air-seasoning.*

Shrinkage. As a log shrinks, the parts season differently. These characteristics are important to know for use of the wood as lumber. The shrinkage from end to end is negligible, but in cross section it shrinks approximately ½ as much at right angles to the annual rings as it does parallel to them. This is important in framing, and flooring: a stud will not shrink much in length, but will shrink somewhat in the 2" and 4" way. Likewise, a green joist will change in depth as it seasons. In flooring, edge or vertical grain flooring, if green, will shrink about half as much as green flat-grained flooring.

Density. The tree undergoes a considerable impetus early every spring and grows very rapidly for a short time. Large amounts of water are carried through the cells to the rapidly growing branches and leaves at the top of the tree. The result is that the cells next to the bark, which are formed during the period of rapid growth, have thin walls and large passages.

Later on, during the summer, the rate of growth slows up and the demand for water is less. The cells which are formed during the summer have much thicker walls and much smaller pores. Thus it is that each year's growth is of two types, the spring wood as it is called, being characterized by softness and openness of grain and the summer wood by hardness and closeness, or density, of grain. The spring wood and summer wood growth for one year is called an *annual ring.*

There is one ring for each year of growth. This development of spring wood and summer wood is a marked characteristic of practically all woods and is clearly evident in such trees as the yellow pines and firs, and less so in the white pines, maple, and the like. Careful examination will reveal this annual ring, however, in practically all species.

Reprinted, with revisions, from *Light Frame House Construction*, printed 1956 by HEW.

Eucalyptus Lumber

The blue gum eucalyptus — *Eucalyptus globulus* — was first imported to California in 1856, along with 13 other species of the Australian hardwood. By the 1870s, the blue gum, due to its tremendous growth rate, was becoming well known throughout the state. In 1877 Southern Pacific planted the trees along its railroad tracks for steam engine fuel. By the early 1900s, there were numerous eucalyptus timber companies aggressively promoting investment in their stock and planting large areas of blue gum in California.

What excited people, and fanned the hype, was that blue gum grew seven times as fast as the best quality of mixed hardwoods — to 12" diameter in 10 years. Early literature claimed it could be used as wharf piling, telephone poles, railroad ties, flooring, and tool handles, as well as firewood.

However, the miracle wood turned out to have many problems — checking, splitting, twisting — and in 1913 the U.S. Forestry Service presented a paper on the subject after studies made at a dry kiln in Berkeley: *Eucalyptus Lumber,* reprinted from the Hardwood Record, Chicago, 1913. A complete copy of the report can be obtained by sending 60 cents (for xerox fees) to Forest Products Labratory, 1301 So. 46th Street, Richmond, Calif. 94804, Attn: William A. Dost.

Bassi Eucalyptus Mill

As lumber becomes increasingly scarcer and higher-priced, the challenge of working with eucalyptus becomes more relevant. Question is, given its drawbacks and peculiarities, can individual builders begin to utilize it in construction? One application, perhaps the only commercial use of blue gum in America, began about 3 years ago at the Bassi Distributing Co. in Watsonville, Calif., which makes eucalyptus pallets. (Pallets are the wooden skids designed for fork lift use.) Because pallets get knocked around quite a bit, the best ones are made of hardwoods such as oak. When oak began to get increasingly scarce and expensive a few years ago, George Bassi and Ken Thayer, owner and manager of the pallet firm turned to eucalyptus and gradually have learned to work with its difficulties. Blue gum surprisingly has about 50% more strength than white oak (static bending modulus of rupture) when green. Here are some of the Bassi techniques:

The logs are heavy and difficult to cut. They start with 12" - 30" trees. After cutting into planks, they are then cut into pallet size — about 4' long. The wood is easiest to work with in narrow and thin dimensions, such as 1" thick, 4 - 6' long. 2" stock is much harder to work with and more likely to warp.

After cutting into 1x4's and 1x6's, the ends are sealed with wax or *Mobilcer*, a Mobil saran product. This seals moisture in, allows the wood to dry slowly, and stops the heavy checking and splitting characteristic of blue gum grown in California. It is then stacked out of the sun. They have found it best to nail it onto the pallets within a month after milling, otherwise it gets too hard and either splits or bends the nails. Six months is maximum time it can be stored before nailing.

Build with Eucalyptus Saplings

The 1913 report was for lumber yards, with the heavy equipment to cut, move and mill large logs. Builders, without the equipment, have tried working with blue gum in various manners: some California boatbuilders have used it, but very carefully in boatbuilding. They say they've used the middle layer — not heartwood or sapwood. I split some shakes out of a 10" log about two years ago, nailed them on the fence. They split a little, but other than that don't show any adverse signs of wear. The shakes have to be short, to split any amount of them you'd probably need a hydraulic splitter (tough fibers), and I'd use them on walls, not roofs, and they just might work. Another, perhaps the best use of blue gum, is to use the saplings as a building frame. Here are bits of a conversation with Val Agnoli about framing with eucalyptus poles:

Blue gum.

Eucalyptus boat.

Bassi eucalyptus mill.

Eucalyptus pole frame.

Val: ...I built a shed of eucalyptus, it cracked a bit, but just the big pieces, not the small ones. It was lightly spiked together....

Lloyd: I framed a shack with eucalyptus I cut across the road. I made the mistake of stripping bark off, I should have left it on to allow slower drying. Poles 3-4" diam., they cracked, but it didn't hurt anything...how would you fasten it?

...drill holes and pin it with rebar, big spike down so things can rotate. You've got to let it twist, rotate. Use the smallest members you can.

...that eucalyptus house in Berkeley, the roof beams twisted and lifted up the roof, they had to go back and remove them. I think their mistake was in squaring them off, if they'd had a round cross section, there wouldn't have been corners to lift the roof up.

You've gotta let it turn. Put a strap over it....Cut trees from the middle of the stand where the wind hasn't been twisting them.... stay with small sizes...you could band them together, a number of small poles, make a laminated pole beam, you could span 30-40 feet. Wire 2" poles together with no. 9 wire, get it real tight with wire puller.

I've been thinking of eucalyptus frame, then infilling with wattle and daub.

If you hold in the moisture it'll rot, eucalyptus rots easily, you gotta keep air circulating....

...maybe use it for siding, saw it up, nail while green. Or eucalyptus shakes. I made shakes out of an 8" section 2 years ago, put them on fence, they're still here. Cracked a bit but basically they're o.k.

Eucalyptus shakes huh?

Yeh, they're hard to split by hand, but if you got a hydraulic splitter. Probably have to nail them up while green, or could predrill holes. Use on walls, but not roof.

You could oil them to make them more stable...basically you gotta use it while green, leave the bark on, it makes them dry out slowly. Because the outside starts shrinking first, the center doesn't, there's great tension, the outside cracks. Wire is a great tool, it's universal, easy to use. Get a big roll no. 9 galvanized wire, you can fasten a lot of things with it...build forms that'll take hydraulic pressure.

Sure, use wire like they use lashing in the tropics.

...use pair of wire tongs for tightening, just keep tightening, wire'll sink into wood you know...easy to work by hand. To prevent sharp ends that'll tear hell out of your hands, take a pair of pliers, turn the ends, make these two little curls...use them architecturally that way. You can retighten them as the wood shrinks. I tied down some sheds about 5 years ago on account of the wind. I just drove rebar stakes into the ground and tied the whole building down and it's still there...another thing about using small poles is they're easier to lift. But if you had say four little ones, make a block in the center to hold them....

LIGHT POLE FRAME

These photos show the Caribs of Surinam building a pole and thatched house. Prepared originally for *Architectural Design* magazine by Peter Kloos.

At the housing site the horizontal frame is laid out, to determine exactly where the posts are to be placed. There is much arrangement and re-arrangement before the builders are satisfied. The trade winds, together with the course of the sun are responsible for the orientation of the house, its ridge-pole lying almost north-south.

The posts are sunk into the sand. The Caribs use a shovel and their hands to dig a narrow pit. The only resistance to horizontal forces (the trade winds) is the posts, dug about two feet into an undisturbed soil. For a house of normal size six or eight posts in two rows are sunk into the sand. The horizontal frame, previously laid out upon the sand, is now placed on the posts, the tops of which are notched. The horizontal beams are not secured to the posts and neither are the cross-beams. Only the beams alongside the beams lying on the posts are nailed to the cross-beams. The roof frame consists of sets of poles which cross at the top: a V-shape placed upon each cross-beam. These poles are placed in a notch and secured with a nail. On top, where they cross, they are nailed together. The frame of the roof is held vertical temporarily by poles lashed to the cross-beams. When they are in place a slender pole, nailed to the horizontal frame, will keep them in place. Then the ridge-pole is laid upon the triangles forming the frame of the roof.

From the ridgepole the rafters hang down, their slender ends pointing downwards. At the trunk-end they are notched and hooked into the ridge-pole, and they are secured by a nail. Usually they are also nailed to the horizontal frame.

The frame of the house is now finished. Note the notched top of the posts and the construction of the horizontal frame.

House is thatched with palm leaves, makes thick, watertight cover.

The eastern part of the roof is thatched first to prevent the trade winds from blowing away the whole roof.

SAPLINGS

Buildings framed with saplings are found throughout the world, especially in the tropics, where they are usually roofed with thatch. Saplings for frame can be cut by hand, often at the site, and do not require power tools. Like other building materials that are cheap, or as in this case, free, poles take *time* to cut and fit. But you have the satisfaction of using resources well (thinning a forest), not patronizing a lumber yard or sawmill, and ending up with a fine looking framework. You could *grow* bamboo or eucalyptus for a light pole building in 4-5 years.

100 year old Barn near Santa Rosa, Calif. Redwood saplings were nailed together. Plate poles were adzed off on top and bottom, but posts and braces were left round.

Nepalese house structure.

Probable frame of Iron Age Barn, Avoncroft Museum

The building shown at the bottom of the opposite page is framed with eucalyptus poles, put together with nails. (The roof is old chicken roosts.) I made three mistakes here: poles should not have been stripped; leaving bark on keeps moisture in, retards shrinkage. The poles should have been wired together, rather than nailed; cracks have developed at each nail, although this doesn't seem to harm much. Last, some sheathing other than wood would have been best, as these poles were not perfectly straight, and nailing on wood, putting in windows was hard. Unless poles are straight it might be best to skin such a frame with plaster, or infill with adobe, or stretch wires between poles. Hang gunny sacks dipped in concrete?

Gutter cut into projecting pole beam. New Mexico.

Taos pueblo, New Mexico

MORTISE & TENON

by John Welles

The *mortise and tenon* method of frame construction, used in early America before sawmills, consisted of hand-hewn timbers with notched and pegged joints. Timber could be gathered and buildings framed entirely with hand tools. (The method was later also used with milled timber.) In summer 1972 John Welles and friends cleared a site in the Connecticut woods, and built the small barn framework shown in these photos. John and two friends worked for about three days cutting and preparing timbers, then people from the town put up the structure in half a day. All materials, including foundation (rock) are from the site. The traditional feast for workers after the frame-raising finished the day.

Making Diagonal brace flush w/ outside of building

Trimming Tenon with Chisel

Where I live, Connecticut, old houses available for tearing down are often hewn timber — pegged joint — lath/plaster — plank floor — shingle roof construction. During dismantling these houses and barns one cannot help wondering at the seeming vast amount of work to produce these frameworks. Hewn timber went out of fashion in our area in the late 1800's with the advent of saw mills. Big sawn timbers are still occasionally mortise/tenoned joined, but rarely pegged. In the course of building my house I collected the various tools required for such work and learned to make the joints necessary to re-erect my house. When my house was more or less livable, I started on a shop which is a 42ft. geodesic — metal struts, corrugated metal on top, various experiments using screen, cloth, wire mesh, wood and cement on the remaining panels, all sprayed with urethane foam from the inside. When this finally got done, I was sick of plastics and chemicals and more than eager to change work. Some friends were clearing a field in the forest, and wanted a barn. Trees were available, and we had the tools, and we thought, the know-how; so a barn raising started. Trees were hewn square (mostly), joints were cut, sections were prefabbed and erected in less than a week. The barn is about 20ft. square.

Hewing Logs—Tools required:
1. Big saw to cut trees down & buck to length.
2. Regular Axe.
3. Broad Axe, 2 preferably, right & left hand.

Several theories on hewing and squaring logs. All work. Personal preference determines method. The log is cut to length and put upon blocks and held in place with dogs, if you have them. We didn't and needed them only on small timbers which shifted with axe blows. The desired beam size is laid out on the end of the log and marked on the log with a chalk line. If the log is bigger than beam you hew one side then go 90° down from that face.

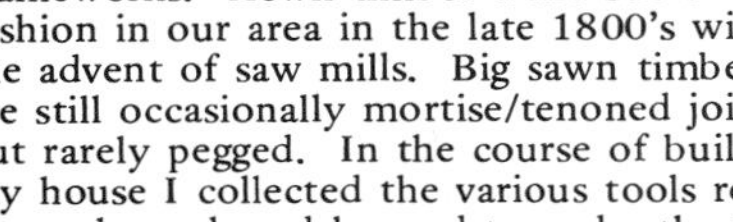

Now comes the choices. Broad axes, as found in antique stores, junk shops etc., often come with short bent handles, if they have handles at all. The blades come in various shapes and sizes depending on age and type, but all have one thing in common, the sharpened part of the blade is beveled on only one side

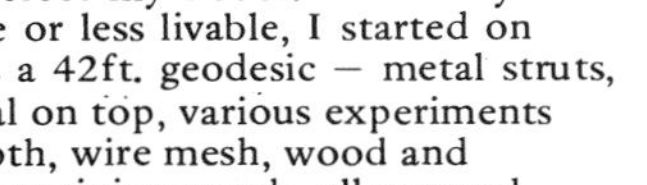

and the side opposite that bevel is flat back to the poll.

The handle can be set into the head from either side making a right or left hand axe. Right or left hand is determined by which side the bevel is on as you hold the axe by the handle. Now a log should be hewn from the top of the tree towards the butt. The reason for this is that a tree grows with a taper, and if you hew from the butt up, your split off chunks would go deeper into the log than intended, whereas hewing from the top down your split off chunks do not remove more than intended.

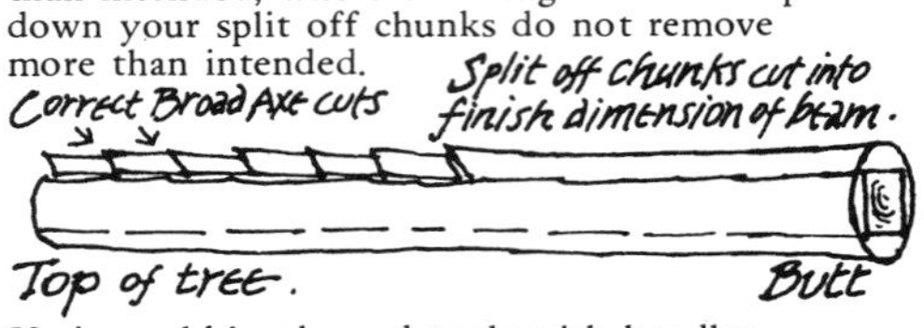

Various old books and tools with handles themselves seem to indicate that they were used with short swings about 45° to the length of the log, cutting on the same side you are standing on. The log is first scored with a series of axe cuts opposite from the hewing cuts about 6" apart. The log is *scored* from the butt up, and *hewn* from the top down. The right or left handed axes are used depending on which side of the log you are hewing down on. The scoring cuts over the wood fibers at six inch or so intervals so that the broad axe can split off those big chips.

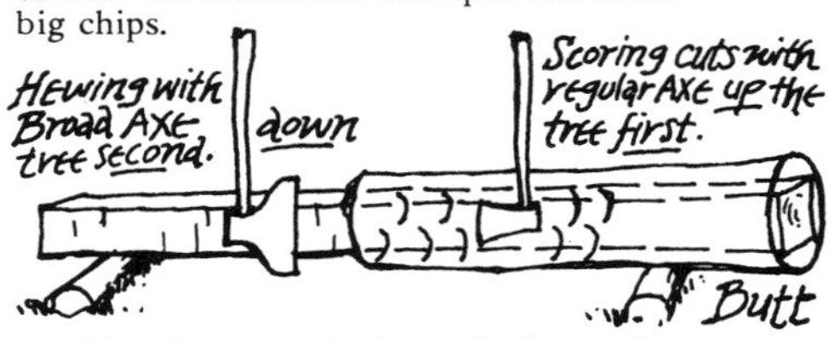

Looking down on the log to be hewn the bevelled side of the broad axe is to the outside of this log being hewn.

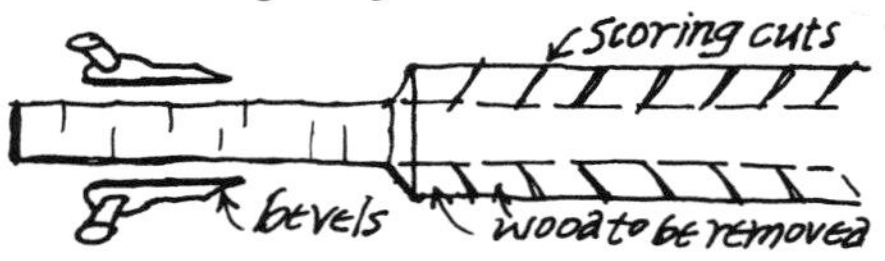

The handles of broad axes are bent towards the bevelled side so that as you swing the handle the blade remains parallel to the timber and vertical to the ground.

The above method is the formally correct way. The log is up on blocks and wedged in position or held by dogs. (See drawing below.)

The dog is a piece of metal, bent and sharpened on both ends, driven into the log and its support, to hold the log from turning.

Hewing is done while standing beside the log, hewing damn carefully next to your legs.

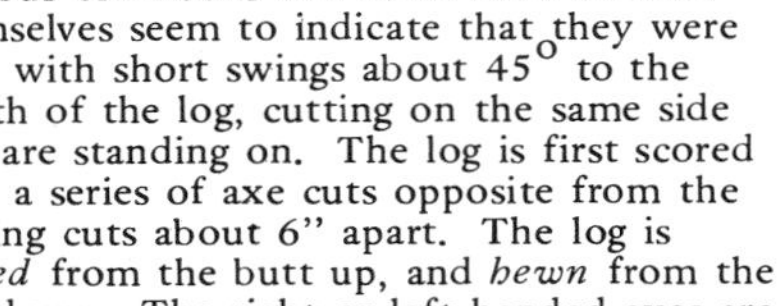

Hewing next to your legs gives one a very uncomfortable feeling. Broadaxes are big, heavy, and should be razor sharp. I do not like using such a tool that close to my legs, in the way that "they" apparently did and "they" say that you should do. So, I offer my personal method which I feel is safer, and I know is quicker. Some of my broadaxes didn't have handles, so I fitted new, unbent standard axe handles to the heads. I then scored and hewed at 45° to the ground, on the *opposite* side of the log from where I was standing. When I was young I was told when using an axe to keep the tree between me and the axe whenever possible, as when removing limbs — trim the opposite side from where you are.

So — log upon blocks, held the same way, but put a 90° notch in your log support. And score and hew on the top opposite side from where you are standing.

With the log between you and the axe, much bigger and safer swings can be done, and the chips fairly fly. When the two top sides are hewn, the 90° notch in the support holds the log for the two final cuts.

When you have all the timbers square, you lay out your mortises and tenons. Timber sizes vary according to intended use, but uprights are usually between six and eight inches square, and sill, (bottom) beams somewhat bigger. Plates about the same.

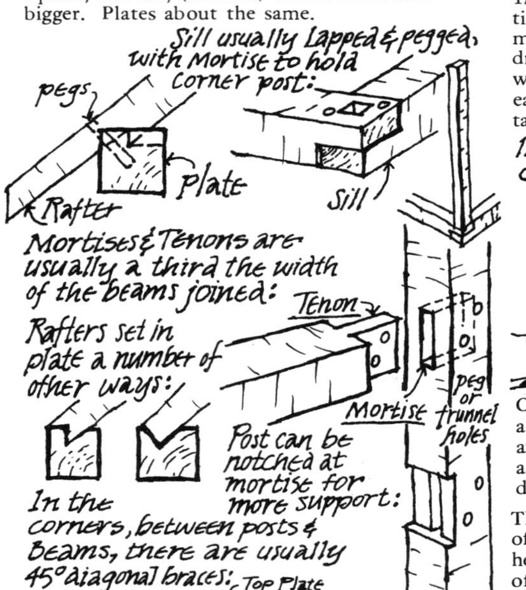

Sill usually Lapped & pegged, with Mortise to hold corner post:
pegs
Plate
Sill
Rafter
Mortises & Tenons are usually a third the width of the beams joined:
Tenon
Rafters set in plate a number of other ways:
Mortise
peg or trunnel holes
Post can be notched at mortise for more support:
In the corners, between posts & Beams, there are usually 45° diagonal braces:

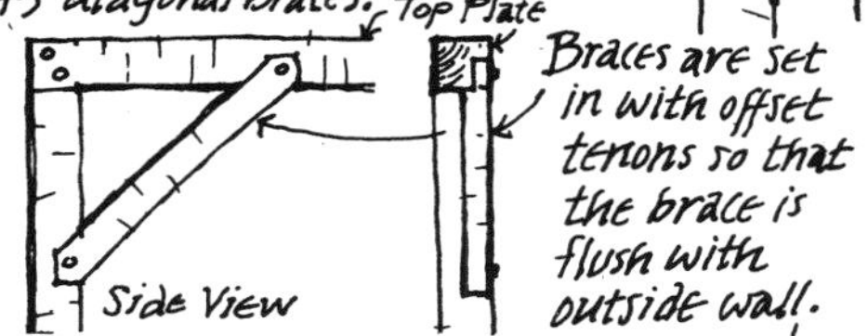

Tenons for horizontal beams are cut full height of beam, leaving 1/3 of wood as tenon. To saw cuts, chisel away wood, champfer edges with chisel or draw knife so tenon will start into mortise easily.

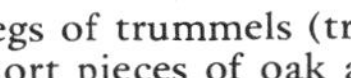

Pegs of trummels (tree nails) are split from short pieces of oak and trimmed with a hatchet or draw knife to a taper. Don't make them perfectly round but leave them rough. Taper them towards, but not to a point.

They should be full thickness where they will pass through tenon.

When you lay out and drill the holes for the pegs, the old people purposely made the holes misaligned so that when the peg was driven in, it pulled the joint together.

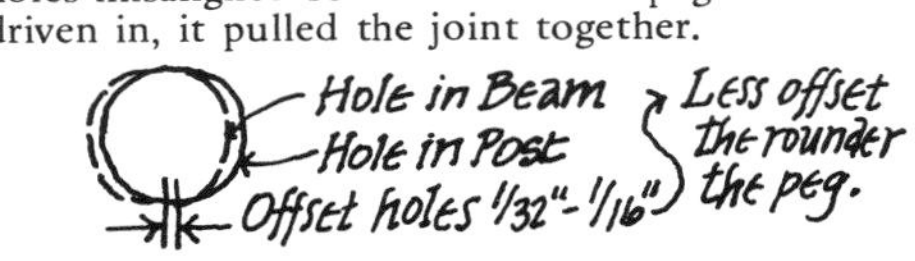

The diagonals are set the same way with offset pegs, but they give their bracing more by the shape of the mortise & tenon. When Joint *A* is driven together, the diagonal should be tightly wedged into position at *B*.

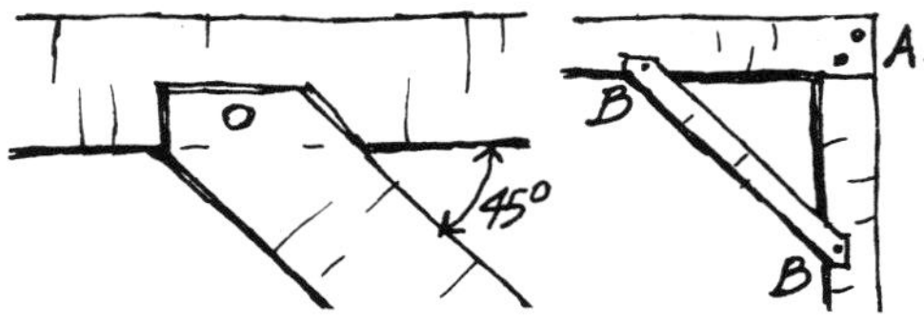

The diagonal must be put into place as joint *A* is assembled, as it cannot be put in afterwards.

The mortises are laid out to correspond to the tenons, and hopefully be only very slightly larger. Again, they are usually about 1/3 the thickness of the timbers. If the timbers are between 6" and 8" square, the mortises are usually 2". 2 or 3 2" holes are drilled down through the beam, and the wood remaining is chisled out. Now that is easily said, but sinking a 2" hole in white oak takes some doing. 3 choices.

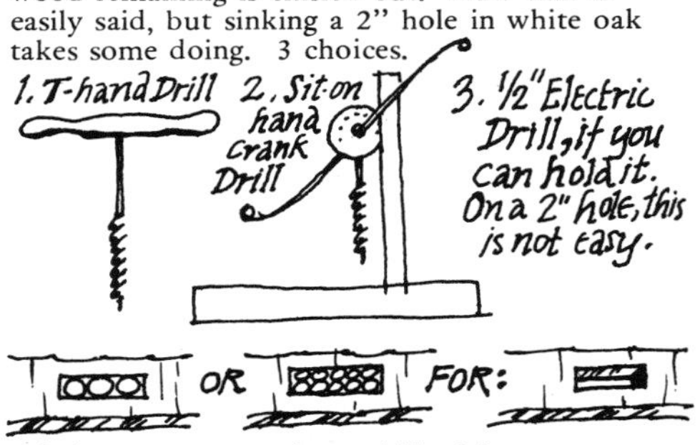

Of the 3, I use the sit on drill. It's easy, accurate and goes as fast as you want. These are not common tools, however, and a brace and one inch bit will do. You just have to drill more holes.

The old people had 90° chisels for the corners of the mortises, but these are scarcer than hens' teeth, and a 2" chisel will do it. Speaking of chisels, if you can acquire the big, old steel, 1½" and 2" chisels, by all means invest in them. Modern chisels, hardware store variety, do not last. They break, and don't hold an edge. The plastic handles get slippery when sweated on but wood gets sticky. The whole chisel including handle ought to be *at least* a foot long.

Once you've hewn your timbers, you can adze them down super smooth, but if you have gotten halfway proficient with the broad axe you won't need to. An adze requires more patience to use efficiently than a broad axe, and is intended more for finish dressing of a timber than the rough hewing.

continued

Second frame assembly being Raised.

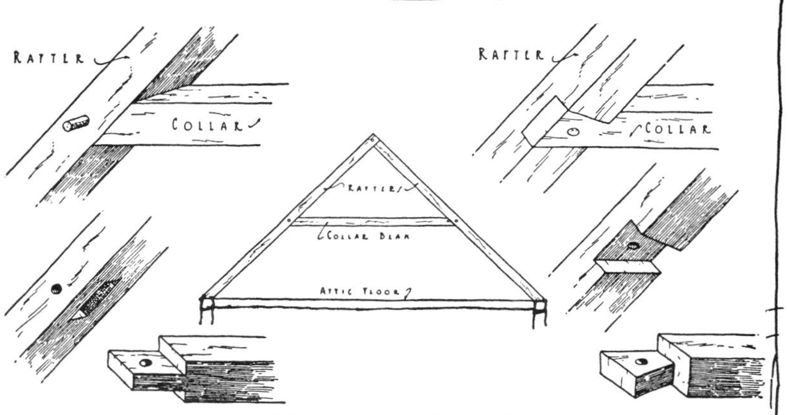

Roof Framing Details
from Early Domestic Architecture of Connecticut

Frames in place – 2 safety ropes, each direction.
Temporary brace is nailed in place before: when assembly is upright in place, braces are nailed to hold it.

As far as choice of woods for timbers, oak is best, and most commonly used. Chestnut was used a lot and is wonderful to work, but doesn't grow anymore. The harder softwoods, fir and spruce, are ok, but not as strong. Avoid super hard woods, elm, ash, hickory, as they twist and check a lot, aside from being ornery to work. Red oak and black oak are best. Hard birches and maples are ok. Use what's strongest of your local woods. Avoid the very soft woods, they aren't strong enough. I'm told the old guys girdled the trees (cut away the bark around the bottoms) to kill them and let them stand a year to season, but I don't know whether that's necessary. Green wood certainly is heavier.

As far as layout and plans go, major uprights were usually put every 8 or 10 feet on a floor plan of say 15' x 20'. There would be 6 major uprights. Wherever doors or windows were to go, 4 x 4 uprights were pegged into place with offset tenons. Rafters were usually 4" or 5" square about 2' on center with about a 45° pitch. On bigger houses, everything was scaled up accordingly. At the rafter peak there were/are several choices.

1. Lapped Joint w/peg
2. Bevelled joint & peg
3. Square or odd shaped beam as Ridge beam with Rafters mortised in.

Beams were built with much more elaborate framing which I won't cover here. See Eric Sloane's books. Floor joists were set into sill beams. Joists were often left round but for one side flattened for the floor planks.

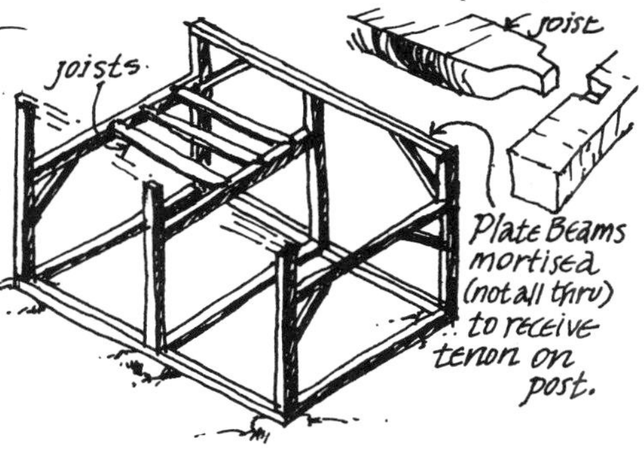

joists
joist
Plate Beams mortised (not all thru) to receive tenon on post.

If there was to be a second story, floor joists were squared and set in as the first floor, again around 2' on center.

Sections are usually prefabbed – 2 uprights – a crosspiece – 2 diagonals – and tipped up. Short tenons on the bottom of uprights hold them in place once upright.

More on Mortise & Tenon:

Two excellent reference books on mortise and tenon construction are *Carpenters Tools,* with details on making the timber connections, and *Early Domestic Architecture of Connecticut,* with many good framing sketches. *Old Ways of Working with Wood* is also pertinent, as is *Heavy Timber Construction.* Details in bibliography.

Moving Heavy Objects:

Much of building is moving things around – rearranging – to rearrange easily then is efficient building.

Moving big beams, timbers, logs.

Square or rectangular.

1. –Jack up one end. Scissors jack, hydraulic jack, walking beam.
   –Lever up one end.
   –Block & tackle & A-frame or tripod.
2. –Insert rollers (section of trees, poles, old propane tanks, lengths of pipe), one as close to middle of beam as possible, and one at end in direction beam is to be moved. Rollers can be angled to change direction, or beam can be pivoted on center roller.
   –Beam then dragged, pushed or levered along on rollers.

Round beams.

1. –Round beams themselves can be rolled, or moved as above – roll onto a center pivot to change direction.

Loading or unloading from or onto a truck.

1. –Bring beam up to back of truck about 90° to the truck. Roll or slide one end up a sloping log onto bed of truck.
2. –Pivot beam 45°.
3. –Insert rollers (piece of pipe), push beam forward.

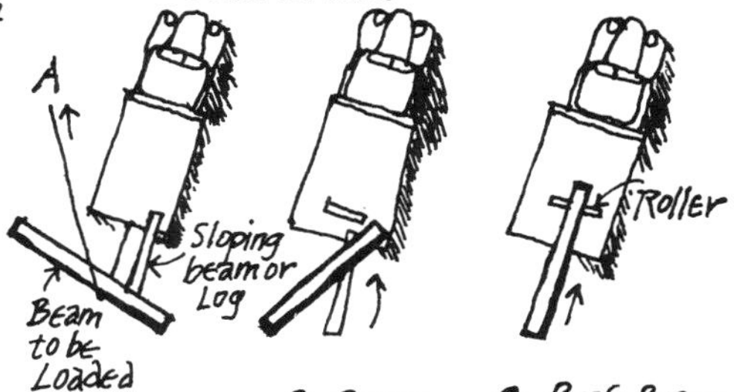

1. Beam 90° to truck
2. Beam 45° to truck
3. Push Beam on truck.

A single turn of rope going under and coming out on top – fastened to truck (see A above) rolls Log with 2:1 mechanical advantage.

4. –Bring beam or tree parallel with side of truck – roll or slide up ramps on side with rope.

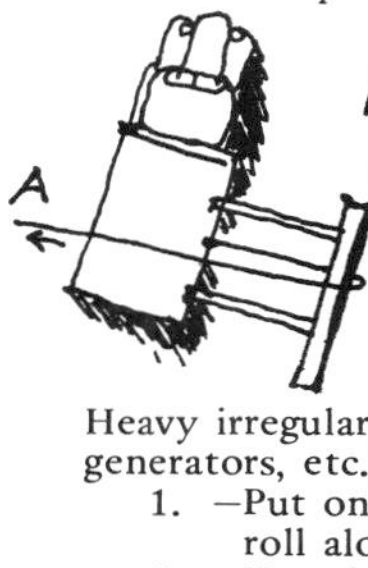

When working with heavy objects, think about Levers. Rollers, pivots at center of Gravity.

Heavy irregular objects (motors, generators, etc.)

1. –Put on a platform, or big plank, roll along, as above.
2. –Use chain hoist or block & tackle or tripod or big tree limb with supports to lift onto truck.

Eric Sloane has written and illustrated a number of books on early American barns, bridges, tools and life. The drawings are beautiful, the information rare. The best of these books for the builder is *A Museum of Early American Tools.* Details in bibliography. Here, courtesy of Eric Sloane and Thomas Y. Crowell Company, Inc., are excerpts from these books of pre-industrial America.

THE JOSÉ DE JESÚS VALLEJO ADOBE, FREMONT CALIF. BUILT 1842. MORLEY BAER, ADOBES IN THE SUN

DIRT

Dirt is the cheapest and most abundant building material known. Often you can get the material for a house on your own site, although earth buildings are best in dry areas. The best book on the subject is *Handbook for Building Homes of Earth* (see bibliography). The basic methods of building with dirt are:

1. Adobe bricks, usually made of dirt, straw and a stabilizer, then set with mortar to make a wall.
2. Rammed earth, where moist soil is tamped into forms to make a finished wall. (A very difficult technique.)
3. Pressed earth blocks, as made with a device such as the *Cinva-Ram*.
4. Wattle and daub: mud, straw and cowshit mixture plastered on a woven branch framework.
5. Cob: stiff mud molded into balls a little larger than a person's head, then piled up to make wall. Slow to construct.
6. Sod: see p. 71.

Adobe walls are thick (for strength), slow to build, heavy, but provide a well insulated house. Whether plastered or left natural, adobe makes a beautiful wall surface texture, and blends with the landscape.

Remember that the best natural soil you can use for making earth walls is a *sandy clay* or a *clayey sand*. If you happen to have such a soil, you have as good a natural building material as can be found. Without the addition of anything more than water, some kinds of sandy clays or clayey sands can be made into walls that will last a lifetime — or even longer.

If you do not have this kind of soil, you might be able to make it. If you happen to have mostly sand, maybe you can find some clay to mix with it, or if you have clay, you might find sand to mix with it.

WHERE TO LOOK — Often you will find a situation like this: beneath the organic topsoil, you will find a layer of sand. Below this is often found a layer of clay. By mixing the sand and clay together you might make a good sandy clay. Also, remember that on the top of rolling hills (not mountains) or ridges you are more likely to find clays, and sands will be most common at the bottom. Probably just what you need, a mixture of both, can be discovered somewhere between....

From *Handbook for Building Homes of Earth*

ENTRY TO ED ALLEN'S MUD HOUSE

CINVA-RAM

This is a steel device with long handle that compresses dirt, cement and water to make a high-quality earth block. The press weighs 140 lbs, costs $175, and comes with wooden inserts to make different size blocks and tiles. Two men can make 300-500 blocks a day and walls built of Cinva-Ram bricks are said to be far superior in strength and water resistance to conventional adobe or rammed earth. Full information from Bellows-Valvair, 200 W. Exchange St., Akron, Ohio 44309.

See p. 71 for more on Cinva-Ram.

RAMMED EARTH

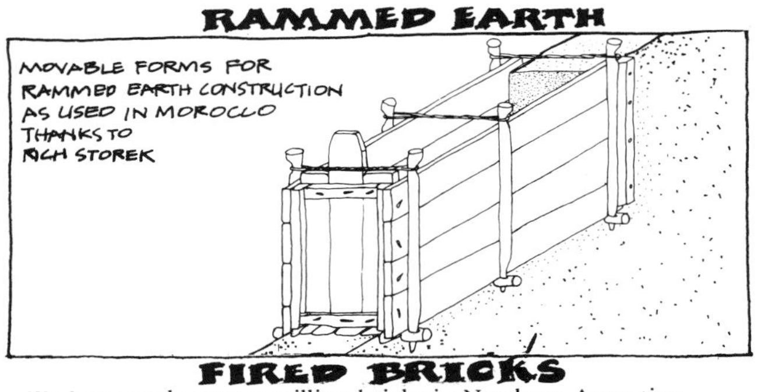
MOVABLE FORMS FOR
RAMMED EARTH CONSTRUCTION
AS USED IN MOROCCO
THANKS TO
RICH STOREK

FIRED BRICKS

...We have made over a million bricks in Northern Argentina and used the age old method which, in brief, was thus: —

First we made a mud bath from top soil with plenty of straw mixed in — linseed straw was best. This was stirred and stirred, using horses to do the work, until the right mixtures and consistency was attained. Then, using wooden moulds and trestle tables, bricks were formed and left to dry. Later they were stacked in lattice-like walls so that the sun and air could dry them further.

The next stage was to make a flat topped pyramid of them with tunnels below. This was plastered all over with more mud and then the whole thing was fired. Log after log was shoved down the tunnels, continuously day and night for three days. The pyramid glowed at night like a large ruby. Then it was left to cool for a week and then pulled to pieces and, if we were lucky, we had 60 per cent "Bells" that pinged.

Oh, the straw was essential to draw the flames into the heart of the bricks so that they might be cooked throughout.

A.F. and E.M. de Ledesma
Guildford
Surrey, England
(From Dec. 28, '72 *New Scientist*.)

HOGAN BUILT IN NEW MEXICO BY STEVE KATONA

The mud-dauber wasp is a solitary rather than a social insect. It builds a pipe-organ nest for its young and provides them with food without the help of workers. Its nest is a long tubular structure of hardened mud subdivided into separate cells. The wasp stocks the cells with an abundant supply of paralyzed spiders and deposits an egg in each. Unlike most other solitary wasps, the female mud dauber gets some help from the male, who guards the nest while she is hunting.

YUCATAN

ADOBE WITH SOD ROOF, BILL GREENOUGH. PLASTERED ADOBE IN NEW MEXICO, JOCK FAVOUR.

MAKING ADOBES IN SOUTHERN CALIF.

DIRT, SAND, WATER, STRAW AND BITUMIL (LIQUID ASPHALT STABILIZER) MIXED IN CEMENT MIXER. 5 PERSON TEAM MADE 700 ADOBES A DAY. 2 PEOPLE MIXED, 1 WHEEL-BARROWED, 2 MADE ADOBES.

IN NEW MEXICO, ADOBES ARE LAID CRUDELY TO PROVIDE ROUGH SURFACE FOR APPLICATION OF MUD & STRAW PLASTER. DOOR FRAMES ARE SET, THEN ADOBES LAID BETWEEN.

ADOBES WERE MADE FOR THE GREENOUGH HOUSE. BRICKS WERE FORM WORK ONLY, WERE NOT TESTED. TEST INFO. IN HANDBOOK ON BUILDING HOMES OF EARTH.

THIS IS BEST WAY TO STACK ADOBES AFTER THEY'VE DRIED.

AT RIGHT, AN ADOBE MAKING MACHINE – FOUND IN SAN DIEGO BY MALCOLM EVERTON

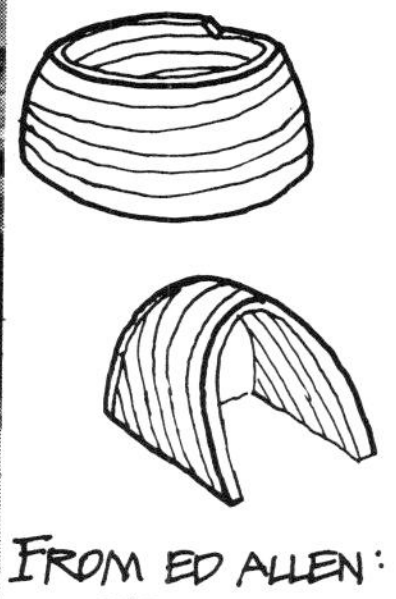

FROM ED ALLEN: ...TO BUILD A DOME OF MUD WITHOUT FORMWORK, WORK UPWARD IN TIGHTENING SPIRAL. IT'S NOT DIFFICULT, EASIER AS YOU CLOSE TOWARDS TOP. TO BUILD VAULTS WITHOUT FORMS, START FROM END WALL, AND BUILD OUT A SLOPE. AT RIGHT, SKETCH OF MUD HOUSE BUILT BY ED AS COLLEGE THESIS PROJECT. WOULD BE PRACTICAL IDEA IF ARCHITECTURE STUDENTS COULD LEAVE BEHIND A HOUSE UPON GRADUATION.

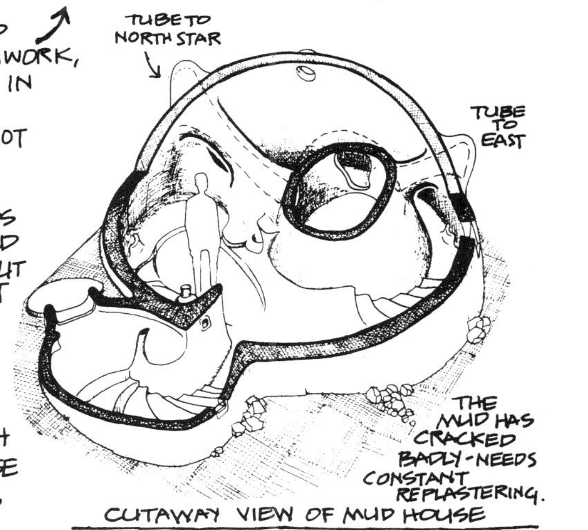

TUBE TO NORTH STAR
TUBE TO EAST
THE MUD HAS CRACKED BADLY - NEEDS CONSTANT REPLASTERING.

CUTAWAY VIEW OF MUD HOUSE

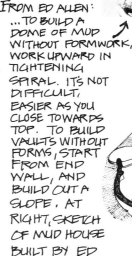

CALIFORNIA ROUND ADOBE *by The Adobe Gang*

ome friends and I built a 900 sq. ft. adobe house using local earth nd wood beams for 300 dollars over a 2½ month period. We tudied the Calif. missions, since they were nearby, and read all the vailable books on the subject. Adobe houses have been built for enturies that survived a variety of weathering conditions. So uilding with adobe seemed the most inexpensive and strongest naterial we could use.

Ve bought some land, tapped a spring, and found a source of good dobe soil. Contrary to popular myth, many soils will make good ricks. An adobe brick consists mainly of a soil that is a little more and than clay in its overall composition. We found our brick soil to e a light color, almost vanilla, and free of twigs and other surface rganic material. We discovered our main deposit on the side of a hill oad where a Caterpillar blade had revealed it. This made our job of ligging it much simpler since the only machine-tool we used was 1954 International pick-up truck.

Ve tossed the soil from the hillside through a screen and into the ed of the truck. We carried over 25 tons of earth from this site to ur building location, a short drive away.

'he soil was dumped at the building site where a three person crew vorked it into mud. One person spread the dirt on a level area, nother squirted water over it lightly before it was shoveled into the vheelbarrow where the third person mixed the mud. A mixture is eady to be poured into the molds when it drops easily from a aised hoe.

Vooden molds should be carefully constructed with tight corners astened with wood screws. Our molds were made of 1x4 interior ir and were 18" long, 12" wide and 4" high. The success of brick nolds is centered on the smoothness of the inner surfaces. We ound metal-lined molds best as we could remove the mold without lamaging the newly-poured brick. We placed the mold on level oards to ensure a smooth bottom surface for the brick. Once the op of the brick is smooth the mold can be lifted off. We found ve had to wash the molds after each use as mud builds up quickly.

Bricks should never be allowed to dry in direct sun. We put as many f our bricks as possible in the shade. Some bricks were allowed to ry under a thick layer of straw but this proved risky. Sun and wind re the foremost enemies of the new brick as both will result in uick drying which almost guarantees cracking. Bricks need rotecting for their first five days, usually, and then they can be tacked for the thirty day drying period. Some climates require onger or shorter drying periods.

During the drying period we prepared the foundation. In spite of he fairly shallow rock-filled foundations of most missions and lder structures, we felt that the foundation was of extreme mportance. Working always with only the materials at hand, we earched old logging sites for lengths of broken logging cable. Ve dug down 3' with the trench being about 1½' wide. In the rench we circled the logging cable as our version of steel einforcement. We mixed and poured the concrete at the site vith the same wheelbarrow we had used for the bricks. We gathered our aggregate-sand from the banks of a nearby small creek. We mixed it at a 6-1 ratio and it worked quite well. We leveled the foundation carefully and placed nail spikes in the drying concrete to tie the first row of bricks into. Most unique was our innovation of using rows of bricks as foundation forms. We placed tar paper inside and poured the concrete into the trench and brick forms.

We did not have the money to follow building codes but we did consider as many technical factors as possible. We built a large foundation on firm ground to avoid any settling (a common problem due to the immense weight load). For earthquake we decided to run two strands of barbed wire around the entire house every third row of bricks. We mortared the bricks together (both on the ends and top) by using the same mud mixture as we had for the making of the bricks. However, we found the bricks welded together better if the mixture was a little more sloppy-wet. As we laid bricks we left spaces for doors and windows by measuring the size of the windows and doors we had on hand.

Before the walls were layed we designed the roof and placed three huge center roof-support beams in a close triangle at the middle of the house. We set these in a half yard of concrete and they reached up 17' from ground level. At the top of the beams we tied them together with 2x12 planks mitered into each other.

The roof was designed so that it would not exert exceptional pressure on the walls. We gathered 15 small trees (5" thick, 19' long) that had been recently knocked over in a nearby logging site. We tied them into the 2"x12" roof plate and ran them out to the top of the bricks and mounted onto long bolts that we set into the concrete bond beam. The bond beam is an important item as it secures the upper level of the wall.

Small 2" branches were nailed across the long roof beams every 2'. This produced an interesting spider web appearance and allowed us to nail a roof of tin onto the limbs. This tin, incidentally, was a gift from a nearby rancher who no longer needed his old barn. The most important factor about the roof is the overhang we built. On most sections the roof juts over the wall 4' providing very good protection from rains of 70" a year (keep in mind that we added no waterproofing to the bricks). The adobe has lasted through two severe winters with virtually no damage.

Adobe Factors:

Average brick: 12" wide, 18" long, and 4" high. It weighs 50 lbs.

We built 2000 bricks for the adobe and used about 1200 of them. Approximately 300 cracked but we used some of them later in building an outdoor kitchen.

Our costs for tools, gasoline, windows, nails, were about $300.

Our average size work crew was 7, both male and female.

NEW MEXICO

...I don't know how much you know about the use of adobe by the Chicanos & new settlers in the northern New Mex. area. Some of the adobes back in the hills there are over 100 years old. The walls grow thicker & thicker since they are occasionally being remudded. In the mountains peaked roofs are used – usually covered with tin nowadays, because of the snow. Lower down they are just flat with big vegas set in the adobe, boarded & covered with mud. This adobe is a wonderful material – cool in the summer, & holding the heat well in the winter. If you are lucky you can find the right earth right on your land – mix with a little water & straw, pour into wooden frames, pull out the frame and let dry in the sun. Hard work making the bricks but the buildings go up fast, with the help of friends and neighbors. There is a puddling technique, if you don't want to make bricks – you pour the mud into board molds set up to shape the wall. Also there is a device for pressing out hard neat bricks which some people use now. The classic Spanish style is based on the rectangle or *L*, but new people are into building maybe circular dwellings and freer forms. Everyone who works with it agrees it is a wonderful material for a dry climate. Spanish people often build with cinder blocks today, maybe because the adobe is too hard work or they are into the prestige of a new *cleaner* American way. Settlers often work with planks, because it goes up faster, sometimes mudding over. And of course there are a lot of new trips people are on in the domes & so forth. I & a lot of my friends feel that for northern New Mex., the old Spanish way is still the best – the most functional, beautiful, & in harmony with the land.

Irwin Klein

CINVA-RAM

by Kelly Jon Morris

In the region where I was working, there were probably at least forty CinvaRam buildings erected in a four year period. Most them were large school buildings built with the help of Peace Corps Volunteers and/or the Togolese government. People were just starting to use CinvaRam for homes as I left in 1972. So a lot of what I wrote is based on big buildings rather than small ones, and tropical rather than temperate climate. But I hope it will whet a few peoples' appetites. It's no panacea but the CinvaRam has already had a great effect on local self-help construction in the developing world and I can't help but think that people interested in low-cost, hand built housing here can put it to good use (especially in the Southwest).

The CinvaRam press is a portable press for making blocks from a dampened mixture of humus-free soil and a stabilising agent, usually portland cement. It is hand-operated and is light enough to be carried by two people. A lever is pressed and actuates a piston which compresses the soil-cement mixture in the chamber into a block. When the lever is pulled to the other side and chamber cover removed, the piston ejects the block from the chamber.

The advantages of construction with CinvaRam:

1. It is cheap. When blocks are made under optimum conditions, the proportion of cement to clay can be as low as 1:15. With 92 lb. bags of portland cement, you can get 95 or 100 blocks per sack. One square meter of wall space uses about 30 CinvaRam blocks, so one sack of cement can produce enough bricks for more than three square meters of wall. When bricks are made under less than optimum conditions (e.g. using wet clay, which inhibits mixing of the cement evenly, thus more cement necessary) the cost advantage is reduced accordingly.

2. It makes use of local material: earth. Clay is ideal block material, but any humus-free soil will do.

3. The CinvaRam is simple and anyone can learn to operate and take care of it with a few minutes instructions. It takes only a couple of hours for a crew to get the rhythm of it and achieve optimum block production speed.

4. The machine is light but sturdy, therefore mobile. Properly oiled and cared for, one machine can produce tens of thousands of blocks with only minor adjustments with an adjustable wrench.

This is the CinvaRam. If you get one, you get a pretty complete manual. However, here are some additional comments from experience.

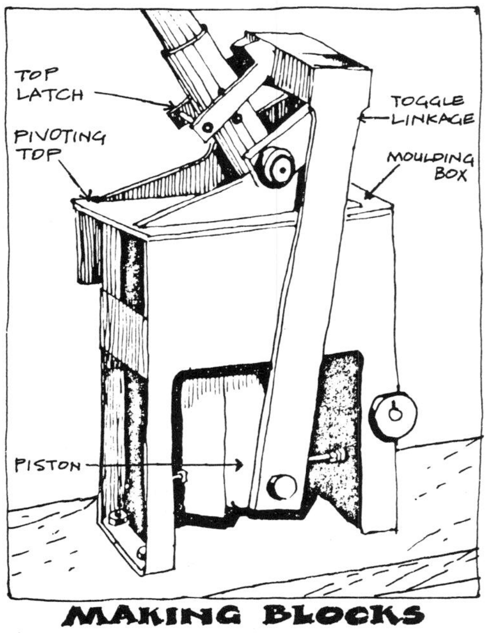

MAKING BLOCKS

1. *Moisture content.* Best way to get the correct moisture content is to start out with dry soil, screen it, add cement and mix thoroughly, *then* add water until you get the right amount. The cement doesn't mix as evenly and thoroughly if the soil is already moist. If you are digging up moist soil try to let it dry a while before using.

2. *Screening.* The method they show is fine, but you might try this one instead:

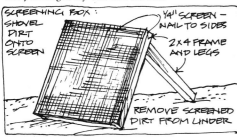

Then shovel the earth on to the screen and remove screened earth from underneath.

3. *Loading the mold box:* Gently tamp mixture into corners and a little in the middle for each shovel-full. After the first few bricks, you can judge how much tamping is necessary. Without tamping you will get lousy corners on your bricks.

4. The press should be thoroughly oiled (crankcase oil, etc.) everywhere inside and out, every day of operation and immediately before any periods when not in use.

5. How many bricks per day? Our normal crews were four or five people – two people mixed, one loaded the machine, one pressed the lever to make the brick, the last one pressed the lever to eject the brick. The first lever-man removed the brick and put it to dry while the loader loaded up, and the second lever-man re-positioned the lever. This was a pretty smooth way of operating and in a 7-hour day they would go through three sacks of cement or about 250-300 bricks. One brickmason was a real hustler – with about 15 people and one machine, he organized three mixing crews and made 900 bricks in a day.

6. Cement and lime aren't the only stabilizing agents. I read an AID manual that said you could use cowshit. Some people are experimenting with CinvaRam bricks with no stabilising agent, especially for use on the upper courses of walls.

BUILDING

1. *Site Selection* is important no matter what you're building with, but particularly with CinvaRam blocks. The blocks are solid and heavy – twenty to twenty-five courses of block weighs a helluva lot. A site where you can build on solid ground or rock is ideal. Sandy or marshy soil is dangerous. Of the five school and other buildings we built, two were built on soil that was fairly sandy. Both were full of cracks (about 25 in each one) and the addition of a grid of 8mm reinforcing rod to the concrete footing of the second one did not help at all – it was still full of cracks. Fortunately for me, the walls were not load-bearing as the roof trusses rested on reinforced concrete pillars, but in most attempts at low-cost housing use of CinvaRam, the walls will be load-bearing and pillars absent, so *beware.*

Another factor in site selection is grade. If the grade is too steep, you need a massive foundation, which would tend to negate the cost advantage of CinvaRam.

In short, CinvaRam is at its strongest and least expensive on a solid, relatively flat site. The farther you go from this ideal, the more you'll have to spend to make up for it.

2. *Foundations* depend on your site and what materials are most readily available. Cemented rock (or "cyclops concrete") is the kind I always used since I always had a large supply of granite or quartz rocks in the nearby hills. We'd dig the foundation hole 1 foot wide and about 1 foot deep at the highest point of the grade. Since there was no topsoil to speak of, this put us in clay or hard laterite already. You may have to go deeper according to the site. We then poured a 3 to 4 inch thick footing of 1:2:4 concrete and began cementing the rocks into place with a mortar of 1 part portland cement to 4 parts medium grain sand. The structural principles of this rubble masonry are simple – the basic thing is to lap your stones and not let joints break on top of each other.

Once you get above ground level, aesthetics may enter into it, so read Ken Kern's chapter on Stone Masonry. (Even if aesthetics don't enter in, read Kern anyway.)

Another thing to keep in mind is to use the largest stones possible – using a lot of little ones wastes mortar like crazy. Also keep a pile of small pieces of stone around to mix in with the mortar where you might have too big of a mortar joint. And don't put stone on stone anywhere without mortar between them.

We brought the foundation 6 inches above ground level at the highest grade point and leveled-off from there. I recommend using a combination of the old masons level and the water level to do it. (See p. 46 for water level.)

WALLS are built just about like any brick wall, with the following adjustments:

1. *How many bricks?* About 2 bricks per square foot of wall area. Add a few more for a fudge factor.

2. *Mortar* is mixed in the following proportions:
   1 part cement
   2 parts sifted clay
   4 parts sand

The "experts" say to use a 1:3:5 mix, but I found that to be a little weak for my purposes. In laying the courses of brick, the mortar should be no more than about ¾" thick.

3. Walls built of CinvaRam block must be reinforced by a corner or a pillar every 10 feet. Whether you use corners or pillars will depend on what type of roof system you plan to use. For large, relatively heavy roofs such as those we put on schools, pillars were necessary. The trusses rested on the pillars and were well attached – thereby transmitting the weight of the roof to the foundation and securing it well to resist the tornado force winds of March and April. However, few CinvaRam homebuilders here would have to contend with either problem. The best solution would seem to be load-bearing walls reinforced by angles and corners, with a reinforced concrete lintel going the perimeter of the building level with the top of window and door openings.

The need for angles in the wall where there is no corner formed by two walls meeting can be put to good use, such as:

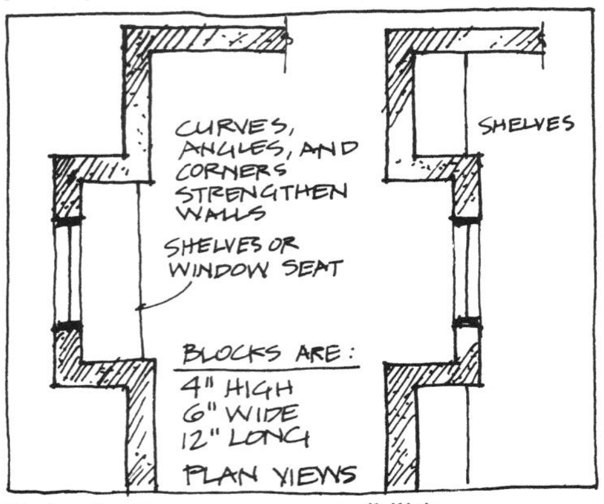

Obviously, there are many possibilities.

As for the lintel, it is easily poured in place on top of the wall once the walls have reached the level of the tops of the window and door openings. It can be poured in sections on succeeding days, so you can use the same form boards over and over. Width of the lintel is the same as the bricks, 6 inches; height is brick plus one layer of mortar of 4¾" - 5". Forms can be built as follows:

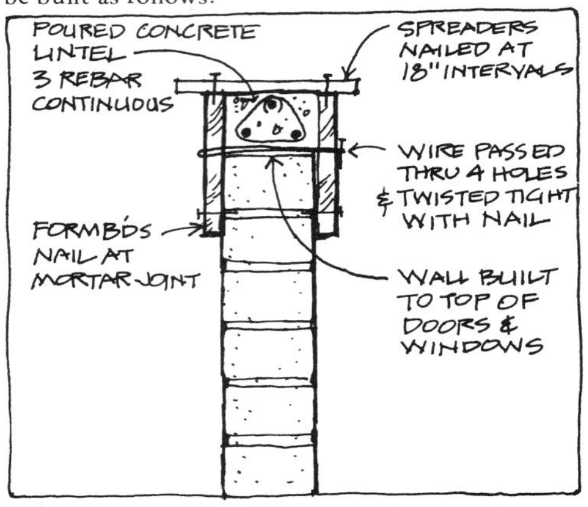

When the formwork passes over a door or window opening, insert board from bottom, nail on both sides, and support by braces from bottom (2"x4"s, tree branches, anything).

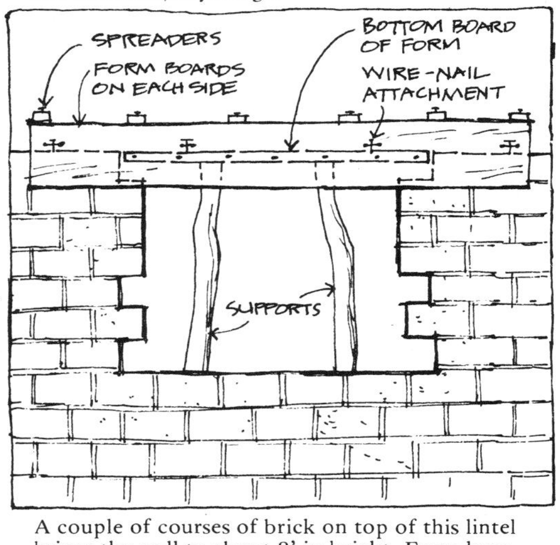

A couple of courses of brick on top of this lintel brings the wall to about 8' in height. From here the roof sits on top of the walls. On a gable roof it is generally more economical to continue the gables in brick rather than setting a truss on the end and covering it with some kind of siding.

To build the gable ends, tack a 2"x4" in the middle so that it sticks up to the proper height, put a nail in the end of it and pull a cord from the top of it to the two corners. Use this as a guide for building the gable end.

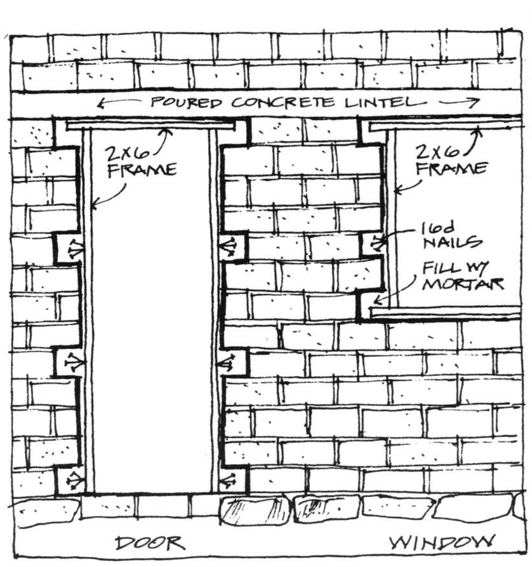

4. *Door and window openings:* The tops of the window and door openings are generally at 6'9" which lets you take advantage of the lintel on the exterior walls. The lintel with 2 courses of bricks (total=about 15 inches) brings wall height to 8 feet. Door and window openings should have half-brick spaces along each side.

When mounting the doors and windows in their frames or the frames separately, the frame should have something sticking out to correspond with the notches in the side of the openings. Blocks of wood or preferably just drive a few 16d nails.

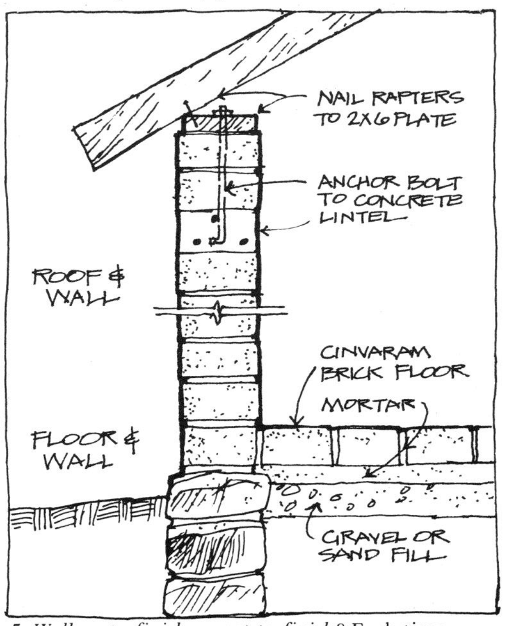

5. *Walls – to finish or not to finish?* Each time we built a CinvaRam building – a school or whatever – we were forced by local opinion to do something we dind't want to do – plaster the walls inside and out. To the local people, a school was a symbol of status, of progress toward "modernity" – and all "modern" buildings they had ever seen were plastered, so theirs had to be plastered, too. It's a shame, because when CinvaRam bricks are made properly they are hard as a rock, with very firm corners and an attractive color. They need no finish provided the tops of walls are protected by a reasonable roof overhang, according to the local climate.

If, for some reason, you decide to finish all or part of your walls, here are some possibilities:

1. Oil of any kind. Linseed oil is usually the recommended thing, but I've heard of using motor oil or even locally made red palm nut oil, which the Africans use as cooking oil and export to us to make Palmolive (Palm + olive – get it?) soap and other things.

2. Whitewash – just mix it up and brush it on – it seems to hold a long time.

3. Plaster – the books say to use a plaster like this:
   1 part portland cement
   3 parts fine sand
   5 parts sifted clay

These proportions are wrong. We plastered our first building with this mix and 18 months later it was falling off. Meanwhile we switched to this and it seems to hold up better:
   1 part portland cement
   3 parts sifted clay
   5 parts fine sand

(A 1:2:4 mix may even be necessary in some cases.)

Trowel it on to a maximum thickness of about 3/8 inch and smooth it off. Anything you can do to reinforce it will help. Ken Kern recommends laying pieces of wire in the brick courses as you go letting them protrude from it a little bit and then using these wire pieces to hold chicken wire in place to be plastered over. Once we had a bunch of 4d finish nails laying around so we drove them into the wall and let the heads stick out a ways, then plastered over them.

4. U.N. experimenters have used a mixture of cement, sand and clay mixed to paint consistency or thinner and then painted on with a 6" brush. It seems to work OK, too, but I don't know the proportions.

END

STONE

COTTAGES & FENCES BUILT OF STONE GATHERED WHILE CLEARING FIELDS IN IRELAND.

KHAJURĀHO, INDIA. TEMPLE SCULPTURE SHOWING AN ASPECT OF THE CONNECTION BETWEEN ANCIENT FERTILITY CULTS AND HINDU RITES.

ROOF OF 4'x5'x1" THICK SHEETS OF SANDSTONE. FARM BUILDING IN IRELAND.

DRY STONE WALL, IRELAND

MACHU PICHU, PERU.

SANDSTONE WORKED & CARVED BY WILL WOOD & JIM WALKER, STONE IS SOFT, EASILY CUT & FIT.

JACK LAVERY, 1922

COTSWALDS, ENGLAND

IRISH BARN

TOMB OF DARIUS, IRAN.

HOPI STONE WORK

There is very little information available on laying up stone. This is partly due to the traditional secrecy of stone masons, and partly because stone differs greatly in different localities. Stonework, like adobe, is free, heavy and takes time. But it does not damage the earth, and houses of local rock are a harmonious addition to the landscape, as can be seen throughout Ireland.

Laying up stone can be meticulous, or funky. Beautiful walls are often built without mortar. In the Cotswalds in England there are still a few masons who can build the high mortarless sandstone walls as shown above. In Big Sur a unique local technique of dry rock walls (not approved by the building inspector) has been developed by Joe Kingswold and other builders. The best way to learn is to work with a local stone mason.

FROM KEN KERN

Through the centuries stone masons...have succeeded in maintaining a respectable, highly paid and somewhat apostolic status in the building industry. Their "trade secrets" are maintained to this day, and include such important items as an intimate knowledge of rock, the correct mortar proportions and use of auxilary materials, the proper selection of tools and organization of work procedure and — finally — an esthetic awareness of the rock in place: The total effect and composition of the finished wall.

With fear of over-simplifying the stone masonry skill it should be stated that the foremost prerequisite of any mason worth his mortar is an intimate — nearly intuitive — knowledge of rock. Pick up a rock. Where the inexperienced observes color, weight and form, the experienced stone mason notices bedding, seams, rift and grain.

INCA STONE

The incredibly tight mortarless joints of ancient Inca stone work has mystified people for centuries. Recent investigations reported in *Science News* have uncovered what could be an answer. The *theory* is that the stones were cut to close tolerances and smeared with a paste made from lichen that typically grew and fed on the stone. After the stones were set, the lichen ate away the irregularities in the stone where oxygen was until the stone was so tight no air could reach the lichen and it died, leaving a perfect fit.

TRULLI

In the 1600's the beautiful stone domes — *trulli* — of the Adriatic coast of Italy were built with no mortar so they could be quickly taken apart when the tax assessor made his rounds.

In 1644, following some years of a mild boom in the construction of new trullo *shelters...Gian Girolamo heard of the subsequent approach of the king's investigator, and his forces in one activity-filled night ensured that only the houses previously known to the crown remained to be revealed by the dawn. The inspector arrived, found nothing out of order, and departed. The Selva was rebuilt from the same stones, again without mortar.*

From *Stone Shelters*

BUILDING A STONE HOUSE

1-Stone buildings should be kept low, because after they reach a height of five feet, the cost of lifting the stone and concrete increase progressively with the height. If a second story is needed, it should be based on dormer construction. 2-Cellar space should be reduced to a minimum and all possible floor areas should be of concrete, laid on the earth. If it is desired, other types of flooring may be used over the concrete. Heating pipes and wires can be laid in conduits or channels. 3-The house should be a unit, with door and window frames of solid material, built into the stone and concrete walls, and without trim. 4-The walls are to be built in movable forms....5-Keep the roof lines as simple as possible, — few if any dormers or extra angles. 6-Make all shapes as regular as possible, eliminating excrescences and cutting corners down to a minimum. 7-Build large enough, because stone walls once built are hard to break down if additions are desired.

From *Living the Good Life*

FROM LAST WEC *by Mark Mendel*...

I worked with a mason of 60 years to learn to lay fieldstone. Practice is the way. Knowing which rock to choose from your pile. Like a puzzle with no two parts the same. Choosing the wrong rock means that your work comes down on your feet....

An old Maine stone mason told me that "Even a round rock has a flat side if you can find it."

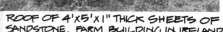

BALED HAY

by Roger L. Welsch

In 1904 new lands in northwestern Nebraska were opened for homesteading under the provisions of a law authored by Moses Kinkaid, a Nebraska congressman. A substantial portion of the Kinkaid lands were the Nebraska Sandhills – a vast, desolate tract of grass-covered dunes. The Sandhills magnified the difficulties of the previous homestead lands: they were even more barren of trees and the weather was even more hostile. Furthermore, the sandy soil made poor construction sod, for if it did not disintegrate during cutting and handling, it would soon crumble after being laid up in walls. The oxen, which were preferred for sod cutting, had been traded for horses, preferred for general farmwork and transportation; the plows for cutting were gone; and, to some degree, the skills for sod construction had been forgotten. Again it was necessary to devise new techniques and materials for home building.

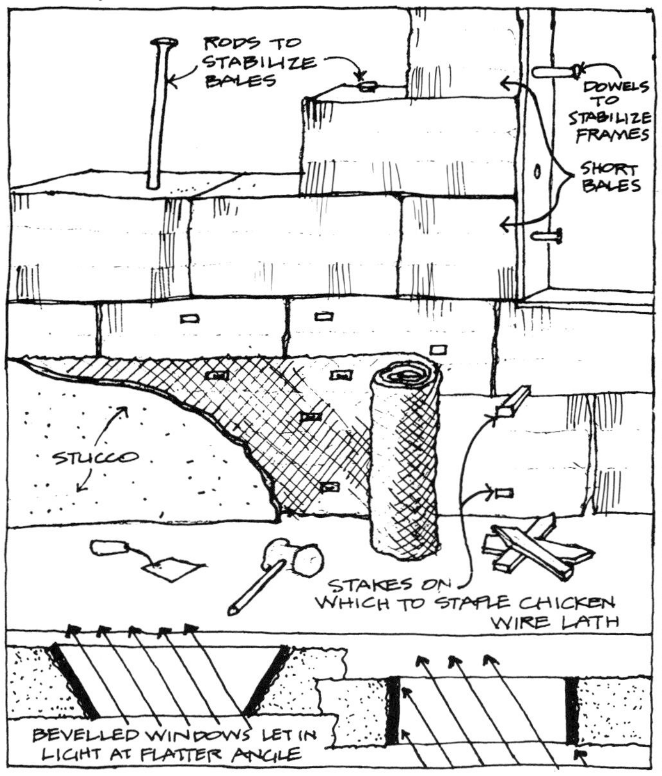

Wild grasses and domestic hays were (and are) the most important crops of the Sandhills. They were cut and stacked for use on the same or neighboring farm, but if the hay was to be moved it was necessary to bale it. Balers were first introduced in the 1850's and by the 1890's hay presses were in general use. Indeed, by the 90's railroads refused to handle loose hay. It was then perhaps inevitable that some settler, desperate for a cheap, available building material, would eventually see the big, solid, hay blocks as a possibility. Soon baled hay was indeed a significant construction material – never reaching the importance, scope, and technology of sod, but widely known and used throughout the Kinkaid Sandhills.

TUMBLEWEED & GUMBO SOIL

Most baled-hay buildings have concrete foundations and concrete or wood floors....

One of the great advantages of the hay house was that the best hay for building was the worst hay – and therefore also the cheapest – for feeding. Late-fall hay that is tough and woody is also the most solid in the bale and made the firmest walls. And the low cost was most certainly an important consideration for the man who had come to the Kinkaid lands because they were so cheap. In one case two problems were solved with one motion: "...My folks bought a farm 9 miles west of Bridgeport, Nebr., in December 1912 – it had a sod house on it and lots of tumble weeds in the fence rows so my father had a hay baler so he baled the tumble weeds and made a good looking 2 room house."

The bales, about three to four feet long and one and one-half to two feet square, were stacked like bricks, one bale deep with the joints staggered. Of the very few buildings in which I could actually see the bales, of those I have seen pictures of during construction, and from descriptions, it would appear about half used mortar between the bales; the others simply rested one bale directly on the other. Where commercial cement was not available or was too expensive, a home-made substitute was used: "They mortared the bales together with gumbo soil, 2 part, sand, 1 part, and enough water to make a real thick paste."

Four to five foot wooden rods (in a few cases iron rods) were driven down through the bales to hold them firmly together. This had to be done even where mortar was used between the bales. The roof plate and roof were also fastened to the top bales of the wall with rods or stakes.

ROOF, DOORS, WINDOWS

The roof itself was frame, covered with cedar shingles or asbestos shingles that were shipped into the Plains in large, wired bundles...

By far and away then the most common roof configuration was some form of the hipped roof – which was also so common to the sod house in Nebraska that it is known in the area as a "soddy roof." This was no accident: the hipped roof permits all walls to be of the same low height – a very important factor, for no matter how square the bales might be and no matter how carefully they were stacked and bound, a high stack at the gable end of a house offered real danger of collapse during the settling of the house.....

Furthermore, unless the builder was generous with ceiling joists, to provide a stout chord for the roof triangle, lateral pressure would be exerted on the walls, and, while the bales could withstand plenty of vertical weight, the walls would buckle and fall if the rafters were permitted to push out against them. The hipped roof, too, required ceiling joists, but if the plate was fastened firmly at the house corners, it served as a closed box and reduced horizontal stress.

Window and door frames were set as the walls rose around them. Baling machines permitted bales to be made in any length, so half-bales were made especially to be used for butting against window and door frames or full bales were unwired and cut in half, rewired, and used at windows and doors. Dowels were driven through holes in the frames into the surrounding bales. Doors and windows were always of commercial manufacture and they were always set at the outside of the casement to provide a handy place inside the house for plants, books, or decorations; if the windows had been placed even with the inside walls, it was explained to me, it would have been far too easy for water to settle on the broad sill, soak through to the hay, and cause rot. The deep walls severely narrowed the angle that sunlight could enter the baled-hay house, so (as was also the case with sod houses) window casements were either beveled or two frames were placed side by side.

FLEAS

The walls were left to settle a few months before they were plastered and the windows were installed. Then, when the hay was dry, short stakes were driven horizontally into the bales, both inside and outside the house. Chicken wire or "hardware cloth" was stapled to the stakes and stucco or cement (for the outside walls) and plaster or plaster-board (for the inside walls) was applied to the wire. One builder fastened the screening to the bales by passing long wires between them as the walls were laid up and then pulling the chicken wire against the wall from inside the house. Depending on family finances, the covering was commercial, or it was home-made from sand and alkali mud that was scraped from the bed or banks of shallow, frequently dry Sandhills lakes. In about one-quarter of the houses I have been able to check in this regard the plaster was spread directly onto the bales, without wire screening or lath....

While several of the buildings I have found are sheds or barns, few of the houses I know have had baled-hay out-buildings. Although it was never expressed to me during this project as explicitly as it was during my sod-house work, this may well have stemmed from the same contrast between the great efficiency of hay (or sod) as opposed to the prestigious but expensive inefficiency of frame construction on the Plains. Many sod-house dwellers and builders told me that wood was good enough for animals but people deserve the comforts afforded by sod, which is warm in the winter (even more important where fuel was such a problem), and cool in the summer (vital where the temperatures soar daily to above 100 degrees in the summer and where there were no trees to shade the house).

The baled-hay house also shared some of the drawbacks of the soddy: unplastered walls, for example, provided a fine breeding ground for fleas and one school teacher who had occasion to spend a night in a hay house owned by the father of one of her students said that it was a busy night and suggested that the ease of heating the hay house was certainly partially the result of the constant exercise hay-house owners got from scratching and slapping fleas. One advantage the hay-house had over the soddy was the relative lightness of the walls as noted by one settler who had both a sod house and a hay house: "The sod house is a little harder to keep up than the baled-hay house because of the extreme weight of the sod, the ground under it thawing and freezing lets it settle, the doors and windows get out of line."

FIRE HAZARD

As might be expected, fire was a particular hazard to the hay house. One of the superior qualities of the sod house was its resistance to fire, and grass fires were a formidable danger to Plains dwellers, for even though the smoke, smell, and fleeing animals warned pioneers of approaching fire, the flames, flying before a wind faster than a man could ride on horseback, would burn out anyone not prepared with a firebreak and gunnysacks to be slapped wet against invading sparks. Even after prairie fires were no longer a danger in the eastern ranges of the hay house, several houses burned; the tinder-dry straw or hay, informants report, burned with frightening rapidity.

The earliest hay building I have found was a Bayard, Nebraska, school built in 1886, and the latest was a dance hall built shortly after the Second World War, about 1946. The latter however, is an anomaly, being the only Nebraska baled-hay building east of the Sandhills, about eight miles west of Lincoln. The latest building within the tradition was built in 1939. The height of bale construction was from about 1900 to 1935. (However, a 1960 issue of *Grassland News*, an organ of the New Holland farm machinery company, reports that a church had just been built in Alberta, Canada, using a New Holland baler).

CATTLE HAZARD

An added, unexpected benefit of baled-hay construction came to light when some of the houses were razed. Farmers were amazed to see their cattle abandon fresh pasture to eat the 50-year-old bales – in some cases eating right into the walls of the abandoned house....

"Nebraska School Buildings and Grounds, a bulletin published by the State Superintendent in 1902, describes a school erected in Scotts Bluff County in 1886 or '87 that had walls of baled straw, a sod roof and a dirt floor. This strange building was 16 feet long, 12 feet wide, and 7 feet high. Two years after its erection cattle, on range in the vicinity, literally ate it to pieces. Few straw school houses were built because of the fire hazard; although State Superintendent Fowler, in 1900, argued that bale hay could be used as a school building for the fall terms and then fed to the cattle in late winter."

BALED HAY DOME

Ranchers in the Sandhills still build hangars for their airplanes out of baled hay, but as early as 1929 (perhaps 1930) Harry Hiles built a round baled-hay hangar near Gothenberg, Nebraska, and published a pamphlet promoting such constructions:

"The wall is a unity of concrete and steel, in which baled hay is used, to replace more expensive material, to afford lightness to the wall and to insulate the wall from penetration of heat or cold. The wall is formed by running parallel bands of concrete and steel about 21 inches apart, one above the other horizontally within the wall from the bottom to the top, all of which is done by using the bales of hay as internal forms one directly above the other within the wall (sic!). Then afterwards the walls are finished inside and out with stucco.

"The circular form of the wall, together with the added strength of the bands of concrete and steel and the perpendicular pillars afford a most excellent roof supporting structure. No interior supports are needed, the roof being of the same material as the walls, being built like a dome.

"No interior supports are necessary, the roof being built in a dome shape, of the same material as the walls. The weight of the bales is cut down to 40 pounds per bale instead of the usual 80 pounds."

PLASTER

Here are two plastering techniques used by Robert Venable. On his shop: 2 coats of a lime and sand mix, 1:3 proportions. Use minimum amount of water, and dehydrated lime. Add white cement for a finished white wall, or paint with lime and water mix. The walls on a studio he built (he framed *and plastered* the walls down on the floor, then raised them), he used one coat of sand, vermiculite and cement, 1:6:2 proportions. Vermiculite is a lightweight aggregate that must be mixed quickly.

Gypsum plaster information on p. 147. Ferrocement techniques on pp. 122,124,125.

SHOP.

STUDIO.

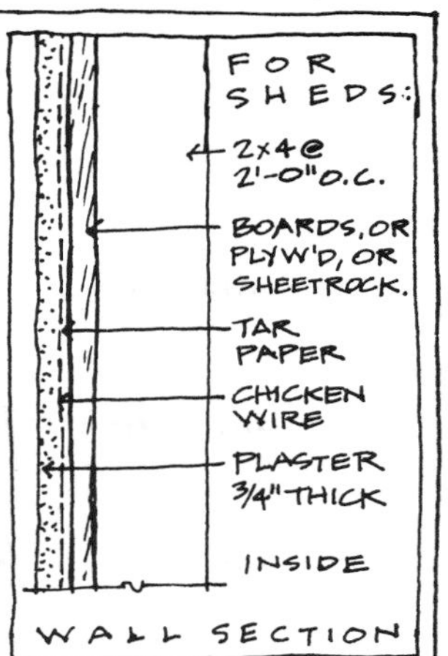

WALL SECTION

SOD by Roger L. Welsch

A visit to the Plains of the north-central United States can be a shock even to twentieth-century travellers, so imagine what it must have been like 75 years ago. Crossing the Missouri River was as substantial a psychological agony as physical, because it marked the eastern boundary of the Great American Desert. Today it is in fashion to laugh off that label, for the land has proved to be fertile, but two factors have changed since the early days of settlement: 1) techniques have been developed to ease harsh prairie conditions: house insulation, gas and electric heat and cooling, rapid and large scale transportation of people and freight, irrigation, and river stabilization, among many others; 2) we who live on the Plains now have come to accept its severity as ordinary. Thus, the Great American Desert is no longer what it was and we are not the people the nineteenth-century settlers were. With its nomadic Indians, treeless wastes, blowing sands, meager rainfall, poisonous snakes and mighty bison, severe winter cold and summer heat, tornadoes, grasshopper plagues, cacti, and ferocious storms, the northern plains fulfilled in every way the pioneers' concept of "desert." His hope then was not to exploit a garden but rather to create a garden from a wilderness.

The settlers came from Northern Europe, eastern America, and the east-central states; they were used to building materials in abundance — if not excess. At their previous homes they perhaps had to pick up stones from the plow's furrow at virtually every step or they had to fight trees as weeds, burning, hoeing, grubbing, cutting. Wood and stone were always at hand for houses, barns, and fences.

But in the north central states area there were no trees. Trees on the Oregon Trail, which cuts lengthwise, east to west, through Nebraska, along the south side of the Platte River, were so rare that they had names and served as unmistakable landmarks. Even along the rivers, where one would certainly expect to find stands of wood, there were only low scrub bushes and willow wands. Only in some canyons could one find groves of ancient, gnarled cedars.

Although good clays were available for brick, there was no fuel to fire them. Stone suitable for construction is found only in the extreme eastern regions and the great distances of the Plains and the lack of commercial transport forbad large-scale moving of cut stone.

Since the acquisition of homestead lands required construction of a permanent dwelling within a limited time, would-be settlers had to develop some technique for home building: sod.

The Mormons, when they first crossed into Nebraska in 1846, found the Omaha and Pawnee Indians living in great earth lodges — heavy timber frames with a sod covering, and since the Mormons also built sod houses during their first winter here, it may well be that they picked up the idea from the Indians, while they were also bartering for food and animals. On the other hand, the Mormons and later settlers may have borrowed the concept from English or European antecedents, low turf and stone huts used especially for temporary field-housing.

Whatever the origin, the technique developed rapidly and spread throughout the region, where it dominated for fifty years....

NEBRASKA SODDY

The site for the house was cleared of grass and brush and leveled with a spade. Holes were filled with loose earth and the floor area, usually a rectangle about fourteen by sixteen feet, was tamped concrete-hard with a fence post or wagon tongue. Very few homes enjoyed more of a foundation than this. The walls were usually aligned straight north and south, east and west, with the help of the North Star on a clear night.

A good stand of slough grass or buffalo grass was found for the wall sods. If the sodbuster had a choice, he built in the fall when the prairie grasses were wire-tough and woody, yet at a time when the ground was moist enough to hold the soil firmly to the sod bricks when they were lifted to the wagon box or stoneboat bed. If the grass was high, like big blue-stem, he might mow it first, saving the cut grass for roof thatch.

Occasionally the sod, about one acre for the average house, was cut in one direction with a cultivator equipped with vertical knives and then plowed in the other direction so that the sods came out ready cut to size. But usually the sod was cut in a long, solid ribbon and then hacked into blocks about two feet by three feet by three or four inches with a sharp spade. The standard

The bevelled windows permitted the entrance of sunlight at a broad angle and the horizontal logs above the window frames bore the weight of the heavy sod.

The Haumont sod house, still standing, about 14 miles northeast of Broken Bow, Nebraska.

walking plow, a turning plow, could not be used, for its purpose was to tumble the soil and destroy the root mat; a breaking plow, or grasshopper plow shaved off the three-inch ribbon, lifted it from the prairie on iron rods, and then gently rolled it over unbroken. Although horses were the most common draft animal, oxen were generally preferred for sod cutting, for they pulled more slowly and evenly. Homesteaders have told me that as the plow slashed through the virgin sod it sounded like the opening of a giant zipper.

At the house site the sods were laid without mortar, grass down, like huge bricks; in fact, in many areas sod houses were called "brick houses." The walls were two or three sods thick and each sod was set behind the others so that the joints were staggered to discourage invasion by unwelcome guests — snakes, mice, or winter winds. Each layer was begun in a corner of the house and each layer was finished before the next was begun. Each sod was also put atop the others with staggered joints. Every third or fourth layer was set crosswise on the vertical piles. This binding course added substantially to the stability of the walls.

As the levels for the window and door sills were reached, the frames — simple, open-ended, plank boxes — were propped in place and the walls rose around them. Sometimes one-inch holes were drilled through the planks and peg or dowels were driven through them and into the walls to hold the frames firmly in place. Frequently the sides of the casements were bevelled to permit more light to pass through the thick walls and into the dark, sometimes smoky, interior of the house.

We dilettantes would probably have just piled our sod directly on top of the window and door casements, but those who learned from tradition knew better. The heavy walls settled six or eight inches in the first year or two, but at the window and door location, where the load was obviously much less, the settling was only two or three inches. To allow for this difference, to prevent the crushing of casements, jamming of frames, and breaking of the precious glass panes — usually the most expensive part of the house, the walls were built up eight inches above both sides of the casement; then logs or planks were set across, leaving a gap between the window and the wood members bearing the overlying sod. This gap was stuffed with loose rags, paper or grass, and as the walls settled around the windows this material was compacted without affecting the windows themselves. These windows were usually commercial, double-hung, twelve-pane frames, and the doors were homemade of three planks with two or three cross-members.

The walls were shaved smooth with a sharp spade and all holes were filled and tamped.

MUD & SNAKES

The roof was "the proof" of the soddy; if it succeeded, the soddy succeeded, and if it failed — and it frequently did, the house almost certainly failed. The following italicized quotes are from various people who lived in sod houses. Detailed source information in the author's book: *Sod Walls: The Story of the Nebraska Soddy.* See bibliography.

*From hardly any rain we soon had more than we needed. Our roof would not stand the heavy downpours that sometimes continued for days at a time, and it would leak from one end to the other. We could keep our beds comparatively dry by drawing them into the middle of the room directly under the peak of the roof. Sometimes the water would drip on the stove while I was cooking, and I would have to keep tight lids on the skillets to prevent the mud from falling into the food. With my dress pinned up, and rubbers on my feet, I waded around until the clouds rolled by. Then we would clean house. Almost everything had to be moved outdoors to dry in the sun. Life is too short to be spent under a sod roof....*Mrs. H.C. Stuckey.

*In June, after the folks were nicely situated in their new one-room soddy, it commenced to rain and continued to rain for four days and nights. The roof was laid with willows, with sod on top of them, and naturally it began to leak. Father said, 'Let's get under the table,' so we did. The long ridge pole on the roof began to crack from the heavy weight of the wet sod, and finally the roof caved in with the pole resting on the table. We were buried beneath the sod and the muck. Finally father saw a little patch of light and he dug his way out...*Mrs. Jane Shellhase.

But many were not so distressed about the rain falling through the roof because they knew that it was also falling on the fields.

*During the last years it got pretty leaky. We were so glad for the good rains that caused us to have to get up and put dishpans, washpans, tubs, skillets under the drips, that we did not mind. My dad used to say those drips played a regular tune with soprano, alto, tenor, and bass, in the harmony...*Mrs. Leota Runyan.

These problems arose from a misunderstanding of the sod roof. Sod is a superb insulator but a miserable water-repellent.

There were three standard styles of roof: in order of frequency, the gable roof with two surfaces, the hipped roof with four surfaces, and the shed roof with one surface.

There were two standard types of roof construction, cut lumber and pole-and-brush. Cut lumber was obviously the best, but it was also the most expensive and, therefore, less common than pole-and-brush. The sheathing planks on a cut-lumber roof were arranged in various ways, running from gable to gable or from ridge to eave, with and without framework. Then an insulating layer of sod was set over waterproof canvas or tar paper.

A pole-and-brush roof was cheaper, but more complicated. One to seven cedar beams were run from gable to gable; then from ridge to eave were run, every six or twelve inches, willow poles. On these was spread a layer of bluestem grass. Next, successful roofs had a waterproof sheet of gypsum plaster or alkali clay from a creek bank or dry lake bed. Atop this was placed the sod, grass up.

Inside, the fancier houses sported a plank floor (these were always the most popular houses for square dances) but this too sometimes caused regrets.

*In the Spring when we had the big rains our roof did its share of leaking and even after the rain was over it seemed like our roof leaked for another day. Mother often wished that it was reversed — the boards on the roof instead of on the floor...*Mrs. Marie Varney.

And so, in two or three weeks, with an investment of two to twelve dollars, a sturdy house was built.

After waiting several months to permit the walls to settle, the inside walls were plastered with a thick mixture of clay and fine sand. Some were then papered with commercial wallpaper or newspapers. Muslin or canvas was hung for a ceiling to add elegance, to make the interior brighter and to keep mice, mud and snakes from dropping onto the bed or into the soup.

COMPLEX

Construction techniques were therefore more complex than is commonly believed. But this I had expected when I began my study. The real surprises lay in the rationale for the sod house. Why did the Nebraska settlers build sod houses? The obvious and best known answer — that there was nothing else — is as incomplete as our knowledge of the building techniques. One by one, new answers came to light.

For example, when I began to work on this project, I advertised in several central Nebraska newspapers for early settlers who remembered building soddies. I was surprised by the hundreds of replies I got — but I was amazed by the number of letters I got from Nebraskans who still live in sod houses. Hundreds still live in attractive, comfortable, nineteenth century and early twentieth century sod houses.

The conclusion suggested by this situation was re-enforced through interviews with former sod-house dwellers and published memories of pioneers.

*I think it was a good thing people in those days had sod houses to live in; with no more fuel than they had, they would have frozen in a frame house. Those that had frame houses hauled coal from Grand Island or Kearney (to Custer County, twenty-five to one hundred miles)...*Mrs. W. H. Hodge.

*Our dugout was so warm that during the blizzard of 1888 we sat in it and let the fire go out...*Mrs. Myrtle Herrick.

Dozens of Butcher's pictures show a sod house surrounded by board fences, heavy wooden windmills, wooden sheds and cribs and huge frame barns. Again the settlers themselves, like Mrs. Hodge in the above statement, verified the fact of the assumption; repeatedly pioneer Nebraskans testified to the comfort of the sod house, its suitability to prairie conditions, and the unsuitability of frame construction.

*...The out buildings were wood. The barn was covered with tin later on, but the cattle and all animals (had) winter coats heavy enough to take care of themselves. Some people thought the frame house was not warm enough for a family to live in...*Mr. and Mrs. Clarence Carr.

The frame house was painfully vulnerable to prairie fires, severe weather and wind. Wood burns, warps, leaks, shrinks, swells, rots and can be eaten by insects and rodents. Sod resists all of these.

In addition to the superior quality of sod, there was another reason for its popularity in house construction even after lumber was available and inexpensive: the momentum of tradition. Plains houses *were* sod houses. When one built a house, he built it of sod, by reason of a regional habit. Just as we continue to build with lumber despite the disadvantages listed above and the ready availability of stone and brick, erecting dangerously flammable cities of wood, Nebraska pioneers continued to build houses of sod out of pure traditional habit....

CANVAS

Natural fiber canvas is easy to work with, cheap, covers space quickly and lights it with a translucent glow. You can buy waterproof and/or fireproof canvas or the untreated material (cheaper) and treat it in place. Tents of the Bedouin, Tuareg and other desert nomads could be studied for good use of a flexible membrane: a very few poles are used and the skin is both weather protection and tensile structure between poles (no rafters). A tent, compared to a conventional house, is more like a bird than a tree.

Canvas can also be used to cover light rectilinear wood frame buildings. In the 1910's, the William de Ronden-Pos family used canvas for the roof of their San Francisco house. The inexpensive and translucent roof gave the dark redwood rooms a luminous light.

Here is information on working with canvas from Carey Smoot:

After a lot of investigation into materials such as vinyl laminates, fiberglass laminates, teflon coated glass (like Frei Otto uses), I prefer cotton canvas which is organic and has a life of 5 to 10 years depending on sun exposure and humidity. The very fact that it is biodegradable like wood makes it a more responsible building material....

I usually use fire & water resistant canvas that only comes in 30" widths. This becomes my maximum width of the panel.

Allotting for seam design and for shrinkage rate is important. All organic cloth shrinks 2 to 5%....

Sewing is the best way to put canvas together....

When sewing 4 ounces or better an industrial sewing machine is recommended. A double or single needle lock stitch with a puller or better yet a canvas chain stitch machine with a puller if you can find one to rent. Look in the Yellow Pages. To save costs, design your work around your home model machine. Light rip-stop cloth (nylon) can be sewn on all home models. Just get the right needle size and thread type.

CANVAS ROOF ON THE DE RONDEN-POS HOUSE. c.1915

It is important to select the proper thread. A polyester dacron thread for all applications is best. For 1 to 3 ounce canvas a no. 16; for 4 to 10 ounce or more a no. 12; for 1 to 3 ounce rip-stop nylon a no. 24; heavier nylons and dacrons are the same as canvas.

When using dacron thread on a canvas seam a natural water proofing takes place. As the needle makes the hole in the cloth and inserts the thread you would think the stitch would leak, but the canvas will shrink and the thread will not, thus closing the needle hole....

Always sew from the top to the bottom in a succession. Sew the last seam inside out.

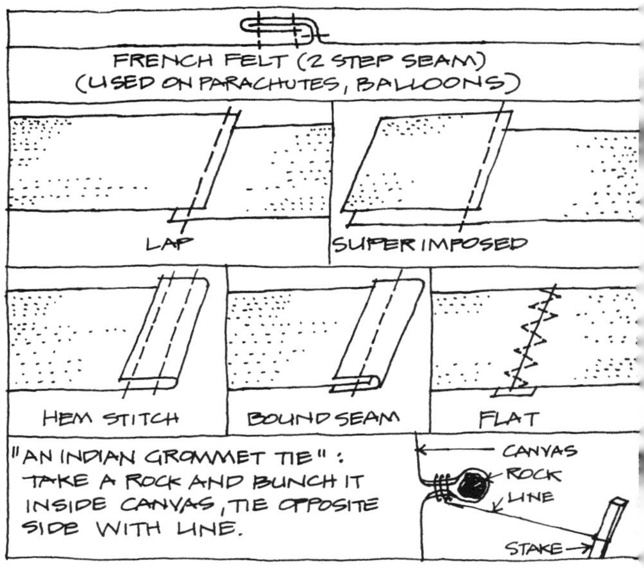

A good waterproofing is Weathermaster, from Champion Products, Chicago. One gallon covers 100-120 feet.

·WINTER TENT· *by Keith Jones, Ambler, Alaska*

Don't underestimate tenting – whole families have lived year round in white wall tents in the Arctic.

Although tenting on the trail all winter is still common, most people who tent now, move into tents in May for spring, summer and fall – 6 months of the year. Each year less people tent, but it's still common to move to camps, – spring camp to hunt muskrats and dry caribou meat, summer camps to net and dry fish, fall hunting and berry picking camps.

After 6 mo. of Arctic winter, no shelter can contain the spirit of man in spring time when in the short space of a few weeks, all the birds return – the snow melts and the river breaks up – the plants burst into bud and bloom and it's light day and night. Then's when tent living is at its finest.

A wall tent is most easily pitched with 5 members; a pyramidal or conical tent with a 3 member external tripod, to give maximum room inside.

- –Remove all or most of the snow cover – an area as large as the tent floor. In winter the earth is a giant heat reservoir as mice and other small creatures well know.
- –Fall 3 or 5 spruce trees of the right size across the floor area or as close as possible. Limb the trees – throw out the coarse branches later. No floor in the tent wanted when you are camping in the forest in winter.
- –As temperature gets colder pitch tent lower. If really cold, -40° to -60°, bank walls with boughs, brush and snow.
- –Bring in wood stove, grub box, bed tarp, caribou skins and bedding from the sled.
- –Cut logs of wood – both dry and green spruce.

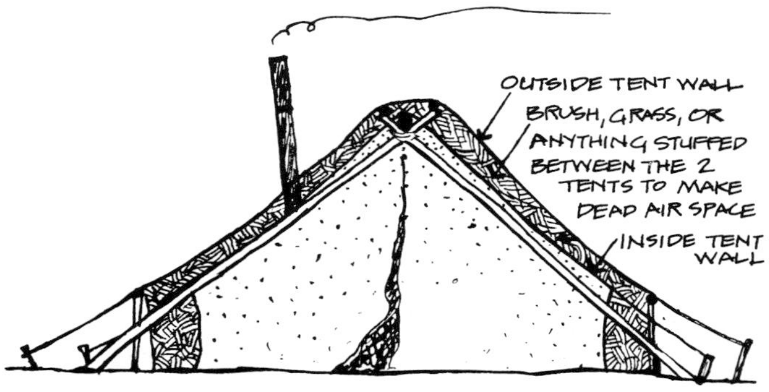

CROSS SECTION OF DOUBLE WALL TENT

It's best to make a good metal or asbestos stove pipe for safety. Blazo tin bottom is most often available. (A recycled item.)

1st – Cut hole in tin – slightly smaller than pipe. Enlarge to fit pipe at angle of tent by beating edge with a handle. There is some stretch in both tin and canvas, but if it's too small – it will pull and wear out. If too big it leaks air and works in the wind.

2nd – Take 8 thin blazo slats and nail tin to canvas – preferably along a seam for extra strength. Do it while tent is down – cinch nails up tight and bend over.

3rd – Cut out hole in canvas. Pitch tent. Install stove and pipe. Try to have stove pipe hole down wind so sparks don't burn tent.

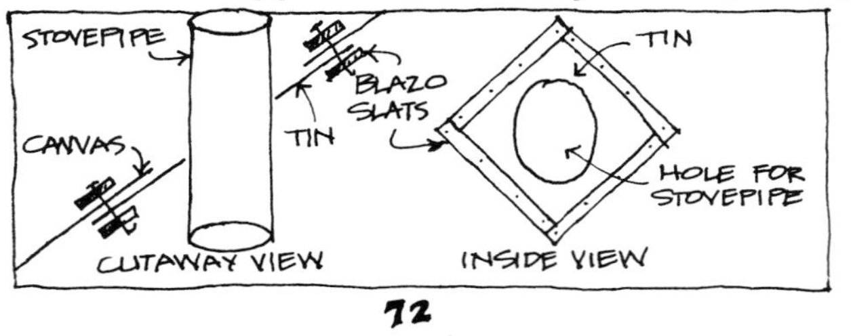

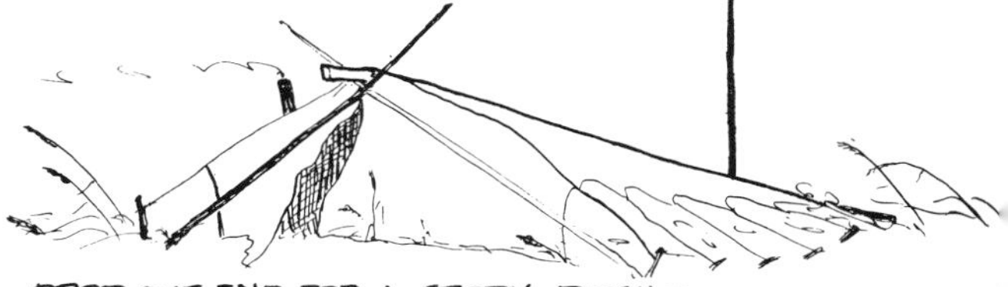

DROP ONE END FOR A STORM PITCH – ALSO TAKE OFF ONE SECTION OF STOVE PIPE

ON TENT FRAMES – More for the coast and windy country.

It's all a matter of degree – how much framing you have, from enough to hold the tent up to a complete canvas covered house.

- –Measure inside seams on *your* tent (don't trust given dimensions, like 8'x10').
- –Build a frame slightly smaller – maybe 1"-3" smaller depending on how big your tent is.
- –Better too small than too big (it won't fit!) But if it's too small the tent will flap in the wind, wearing itself out and driving you bugs.
- –If it's too tight and you have to force it onto the frame it will pull the tent weave and make it leak.
- –A new tent stretches somewhat and gets to fit its own frame.
- –Frame inside – handy to hang things from, build shelves, etc. into, lean against – and supports tent better.
- –Frame outside – more room inside. Doesn't have to be as precise.

Most wall tents come with only a 6" overlap and 3-4 ties to tie the two flaps shut; both are inadequate. Mostly the overlap needs to be 3'-4' so it can be really wind tight. Also, a large flap can be controlled with stakes and sticks to make an entrance way.

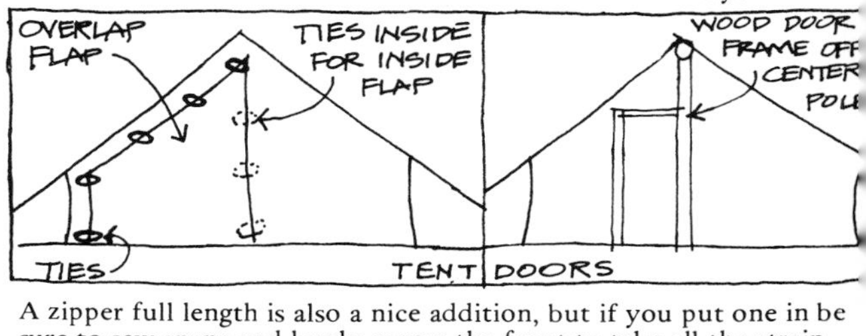

A zipper full length is also a nice addition, but if you put one in be sure to sew snaps and hooks across the front to take all the strain off the zipper or it won't last long.

Wooden-frame doors are nice – usually with a tent frame but not always. Make any size and use the same construction as for stove pipes.

- Do it on floor – tent not pitched.
- Nail 14 wood boards to tent (4 each side of door and 3 each side of door frame.)
- Cut tent, fix latch – canvas can be hinge or you can use leather, or metal hinges.
- Hinge side must be away from center so you have something sturdy to latch to.

GABLE ROOFED HOUSE ON MAUI, 1915

HAWAIIAN LASHING

Te Rangi Hiroa (Peter H. Buck), son of a Maori chieftess and an Irish father was a scholar who concentrated on Polynesian arts and crafts. He wrote *Arts and Crafts of Hawaii*, a book published in 1957. In 1964 when it was about to go out of print, the Bishop Museum Press of Honolulu decided to reprint the work in a series of separate booklets. Here, thanks to Gary Snyder, are some details from the section *Houses*, on the art of Hawaiian lashing. Details on obtaining booklet in bibliography.

The sketches below show the lashing techniques for both gable and hipped roof structures. They are both 12'-6" long, 10'-3" wide, 8'-6" high. The lashing material was generally a three-ply braid of *'uki'uki* grass. "They make little use of these dwellings, except to protect their food and clothing," wrote a member of Cook's expedition in describing the small (4'-6' high) houses of Hawaii, "and most generally eat, sleep and live in the open air, under the shade of a *kou*, or breadfruit tree."

THE GABLE

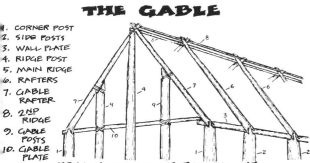

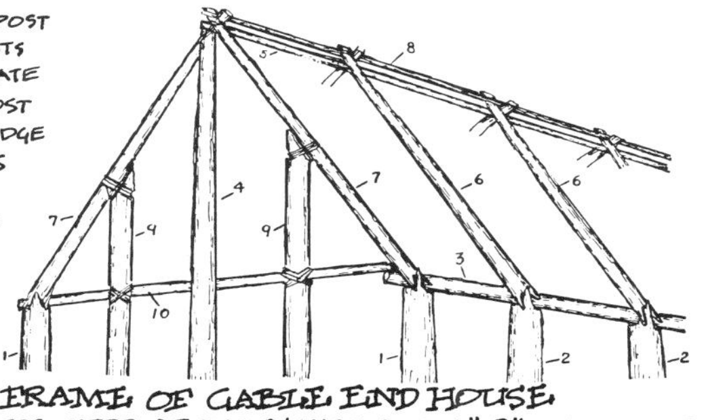

FRAME OF GABLE END HOUSE

SIDE POSTS WERE ABOUT 4' HIGH AND 6"-8" IN DIAMETER. POSTS WERE HARDWOOD (UHIUHI, NAIO, PUA, OTHERS). SET POSTS IN HOLES, MAKE FIRM WITH STONES.

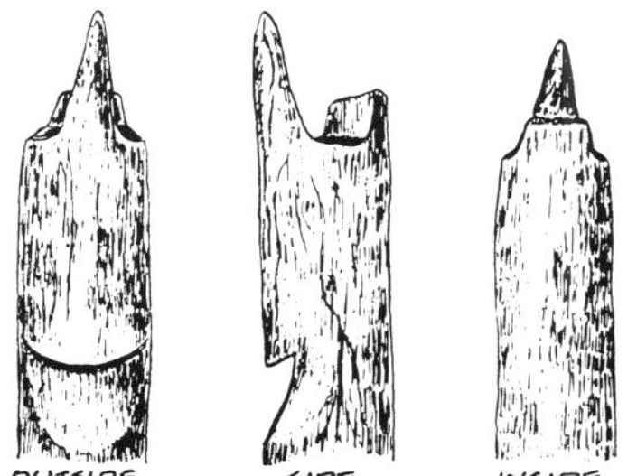

LASHING WAS DONE WITH A 3-PLY BRAID OF 'UKI'UKI GRASS. COCONUT FIBER WAS RARELY USED.

TOP OF WALL POSTS

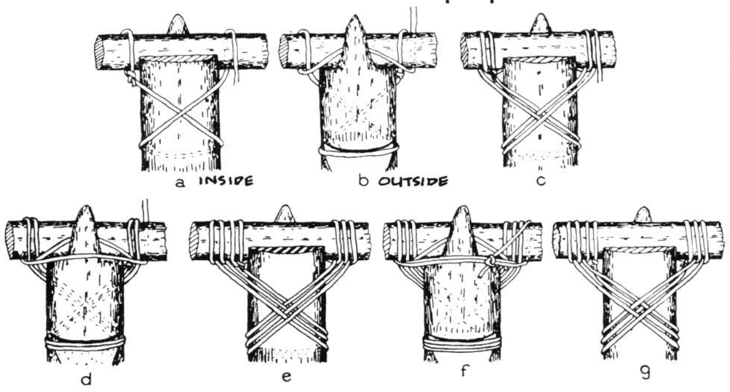

LASHING WALL PLATE TO WALL POST

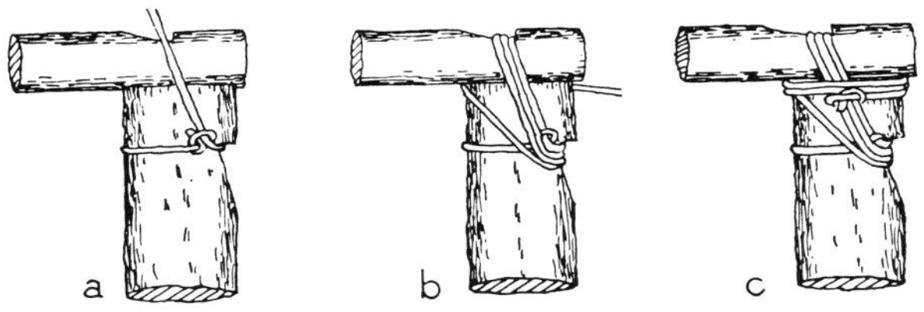

RIDGEPOLE TO RIDGEPOST

HIPPED ROOF HOUSE

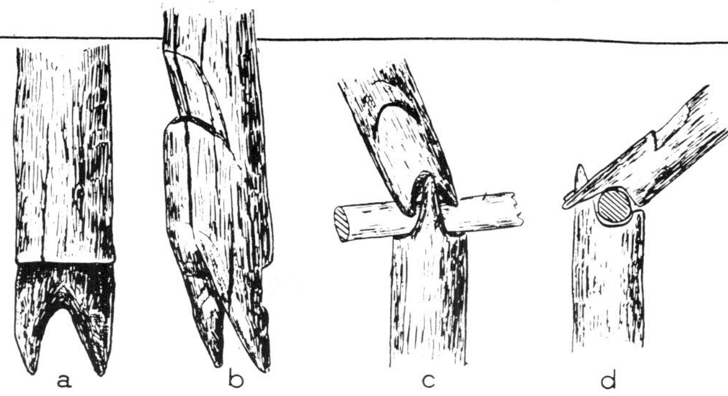

RAFTER END — POST, WALL PLATE, RAFTER

LOWER RAFTER END, AT POST AND WALL PLATE

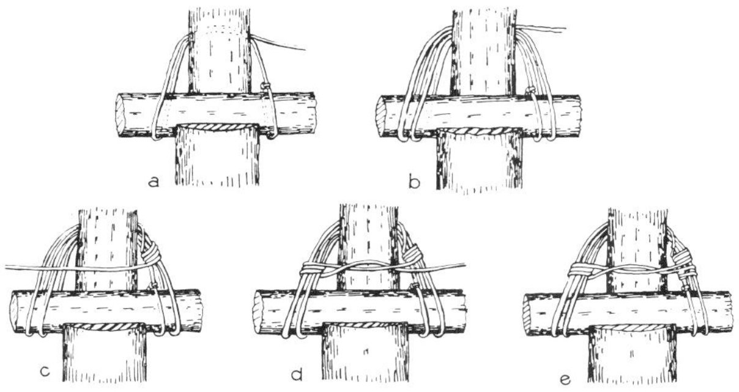

LASHING RAFTER TO WALL PLATE

AT (d), PULL TAUT BEFORE KNOTTING (e)

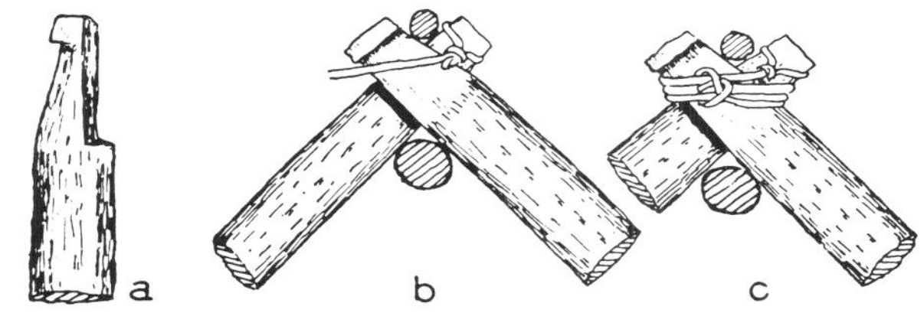

LASHING RAFTERS AT RIDGE

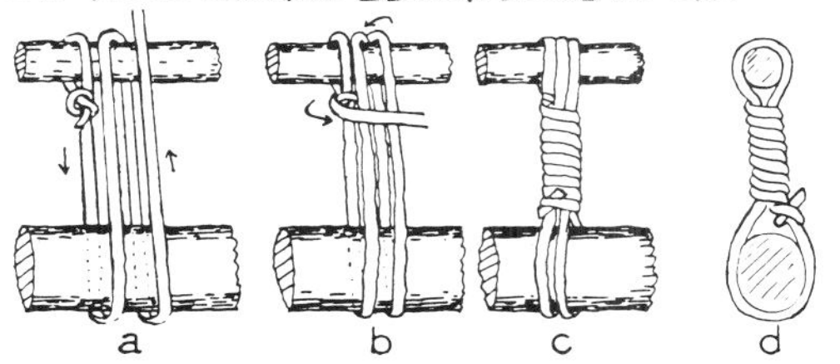

LASHING RIDGEPOLES TOGETHER

LASHING GABLE POSTS TO GABLE RAFTERS

NOW BASIC FRAME IS COMPLETED. THE PURLINS ARE ADDED TO THE FRAME TO ATTACH THE THATCHING. SEE THE DRAWINGS AT THE ABOVE RIGHT FOR LOCATION OF PURLINS.

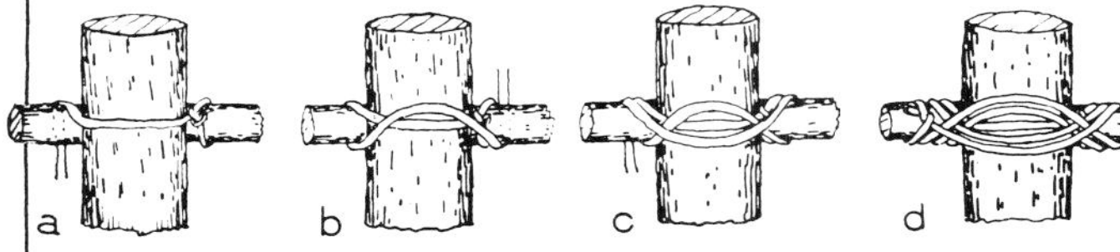

LASHING MAIN PURLINS TO RAFTERS AND POSTS

ORNAMENTAL LASHING VIEWED FROM INSIDE.

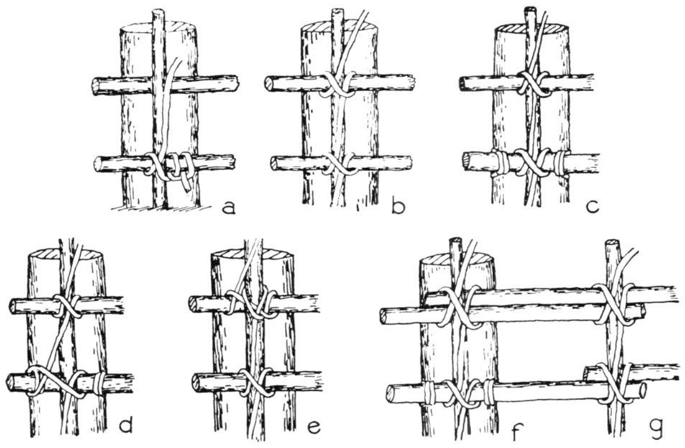

LASHING PURLINS TO SUPPORTING ROD

Lashing purlins to supporting rod: main purlins ¾-1" thick; thatch purlins, ½-¾" thick go between main purlins on both roof and wall. They are not lashed directly to rafters or posts but to ½" vertical rods.

If all was completed perfectly according to the rule from beginning to end, the householder would live to be white-haired, bent with age, dim-sighted, to crouch before the fire with wrinkled eyelids hanging down upon his cheeks or held up with sticks, to lie feebly down and be carried about in a net, and to go away from the world of light as gently as the wafting of a zephyr.

HIPPED ROOF FRAME IN BISHOP MUSEUM, HONOLULU

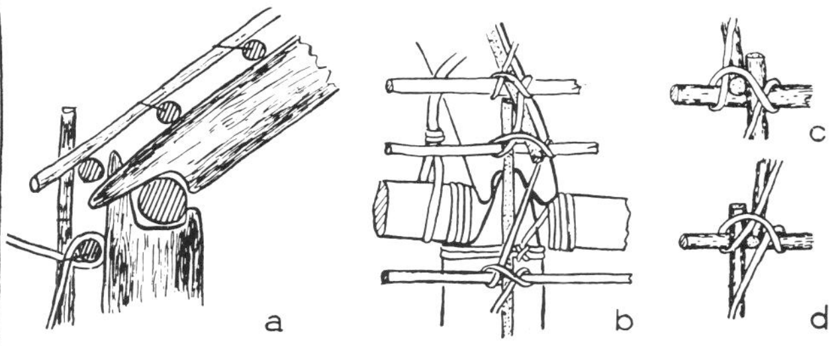

JUNCTION OF ROOF AND WALL RODS

HAWAIIAN NAMES: *

- POU KIHI: CORNER POSTS
- POU KUA: WALL POSTS, BACK
- POU ALO: WALL POSTS, FRONT
- LOHELAU: WALL PLATE
- POU HANA: RIDGE POSTS
- HALAKEA: TEMPORARY RIDGE POSTS
- KAUHUHU: RIDGEPOLE, ALSO KAUPAKU
- O'A: RAFTERS
- KAUPAKU 'IOLE: 2ND RIDGEPOLE
- KUKUNA: GABLE POSTS
- 'AHO PUEO: MAIN PURLINS
- 'AHO: THATCH PURLINS
- 'AHO HUI: THATCH PURLINS SUPPORTS.

* PRONOUNCE EACH SYLLABLE EVENLY. 'AHO SOUNDS LIKE HAHO.

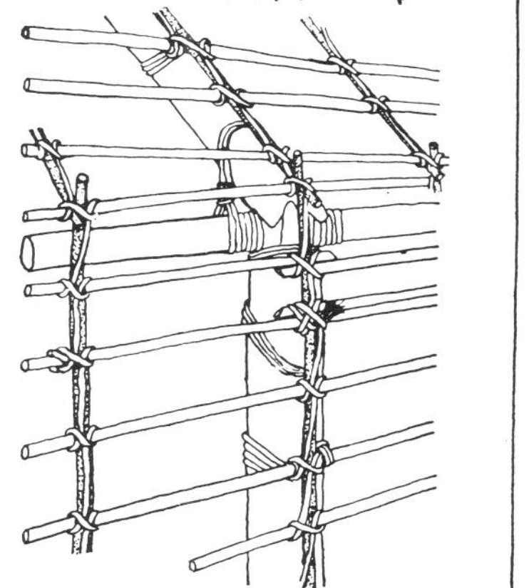

CLOVE HITCH TECHNIQUE W/ PURLINS

THE HIP

The hipped roof differs from the gable roof in that the triangular section above the wall plate slopes inward and upward instead of being vertical. The hipped roof house did not appear in the Islands until after European contact.

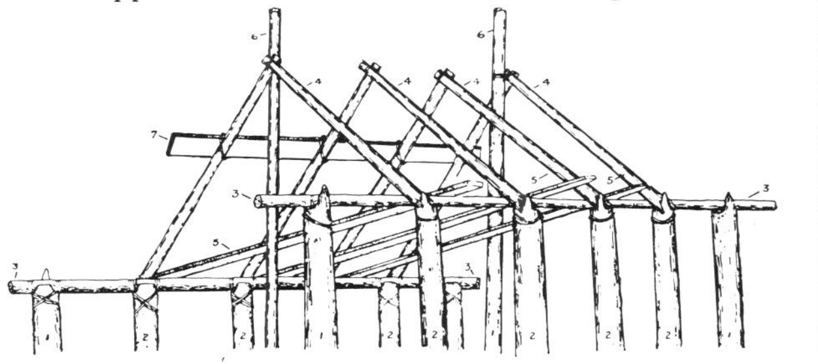

PUTTING UP HIPPED ROOF

INTERIOR POSTS ARE TEMPORARY

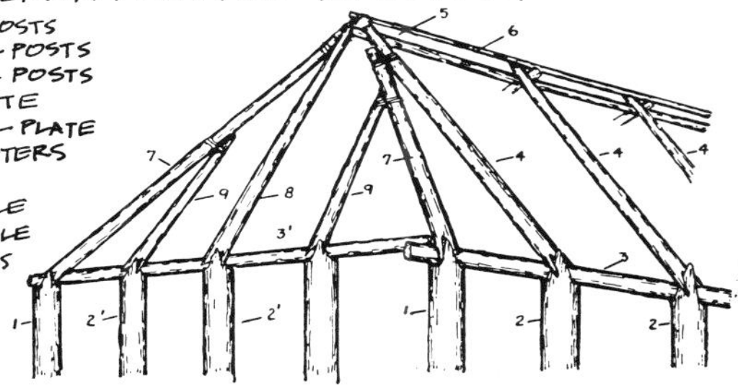

FRAME

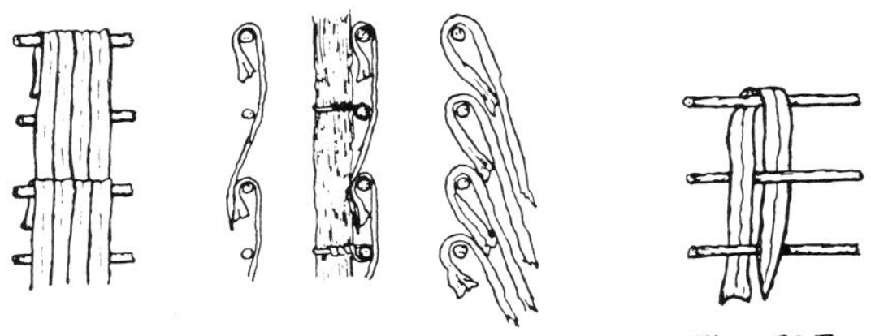

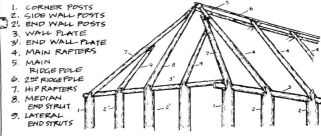

PANADUS LEAVES FROM BIG ISLAND — TI-LEAF

THATCHING

HOUSES ON NIIHAU, BUILT WITH NO WALLS. SKETCH BY DR. WM. ELLIS

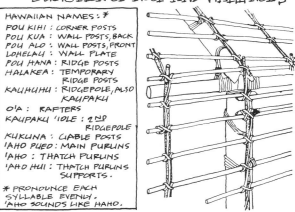

The building above is of a *mudhif* built by the *Ma'dan* or Marshmen of Southern Iraq. It is a 6,000 year old building technique, involving use of the giant reed (*fragmites communis*) that grows to a height of 20' along the lower Tigris and Euphrates. It is bound into bundles, which are then stuck into the ground opposite each other in two rows. A man then climbs onto a reed tripod, and as others pull the tops down, binds them together. When the arches are in place, horizontal ribbing is fastened, then mats tied in place. This structure is a restaurant outside Baghdad, Iraq.

Dale McLeod

REED

Reed has been used in various rural areas of the world for frameworks, walls and roofs. It insulates well, and is light and easy to work with, but is also flammable and makes good insect nesting. Reed does not last long and must be replaced often, but is a material that grows naturally and needs no processing.

Sitting in the Euphrates mudhifs, I always had the impression of being inside a Romanesque or Gothic cathedral, an illusion enhanced by the ribbed roof and the traceried windows at either end, through which bright shafts of light came to penetrate the gloom of the interior. Both on the Euphrates and the Tigris the mudhifs represented an extraordinary architectural achievement with the simplest possible materials; the effect of enrichment given by the reed patterns, came entirely from the functional methods of construction. Historically, too, they were important. Long familiarity with houses such as these may well have given man the idea of imitating their arched form in mud bricks, as the Greeks later perpetuated wooden techniques in stone. Buildings similar to these mudhifs have been part of the scene in Southern Iraq for five thousand years and more. Probably within the next twenty years, certainly within the next fifty, they will have disappeared for ever.

Wilfred Thesiger

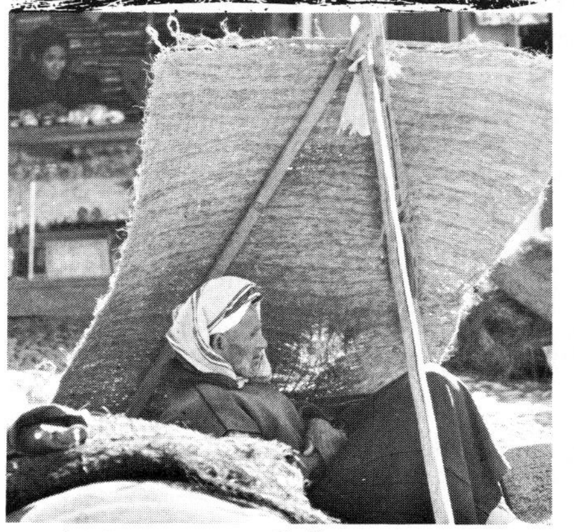

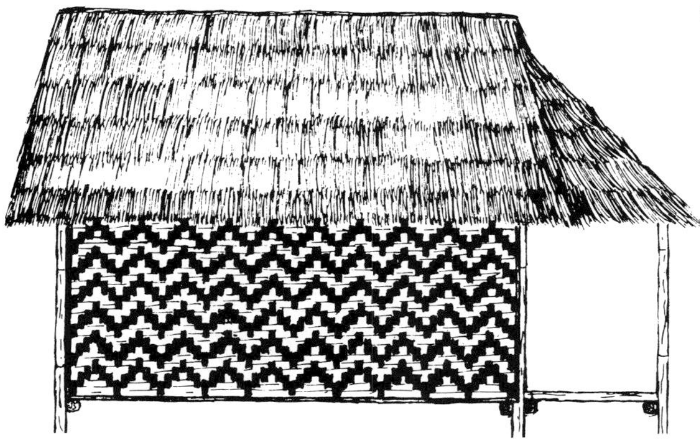

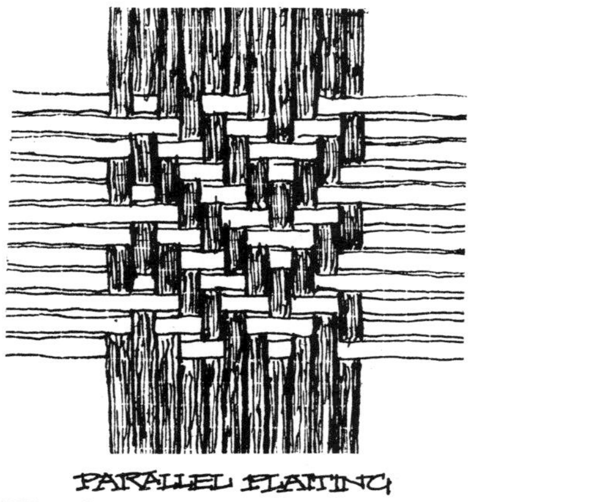

PARALLEL PLAITING

SWAMP REEDS, SPLIT BAMBOO, SPLIT WOOD MAY BE WOVEN INTO MATS AND HUNG ON THE SIDES OF BUILDINGS. THIS WOVEN PROCESS IS CALLED PLAITING. IT MAY BE DONE WITH 2, 3, OR 4 ELEMENTS. THE 2 ELEMENT PLAITS MAY BE WOVEN PARALLEL OR DIAGONALLY (AT 45°) TO THE BORDERS IN EACH CASE THE THREADS RUN AT RIGHT ANGLES TO EACH OTHER.

John Bradbury

• FIRE-RESISTANT TULE-ADOBE SHINGLES •

Details of a device for making tule-adobe shingles used by the Rall family in Santa Barbara in the 1920's: tule 8-10' in length was cut and laid out on table while still fresh (otherwise wetted down for flexibility).

TULE SHINGLE WORK TABLE

Tule is layered to depth of 3'' from tapered end of stalks.

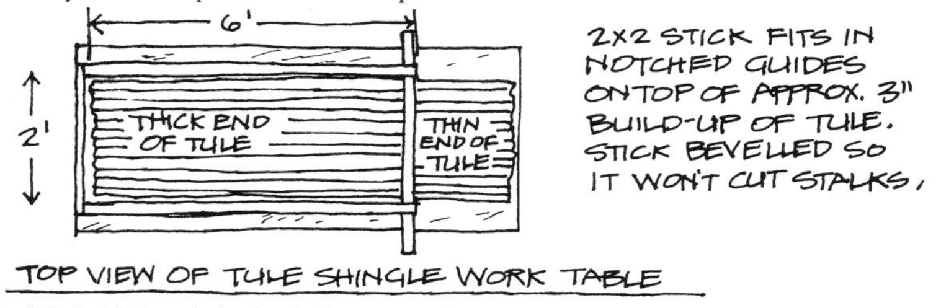

6'
2'
THICK END OF TULE
THIN END OF TULE
2X2 STICK FITS IN NOTCHED GUIDES ON TOP OF APPROX. 3" BUILD-UP OF TULE. STICK BEVELLED SO IT WON'T CUT STALKS.

TOP VIEW OF TULE SHINGLE WORK TABLE

2x2 stick (with beveled edges) then pressed on tule, eased into notches. Next a liquid adobe mixture (any sticky clay type soil with water, mud-pie consistency) is poured over a 2 sq. ft. section behind stick. Then projecting ends are folded back over stick and more adobe is worked into the tule from the top.

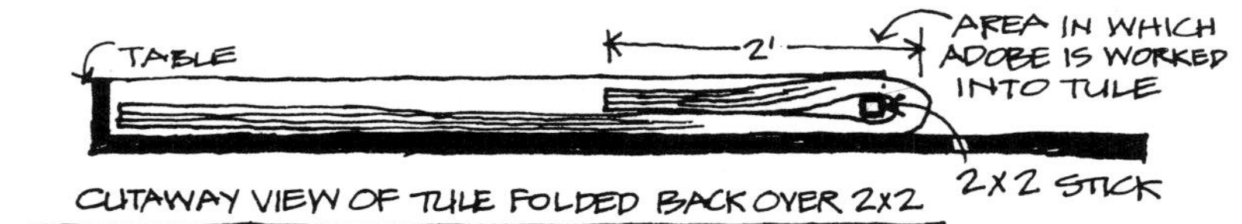

CUTAWAY VIEW OF TULE FOLDED BACK OVER 2X2

The shingle is then removed, placed on the ground, adobe side up to dry. The shingles are then wired onto 2x2's set into the rafters of a minimum 45° angle roof.

Extracted from detailed article by Marsha Zilles, available for $1.00 from Marsha, R.D. 4, McElroy Rd., Ballston Lake, N.Y. 12019.

WATTLE & DAUB

At the beginning of the 13th century a Royal Ordinance insisted that all London houses then covered with reed and rush were to be plastered, as a result of a serious fire in 1212. This so-called *wattle and daub* construction, known by many other dialect names lasted up until the Renaissance time.

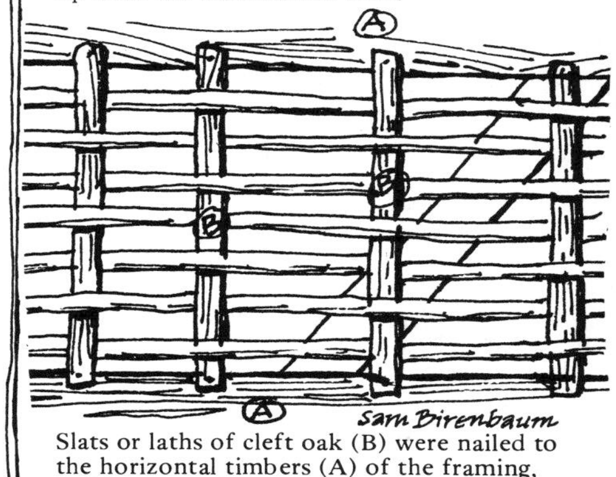

Sam Birenbaum

Slats or laths of cleft oak (B) were nailed to the horizontal timbers (A) of the framing, and between these were interwoven *wattles,* hazel sticks ½ in. to 1 in. in diameter, with the bark left on. To this primitive lathing was applied a thick layer of clay, mixed with straw to toughen it. The *daubers* worked on each side of the wattle throwing on the clay until the desired thickness was reached. The next development was the addition of a plastered surface on each side of the clay, the latter material being liable to expansion and contraction under the influence of the weather.

Next came the introduction of laths, which in the Middle Ages were of cleft oak or beech, classified according to their strength. Often they were *sprung* into grooves in the timber framing. Lath and plaster walls were usually hollow, but were often filled with straw or bran. Lath and plaster ceilings only became general in the 16th century.

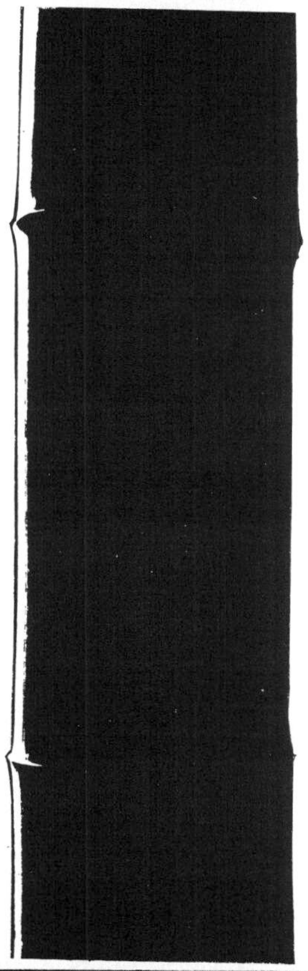

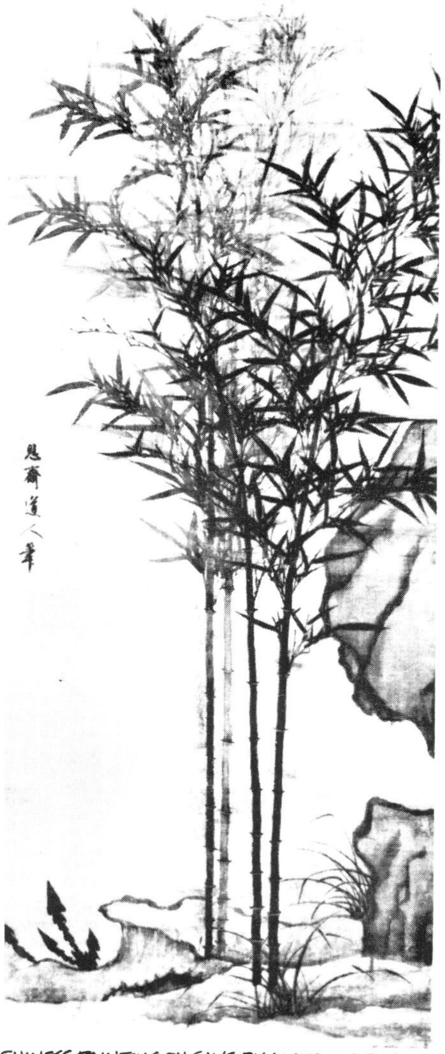

CHINESE PAINTING ON SILK BY LI K'AN AD 1300.

Kozo Kitahara has been making the Japanese vertical bambook flute, the *shakuhachi*, for 25 years.

He brings to his work a sense of humbleness and dedication which is immediately communicated. Loving the bamboo itself, he cannot even bring himself to eat its shoots. In the winter, he goes up alone into the mountains, where he selects each one of the stems he will work. To stand thus in silence in the snow of a bamboo grove is his only holiday.

RONYOON, BURMA. BUILDING A TEMPORARY THEATER.

THE BAMBOO MEN

A typhoon once struck two tall Hong Kong buildings, both protected by scaffolding: one of bamboo, the other of steel. The bamboo held firm, but the steel collapsed in a contorted heap....

Foreign visitors usually stand aghast when they see a half-completed skyscraper parcelled up with bamboo strips, without a single nail, nut or bolt in the whole structure....

You can sit by a 20-story office window and watch the Bamboo Men operate. Suddenly you will catch a glimpse of a waving pole. Then a little man will appear, scampering up to the top. After tying more poles, he will disappear skywards. These are the men who have recently built a 4000-seat theater in just over a week....

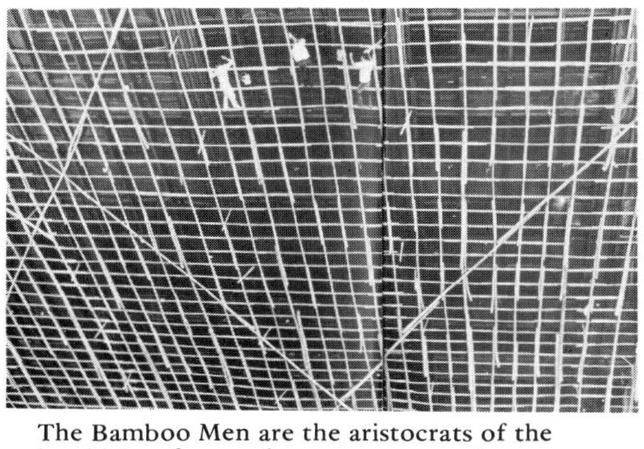

The Bamboo Men are the aristocrats of the local labor force...they can command over $500 a month....

In a typical operation – like repainting the exterior of an office building – trucks move in first to dump their loads of weatherbeaten bamboo poles on the sidewalk.

The bamboo has to be at least three years old before it is seasoned enough to use, and the workers pick through the poles carefully, rejecting an occasional piece because it is cracked or weak – or simply because they don't fancy it.

Next the heavy base poles are swung up and lashed to the window ledges or other projections.

Soon the framework is towering over the street, and the workers are swinging along it, secured only by the tough bamboo strips and their own confidence.

By Ivor Smullen
Chronicle Foreign Service

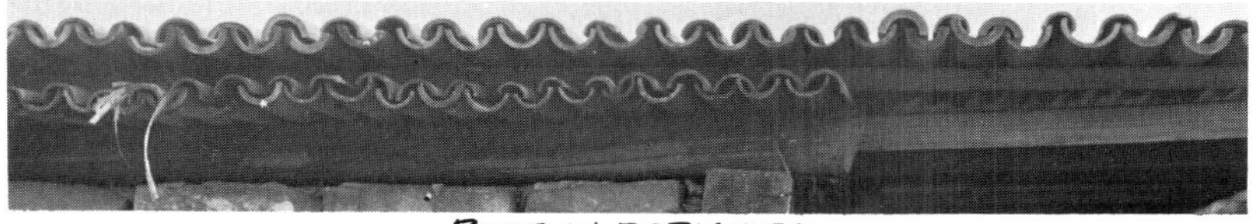

ROOF IN FORMOSA.

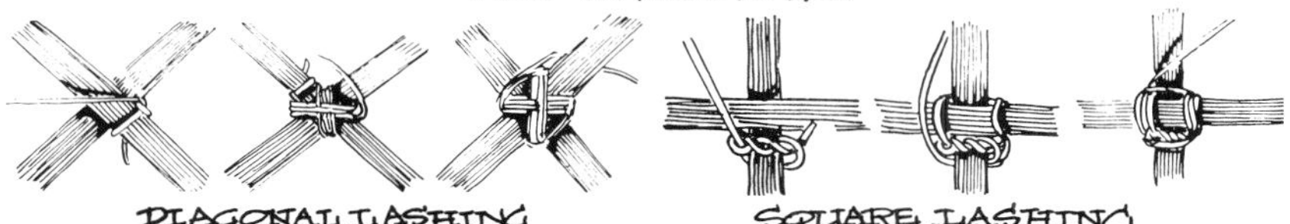

DIAGONAL LASHING
START WITH TIMBER HITCH — 4 TURNS AROUND — 2 OR 3 FRAPPING TURNS & SECURE

SQUARE LASHING
START WITH TIMBER HITCH — 4 TURNS AROUND — SECURE END WITH HALF HITCH

from Bushcraft

BAMBOO HUT BUILT IN INDIA BY G.A. RUDA.

COLOMBIA.

PLASTERED BAMBOO FRAME HOUSE, COLOMBIA.

INTERIOR, RUDA HUT.

BAMBOO

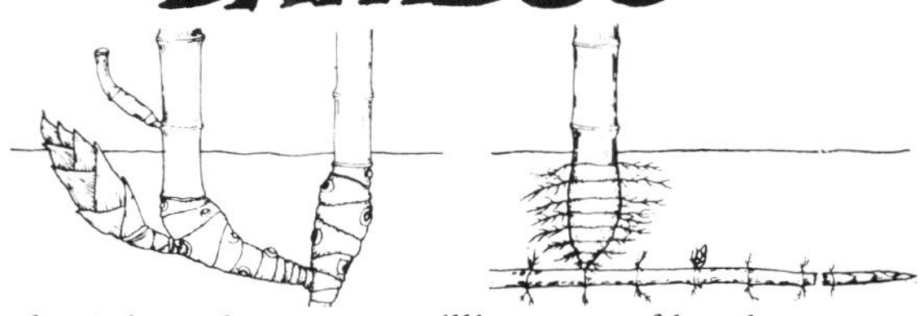

In the Orient almost two million tons of bamboo are used annually in housing.

Regarded as a material to work, bamboo shows itself "grateful" – to use the artisan's term. It is flexible yet tough, light but very strong. It can be split with ease, in one direction only, never in the other; it may be pliant or rigid as the occasion demands; it can be compressed enough to keep its place in holes; after heating, it can be bent to take and retain a new shape. It is straight and possessed of great tensile strength.

The speed of its growth is phenomenal: during a single day, in observations conducted in Kyoto in 1956, a shoot of timber bamboo grew 47.6 inches. Bamboo is useful in controlling erosion, can grow on steep hillsides and riverbanks, has an interlocking root system, produces mulch from its leaves and propagates without attention. Grow your own building material.

The best source of information is the beautiful book *Bamboo* with fine photos of this elegant plant and its diverse uses: tools, buildings, flutes, bridges, fans, food, and a 30-story high lashed bamboo scaffold on a Hong Kong high rise building. Details in bibliography.

The adoration and utilization of towering weed.
Civilization as seen by a material.
Every single thing that plastic isn't.

—Stewart Brand

Growing bamboo: there are two types: clumping (sympodial) and running (monopodial). Bambusa Oldhamii (clumping), which grows to 20-40', 3-4" diameter is available in California. How fast it grows depends upon age of the plant; growth can be speeded by lots of water and nitrogen fertilizer such as blood meal and cotton seed meal. Timber bamboo (running) – *Phyllostachys Bambusoides* – is difficult to obtain in this country; it grows to a height of 25-40', 6" diameter.

THATCHING

NORFOLK REED ROOF ON BOAT SHED, NORFOLK, ENG.

SAMOAN ROOF FRAME LIFTED IN PLACE BY 4 MEN.

POLE ROOF FRAMES AT NETHERLANDS OPEN AIR MUSEUM, ARNHEM, HOLLAND

Thatching, perhaps the world's most common roofing material is virtually unknown in America. As with adobe, stone, and other un-processed materials, there is nothing to *sell,* therefore no entrepreneurs advertising the virtue of thatched roofs. Thatching *is* time-consuming, and it will burn, but when the western world begins running out of processed materials, we may discover that reeds, straw and fronds can provide a waterproof, insulated, and biodegradable roof.

The most elaborate, sophisticated thatching is practiced in England, where there are currently said to be about 800 thatchers still at work. There are three basic types of English thatch: Norfolk reed, which lasts 50-60 years; combed wheat reed, 25-40 years; and long straw, 10-20 years. An excellent book, with 500 photographs, is *The Thatchers Craft* (see bibliography.)

The thatching of Irish cottages is simpler, and requires less skill. Often a layer of sod is placed under the thatch for insulation. Much of the straw used in Irish thatching is what is left over after harvesting grain. A good description is in *Irish Folk Ways.*

The best source of information for simple thatching is *Bushcraft,* a survival guide written by a WW II jungle rescue leader. (See bibliography.) Waterproof thatching requires a steep pitch, and the *overlap* principle. Suitable materials range from long reeds and grasses (best used dry), long stalk ferns (bracken), palm leaves, to straw.

SIMPLE THATCH
from Bushcraft.

SEWN THATCHING

Stitch at bottom of first thatch on lowest thatching batten. The second layer must overlay the stitching of the first row and include the top section of the underneath layer in the actual stitch. It is better to have each layer held by three rows of stitching. The stitching of every row must be completely covered by the free ends of the next layer above it.

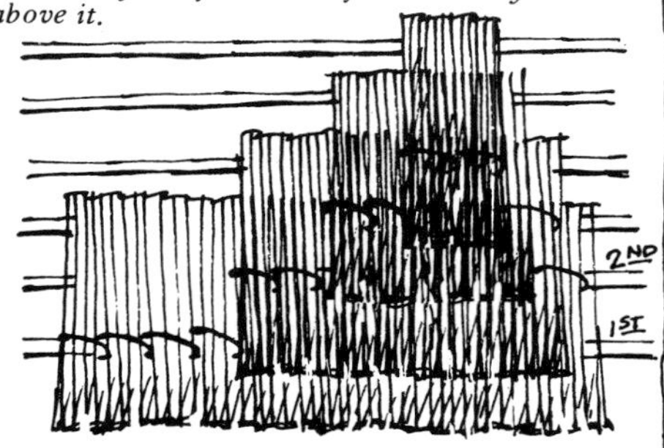

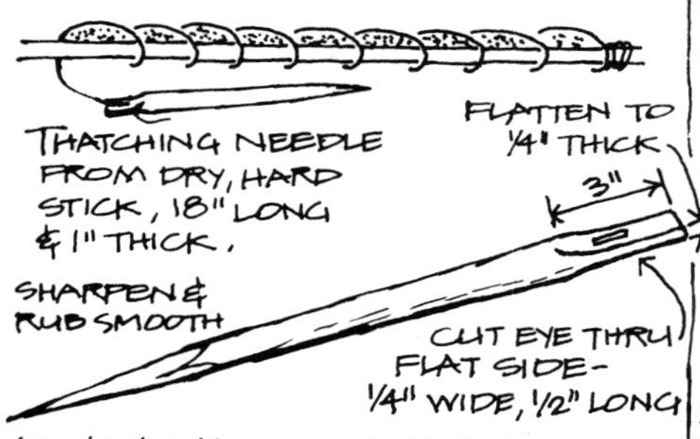

Lay the thatching material with the butts towards the roof and the lower end on the lowest batten. Secure one end of the sewing material with a timber hitch to the thatching batten, thread the other end through the eye of the thatching needle and sew in the ordinary manner to the thatching batten. To avoid holes where the sewing may tend to bunch the thatching together, pass the needle through the thatch at the angle indicated in the sketch and push thatch over the crossing of the stitches.

STICK THATCH

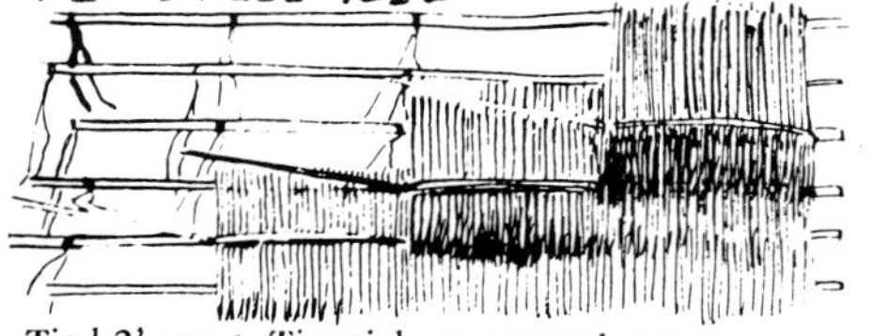

Tied 2' apart. Tie stick at one end, put thatch underneath, tie other end. Follow same principles as with *sewn thatching.*

TUFT THATCH

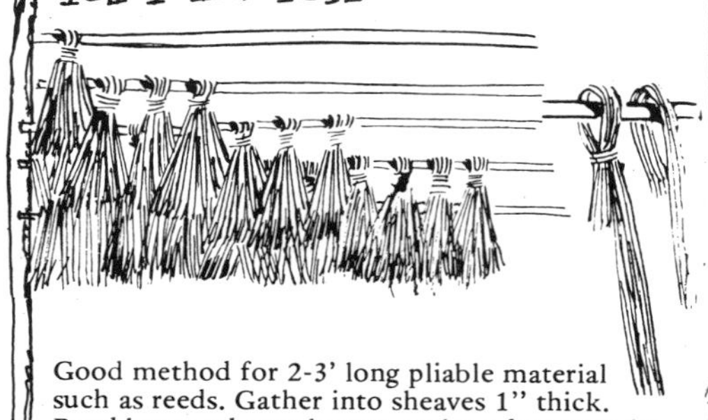

Good method for 2-3' long pliable material such as reeds. Gather into sheaves 1" thick. Bend butt end over batten, twist a few strands around the sheaf to hold tight. Slide along batten. This looks good from inside, is good weather protection. Important that long free ends overlap 2 or 3 rows below. Do not bunch tightly — leave ½" between bent-over ends.

STALK THATCH

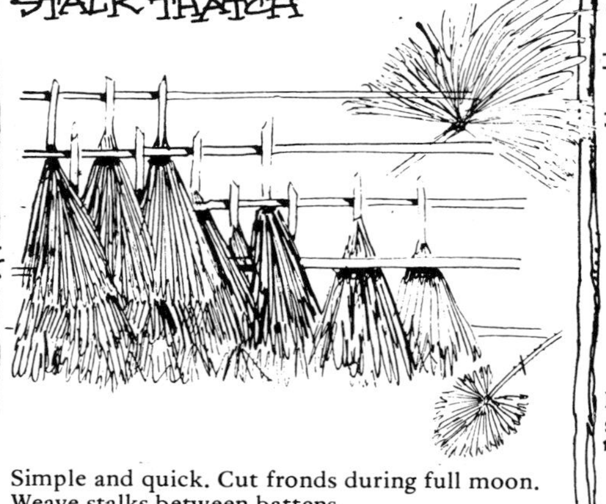

Simple and quick. Cut fronds during full moon. Weave stalks between battens.

SPLIT STALK THATCH

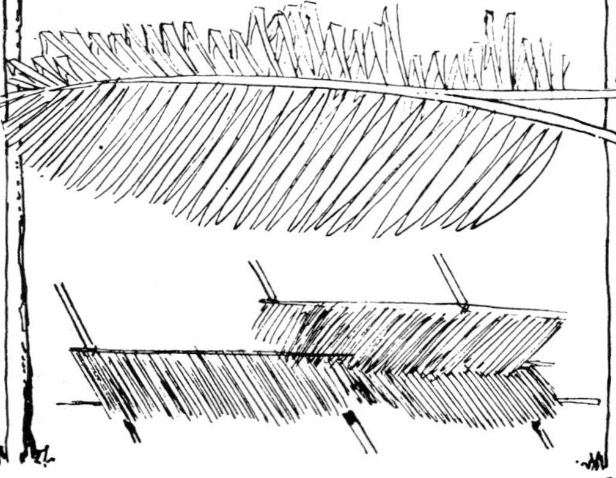

WOVEN THATCH

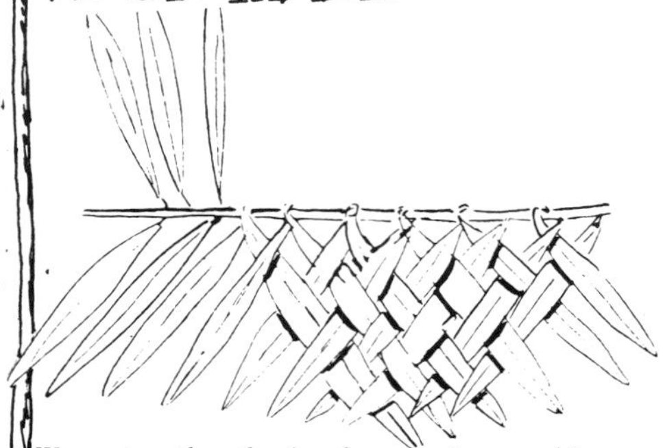

Weave together thusly, then overlap as with other methods.

SEWN BATTEN THATCH

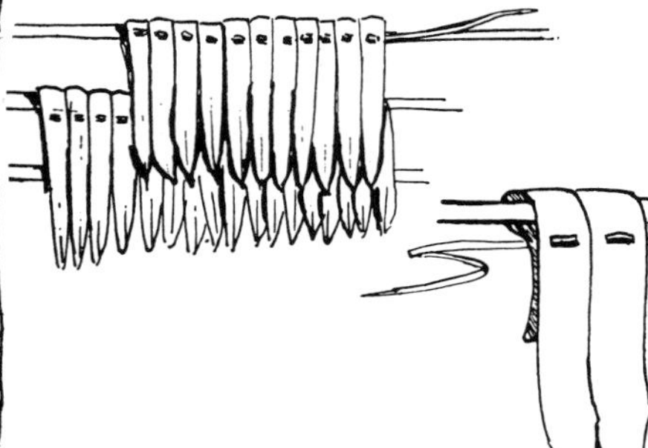

Neat and efficient for certain materials. Make sure material will not curl when it dries.

THATCHING THE RIDGE

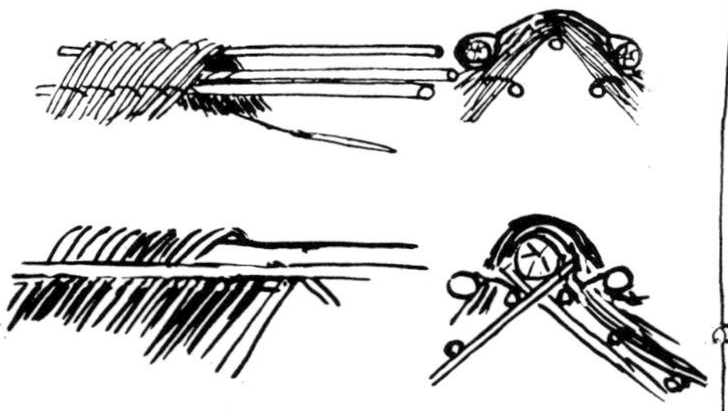

SEWN RIDGE THATCH

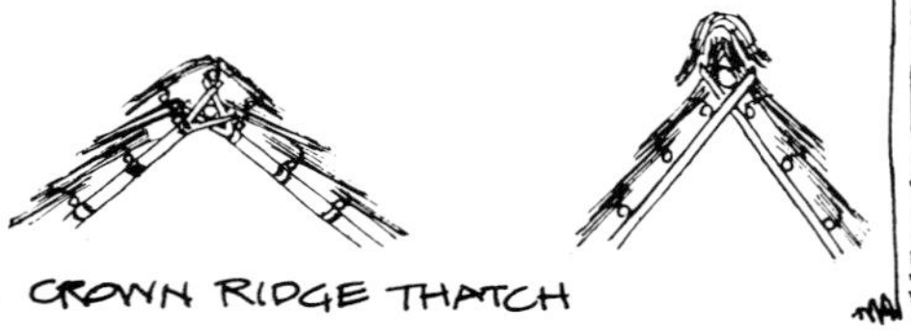

CROWN RIDGE THATCH

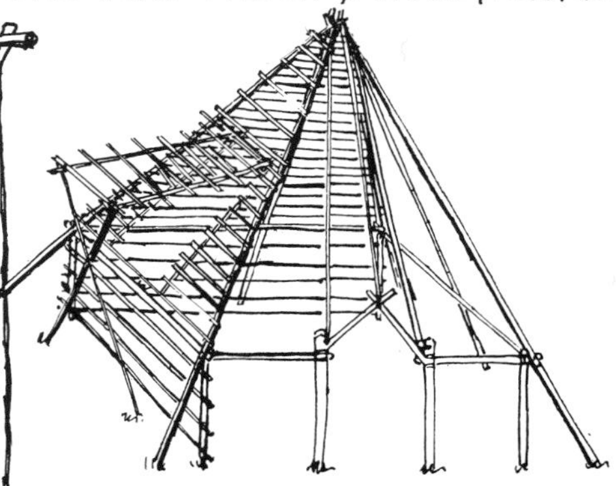

PYRAMIDAL HUT

NOTE X BRACE DORMER WINDOW

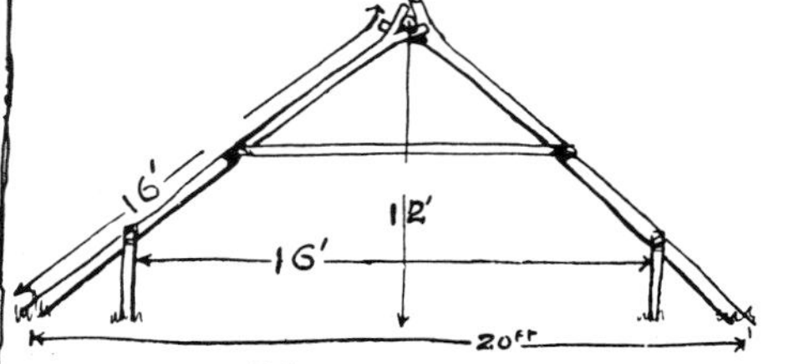

CUTAWAY VIEW

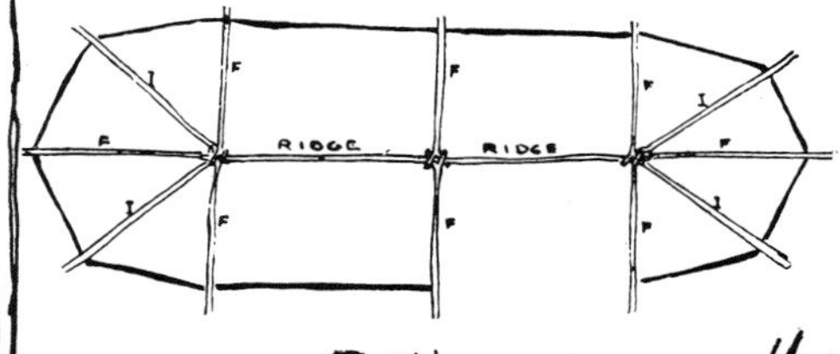

PLAN

MAXIMUM POLE SPAN IS 6'

LONG HUT

THE TIMBER HITCH

START LASHINGS WITH THIS KNOT

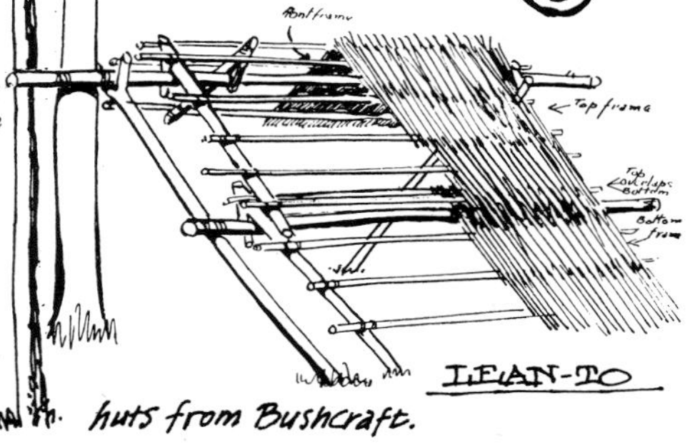

LEAN-TO

huts from Bushcraft.

RECONSTRUCTED DUTCH FARMHOUSE C.1700'S. STRAW THATCH. NETHERLANDS MUSEUM.

INTERIOR, CHIEF'S HOUSE IN WESTERN SAMOA

SIMPLE IRISH THATCHING METHOD. ROPES HOLD STRAW DOWN. HORIZONTAL ROPES WEIGHTED BY ROCKS ON EITHER SIDE.

BUNDLES OF NORFOLK REED AT STORAGE YARD, NORFOLK, ENG.

DOME BUILT BY ZULU WOMEN

Where the construction allows the thatch to be gathered in a finial, great care is lavished on its finish and decoration, both to ensure waterproofing at a crucial point where the thatch lies horizontally, and because the finial houses the *abaFana,* ritual thundersticks invoked as protection against the lightning.

Final thatching is carried out in the last decades of the occupant's lifetime. The decoration now attains a baroque exuberance. Bands of criss-cross ornament, reminiscent of bead-work patterns, alternate with the swags and festoons more appropriate to the medium of plaited grass....

Shelter in Africa

HOUSE IN PADANG-PANDJANG, SUMATRA

SHEEPCOTE AT NETHERLANDS MUSEUM. HOUSED SHEEP & COLLECTED MANURE FOR FIELDS.

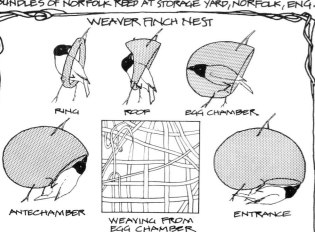

WEAVER FINCH NEST

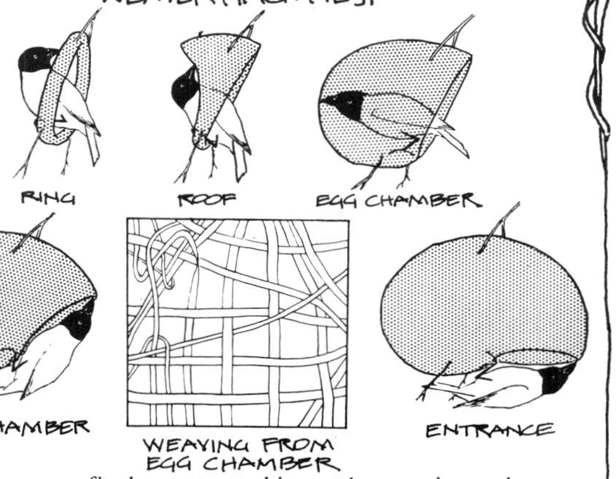

The male weaver finch constructs his nest by weaving and thatching. The female does none of this. After the nest is substantially completed, the female lines the egg and brood chamber with non-woven materials.

The male works deliberately: he collects the long grasses or strips from palm leaves: he attaches them to a chosen spot; using his body as the radius, he forms the ring, the egg or brood chamber, the antechamber and entrance. This is all accomplished by *weaving* and *"hitch-knots."* All the weaving is done by the bill, although the finch often uses one or both feet to help hold the strip. In a sense, weaver finch weaving is less "stereotyped" than classic human basket weaving. As the pictures show, the pushing of the grasses is flexible, almost free form. When working on the ceiling, for instance, the male weaves until the mesh size is so small, he can no longer thread. At this point, he starts *thatching.* He switches from long, narrow strips to short, wide strips. He switches from pulling the long strips in and over others. Instead, he pokes first one end, then the other into available gaps. When there is no light entering through the woven roof and thatched ceiling, he stops work.

The *entrance tube* is added only after the female accepts the nest. If she doesn't, the male simply unhitches the hitch-knots and destroys the nest. If she begins to line the egg chamber, he no longer can enter the nest. He stays on the outside completing the entrance tube. Many times the male appears to be unable to stop weaving. He keeps strengthening attachments, tucking in loose ends and adding new, fresh green strips.

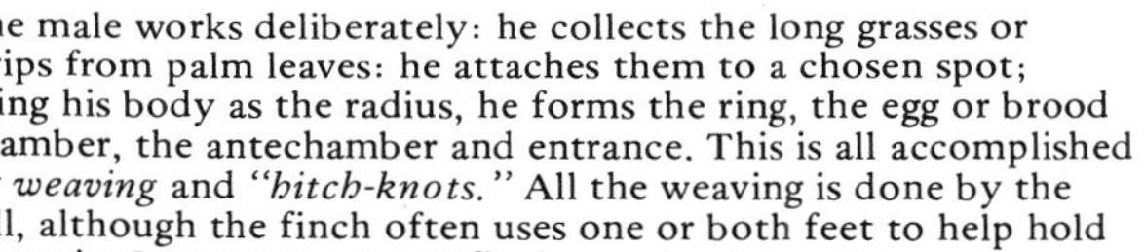

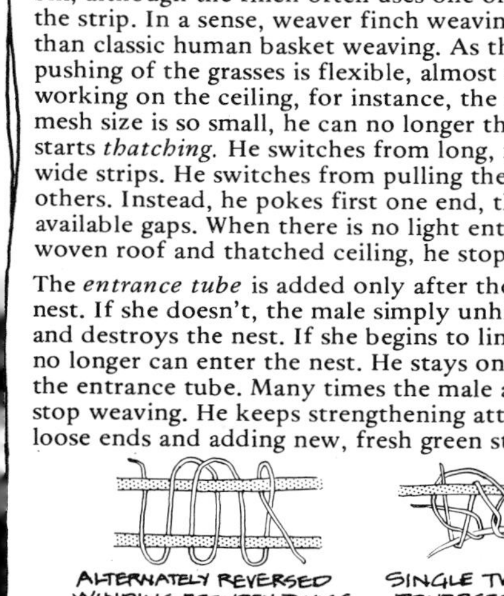

WEAVER NEST

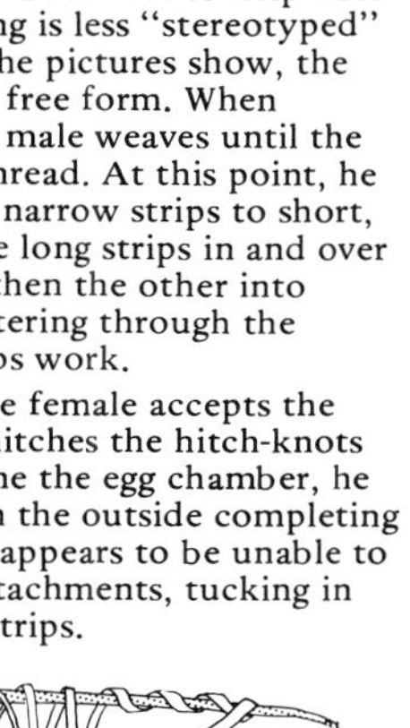

BY PETER WARSHALL FROM STUDY BY NICOLAS & ELSIE COLLIAS

TRADITIONAL STYLE ONE ROOM HOUSE OF HIGHLAND MAYAN OF CHIAPAS, MEXICO. WATTLE & DAUB WALLS

Carpathian mountain house. Smoke from fire escapes through thatch.

Shag roof house in Rumania, built half underground. The haystack roof is easier to build than a thatched roof.

Natural fork branch, used for pitching hay. Middle tine is added.

There is a lyrical decree in the art of building, an edict, an understanding waiting available to be used by those builders with a sensitive hand. It may be a method of material, a form and appearance, an approach to the environment or just a light touch, or it may be the way the parts combine to make a whole-building by form and fit.

Christopher and Charlotte Williams and their four year old son took a 15 month trip in a VW bus, studying architecture, crafts, and the utilitarian arts of indigenous societies in North Africa, Syria, Turkey, Bulgaria and Romania. To find what they were looking for, they would approach the tourist offices with a card stating [in the native language] that they were interested in indigenous people and their work. As a rule they found that it took some time to find someone who understood: urban people couldn't understand why Americans would be interested in primitive culture. However, sooner or later someone would understand, and say: "Oh, you want to go to...."

Usually they would find a guide. When they did get out to the rural areas, the people readily understood the Williams' interests, and they found it easy to communicate to the people their appreciation of local art and ingenuity.

They found the VW bus an excellent and cheap way to travel, and felt that being a family unit (with child) made human relations easier than traveling alone or as a couple. The record of their trip will be published in spring 1974 by Random House in the book *Craftsmen of Necessity* and following, courtesy of the Williamses and Random House are some excerpts from this excellent book.

For years the Norwegians have used the organic roofed house. When first constructed, birch bark is underlayed as a sealer. Birch bark is weather tight when overlapping and almost completely resistant to rot. Then squares of pasture dirt about five inches thick with roots and grass are laid over the bark. As the seasons pass the mat perpetuates itself, root intertwines root and the roof becomes a solid whole, rain and weather only strengthen it. In the winter the dead stalks of grass hold the snow for effective insulation. The spring rains beat the grasses down to shed the excess water, then bring the roof to life again. In summer the grasses grow long and effectively reflect the sun's heat. As the years pass the roof renews itself from season to season and needs little or no maintenance.

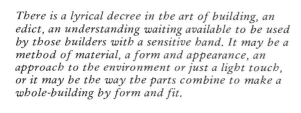

OF NECESSITY

Gate in Rumania, made from sycamore tree, is counterbalanced to swing easily.

Notched wall planks are supports for gate and bannister.

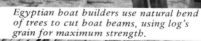

Egyptian boat builders use natural bend of trees to cut boat beams, using log's grain for maximum strength.

Most of rural Egypt is spotted with large free standing pigeon rookeries. The pigeons provide fertilizer and pigeon is a luxury of the Egyptian diet. Only the young pigeons are for eating and not all of them can be consumed, some must be left for breeding stock. Each of the hundred odd holes in the tower provides a brooding nest, the holes extend through the walls of the structure. Each can only accommodate enough young to maintain the rookery, the unfortunate extra, usually the smaller ones, fall inside the tower and to the bottom. All that is left for the owner is to enter a door through the base of the tower and pick up his daily harvest of squab. Self-regulating, self-maintaining and the broods tend to build a larger and healthier stock.

The blacksmith in some places is still regarded as a somewhat fearful person. His art is difficult to understand, his materials come from the depths, he moves about in a darkened recess, he builds and tends pits of intense heat, and pulls from them irons white with fire that he violently renders. But villagers know that the blacksmith, this outcropping of industrial technology, is highly essential to them as a bulwark against the realities of their difficult existence.

Most all blacksmiths work no wood, those tools requiring iron working parts and wooden handles have to be formed from a collaboration between two artisans. Most blacksmith markets are closely annexed by the handle makers. Usually the customer will select the hoe, shovel, or chisel then take it to the handlemakers to have a handle fitted.

The rhythm of the woodworker making a piece of a tree into a man-designed form is an action and atmosphere very removed from the environment of the blacksmith. Both require honesty and understanding, but the blacksmith works swiftly with forceful movements necessary to shape the red hot but still tough iron before it cools to require another heating. He is urgent and aggressive. The woodworker must be more methodical and sympathetic, quiet and alone.

The woodworker is a lonely man, alone with his hands. Behind his hands he contemplates their work. To his sides lie his tools, ready for their turn. Long bladed knives to cut long thin wafers of wood, choppy hatchets and adzes to cut chunks, curved shank gouges to cut spirals. By following the edges and flutes of a tool with thumb and eye the wood cutter can determine the character of the finished piece by choice of his tool shape. His tools lie before him as a vocabulary, each one possessing a subtle inflection of meaning. With this language of tools and the motions by which he uses them a conversation is conducted between worker and material. The wood argues in knots and agrees in smooth grain.

WRECKING & SALVAGE

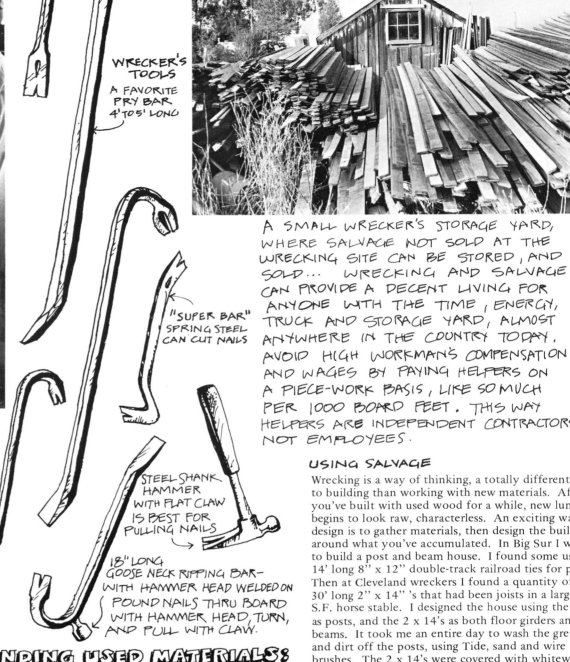

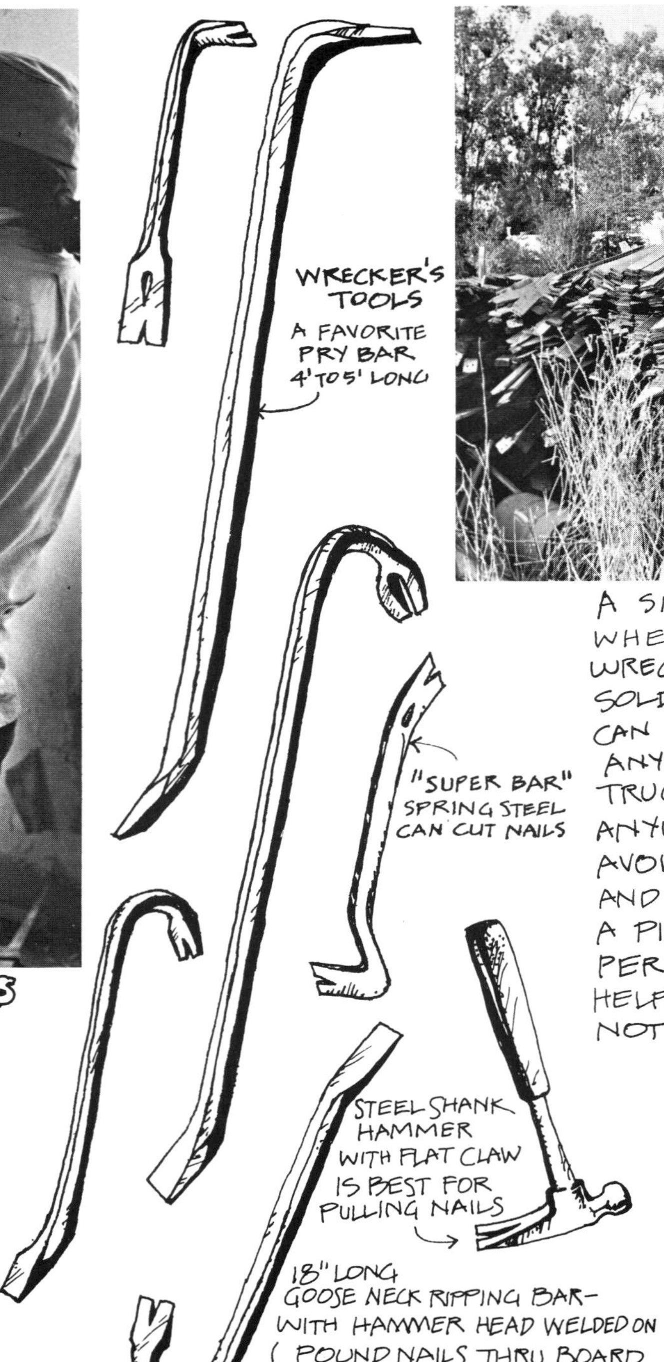
WRECKER'S TOOLS
A FAVORITE PRY BAR 4' TO 5' LONG
"SUPER BAR" SPRING STEEL CAN CUT NAILS
STEEL SHANK HAMMER WITH FLAT CLAW IS BEST FOR PULLING NAILS
18" LONG GOOSE NECK RIPPING BAR- WITH HAMMER HEAD WELDED ON
POUND NAILS THRU BOARD WITH HAMMER HEAD, TURN, AND PULL WITH CLAW.

A SMALL WRECKER'S STORAGE YARD, WHERE SALVAGE NOT SOLD AT THE WRECKING SITE CAN BE STORED, AND SOLD... WRECKING AND SALVAGE CAN PROVIDE A DECENT LIVING FOR ANYONE WITH THE TIME, ENERGY, TRUCK AND STORAGE YARD, ALMOST ANYWHERE IN THE COUNTRY TODAY. AVOID HIGH WORKMAN'S COMPENSATION AND WAGES BY PAYING HELPERS ON A PIECE-WORK BASIS, LIKE SO MUCH PER 1000 BOARD FEET. THIS WAY HELPERS ARE INDEPENDENT CONTRACTORS, NOT EMPLOYEES.

OUR GROWING RESOURCES

You need *time* if you're going to work with used building materials, especially if you do the wrecking yourself. But with the rising costs of lumber and other supplies, wrecking and salvage are becoming increasingly important builders' alternatives.

The most difficult type of wrecking is to take down an entire building, especially if you must remove the concrete foundations. Hauling unsalvageable material is expensive, and enthusiasm usually dwindles during the project. However, given the manpower, persistence, and tools, it can be done, and the rewards are considerable.

A good alternative to the responsibilities and time inherent in wrecking an entire building is to somehow work out an agreement with a professional wrecker to remove salvageable materials before demolition. Wreckers generally smash up a building with a bulldozer and haul the splinters and rubble to the dump. Thus if you can get in ahead of the tractors, you'll save the contractor work and dumping fees. Minor Wilson worked out such an arrangement in San Francisco, paid the wrecker $35 per 1,000 board feet of lumber removed from an old school. He also got doors, windows, toilets, hardware, etc. It was enough for Minor to build an entire new house, with little new materials required. Go around and talk to wreckers. The biggest obstacle will be workmen's compensation insurance, but it's possible to work out a "hold-harmless" agreement, saying that you'll not sue the wrecker if injured.

An interesting aspect of wrecking, and the reason that the big wreckers salvage little material lately, is the high rate of injury. California wreckers must pay $35.00 per $100 payroll for workmen's compensation insurance. Thus the employer can hardly afford to have a man pulling nails. It's far cheaper to smash and haul.

Professional wreckers have heavy equipment, and big payrolls; they're geared for large jobs and their bids on home-size buildings are often high. Thus your best chance will be to find a relatively small building. Watch the want ads, or if you see an adandoned building, trace down the owner.

FINDING USED MATERIALS:

- *lumber yards and mills will usually sell twisted and weathered lumber cheap – ask to scrounge thru their scraps.*
- *look under "wreckers" in the yellow pages. Most maintain a yard and sell materials.*
- *dumps are full of lumber and other stuff. Try to get the dump operator to let you scrounge. If it's one of the big city dumps where they drive right along with bulldozers, get materials from the people dumping before they go into the pit.*
- *cruise with a pickup. Be a "perceptive scavenger". There is always an unbelievable amount of stuff in city trash boxes – the large metal kind that sit outside remodel jobs. Doors, 2 x 4's, wooden windows replaced by aluminum. I have always meant to take my truck, two friends, work from dawn to dusk in S.F., have a place to stockpile, and see if we could get enough in a single day to build a small building.*
- *cruise alleys, look in front of Chinese stores, in back of any kind of business that receives goods in wooden crates. Glass companies, furniture stores, docks, railroad yards.*
- *construction sites are notorious for waste. Ask the foreman.*
- *driftwood on the beach*
- *cleaning up after disasters like tornadoes.*

WASHING SALVAGED BOARDS:

- LAY OUT ABOUT 4 BOARDS, THEN WITH HOSE SQUIRTER, WET BOTH SIDES – DIRT WILL SOFTEN
- THEN WIREBRUSH ONE SIDE OF ALL FOUR, THEN TURN, AND GET THE OTHER SIDE.
- SQUIRT AGAIN, STACK AND LET DRY FOR A FEW DAYS.

USING SALVAGE

Wrecking is a way of thinking, a totally different approach to building than working with new materials. After you've built with used wood for a while, new lumber begins to look raw, characterless. An exciting way to design is to gather materials, then design the building around what you've accumulated. In Big Sur I wanted to build a post and beam house. I found some used 14' long 8" x 12" double-track railroad ties for posts. Then at Cleveland wreckers I found a quantity of 30' long 2" x 14" 's that had been joists in a large old S.F. horse stable. I designed the house using the ties as posts, and the 2 x 14's as both floor girders and beams. It took me an entire day to wash the grease and dirt off the posts, using Tide, sand and wire brushes. The 2 x 14's were covered with whitewash and horse piss stains, which were easy to brush and hose off. Under all the dirt was beautiful wood, with the patina of age that new lumber will never have.

George Taylor is the most organized wrecker I've met. Working basically by hand, without machinery, George takes his time, cleans material as he goes, and salvages almost everything, including nails that he picks up with magnets. Here's part of a conversation with George about his operation:

George, we think wrecking is going to be more and more important for someone trying to build his own place. What do you think we should tell people about it?

Well it's becoming harder and harder to procure a place to wreck. You can't just give the state $5 for a little building they have to move for a freeway, they want a $500 bond. They had too many people get buildings from them, just take out what they wanted, and leave them with the mess.

There seem to be a lot of carpenters out of work. If they could get a bond, how would they go about finding good buildings to wreck?

Government buildings, there is still a great surplus of these temporary buildings that were put up during WW II. If they approach either the Army Corps of Engineers or any district Navy real estate branch. There are an awful lot of buildings going down that people aren't touching, just being wasted – bulldozed.

How do you find out about government buildings?

They've got a mailing list.

Would the government have mailing lists everywhere in the country?

Yes. Each of the Naval districts will. The Army Corps of Engineers. Just as they would if there were going to be bids to do work for them.

What about a guy in Kansas? Would he contact an Army Base near him?

...and ask who their Army Corps of Engineers are, whether they'd be Air Base, Army Base, or whether it would be the Navy.

Each branch of the service has got a corps of engineers?

The Air Force sort of lets the Army take care of theirs, but it would be their Corps of Engineers.

Do you always clean all your lumber as it is taken down?

As it is taken down. Even before it goes from here to the ground. Because all of the nails that are in it get bent getting it from here over there. Somebody has to straighten those nails, they rust, they pull harder.

So you take nails out right away. What about the plumbing, the electrical and the rest of the stuff?

It's all good salvage. Except at this time I don't salvage too much of the cast iron pipe because everybody's gone to PVC, so the only value in cast is junk.

Do you sell that scrap, the nails and the cast?

Yeah, all of that is good and saleable. The first sets of buildings I had I saved 16 tons of nails and small stuff. Just small scrap, I sold all of the pipe and things as useable.

How do you take a building down?

Find out the way it was built and reverse it. Take the nailing of all the sheeting, the rafters and the joists and just reverse it. Often you have to start from the center. Two men would work from the center out or for one reason they would want to shelter the one side first. Even on roofing. Find out how it was nailed when built.

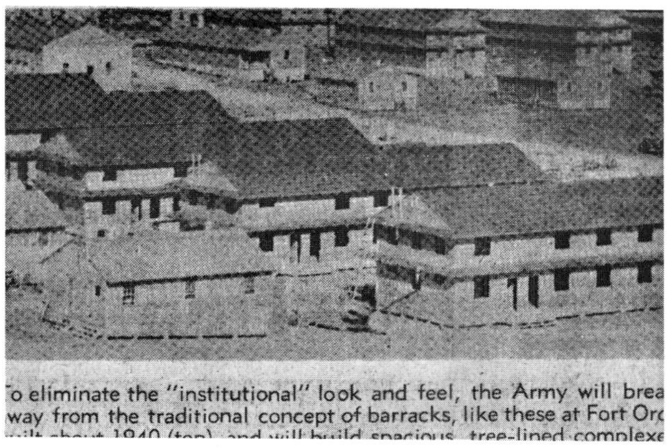

o eliminate the "institutional" look and feel, the Army will brea
way from the traditional concept of barracks, like these at Fort Orc

Professional wreckers don't save lumber any longer. That's really ridiculous. It must be because they're sloppy, careless. That must be the highest rate of injury of all the trades.

It is. If you keep debris around, people have to walk over it, nails sticking in it. We take all nails out right away so we aren't tripping or falling with a knee on them. Those are the little things that could hurt a man. There's a lot of lost time. If you're going to clean the nails, it's no harder to clean them right away.

Actually, you're not using any machinery, are you?

No, most wrecking injuries come from wrecking with machinery and still having men trying to work in there by hand. Say they'd want to pull down this overhead here and salvage the timbers away from it or grab a couple of 6 x 6's. So people are continually getting hurt. Going down and just hurriedly getting in to drop the roof trash to get a few joists out. When you start and dismantle it from the top to the bottom, you're just as clean as the carpenter that put it up there.

How much do wreckers bid on typical jobs?

30-50 cents per square foot of floor space. They count each story separately. Another way to figure is it costs a wrecker about $30-50 per 1000 board feet of lumber in a building, to smash it and carry it to the dump. So you could bid on a building for the lumber that's in it. Just make sure you get enough time. Another thing, each man can salvage about 1000 board feet a day....

So you've been doing it for 12 years? And you've been supporting yourself by doing that?

Oh yeah.

Did you have any building experience before this?

None whatsoever. I didn't know a board foot from a penny nail. I was a chief pharmacist mate in the Navy for 23 years.

ROLLING SEATS FOR COMFORTABLE FLOORING REMOVAL. FIR T&G BEING REMOVED WITH RAM NAIL PULLER.

TOOLS

TWO TOOLS FOR PULLING NAILS WHERE YOU CAN'T POUND NAILS THRU FROM OTHER SIDE: CATS PAW →

GEORGE HAS 3 DIFFERENT WIDTH CLAWS FOR DIFFERENT OPERATIONS – HAMMER THE CLAW IN AT ANGLE UNTIL IT GRIPS THE NAIL HEAD, THEN PRY UNTIL HEAD POPS UP. DO ALL THE NAILS ON BOARD, THEN PULL NAILS WITH A CROWBAR.

← RAM NAIL PULLER – DOESN'T SCAR WOOD AS BADLY AS CATS PAWS. PLACE JAWS OVER NAIL HEAD, PUMP HANDLE – WHEN JAWS SINK INTO WOOD, YOU PRY BACK AND NAIL LIFTS OUT. WARDS HAS BEST ONES THESE DAYS.

NIPPERS → FOR CUTTING OFF NAIL HEADS, AS ON METAL ROOFS, WHERE YOU CAN'T PRY.

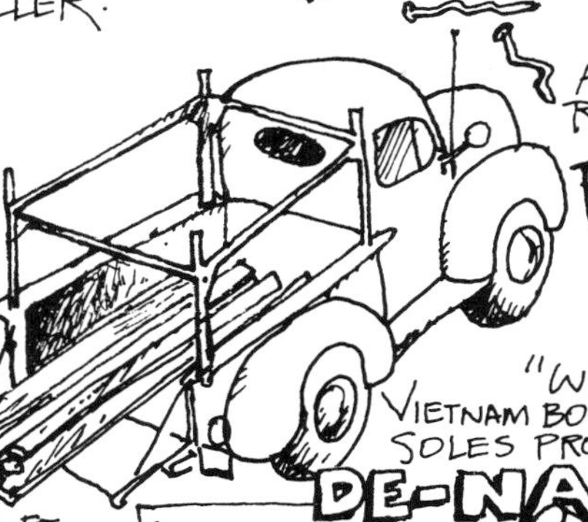

PICK UP TRUCK (PRE-'65 CHEVYS) WITH RACK

CANVAS BACKED LEATHER GLOVES "WHITE MULE" BEST

VIETNAM BOOTS WITH STEEL IN SOLES PROTECT FEET FROM NAILS

GEORGE'S TOOLS FOR THE PRO, STEEL SAWHORSES THAT CAN HOLD 2000 BOARD FEET OF LUMBER, AND SPECIAL TOOLS TO STRIP 1x6, 1x8, 1x10'S WITHOUT BENDING OVER. TWO BIG BARS OF CHROME NICKEL STEEL THAT CAN BUST CONCRETE.

EXPLANATION OF BOARD FEET: LUMBER IS SOLD BY THE 1000 BOARD FEET. ONE BOARD FOOT IS ONE LINEAL FOOT OF A 1x12, 2x6, ETC.

$$\frac{\text{Thickness} \times \text{Width}}{12} \times \text{Length} = \text{Board Feet}$$

FOR EXAMPLE: 8' LONG 2x4 = 5 1/3 BD. FT.

WEIGHT OF WOOD: DRY WOOD AVERAGES ABOUT 3 LBS PER BD. FT.

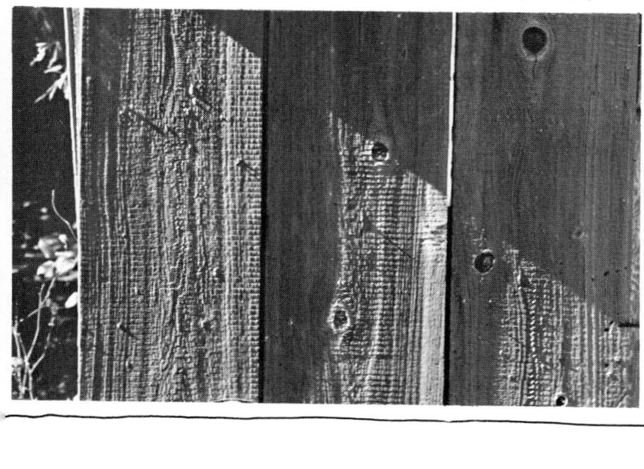

DE-NAILING LUMBER

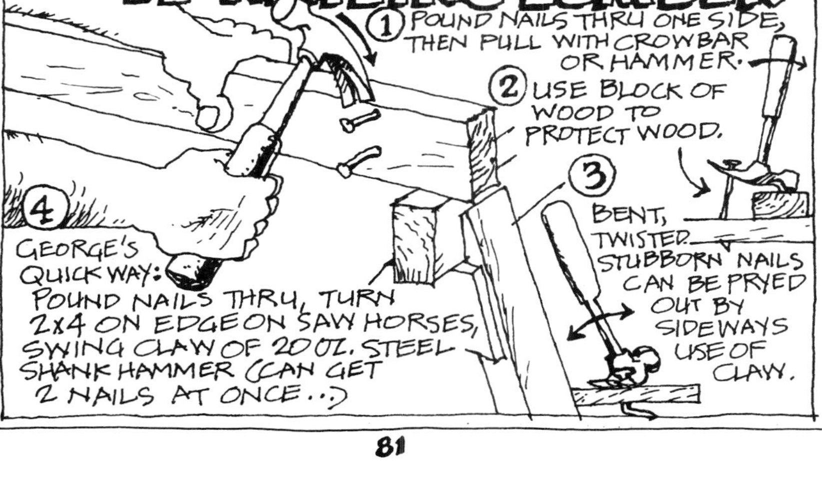

① POUND NAILS THRU ONE SIDE, THEN PULL WITH CROWBAR OR HAMMER.
② USE BLOCK OF WOOD TO PROTECT WOOD.
③ BENT, TWISTED STUBBORN NAILS CAN BE PRYED OUT BY SIDEWAYS USE OF CLAW.
④ GEORGE'S QUICK WAY: POUND NAILS THRU, TURN 2x4 ON EDGE ON SAW HORSES, SWING CLAW OF 20 OZ. STEEL SHANK HAMMER (CAN GET 2 NAILS AT ONCE..:)

Scrap sculpture, of short 2 x 4's and nails by Jerry Thorman and Bob de Buck. Although most of the wrecking information here pertains to wood, there are countless other materials that you can scrounge for building. Bricks, old adobe blocks, glass, broken up concrete, etc.

DEMOLITION ADDICT

by Eric Park

A time to build up, a time to break down: I began my career in demolition in late '71, during a break in the autumn rains. The sunlight that poured through the southern and western windows lit up my problem: a gaping hole in my floor caused by the leaking water heater that teetered now, with its quarter-ton bulk, at the brink of the precipice it had created. If I could not fix the floor, our remodeled Springfield home [see p. 52] might well prove valueless; and yet no mere patching operation would do, with the remainder of the floor in so jaundiced a condition. What was there for me to do but to leap down into the crawl space, taking my stand amidst the cobwebs, old marbles and wood scraps, and to let fly with the crowbar?

It was oddly satisfying work, draining and removing the heater, and exposing some eight feet of damaged, but reparable joist — the very skeleton of our home. I carved away the unhealthy portions of the joists, bolstering and levelling the remainder with new two-by-six planks — and thus found myself well advanced in the second phase of a reconstruction. Demolition had been the first. Five days later, we rolled our new heater into position over the resurfaced floor, and celebrated by taking a shower.

In the months that followed, I plastered my walls by first tearing out the loose plaster bordering the gouges and cracks; I tore out the old bathtub, and yards of its plumbing, in revamping the bathroom; and the old linoleum flew from the kitchen and utility floors like quail before one of my father's old hunting dogs.

As I took an axe to the two maple trees that had taken root beneath our foundation, I began to wonder if I hadn't gotten out of hand, become a demolition addict. But my wondering only led me back to my reasons. I blamed the wind that had blown them there, the maples, where they could only crack my foundation; the house would have quickly followed, and a forest would have been ravaged to replace it with a duplex.

I battled back the black berry bushes, out by the alley, and found myself face-to-face with a demolitionist's dream: a 20-by-10 unused driveway slab, abandoned when someone erected a new, two-car garage. Debby, my wife, thought I had lost my senses as I prepared for what proved to be a three-day battle. Was I really going to bust rock, she wanted to know, without being first sent to prison? With sledge, pick and shovel — crack it, break it loose, haul it off — I set out to lower the surface of the earth by six inches. By now it was August, for sane men the time of harvest — but for half an hour I swung the sledge blindly, beneath the full fury of the afternoon sun. Had not that earlier Renaissance begun with the dissolution of the monasteries, the trampling underfoot of the old order?

Soon I was able merely to lift the sledge, and let it fall of its own accord. I discovered that this was all that was needed. Two, three blows a foot from the edge, as many more with the pick, and a square foot of concrete would be dislodged from the rest. Children gathered to watch, begging for a turn. I obliged as my condition worsened. A ten-year-old named Greg proved his mettle with the sledge; I followed with the pick, and the rest of the squealing mob raced to pick up the chunks and chips. Six and seven-year-olds dragged, staggering, lugging and hauling blocks as heavy as themselves to the pile. A neighbor, excited by the tumult, materialized in a jeep with trailer, to haul the stuff away for fill. With my blessing. Here was neighborhood solidarity through demolition!

And yet, this is probably not the place for the discourse on demolition that has moved me to the typewriter: Demolition as a Lifestyle, as a Way, as a Path. I have had dreams of this country's Marines, hardened to the point of derangement by thirteen years of war, being set peacefully to work in tearing up the freeways, the high-rise apartment buildings, the bull-market stock exchanges with their ticker-tape displays of profits in war materials. Hippies and hard-hats levelling all the oil refineries, dismantling all the automobiles, trucks and busses, and leaving nothing but bicycle paths. And when the paths had been completed, and each issued his ten-speed; when the last filling station had bitten the dust, and been converted into a corner minipark; when the last concrete three-story parking garage had been reduced to rubble, and buried, and a meditation garden erected in its stead; when the last television set had been melted back into metal and glass, the last *Playboy* pulverized and pressed into bathroom tissue, then a generation might look back upon its brief traffic upon the stage with a smile — I was going to say "pride," but such a generation would have become too wise for that.

This week I've torn down the shed in my back yard, using the recycled lumber from the roof to begin a fence. We're planning to turn the reclaimed space into a Japanese winter-garden, with shrubs and stones. The concrete walk along the side of our house has got to go, as does the plastic breakfast bar in our dining room — they'll make room for an archway, broom-closet and vegetable garden. And even Debby has gotten into the act, stripping the half-dozen layers of paint with which some fool had coated the wood panelling in our kitchen, and restoring the original surface.

Thoreau tells the story of the deacon whose dried tapeworm was auctioned off, in the settling of his estate, when it might better have been burned, together with the rest of his effects. Better that the deacon had been into demolition! What but demolition was the Flood? The destruction of Sodom and Gomorrah? Might not the Hebrew prophet have had in mind, as he led poor Lot into the hills, that each of us could do with a nice moral purge now and then, and no looking back? Even Milton, in his *Areopagitica*, as he calls for the "liberty of unlicensed printing," recommends that most books be burnt — asking only that the printing of them be unhindered, and the public given its rightful choice in the burning.

Let us, then conceive of demolition not as a means toward worse ends yet, but as a cleansing, a putting the stop. Let us have no more Urban Renewal, but Urban Termination. No more crap, no more skyscrapers. This one's patched together with Scotch tape and bailing wire: let's pull it down, and go looking for the earth.

BASHO DEMOLITION

by Martin Bartlett

Martin Bartlett, who built the Pod described in Domebook 2, *more recently became involved in a wrecking project with friends in Vancouver, B.C. Here is Martin's story of Basho Demolition:*

Ed Miller is a junkman. He has an eye that can spot a brass coupling in a heap of rubble the way a biologist spots rare mushrooms in the forest. Years before the word recycling was coined, Ed was making a living out of trash. Now he presides over Ed's Bottle Depot on a back street in Vancouver, a local potentate of junk.

Mo van Nostrand was trained as an architect. Over the years a series of reality-experiences [called by some disillusionments] have made him value the company of junkmen as much as that of his fellow professionals.

Shooting the breeze around the Bottle Depot, Ed told Mo of a demolition contract that could be had: three old houses and a five-suite apartment building to be torn down to make way for Brutopia Towers. Mo bid on the contract, which involved a lump sum payment on completion of the work as well as total possession of all lumber, fixtures, appliances, appurtenances, relics and the like, the site to be levelled to bare ground within a two-month period. Mo was in the demolition biz.

Workers seemed readily available, attracted by the idealistic and aesthetic aspects of the game as well as the prospect of making a few bucks. We knew, of course, that certain physical hardships were involved, but felt confident that we could cope. Out at the site the first day, the undertaking looked quite large. We fell to with crowbars and sledgehammers, occasionally pausing to watch the flight of birds over the Fraser River and feel the chilly November easterlies blowing down the river from the mountains. During breaks we made coffee in a vacant apartment and read Chuang Tzu aloud. The Mysterious East sensibility seems to be with us a lot in B.C. [Mo's wife, Sonja, is a scholar of ancient Japanese poetry], and when Gordon coined the name "Basho Demolition" there was general applause. Not to mention that the contractors for whom we were demolishing were two unlikely Sicilian gentlemen named Zen and Aquilini.

You might think that such an airy enterprise might not be as efficient as a professional wrecking outfit, and it is true that as the weeks went by a certain lassitude slowed our progress, coffee breaks becoming longer and the time spent with wrecking bars less aesthetically appealing.

Occasionally Ed would come out to the site and poke around, pointing out to us that this or that item was worth a buck or five bucks, showing us how to get the lead out of joints of sewer pipe, how to burn the insulation off wire to redeem the copper and other tricks of the trade. As we tore the buildings apart we stacked the lumber on the site and put ads in the paper advertising it for sale, though during the December drizzle buyers were at first few. The amount of trash, broken shingles, split siding, and heaps and heaps of lath and plaster was amazing and appalling. The nadir of the experience came for me when after several days of rain it froze, freezing masses of soggy plaster to the floor from which it had to be laboriously chipped.

Certain laws of the business became apparent: firstly, clean up as you go. At the beginning we were taken up with the excitement of felling walls and toppling chimneys in the Cecil B. deMille manner [Samson and Delilah...] Later we would have to cope with the totally tedious drudgery of cleaning up the mess.

Second law: have a reliable truck and a place to put the stuff — as the time for the contract ran out there was still a great deal of lumber unsold which had to be removed from the site and is now filling up selected Vancouver backyards, somewhat to the inconvenience of owners of the premises.

So, hampered by inexperience and a tendency to goof off, we did not make as much in hard cash as Ed had glowingly predicted. Nevertheless we did salvage an amazing amount of material; good well-seasoned fir cut from virgin timber sixty or seventy years ago, doors and windows of all shapes and sizes, refrigerators, stoves, toilets, fireplaces, much of which has already found its way into the habitations of people building their own funky alternatives to Brutopia Towers. It is spring and Mo is no longer in the junk business, but is in Japan, having made his contribution to our local mythologies. Ed still presides among his bottles and bits of plumbing. Among the ruins a cherry tree is in bloom...Basho, where are you now?

And we was loading hay on the sleigh and all of a sudden I looked back and there was cossacks. That's the fiercest thing you can see. They had those whips and if they hit you, they could knock your head right off...

CAPTAIN BILL

For some time people had been telling me about Captain Bill, an 82-year-old house-wrecker in a nearby farming town. Finally, looking for some used redwood, I took a trip to see him. Out in the yard of piled lumber, old bathtubs, used windows, doors, plumbing supplies was a lean energetic old man with sparkling eyes, feeding chickens. Within a few minutes, as we loaded my truck with nail-studded lumber, he was telling me how he sank a Japanese submarine off the Golden Gate in World War 2 and I was listening, amazed at his vigor, astounded at these tales of his adventurous sea-going life. In subsequent visits I learned that Captain Bill was born on Christmas Day, 1890, on the Estonian island of Muhu, had been at sea since his teens, and in his early years sailed on the big clipper ships.

After he retired from the sea, Captain Bill became a wrecker, tearing down chicken coops in the country around his home. The town where he lives used to be the egg center of the West, with chicken farms throughout the surrounding countryside. But in recent years, as the hormone-fed automated chicken factories replaced the farms, chicken houses and coops have been abandoned. (Now, under electric lights in Oakland, chickens' eggs roll from their small cages onto a conveyor belt; that's why Safeway eggs taste the way they do.)

His storage area consists of a large yard, and about 8 chicken houses. In one are doors, in another plumbing supplies, another windows, etc. Sheep wander in and out, there is a thriving garden, and chickens lay eggs in old toilet bowls and bathtubs.

One day we went to interview him, and I tried to get him to tell Jack and Rod the story of the sub, but he was off into his earlier past. The submarine incident happened on Dec. 21, 1941 when he was captain of the ship *Tahoe* which took Oakland's garbage out to dump at sea. About 15 miles out of San Francisco Capt. Bill was at the wheel and noticed a troopship with 3,000 men heading out to sea. Between the *Tahoe* and the troopship he noticed a branch sticking up from the water, then suddenly his eye caught the reflection of glass in the branch. Capt. Bill said it looked like a sub waiting for the transport (Japanese subs were suspected responsible for the sinking of a Matson liner on its way to Hawaii the same year). He called *full speed* to the engine room and steered for the periscope. It disappeared, and as the ship passed over it he heard a thud, the *Tahoe* stopped, then started up again. The ship was put in dry dock in Oakland upon return, as she was leaking. Naval intelligence discovered that 40' of the keel's shoe had been scraped off. Later a torpedo was found on Baker's beach, and Capt. Bill speculates the sub was firing at the time of collision. Three days later three drowned Japanese sailors were found off Pt. Reyes. He says there was no publicity at the time because the military did not want to alarm people.

The day we went to see him he told us he'd just sold the last of his doors and windows, and was retiring from wrecking, and was going to do nothing but gardening from now on. I asked if he worked all day, and he replied: "No, not after dark." We turned on the tape recorder part-way through the conversation, standing in the garden on a sunny morning. Each of Capt. Bill's stories has off-shoots and diversions, but in the end you realize they were all relevant. William Vartnaw, now 82, and still alert, saw Hawaii before it was ruined by white men, steered a ship through a hurricane in the Bay of Biscayne, and as recalled in the rambling story below, settled a revolution between Russia and Estonia when he was 15 years old.

Capt. Bill: You read the papers, do you?

Lloyd: Yeah.

I been at it since I was 15 years old.

Reading the papers?

Reading the papers. And I settled that revolution between Russia (*pronounced Roo-see-a*) and Estonia when I was 15 years old.

You settled a revolution?

Yeh.

How'd you do that?

That's another story again. We're going from one story to another and never finish any of them. But that's what it is. Now, when Catherine the Great was Empress of Russia, she was a German Duchess, you know. And when she married Peter the Third she was the same as all Germans. They thought they was superior to any other race. *Ueber Alles,* they call it. Over everybody. She was married to him and he was a good one, a good emperor like grandson of Peter the Great. He want to bring Russia and Peter even with the others. And that's the time they occupied Estonia, Latvia and Lithuania. Those people were from the West, west of the Slavs and they, like Estonians and Latvians, they were the most educated in the world. And they still are, excuse me telling you that. (Laughter.) Estonia University is built before the United States was discovered. And now they are taking out your heart, putting it on a table, and put the new parts in and put it back again and you wake up and you don't know the difference.

What's that?

Heart operation. You can take it out. There's a woman specialist. I guess you read that was in the paper here. Well, that was in Estonia. Estonia is a small country...one million and a quarter in population and they run the Russians, see?

Oh yeah?

Oh *yes.* One million Estonians run 240 million Russians.

Well, what about the revolution?

Revolution? Now that's it. That was good idea you remind me, otherwise I would tell you about Estonia running the Russians.

I graduated from a church school. My father was curator for the orthodox church, that was, the government church. Catherine the Great, she was Lutheran, that was the German revolution. And she ordered all Army officers, Navy officers and Merchant Marine officers, all from Germany. The Russians were layed out just (*pronounced yoost*) like a waste. But again she just took it for a nationality because in Russia there was very few high schools. But working people had no chance at all. Well I got into the church school. There was everyt'ing taught in Russian and Latin. I was reading all the Lent, six weeks before Easter, every Wednesday and Friday in the church. The country was occupied by Russia, you know, and they get day off if they go to church. So the church was full every Wednesday and Friday. That was the days they got off.

Everybody read in church?

No, only the ones that are going to graduate that Spring for the priesthood. Only three. Thirty start and three are graduated. No matter how smart you are and how modern just the three very best ones. Because they don't want to have too many priests, you know. For the priesthood you have to have a diploma and three of us was elected. And, of course, I was always the best. Everywhere I ever done anyt'ing, I been the best. All my life.

continued

Did you become a priest?

I was gonna be. But I tell you how I change. I had photographic mind. When I see somet'ing written, it is just like I make a notebook. When I was sailing Captain on the coast there and all the officers' telephone numbers you have to know and the lumber yards you have to know them so you can call them up and I never had a pocketbook for the addresses or the telephone numbers or nothing. I even could tell the people who want to go and telephone to my friends, I could tell them what the telephone number is.

Anyway, every grade I was in, I was always the best. And that's the way that I get the chance to become a priest. But I was too far ahead. I graduated when I was 14 instead of 16. See? So I couldn't get in again. My father took me to the city and they told him he's too young. I got to wait a couple of years. I couldn't get in there, you see. You see, religion is the biggest kind of fraud in the world. There is nutting that is as dirty as religion is. Telling people a story written by some old Jews. And Jesus Christ or the angels or cherubims or seraphims and holy ghosts, all that is the biggest imagination they can put in the human brains. (Laughter.) Well, you take it or leave it, you know, it's up to you, to decide. If you're religious I won't say if it's good on the one part. But it's bad for other part because they take you for a sheep and lead you. And that's it. And if you don't do it, you've got to pay. Like the Catholics, they're taking 10% off. In war time they took it off my son-in-law, Mike. He was a Polish descent and he was a Catholic. And they just tell him well, you're making good money now and we want at least one dollar a day. And he paid it. And all the Catholics that were working in the Navy Yard. They all paid thirty dollars a month, for the Christ.

Well, anyway, I got to go back to that reading in the church again: When Lent came we got the books. It's all Latin. We didn't understand not a word about it. But then the teacher, he just told us what it meant. It was not'ing holy about it whatsoever. (Intoning): *Scode, Kadi, Ipernate*. And all this. *Scode* means the cattle herds on the hillside and the kittens are playing in the sandbox under the sun and you have to then make your voice that way that when you go with that *Kadela,* that incense. And you go to the altar. That was my job. And then you pray. I can pray from morning to night without thinking. I tried it yesterday. (Laughter.) I didn't go all the way, but I still remember.

That's the first part. That's for the God. And then after you got with the God then comes the Jesus Christ, it's on the right of him, you know. Then the Holy Ghost is on the left, those three. Then comes Saint Mary. But they forget Joseph...they never put him anywhere. (Laughter.) And then you have to make your voice that those fellows will cry and that's a job, and you do it. You've got to change your voice so that you're almost crying. That the God is right there and we are nothing but pagans or criminals. And you beg them to forgive you everyt'ing just like you forgive those who do unto you. See? That's translated. And it works.

And in the Spring we have to go to pass the examinations but that's just like passing here. Same as they do it in Air Corps and everywhere else. Till some of them get caught cheating and then most of them...I don't know how they do that but I know some of them got kicked out because they're copying, looking over someone's shoulder. I was a teacher in navigational school in Christinson Navigational School on Market Street in San Francisco 1918.

How did you learn to be a navigator?

Well, now...well, we're nowhere again.

Well, we're never going to finish that one anyway.

I tell you I went three years to Navigational School in Estonia.

After trying to be a priest?

No, I couldn't be a priest because I had to wait two years. I got to stay home. All the rest of them went to school and had a good time and I had to stay home and make nets. For the fish. We do all our own fishing. A little island in the Baltic Sea and we do all our own work, you know. The net and everyt'ing is made, shoes, clothes, everyt'ing is knit at home and there is not'ing imported except sugar and coffee. The rest is all grown on the islands and there is no doctors, no banks.

Is any of your family still there?

My family most got burned up, you know. When Germans occupied it. My nephew I got a letter from two years ago. But I wrote another one, but I didn't get no more answer so I guess they're either here or there, see. Russians don't want Estonians to communicate too much with Americans. And Americans don't want us here to know how the people do there.

Why?

Because we are so far ahead.

In Estonia?

In Estonia and Latvia, too. We are the two most educated nations in the world.

You mean ahead in education. And what else? The standard of living? Quality of life? They don't have as many cars and electricity.

Well, I don't know about it, because I just got out here. Russia is exporting 150,000 automobiles every year that's manufactured in Russia. But they're supplying all these countries like India, Pakistan and those countries. They are their customers. They are supplying them but they don't want it themselves for two reasons. They got an apartment system. Where it is all pre-manufactured. They are just putting them on top of one another and go up to 14 stories. But they start it from the top and come down and the apartment is already 13 stories and there is not'ing in the basement. I got it on a picture here. I can show it to you. But it takes us all day if I talk at it. Well, anyway, where I was now?

Well, when you were making nets. You couldn't go to the church, and you were making nets....

Yeh, well my father and two brothers and two sisters they were accordian players and dancers. My father could sing better, I bet anything, than Caruso. He was a County Clerk and he had a big farm, old farm. And he was a County Clerk and he was a curator of a church and practically, he could just tell all about the island, what's going on. But he like...he drunk my share, too, you know. (Laughter.)

And then he'd sing?

I have never drink beer or whiskey, and he never drink water in his life. Not a drop. Water is only for washing, as far as that's concerned. He was a County Clerk and every Wednesday have to go to Courthouse. He was not the secretary. He couldn't write Russian, but he could sign his name for a passport, that's what the job he had. And on the way home there's a roadhouse, so he went in there and by the time he got through the horse outside with sleigh was cold and he had to get home. And all of the horses he bought at the county fair. He bought the finest horse there was. And, of course, that fancy horse when he got cold, he showed him how fast he could go and the sleigh overset. So he was in a ditch and his arm was out of joint and he was singing and you could hear him miles away. So that's my father. He never worked a day in his life. (Laughter.) That's right.

He had so many jobs and a big farm. We had a dutch mill. We had an up-to-date blacksmith shop. All the people around there didn't have it and then we had a grain supply of our own. For the years to come, you know, sometimes like last year, say all the crops all went and the big hail storm came and drove through Russia and killed all the crops. And that has happened before. So they always keep the one year's supply ahead. And we had them and then when we didn't use them the people around there they just come and help themselves with a sack full, take it to the mill and in the summertime when the harvest time, then they came and do the harvesting. Old man was in the cellar drinking beer. But then, I was going to tell you how I settled the revolution.

Oh yeah, right.

Catherine the Great, she was a German princess and she liked Germany. And when she got all the officers and all the army training and all that done by Germans, she went to work and did away with Peter III. Nobody knows what happened to him, but he was dead. And she became a real prostitute. She had new ones every night. Even a band leader, he was nice looking and had a good uniform and he got a job for a week in the palace and when he came out he was a captain. And that's the way all this

was done and then she was going to give Estonia and those islands in the Baltic to Germany. And that's where the trouble came. All those people, those officers and Army and Navy, they got thousands of acres of land as their homesteads. And free. And took almost all of the islands where we were. We only had a little island left and the poorest part. But the best part, all around the Lutheran church and all the German barons, they had the best part of the land and then, of course, the people had to do the work for them for damn near nothing.

And then Catherine says this way. "You Estonians you have no title, none of you get title for your land." They done like they did to Indians here in America. You people just went and moved in a little creek and a little more and more and that's the way they start. And nobody had any deed for their land. My father had about 2 or 3 thousand acres of land, too. But he had nothing to show.

So Catherine says this way, "Everyone of you that wants to have a title to the land got to pay 50 rubles a year for 100 years." Then you can get a title for your land.

Now the time went and there was many czars already and when Czar Nicholas the Second, Alexandrovich and he was the king then and the hundred years was over then they wouldn't give us the deeds. And people say, well, what the hell we're suppose to pay a hundred years and it passed three years ago. And they wrote to the czar and the county and everbody and no answer. Just keep on paying. So anyway then they came out that they give us free land in Kamchakta or Sahalin. Out in the far east. Anybody want to go there they get free fare to go over there and settle down. Then we know what was going on. They want us out of here so the Germans could have the islands. And the islands is one of the best places in the world. All the Europeans go there for a summer resort. The mud baths are there that heal all the kinds of rheumatics and everything. They are there all summer. And a nice place.

So anyway, there was a thing coming up. Times just like we got here. Russians had no work. Nothing to eat. 1905 in January, the priests organized it and they was going to see the czar, to do something about it because people are just starving. That they could at least stop taxes. They built the St. Isaacs church in St. Petersburg and used eight tons of gold for painting the cupolas on top. And people are starving and you just got to go and cross yourself and starve. And they are putting gold all over the inside every day. I got a picture of everyt'ing. And the priests went together and they had the czar's picture and they carried that with them and the other one with the cross, you know. They went down to the winter palace. And the czar ordered his body guards out and shot them 150 priests and wounded 350. Then the Russians got mad. And they went to work and shut down every god damn thing in Russia. And that was in the middle of January and then Estonians wake up too. Estonians went to work and those German landlords they had all the thousands of acres of land and all the purebred horses and cattle and they went and burned all the god damn barns down and they had not a bit of food for their horses or cows. And then you hear them coming. Holy Christ!

One morning, we was loading hay for my sister, she had a house, her husband was an engineer and he was out most of the time, but her and the son they were home and they had a cow and no hay, so we supplied the hay. And we was loading hay on the sleigh and all of a sudden I looked back and there was cossacks. That's the fiercest thing you can see. They had those whips and if they hit you, they could knock your head right off, you know wires, sharp wires with a weight on it, and we all know that when they are there, there is going to be trouble. My father was on the other side of the hay and I see him disappearing, but I couldn't go nowhere because they were right there, you know, they could have swing around and that would be the last of me. Then one of them came off the horse and say to me, "Can you talk Russian?" I say, "Yes, I talk Russian better than Estonian. I graduated from church school in Allam." He says, "Yes, I was told that you talk Russian because nobody here in Estonia hardly talk Russian." They talk German and Latish. But no Russian. Only the religion forced, you know. And then he says to me, "Where is the revolution?" Well, I was thinking for awhile what to say but comes some kind of idea in my head, that helped. I told him this way that we have no revolution to speak of. But I tell you we've got trouble here to straighten up the land deeds. Catherine the Great put on the land tax for 50 rubles a year for 100 years and it's already passed now 3 years. And is no relief, we are still paying it. And agreement is that it should be paid in 100 years and then freed. Then they give us land title. Well, he says, "You wait." And he gets on his horse and he get the leader, you know, he come back and he's got a book and "What's your name?" I say my name, and he wrote it down and then he said that, legally what we are demanding is the right thing. We demand the title of our land that is paid for already, three years ago. Then when that fellow, he took it all down and he says, "I think you people are right." He says, "Right is right."

Then I tell them about the Germans: They've got hunting dogs, they've got horses. They've got ones that chase the deer and all the wild animals and they shoot them and they are tramping our crops all to pieces. Bugles are going all the time. (Laughter.) Yeh. They're having a hell of a time there, you know. They want us to get out of there, you know. The idea was that they were going to take over the island, maybe all Estonia. Anyway, the Cossacks say, "That's all right." But one thing more, he says "Have you got any ham?" I said, sure, we've got lots of hams. So he came into the storeroom and we had lots of smoked hams. Killed the animals in fall and then smoked everything. Salt them and smoke them, ready to eat. And by gosh he said, he took four big hams. He said there are no restaurants here and I said, "All right." We have plenty to eat. By gosh, a month later, we got the deeds.

•

When I was in Valparaiso I went to work in a sail loft sewing sails. An American clipper ship came in, beautiful. *White as snow*, the bottom was this color of paint (red), really looked good. And then they say that Capt. was going to take a new crew. So everybody, the beachcombers they was in there, they had their books on sailing and all came to the captain and he looked at them and "very good, very good" and the man that owned that sail loft, he wanted me to get a job. He went and told that Capt. Peterson, he's a good sailmaker. He looked at me and said, "too small." Oh, I got *disgusted.* I was disgusted that they wouldn't sign me on the American ship. But then he told me again, that fellow, don't be discouraged, there's two more coming. Then about four days later I go to work and there comes a fellow. I would say he was an American because the Americans they had their hats sharp pointed on top. And he came and the boss pointed at me. Aha, I say, I think that's an American all right and he wants a sailor. And he says, "Come over. You a sailmaker?" I said, "I was assistant sailmaker on the German boat." He said, "That's just what I want. I send my mate here tomorrow morning and you bring your clothes and you go on board the ship and that's that." Oh gosh, I put my needle right away and go home and get ready to go. Next morning I went there with my clothes and everything and I tell you, you would never believe how I feel when I get on that ship. The people. The captain and his wife and his little girl, eight years old, Catherine. And, of course, I get on board. I was well dressed. Now, I got to tell you another thing again. When we got wrecked in that ship there in the Bay of Biscayne....

•

...the Hawaiian Islands were really paradise, was very nice in those days. Almost unimaginable such a nice place it was.

What year was that?

1912.

Every night we were in the bar, there was American missionaries telling about America. Between every word the Hula Girls came out and the band played every night like that. Oh, it was beautiful, it was really...I was gonna stay there but the captain's wife wouldn't let me. Captains always had their wives along on the sailing ship. She wouldn't let me. She said, gosh she said we want you to go back to America....

Peter Warshall, who prepared the next three pages, is a naturalist who has lived with rhesus monkeys on an island in the Caribbean, studied baboon language in Kenya, and helped set up a wild horse sanctuary in the Pryor Mountains in Montana. Lately he has been working on environmental impact in Marin County, California and on recycling sewage systems and land retention.

EARTH SHELTER

by Peter Warshall

The monarch butterfly on the right is mimicked by the viceroy on the left.

ANIMAL SHELTER

An old Australian aborigine woman lives in the desert. She wears no clothes and has about fifteen pet dogs. Together they wander from waterhole to waterhole. At night, when temperatures drop close to freezing, she lays down and the dogs pile on top of her. In the morning, she is sleeping in among them and her smiling face appears as each dog gets up, stretches and walks a short distance away.

Place this book upright and back across the room. The more disruptive design on the "real" zebra makes the zebra disappear sooner than the "pseudo-zebra" with non-interrupted stripes. This is one basic principle of animal coloration.

The surface area of the body is the best way for mammals to control their body heat. In rabbits, the ears act as radiators of heat. By perking up the ears or keeping them close to their body, rabbits can rapidly increase or decrease the amount of their body surface surrounded by air. Northern rabbits want to keep all the warmth they can, so their ears are small (above right). Desert rabbits like the jack-rabbit on the left want to get rid of as much body heat as possible. They have very long ears. The limbs of the jack-rabbit are also long and thin compared to the arctic hare. The middle rabbits come from latitudes between the American Southwest and Alaska.

Varieties of shelter.

Consider the fawn in the New Hampshire woods. Late spring. I'm stumbling around the fallen branches, trying to reach the blueberry scrubland. (Mt. Sketukechee.) I'm looking among the yellows of the leaves and the new green of the winter and there's the fawn. Now, the thicket is definitely shelter, but so, equally is the stillness of the fawn shelter. Even the lowered head flush to the ground and tucked-in limbs and the flattened ears are shelter. Here, habit is habitat and habitat, the inhabitant and the fusion means shelter.

Besides the fawn, there are probably 4,000 other species in the same woods, and this number would not include some of the unidentified parasitic wasps, saw-flies or butterflies. The diversity and effectiveness of shelter makes most woodland creatures totally invisible. A forest is a series of curtains with animals hiding behind bark, inside the gills of mushrooms, under the wings of birds, inside the stems of plants, among the leaf litter, inside dead animals and cow pats and underground. Almost any natural object will show itself to be a "dwelling" — a chrysallis, a crack in a rock, an abandoned bird's nest, an old tire.

On the same mountain, reproducing on milkweed plants, are monarch butterflies which taste bitter to birds because the monarch caterpillars eat the milkweed and ingest a bitter-tasting chemical poison. Also, on the mountain are viceroy butterflies which do not eat milkweed, taste delicious to birds, and look just like the monarch. The viceroy's masquerade as a monarch is its shelter. Although a blue jay's gourmet delight, the viceroy disguise means that most birds will avoid it, expecting a bitter monarch.

Nearby, in a cold storage plant, live house mice that are much larger than mice of the same species living in the house I lived in and the silos around the town. The temperature in the cold storage plant never rises above 10^{0} C. The large body size of mice in the cold storage plant is a shelter built into the body form. (The increase in size means a relative decrease in surface area which reduces heat loss and maintains body temperature.) Thus, these mice don't have to spend as much time insulating their nests, shivering, and eating for warmth because their body form shivering, automatically gives them what a shelter would provide. Similarly, in hot places like my house, the skinnier mouse dissipates heat more quickly and doesn't have to waste time opening and closing the entrance to its nest and licking its fur for proper air-conditioning.

Animals find shelter by masquerade or mimicry, by body shape and size, by camouflage and fur, by hiding in the complexity of habitats, and by habits themselves. These various ways in which animals are sheltered serve them: shelter is safety (hiding); shelter is a place to reproduce; a place to sleep (dream); and a way of economizing (conserving energy).

To keep heat or lose it easily, fat can be stored in special places on the body. In the camel, for instance, all the fat is stored in the humps. This allows heat to easily escape from the skin in all the other places. Cold dwelling mammals like the whale spread the fat equally over their whole body. This fat-blanket keeps body heat inside the body. In humans, the bushman with large fatty backsides store fat like the camel. The fatty fold that can be found even in the eyelids of Eskimos (the "Mongoloid" fold) is a human example of the "fat-blanket."

Dream shelter.

Porpoises in the sea and antelopes on the savannah do not have a shelter to sleep in. They rarely close their eyes and alternate short periods of sleep with longer periods of wakefulness. They live in groups so that at least one of the group will be awake at all times. Only the great apes (chimp and gorilla) and humans show a single, long period of sleep followed by a single, long period of wakefulness. The great apes make a simple, new nest of bent branches each night, but no permanent structures. The similarity between man and apes seems to indicate that even before man built fortresses, he enjoyed a certain nocturnal safety. This allowed longer dreaming patterns and re-arrangements of thoughts, speculations, fears, and hopes. Besides these three great dreamers, the sperm whale is the only mammal to have the same sleeping pattern as man and the apes. Notably, the sperm whale has no shelter in the sea besides its group and size.

Below is a short bibliography for this and the next two pages:

The Biosphere, A *Scientific American* Book, 1970, 134 pp. $3.25 paper. W.H. Freeman and Co., 660 Market St., San Francisco, California 94104.

Continents Adrift, Readings from *Scientific American,* 1972, 172 pp. $3.50 paper.

The Earth, Life Nature Library, by Arthur Beises, 1962. $3.95. From Time-Life, Inc., Time-Life Bldg., New York, N.Y. 10020.

Adaptive Coloration in Animals by Hugh B. Cott. $20.00 from Barnes & Noble, Inc., 105 Fifth Ave., New York, N.Y. 10003.

The Natural History of Mammals by Francois Bourliere, Third Edition, New York, 1964, 387 pp. $6.95 from Alfred A. Knopf, Publisher.

Desert Animals by Knut Schmidt-Nielson, 1964. $8.00 from Oxford Univ. Press, 200 Madison Ave., New York, N.Y. 10016.

Energy and Power, A *Scientific American* Book, 1971, 114 pp. $3.25 paper.

Creatures of habit.

Notice how few animals' ways of shelter necessitate building. In summer, a deer will molt and in winter, grow a thick coat. Most animal shelters are neither strictly "inside" nor "outside." The harsh inside/outside distinction is modern man's. Instead of putting on a sweater or even a coat inside the house, we would rather turn up the heat. We are all junkies on petroleum and "shelter" has lost its vivid power (think of tipis or hearths or yurts or longhouses) because of this oil addiction.

Rather than take siestas in the hot seasons (a habit in most hot countries), we insist on the nine-to-five working habit and therefore "need" air conditioning. While living on an island in the Caribbean where water had to be imported, I found I could conserve an extra gallon of water a day by working from pre-dawn to about 11 A.M. and from about 4 P.M. to after dark. This is how migrant laborers in the San Joaquin worked and I see no reason why insurance or advertising executives need other hours. We have addicted ourselves to time schedules that not only are unenjoyable and inflexible, but are wasteful bio-energetics.

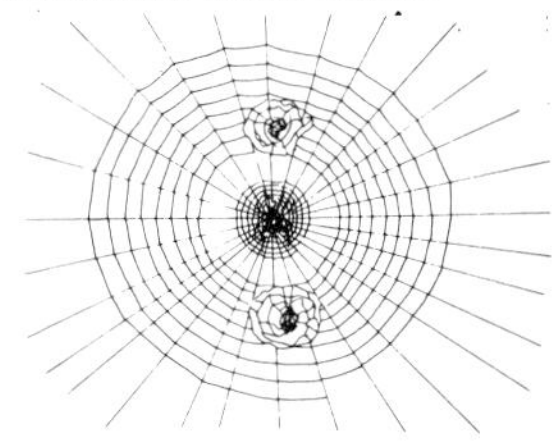

Many spiders fool their enemies by building "mimic spider sites" into their webs. This decreases the spider's chances of being hit during an attack from 100% to 1 in 3.

No natural shelter is an enclosed capsule. Shelters are more like active membranes which filter and select, screen and balance. The best illustration of the outside that is inside is the human lung. Here, "inside" the body is the exchange of "outside" oxygen. The exchange happens across a one-celled thick membrane and is an intricate balance between the amount of carbon dioxide in the blood and the amount of oxygen. Man has created *himself* as a specialized shelter for the inside exchange of outside air. Take the nose. The nose is shaped to help the lung-air interchange. The best rate of interchange (bio-energetically) occurs when the moisture and temperature of the "outside" air are identical with the "inside" lung. So, an African inhales in the Congo. The temperature and humidity are just about as warm and wet as the lung. So, the nose is short, and wide-open nostrils allow easy interchange.

An Eskimo inhales in Alaska. The air is dry and cold. His nostrils are tiny compared to the African so that minute amounts of air enter with each breath. As the thin stream of air passes through the nose, it is warmed and moistened by the membranes inside the nostrils and by the longer passage through the nose and mouth. An Arab inhales in the Sahara. The air is dusty, hot, and dry. The nose is long, as is the Eskimo's, in order to wet the dry air. The nose has large wide openings because the desert air is already heated. The nose is large to create an expansive surface area where water can be used to cool the incoming hot, dry air.

The body, in its shape, provides a shelter for the lungs and this shelter is made from the passage of the easily defined outside to the ambiguously "inside" lung.

A place to reproduce.

A beehive is the meeting place of one queen bee and as many as 40,000 other bees all trying to help her reproduce the colony. These other bees are totally dedicated to keeping the colony unpolluted and well-fed. Bees ventilate the hive with their wings and bring in water droplets to reduce the hive temperature during summer. They store pollen and honey by color in order to quarantine any contaminated food source into a few cells. They shit outside the hive and remove shit on any bee that shits inside. They control parasites and pathogens by adding bactericidal enzymes to the honey and by "ripening" the nectar. This reduces the water content of honey below the level in which yeasts grow (*Scientific American,* April, 1972). All this and more keeps a reproductive shelter reproducing.

These are the ways animals are sheltered. Meditation on almost any phenomenon will demonstrate its sheltering abilities. We have not discussed plants or the psyche, internal organs, body cell walls or English grammar — all of which have sheltering properties. continued

Moving **EARTH** *shelter*

There is a yoga of survival on this planet, in this universe, where you are. You may lead a "sheltered existence" with all the luxuries of twentieth century America, or you may not. Wherever you dwell, right now, it is the bio-sphere of Earth. The bio-sphere is a global envelope surrounding the enveloping crust of the Earth and surrounded by the protective envelopes of the Earth's sky. The bio-sphere is a House of Life for you as a particular confluence and arrangement of the sun's energy. This House of Life provides your essential shelter — the shelter that is there when you're *outside* the shelter you built, borrowed, or bought; *outside* the psychic shelter you imagined. fantasized, created, or thought.

The yoga of survival takes place in this outside shelter. To understand, to be aware of this shelter, to live in peace with this shelter, requires a tough meditation on the bio-sphere's swirling energy movements and grandscale cycles of transformation that are woven into our bodies. The return for this meditation is a certain fearlessness...you *are always* sheltered. There is a story of Tibetan monks who used to meditate on the energies of the Earth while sitting, wrapped in cloth, in the snow. Their meditation, when working, would make the cloth steam.

"Structure" of the planet.

First, I will describe the comfortable view of our planet. A view that sees us as the permanent residents in a stable architecture with solid foundation and immovable floors, living beneath non-leaking ceilings that don't warp unreasonably. This is the view I learned in the Hayden Planetarium in New York City and the view taught by Mr. Zwerling, my Junior High science teacher...This is *not* the tough meditation on the energy forces of the Earth but a kind of relaxed suburban attitude towards our planet that sees *form* as most important and makes the form rigid.

The Earth can be visualized as a series of globes, one inside the other. If you slice through the sky enveloping the Earth, through the Earth's crust enveloping the core and through the core surrounding the Earth center, the cross-section will always look the same. This is called radial symmetry. The spheroid crust of the Earth and the globes inside the crust's globe can be seen as the floors of the bio-sphere. The bio-sphere can be visualized as a room between the crust and the first global envelope in which no life is found. The bio-sphere houses all the Earth's living creatures.

The first ceiling with no trace of life is a region of the sky about 6 miles above the crust. Beyond this first global ceiling are a long series of surrounding spheres that make the bio-sphere look like one "box" in the chinese puzzle. These globes-of-sky surrounding globes-of-sky extend out 52,000 miles into space from the Earth's crust. These surrounding sky-spheres around the bio-sphere act as protective ceilings to this room where life goes through its changes. (See chart.)

The Earth's "onion" architecture would have no "top" or "bottom" if we did not conceive of the "axis" of the Earth and give names to the two ends of an imaginary, astronomical pole piercing the sky and Earth like a toothpick through a cherry. This great axis through the sphere lets us comfortably visualize the Earth as having a top and bottom and spinning evenly on an easy tilt.

"Structure" undulates.

The comfortable relation of the parts of the Earth as quiet global floors and solid ceilings with a single north-south axis has been described. Yet, actual measurements and intimacy with the Earth, its sky and interior deny this rigid conception and make the vision of our planet appear so limited, abstract and static as to be almost erroneous. *Now,* visualize the stable onion architecture begin to move, waver, shimmy, and undulate.

Like a peyote vision, the floors of Earth slide about and crack; the ceilings oscillate wildly and fold in upon themselves; the oceans heave as one huge mass of water, then subside. The whole surface of the planet, the bio-sphere and the sky-spheres are bulging and contracting like a breathing lung. The Earth is now a swirling mixture of energy masses, continually in motion, moving erratically, sending off and swallowing vibrations from its interior and from outer space.

Take a specific example, the single north-south axis that was described as a simple toothpick piercing a cherry. With the new measurements and intimacy, we find not one but many axes and not a stable axis but axes that quiver and shake as the Earth itself wobbles around.

The first axis we might focus on is called the axis of rotation. Right now, if we project the axis of rotation onto the heavens, it will point to Polaris, the north star. But, the axis of rotation did not always point to Polaris and will not point there forever because the axis wobbles. As the axis wobbles, its extension into the heavens shifts. This wobble, the *greater wobble,* is caused by the ever changing pulls of the Sun and the Moon on the Earth's equatorial bulge. The Earth spins like a slowing top. Another wobble occurs inside the greater wobble as the Moon pulls on the Earth's liquid innards. This is the *lesser wobble.* Both the greater and lesser wobbles show additional strange jumps in their wobbles as the wobbles themselves get jolted by earthquakes.

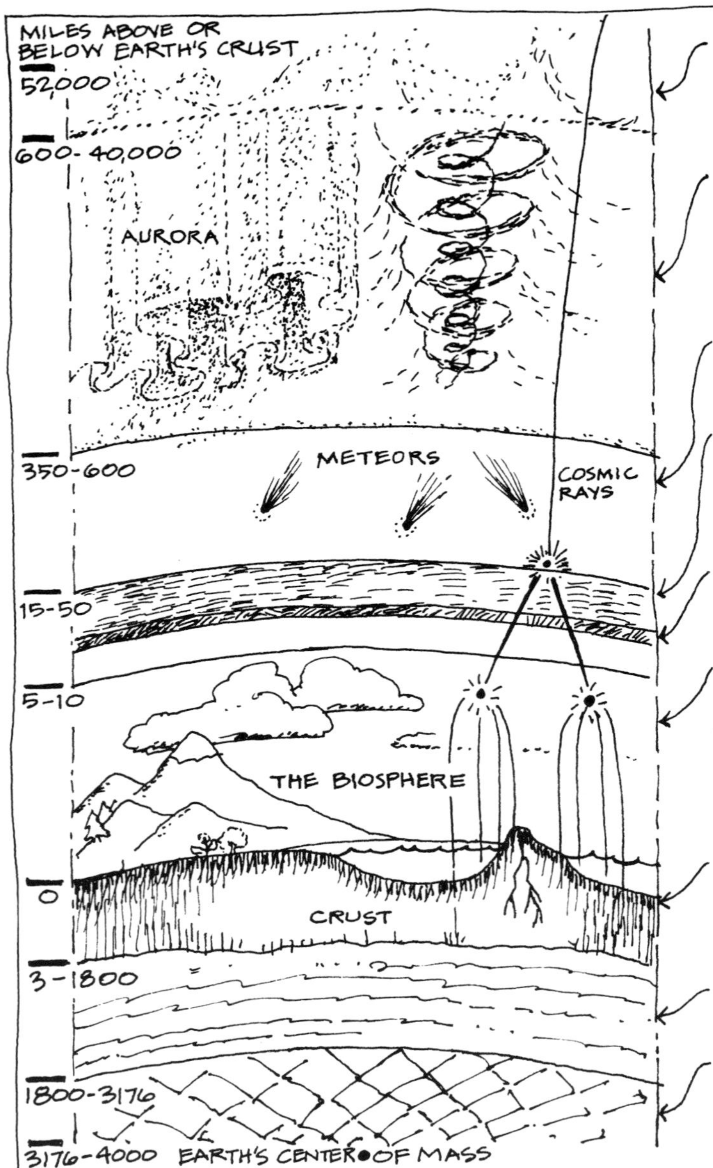

The outermost envelope, the last encompassing globe around the Earth, is called the TURBULENT ZONE. Too far from Earth's dynamo to harmonize with our magnetic field and electric currents, all the lines of force are out of line and sporadic. This is the enveloping roof where Earth's personal influence vanishes into interstellar space.

MAGNETOSPHERE: great arranger and protector. Trapping huge clouds of protons and electrons from the sun in petal-like lotus patterns of magnetism. This radiation (that is lethal to life) spirals back and forth along the lines of force. Only along the magnetic lines going down the axis poles can electrons and protons enter the bio-sphere. There, they create auroras. The celestial poles are embellished.

IONOSPHERE: bouncer of radiowaves, incinerator of meteors, and medium for the closest auroras and noctilucent clouds. The ionosphere is the global envelope that shelters the bio-sphere from extensive cosmic bombing.

The OZONOSPHERE veil screens out ultraviolet light which, unfiltered, would wipe out the creatures in the bio-sphere. The ozonosphere allows enough ultraviolet to filter to the Earth's surface and to kill many bacteria. The CITIES, with so much dust, stop u.v. light from entering in the neither too-much-nor-too-little amount. More bacteria. More disease.

SULPHATE CEILING: a thin envelope that makes rain. The sulphate particles attract water vapor and condense it. Otherwise, water vapor might leave the earth and be lost to life. This sulphate layer also collects fallout, stores it, brings it back down, when it rains.

TROPOSPHERE: clouds and carbon dioxide trap the radiation from the Earth, keeping this heat from leaving our planet. This creates the greenhouse we live in (See *Transformation II:* warm energy moving in masses). Typhoons, jet streams, ocean currents, winds and glaciers come and go, keeping the temperature of the planet within liveable limits: the heat engine of the bio-sphere. Cosmic rays are reduced from primary killers to secondaries. These reduced cosmic rays pass through our bodies ceaselessly and benignly. The atmosphere acts as the *cosmic ray shield.*

Molecules of the CRUSTS' top layer become excited by solar radiation and release longer waves that are the heat source of the bio-sphere (See *Transformation II*)...A thin layer of green-blue algae and land plants produce all the oxygen in the bio-sphere... The movements of the floor "tiles" cause oceans to grow and shrink; continents to crash and separate; plants and animals to diversify and wipe each other out.

The 4,000 to 8,000^{o} F temperature of the Earth's center does not burn our toes because the MANTLE and CRUST are incredible insulators. Earth's warmth is from the sun.

The Earth's magnet creates the magnetosphere that protects all living creatures from electrically charged solar radiation. The dynamo or electric generator of the Earth's CORE produces the magnetic field. This follows a law that electric currents always are accompanied by magnetic fields. The currents of the Earth's cores are started by a weak battery action and energized by great circulations of the Earth's liquid core.

Life can exist on Earth because the sky is layered and each layer shelters us from *too much* ultraviolet light or *too many* meteors or *too strong* cosmic rays or *too powerful* solar radiation. All these global ceilings are not solid shields but veils allowing only small amounts of radiation to enter the house of life. These regions of the sky are selective windows, screens, filters, sheltering membranes.

Life can exist on Earth because the bio-sphere contains "permanent" supplies of liquid water and "permanent" supplies of matter in all three essential states: liquid, gas, solid. The planet can be seen as a great fluid mixture which like all fluid mixes, has its heavy parts settle. As a rocket enters from outer space, we plunge into the denser liquids. It is a journey into an alchemist's retort: regions of the sky change from ether to air to water to earth to fire to dazzling crystal. For instance, within the 3½ miles of Earth's surface is 90% of all our planet's water. (*See States of Mass.*)

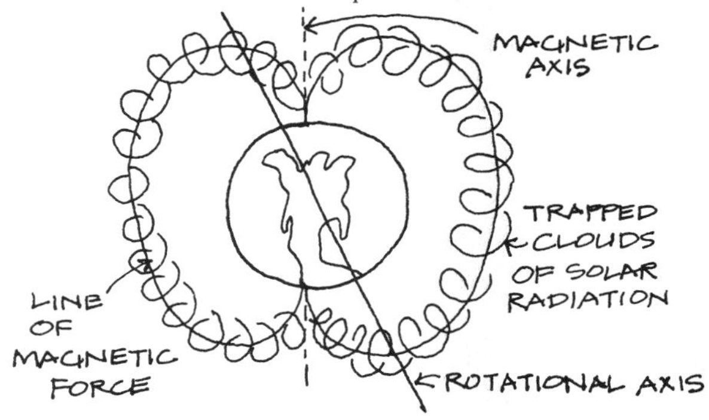

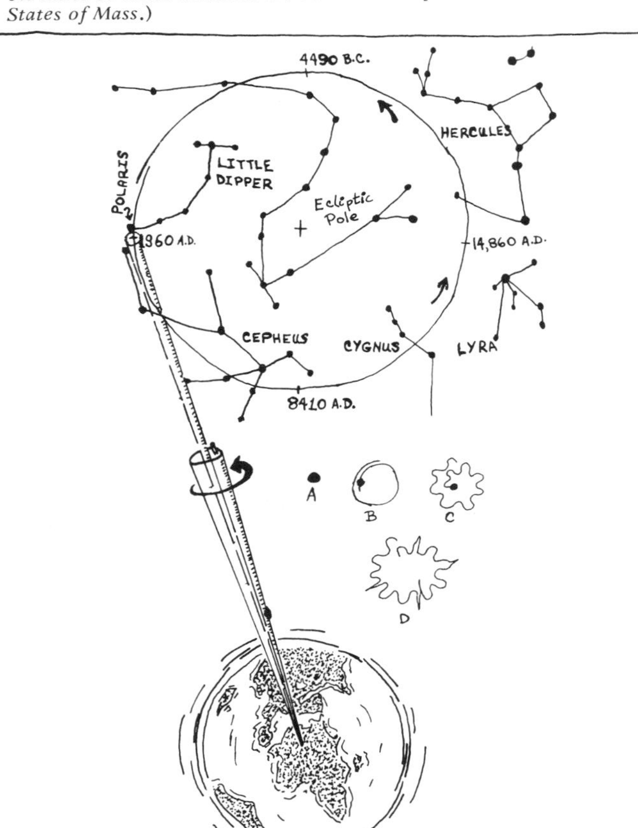

The greater wobble *of Earth completes one "circle" of its wobble every 26,000 years. The picture shows how the extension of the Earth's rotational axis onto the heavens causes the North Star to change as the axis wobbles. Egyptian astronomers (5,000 years ago) found the North Star was Alpha Draconis, not Polaris. "A" shows a stable axis; "B" shows the* greater wobble*; "C" shows the* greater wobble *plus the* lesser wobble *caused by changes in the moon's orbit; "D" shows the* greater *and* lesser wobble *plus irregularities caused by earthquakes.*

Add to this shakiness of the spinning Earth, a rolling or nodding as if the "hole" for the axis were too big for the axis. This "loose" additional wobble is totally independent of the first two. It occurs because we are not a rigid sphere but a deformable planet: the Earth's inner mantle is elastic, the oceans and the outer core of the Earth are liquid, the winds change the atmospheric density and glaciers pile up unevenly. Any object that changes shape so freely has a hard time keeping a steady spin and the erratic speeding up and slowing down of the Earth changes the location of the axis as the weights of Earth shift about.

Next, we notice that a compass will not point to any of these shifting axes. The compass aligns itself with the magnetic field of the Earth, not the rotation. It points to the magnetic north, not the center of spin. Every once in a while, in fact, the magnetic field of Earth reverses itself, changing magnetic north to south and south to north. When this next happens, it will spin your compass arrow 180 degrees.

Finally, there is a geographical axis that does not coincide with either the rotational or magnetic axes. There are other axes as well, such as the axis of angular momentum (see Dec. 71 *Scientific American*). Vainly, we tried to freeze the erratic spin and waverings of Earth in space. But, joyously we found similar quivers and shimmies, similar energy movements everywhere in the bio-sphere and within the interior of the earth.

Knowing that nothing, not even the Earth you are standing on, is standing still, is part of the Earth-shelter-yoga. The more you feel these vibrations (the whole bio-sphere breathing as a lung and exchanging energy like the breath), the greater joy this more accurate and truthful Energy Earth will bring.

· atmosphere is neither too hot nor too cold and fluctuates thin remarkable limits. Locally stormy and unpredictable, the th's overall temperature control is a model of Balance. At st, we've had a few reversible glaciers and the yearly typhoons. npared to the other inner planets (Mercury, Venus, and Mars) heating, thermostat and distribution system of the Earth is most grandscale delicate self-adjustment shelter in the solar em. Above, Hurricane Gladys stalled off the coast from les, Florida. Hurricanes are one of the best heat distributors Earth.

Transformation I: states of mass.

nding on a cliff along the Pacific on New Year's day, oked down at huge waves crashing 100 feet below. ocean was the color of a coffee malted as the wind-pped ocean made its mix of cliff and sea. The pleasures erosion had never been so delightful. Thirty million rs ago there was no fog and the sun shone on Pt. Reyes insula that was in Baja and, 50 million years from v, will hit the Aleutians. All these land masses sliding crashing over the Earth's interior mantle. The Pacific lf was getting bigger.

ocean's atmosphere (and the atmosphere's ocean) is great interior decorator of the bio-sphere. It is climate ostly the ocean/atmosphere's) that switches water from d glacier to vapor and back to liquid. It is ocean/ osphere that creates by erosion the soil that holds er. Especially compared to Mars, which shows no sion by water, no folded mountains and little shifting wind, the terrain of Earth is delicately crafted. There o living creature that exists without dependence on three states of matter and even where only one exists id-only glacial ice or solid-only hot desert of liquid-y sea bottom), there are dormant forms of life ting for the slightest transformation of matter to w life to happen. These dormant forms include gi spores and protozoa encasements and rare spiders springtails.

find happiness in the fog, in mud, and dust. This is ea not to indulge in criticism of the weather. As shores sluff away and deserts turn to meadows, we being entertained by the three states of matter that e the conditions for our life.

Transformation II: energy, mass.

rt, fast waves of the Sun strike the Earth, causing lecular excitement. The Earth, in its excitement, s off longer, slower waves. This is called the *nhouse shift.* The long waves cannot escape back Space. To our great fortune, they are absorbed by dust, carbon dioxide, and water vapor in the osphere of this ground floor apartment. What might e been the simple bouncing of solar radiation on and the Earth becomes the heater of the House.

start again. The warmth you feel on a cool, autumn ning from an outdoor fire is a warmth transmitted ct from the fire's body to your body — without nging the temperature of the air. (This is the heat of *iation.*) Radiation is the way the Sun's heat reaches cool Earth. It is a beaming, bouncy radiation with way of dispersing its heat energy on Earth. Its effect our warmth is small, though delightful.

v, if you poke in the autumn fire with a piece of , the iron will slowly get hot and finally burn your d. (This is *conduction* of heat by exciting molecule r molecule.) This is what the Sun's radiant light does he lowest molecular layer of the atmosphere: the l and water surface. The hot excited floor, in turn, ites the water vapor and air resting on it. The air and er vapor loosen up, get lighter, and begin to rise ically. (Getting high, in the House of Life, by an th in heat, is called *convection.*) Heavier air sinks to ace the rising tides of warmer air. A big circulation ts. Convection becomes the humidifier and air-ditioner of the House of Life — popularly known as ther.

say it all another way: The groundfloor envelope-m surrounding the Earth is a window to the Sun and eater for Earth.

The Earth is only a temporary shelter for living creatures.

The Solar System and our galaxy the Milky Way are just a temporary architecture of the Universe, allowing Earth's experiment.

The shelter of Earth is directly related to the sustaining Energy of the Universe.

Most scientists believe that the Energy of the Universe is being dissipated and that's why it's all so temporary. Specifically, scientists believe that the Universe is whirling to a state of quietude. The Earth will stop rotating. The Sun will burn out. Gravitational collapse will make the Universe fall in on itself. The best they (scientists, speakers of Universe Truth) can do is understand what slows down the collapse and channels energy around and around — postponing Energy dissipation.

Maize uses Earth. Squash uses Earth. Beans use Earth. Earth gets used up. The Hopi essentially held the scientist's attitude but believed that dancing (stamp on the Earth) and singing (adding energized, harmonious atmosphere) could renew the Energy of the Universe, and revitalize the Earth as it spent itself. In other words, they felt responsible for the Energy sustaining the Earth Home in the Galactic Home.

Making the vital energy that sustains our shelter was Hopi Faith. And so they travel together: Faith, Energy, Life, Shelter, Earth, the Milky Way, dancing, chanting, singing, and Breath.

We are average and that is our blessing. The sun is neither young nor old for a star. The Earth is neither too close nor far away for a planet. Our size is large enough to have remelted and formed a core of molten liquid, creating our magnetism, without being so large as to collapse the brittle shell. Our atmosphere is neither too hot nor too cold. The Earth may have evolved from a cloud of dust pictured above. The pattern of turbulence produced eddies where dust coagulated into the planets.

Blueprint for building the earth.

The Origin of Earth
The Origin of the Crust
The Origin of the Oceans / non-interstellar Atmosphere
The Origin of plants using Fermentative Metabolism
The Origin of Photosynthetic plants
The Addition of Oxygen to the Atmosphere
The First Metazoan Animal
The First Land Plant
The First man
The closing identity between human consciousness and the biosphere

When the Sun formed, a dust cloud condensed, now called Earth. About 6 billion years ago. Contraction of the dust cloud and radioactive heat cause the dense dust cloud to melt. Heavier metals sank toward the center of the Earth, leaving the lighter elements nearer the surface. The lighter elements solidified, trapping the hot molten elements in the middle of the Earth. So a billion years ago, the earth's floors became the crystalline core, the molten iron core, the warm silicate mantle and the cool newborn crust — separate floors though still (as today) burning and moving and sinking and rising.

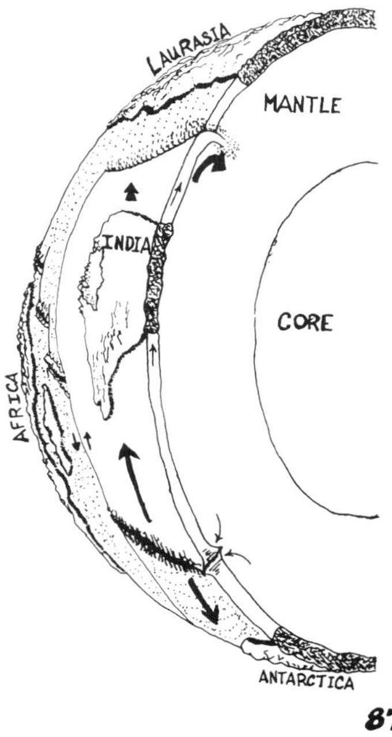

The crust of the Earth is made of a series of "tiles" that move all over the surface. Above, India on a rectangular tile has broken away from Antarctica. The "tile" with India on it drifted northward and crashed into the southeast border of Asia. The tile with India dove under Asia creating the Himalayas, which are two tiles thick. The tile with India was moving about 2½ inches a year. The leading edge of the tile is "eaten" by the mantle. The trailing edge is being pushed by molten basalt erupting from the Earth's interior.

The earth spun faster than now, sloughing off much of its interstellar gasses (hydrogen, helium and neon). A second atmosphere formed (scientists say) which no one knows much about. Then about 3.4 billion years ago, volcanoes, cracking through the mantle began to heat water containing rocks called hydrated silicates. The volcanoes released steam that became the water of the ocean; released the nitrogen, ammonia, and methane that would form the first protein-like compounds; released the carbon monoxide and dioxide that would later make even more carbon compounds and give plants the fuel for photosynthesis. So the ocean and the atmosphere were born together. The ocean's atmosphere and the atmosphere's ocean became the maternity ward for life.

Deep enough in the new-born ocean to be somewhat protected from the bombardments of ultraviolet light from outer space, the dissolved volcanic gases and the new compounds made from the new atmosphere and the new crust experimented combinations on combinations.

Perhaps attached to a clay particle, a bacteria-like organism began to ingest those organic compounds floating in the sea. A process called fermentative metabolism, a process that does not need oxygen to make sugars and proteins, established hold in the Oceans. This giant step in nourishment was fickle in that it depended completely on locally, uncontrollable sources of energy for food.

A billion years later...

The greatest revolution since the first fermentative being: photosynthesis, the ability to use a constant source of energy (the Sun) to fix nitrogen (that is, make one's own food). As a by-product of self-nourishment, the photosynthesizers exhaled oxygen. More and more oxygen began to be added to the atmosphere. The fermenters and other unimaginable organisms found oxygen to be a deadly poison gas. Photosynthetic pollution changed the composition of living creatures on the Earth drastically. The photosynthesizers pushing the fermenters into small anaerobic corners of the Earth.

Maybe two million years ago, our form emerged as the latest result of the oxygen revolution....

1973

We have entered the noosphere where man's mind or mindlessness will shape evolutionary history within the bio-sphere. A transition as great as the first photosynthesizers is underway. Previously, every invention or attitude in human history changed the bio-sphere BUT the bio-sphere could balance the change by changing itself. Now, any act or change in consciousness by humans directly affects the bio-sphere in, perhaps, an irreversible way.

The Aye-aye, perhaps the strangest living mammal, is on the brink of extinction. Something like a monkey-lemur-rodent, the aye-aye lives (?) only on Madagascar. It is strictly solitary, nocturnal and builds a complex nest.

This irreversible upset in the atmosphere, ocean, and Earth crust is caused by the *scale* of human works and the *values* of the twentieth century. You move in a car fueled by Carboniferous ferns, eat corn flown from Iowa, take a cruise to Istanbul, shit in the ocean, pet your hungry cat, see the last aye-aye in a Paris zoo....hundreds of millions of other humans are acting the same way. Result: complex habitats are reduced to simplified agricultures, oil spills change the evaporation rates and reflectivity of the oceans, industry changes the gases of the atmosphere....

Think of your breath. You inhale atmosphere so thoroughly mixed and rapidly recycled that the next breath you inhale will contain atoms exhaled by Jesus in Palestine, by Ford in Detroit, and by Goebells in Munich. Add various molecules of Strontium 90 from bomb tests and the gases from chimneys and exhaust pipes. All this enters your lungs and helps your brain think: "How much energy that runs the biosphere can be diverted to the support of a single species: man?"

The question can no longer be treated as out-there in the bio-sphere separate from everyday habits of Mind. Because we breathe, the question is not a debate between your expanding "inner" consciousness and "outer" awareness of the Earth. Mind process and Planet process have become too close, too inter-mixed, too woven to separate thought and atmosphere, action and ocean, attitude and Earth. end

SHELTER

The best shelter for light traveling i know of is the *one-tarp-tipi*. Fully adaptable – from a simple ground cover to a lean-to to a one pole tipi to a full tipi [for a longer camp].

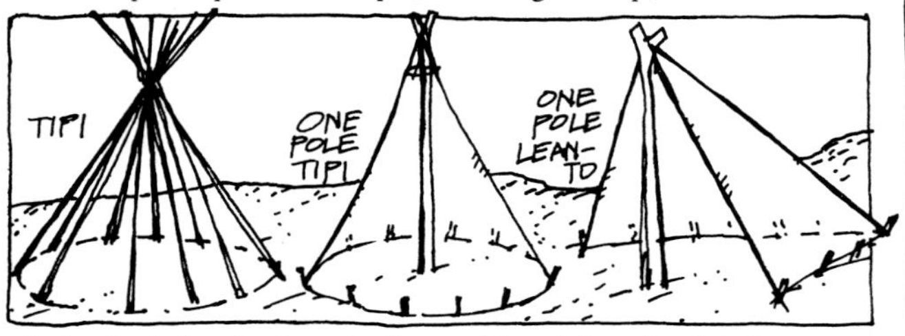

A simple shelter for horseback travel is a tipi made of dark green, waterproofed, 8 to 12 ounce canvas tarps.

To make this tipi, obtain 2 tarps 14' x 14'. Place the tarps next to each other, one overlapping the other by one inch. Cut the tarps into a full half-circle, then sew up the middle where the overlap is. From the scraps, cut 2 pieces 3' x 2' [A]. Fold one corner of each & sew these three pieces to the main body as shown. On what is to be the outer surface of your tipi, attach 4 cords to the outer edge of the center piece [B]. Fold the tipi in half, so it is a quarter circle. Directly opposite the cords, sew on 4 rings or buttons one foot in from the edge, also on the outside of the tipi.

Make about 10 stakes. You will also need 10 pieces of strong cord 18 inches long to attach the tipi to the stakes. The other necessary item to complete this tipi is a 25 foot rope.

When finished this tipi weighs approximately 30 pounds. For backpacking, a lighter tipi can be made similar to this one from a parachute folded in half. It weighs only a couple pounds.

I've found this type of shelter to be the most practical for horseback travel, it is light, waterproof, camouflaged, easy to set up, and comfortable to live in. You can build a fire inside, a necessity for any good shelter, and there is still ample space for 3 people and gear.

There are a number of ways to set up this tipi: **ONE POLE TIPI W/ ROPE:** Lay the tipi out flat, outside up. Under the center seam, lay a pole about 16' long, the thick end flush with the curved edge. Tie the tipi to the pole with a piece of cord at the point where the pole extends beyond the center of the flat edge. This will be the pointed top of the tipi. Now tie your 25' rope to the length of pole sticking out, and pull the rope up and back till the pole forms a 60° angle with the ground. Tie the rope to a tree or stake. This takes a little practice to get it balanced so it won't just fall to the side. Next, stake it down, starting from the door, making sure the tipi is taut. Now tie the 4 cords to the 4 rings and it is finished.

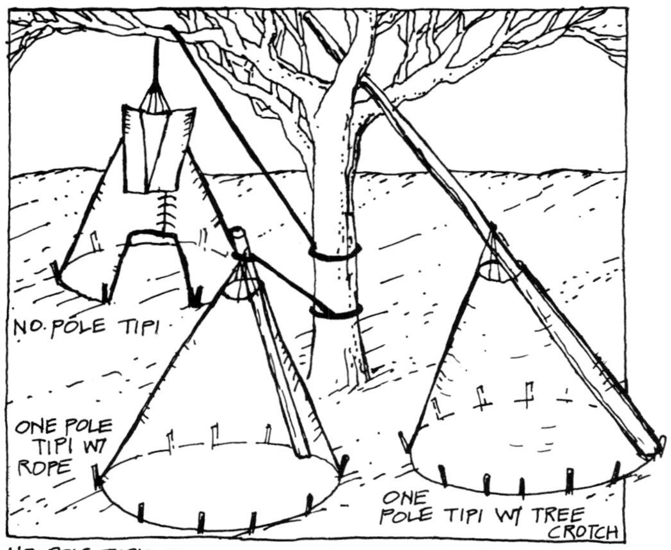

NO-POLE TIPI: Tie your rope at the center ["top"] of the tipi. Throw the rope over a limb & hoist straight up. Tie the other end to the tree trunk. Then stake down and tie center piece.

ONE POLE TIPI W/ TREE CROTCH: Lay out the tipi and pole as with the one pole tipi w/rope. Tie pole to tipi, then lift pole & set protruding end in a tree crotch. Stake down and tie center piece as before.

TRAVEL

Your method of travel would also determine what you could or could not carry. The first and obviously most *self* sufficient is on foot, which would limit your shelter to the small tent or tarp. Your second choice being horseback – giving you a wider range; either the one-tarp-tipi or if a pack-horse is added even a full size tipi [without poles] could be carried, assuming you were planning to establish a fairly long camp [one season, winter etc.]. Your last choice being motor vehicle which is severly limiting as far as wilderness is concerned. Wagons are also useless away from dirt roads & population.

Unless you are in an area where horseback & horse-packing is impossible – heavy timber, no graze, etc. this is your best bet for nomadic *life* because now you can carry your full material needs rather than mere survival needs. And your possibilities are

CON'T TOP RIGHT

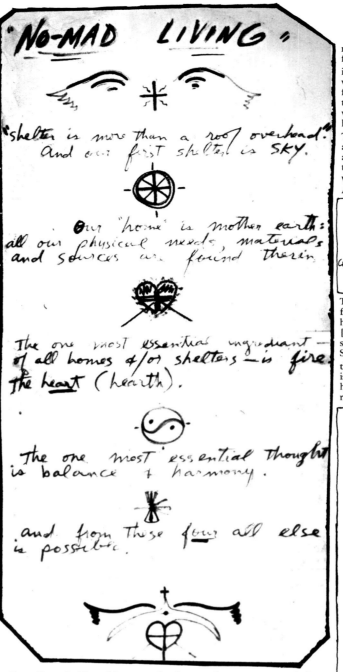

"NO-MAD LIVING"

"shelter is more than a roof overhead". And our first shelter is SKY.

Our "home" is mother earth: all our physical needs, materials and sources are found therein.

The one most essential ingrediant – of all homes &/or shelters – is fire: the heart (hearth).

The one most essential thought is balance + harmony.

and from these four all else is possible.

Dear Bros & Sist...i just received one of your post cards in the mail from a bro. (Wavy Gravy) in Cal. Me & ol' lady been living nomad/mountain/horseback past 4 years – tipis, tents, wigwams, log cabins etc. Wild food (plants & meat) survival etc. been 3 winters 35-40° below spells in canvas tipi & wigwam...the enclosed prepared by 3 members of our clan – myself, Chipita & Jade.

i figured i'd start my information on no-mad living with this small visual/prayer. if you're building a *house* you start at the foundation & if you're building a *way of life* thats where you start also; and nomad living is truly a *way* of life.

The first thing is to be comfortable and at ease in your surroundings. In a house you create your own environment – with your own (more or less) conveniences. *The parallel in the wilderness is to know what exists.* That means how to survive – food, herbs, materials, etc. This is learned by starting slow – step by step accumulation of knowledge – through experiment, books and informants.

To want to *live* in the wilderness & not learn all edible and useful plants would be ridiculous. To want to *live* in the wilderness & not learn the skills of hunting & tracking would be as equally ridiculous. If you cant live *with* your environment, then you are merely *lost* in it.

Another element of survival is shelter – which means everything from the totally improvised to the more sophisticated; from the one nighter to the *home*.

The totally improvised: brush hut, cliff overhang, log hollow etc.; the partially improvised, using something you carry – usually a tarp or poncho: leanto, simple tent, covered wigwam, etc. and the more complete: tent, tipi etc.

the amount of time to be spent in a shelter would also determine its complexity. For instance, a totally improvised shelter could be anything from a few branches to an earth lodge – dug in for the winter. A carried shelter anything from a rain poncho to a full sized tipi.

A nation is coming, a nation
is coming.
Over the whole earth they
are coming.
A nation is coming, a nation
is coming.

Keep the faith.

Ben Eagle

more permanent than a short trip somewhere [which is the usual for back-packers].

in the several years of living this way, the longest straight ride we've taken is two months [without more than a few days stop for graze etc.]. We carried everything we need &/or use for our total existence – that included winter gear etc. [it was spring but we were not returning]. We used one riding horse & one pack-horse per person – the best ratio for this kind of travel.

Then there are the medium-long rides of a month or two from a base camp [leaving out-of-season essentials etc. behind] using a pack horse [maybe one per 2 or 3 people] or the several days to week [or two] ride with saddle bags & riding horse only.

When pack-horses are used you will need special equipment. A good design for pack-saddles, fairly simple to make are:

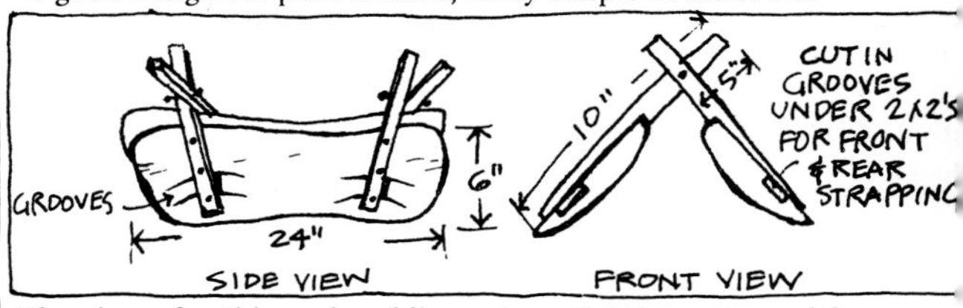

The pieces for this pack-saddle are two slightly contoured forms fitting on either side of the horse, from 18" - 24" long and 6" high [can be padded or used with saddle pad] and four 2 x 2's [or saplings bound with rawhide] about 10" long – fastened so that about 5" comes above the frame with 2" above X. Size according to horse. L – M or Small.

the harnessing needed [made from either leather or strong webbing] is one front chest strap. one or two belly cinches & one rear halter. These can be attached in various methods – using metal rings and/or fastened around the 2x2's.

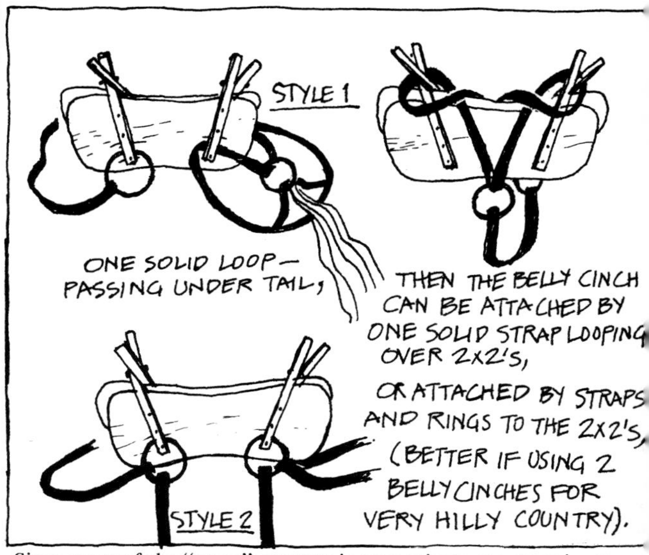

Since most of the "open" country is now private etc. travois are not as practical as pack-saddles, which are better suited for mountains. Old army cavalry saddles are also easily adapted for packing.

The pack saddles shown above can either be used as frames on which to tie equipment and soft rolls (bedding, tents, canvas packs, etc.) or with *paniers* – special made boxes, leather bags, etc., also fairly easy to make.

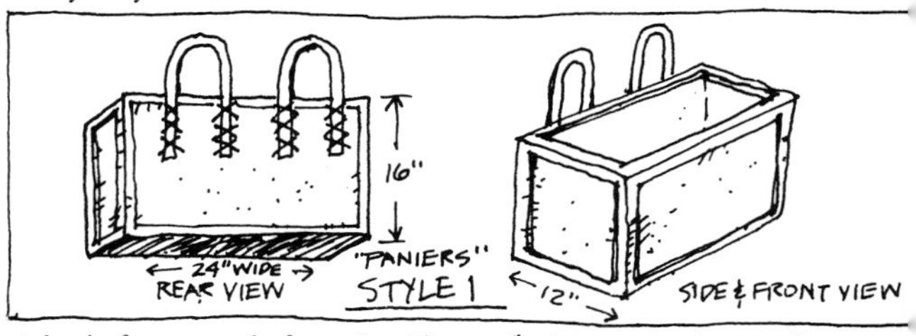

A basic frame made from 1 x 2's notched & screwed or saplings bound by wet rawhide strips also notched and fit, measuring around 2 ft. long and 1 ft. wide and about 16 inches deep. This frame is then completely covered by wet rawhide – stitching pieces together [either cow or horse hides are good]. Two straps can be attached to rawhide, by stitching or grommets, with long enough loops so that the boxes can be hung from the X of pack saddles – restring against horses sides, one on each side. [each one being about as equally loaded as possible so weight isn't off balance and hurting horse.]

Plywood boxes about same size as 1 – also covered with rawhide [for strength & endurance]. But these can be hung differently.

Using two pieces of wood about 6" long, 1" wide & 3" high, one on each end to secure boxes to saddle by passing heavy cord loop under them and over X of pack frame – each loop passing over opposite side.

Make leather bags [not rawhide] about same size as boxes only deeper.

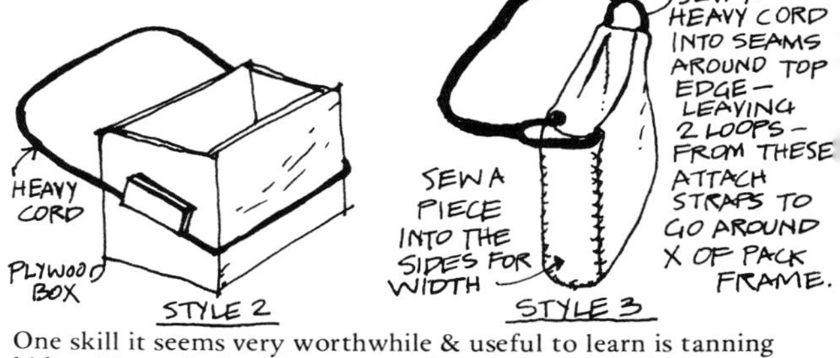

One skill it seems very worthwhile & useful to learn is tanning hides. Not only for horse gear and equipment but clothing, moccasins & other necessities. Also anyone living off wild game should be able to use the hides as well as the meat.

I know this isn't much information, but it may help. And if someone is really into it – they'll learn whatever else it is they need to know.

HOUSECAR

by Kelly Hart

have a 1952 Southern coach with a Waukesha engine built riginally for the Army and subsequently used as a school us. I bought it in operating order five years ago....

was especially suited to conversion as a housecar because f 6½' standing room, the fact that the engine lies under ne floor (pancake style), and the body is composed of two yers of aluminum with insulation in between. All window ames are also aluminum.

built the interior with no particular plan, except that knew I wanted a permanent double bed in the back; therwise I just put things where they needed to be. The us sort of formed itself: the stove needed to be near enter to distribute the heat and there was a convenient ent in the roof nearby for the stove pipe. We wanted the itchen table near the stove so we could sit and watch the re and be warm on the coldest night. We wanted the sink ear the table for efficient movement of dishes, and so, you see, the design was pre-ordained.

hat really transformed the bus from a tunnel into a castle was the ddition of six, 2½ ft. square dome skylights.

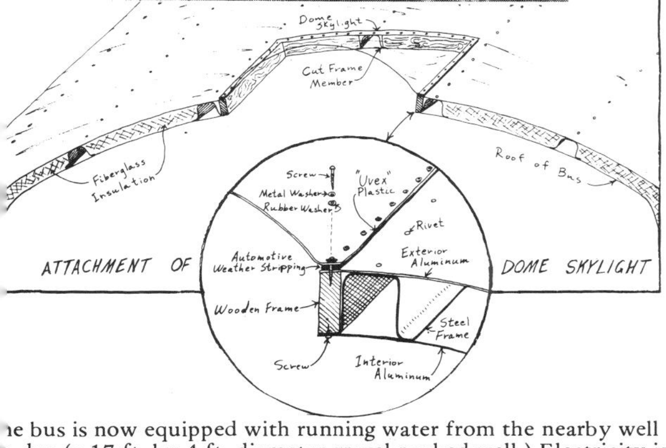

ATTACHMENT OF DOME SKYLIGHT

ne bus is now equipped with running water from the nearby well e dug (a 17 ft. by 4 ft. diameter gravel-packed well.) Electricity is ovided via a ground line from my neighbors house (upon whose nd the bus is parked). There is large semi-circular deck adjacent to e side of the bus with the emergency exit, providing open space r kids to play and me to do my morning kung fu. We have a phone stalled in the closet and a stereo for added convenience and easure. My ultimate luxury is a full sized upright piano from which stripped all the cabinetry but the essential strings and sounding oard, key board and action. The key board is detachable and hinged that it will fold up to conserve space. Another board will fit into e place of the keys to serve as a work table.

s with the piano/work table combination, there are many such ulti-use facilities. The kitchen cutting board reverses to act as an ning board. The kitchen table is a three-inch slab of solid laminated , sturdy enough to knead bread or wedge clay. The driver's seat is w converted with a little typing table in the place of the steering heel (with adjustable seat.)

CONTINUED TOP RIGHT

KELLY'S ABOVE

DENNIS' BELOW

Junk cars have been ignored too long. They have roofs, windows that open, and doors, all of which are most expensive in house building. Old trucks can be a semi-portable foundation for a house. I put my old Peugeot roof with its sun roof on stilts above the bed of my '49 Ford 1 ton, high enough so I could stretch in the morning and added on till I ended up with this Transient Home called outrageous, turtle, ingenious, satirical, etc. by admirers.

The whole thing has hung together through 4 mo. of erratic (schizoid engine) touring from Mpls. to Mexico City and up to Seattle. Mex. border officials thought we were part of the Apollo 18 mission. After giving my hood and fenders to a Mexican trucker I found the engine easier to get at and the overall appearance racier.

The interior is all one space because I cut away the center 3 feet of the cab from the top of the bed to a foot from the windshield.

Now when you unite the cab space with the back you're committed to making the whole truck rigid one piece. Pickup campers can't extend to the front bumper because the whole camper unit moves independent of the cab. To have your upper bedroom over the engine you must have a strong link to the frame in front of the engine.

This way the cab is the center of strength since all the main beams come together and bolt onto the edges of the cab roof and back. The Peugeot is secured by beams running inside of the door handles. Plywood sheathing on the sides and back braces laterally against sway.

The stove is built from the doors to fit the space in the rear overhang. A flat top and full width door make it good for cooking. While driving it burns better and cooking while driving smells real good. The tailgate still opens so I can carry large trees, tires, beds etc. that you find in ditches everywhere. Side windows are not very influential because of its effective streamlining and cops everywhere have only waved and laughed.

Dennis Turnguisy
Prior Lake, Minn.

CONTINUED FROM LEFT

I have lived in this bus for over four years and find it at least as comfortable as any house I've lived in. There were two main reasons why I got the bus in the first place: I was sick and tired of fixing up a rented apartment or house to suit my needs, only to be forced by circumstances of ownership to look for another place. Secondly, I was planning to build a cruising trimaran and I wanted to experience living in a small space so I could find out what is truly essential to the maintenance of my life.

Although I have lost interest in the trimaran, the experiment in bus living has proved to be highly beneficial and instructive. In fact, I often have a feeling of claustrophobia when encased in a conventional dwelling, because the bus has 360 degree visibility plus all the light from the dome skylights. The advantages of bus living are manyfold; to list a few of them:

Mobility...The fact that it is on wheels means that I am not permanently stuck in one locale...this gives me great freedom.

Durability...I was lucky to find a bus that was constructed almost entirely of aluminum so I don't have the problem of the thing rusting away, it is less vulnerable to fire...It's also practically impervious to earthquake damage....

Building codes...The bus is a vehicle, not a house, so it is properly under the jurisdiction of the motor vehicles department, not the housing authorities. When I was driving it I had it registered as a housecar and the registration fees were about $18 a year, but since it has been stationary I have not even paid this. You can't beat that for taxes!

Efficiency...Living in a compact space brings about efficiency in many spheres:

Heat. We have a small wood stove that keeps the bus cozy in any California weather. An armfull of wood will last all night. My yearly expense for heating has averaged about $20...

Ventilation. With so many openable windows and doors (the back opens up completely and there is an emergency exit in addition to the front door), the bus can be thoroughly ventilated in any desired cross direction....

Movement. The inside of the bus measures about 28'x8', or 224 sq. ft., so one need only move a few steps to arrive at any point within. It's amazing how such a simple fact can make life so much easier. Standing in the center of the bus (which is essentially the kitchen), I can reach any implement, commodity, or facility by taking only one step in the appropriate direction....

Organization. I feel that the real crux of successful life in a bus, or any other small space for that matter, lies in efficient organization and use of space. It would not be practical for some people to even attempt such a life because their habits and life styles do not blend with such a highly organized way of living. So if it pleases you to have a place for everything and everything in its place, then perhaps you could be happy in a bus.

Furniture and fixtures. Once I got the bus set up for living (this took about four months) and then chose the belongings I really needed and eliminated the rest, the temptation to shop for new furniture or pretties was severly curtailed. If I buy something and bring it home, I then have to find a place to put it. Every nook and cranny is utilized and has its proper function, so it is no mean accomplishment to come up with new space for new objects. This has definitely curbed my spending urge.

Housework. There practically is none! It takes just a few minutes to sweep, the table can be cleared in a matter of seconds, and there just isn't very much area to keep clean.

Of course there are some disadvantages to bus life that parallel life in any small space. Primary among these is that it gets pretty crowded when you get more than a few people spending very much time indoors, or if you have small children, the confinement during inclement weather can drive you up the wall.

I find this vehicular life so appealing that I now have a 20 ft. semi truck trailer that I am converting to use as my film studio. It was a Navy radar unit and is equipped with a hundred drawers and cabinets, making it perfect for my work.

Kelly Hart
Jenner, Cal.

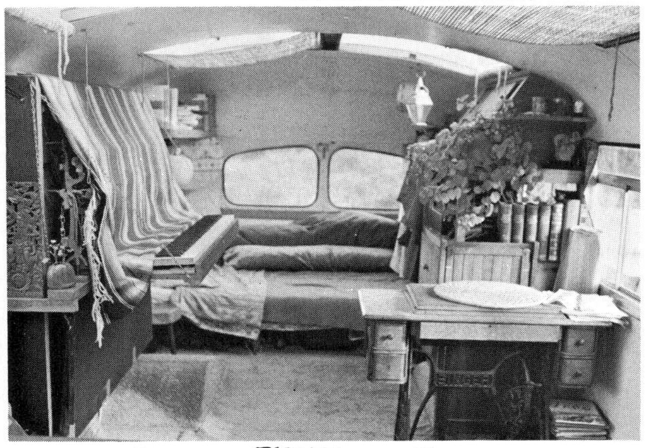

PIANO

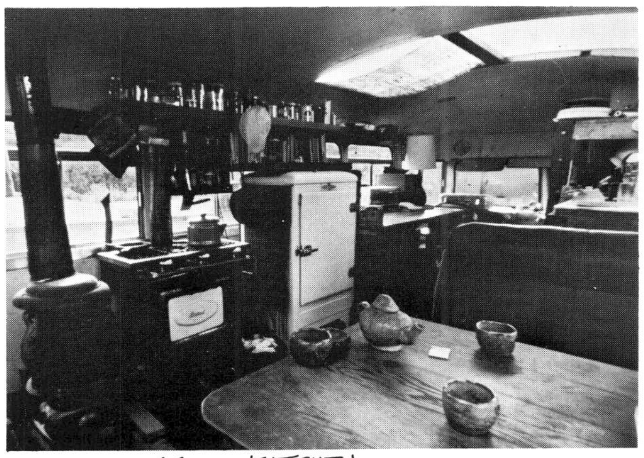

KITCHEN

TIN LIZZIES

ime de Angulo — cowboy, anthropologist, doctor, poet — ed with the California Pit River Indians in the early 00's. "Real Primitive People," he wrote, "not like those ultured' Indians of the Southwest...real Stone Age men... y Indians in Overalls." The following selection, courtesy The Hudson Review *is reprinted from the just-published* dians in Overalls, *Turtle Island Foundation*

was up again, the next year. This time I had a jalopy, yself. Progress. You can't defend yourself against progress. o this time I came up following the Pit River from Redding o. I was meeting more and more Indians after Montgomery reek. I had never been through that territory before, only nce, years before with a drove of horses, through a snow orm. I didn't even recognize the country.

got into the upper land. It was getting dark. I had a olfish-looking bitch swaying on the back, on top of my camping stuff. I was getting tired driving that damned car. I hate them. When I got to Big Valley I couldn't stand the driving over the rough road any more. I saw a campfire a little south of the road. I was awfully tired. I thought: They must be Indians.

I got out and walked over, being careful to make a noise. There was no need for care. There comes Sukmit: "I have been watching for you. I got lonesome for you. I sent my poison after you. The old lady is there, in the camp. Old lady Gordon died. The other woman, my uncle, she is dead. We are all going to die. We can't help it. What have you got there? Is that a coyote? Looks like a coyote. Don't growl at me, you son-of-a-bitch, I am Indian doctor. I ain't afraid of you. Want to be my poison? Say, Jaime, did you get my message? I got lonesome for you. You want to be a doctor? I teach you. I am Indian doctor. I teach you, pretty bad, get scared, I teach you, you no white man...."

Under this avalanche I was being dragged across some wasteland toward the campfire I had seen. There were no introductions of any kind whatever. Nobody paid any attention to me at all. I sat in a corner. I said nothing. Then a little boy brought me a basket full of some kind of mush, and it had salt in it, too. I was reserved and very careful, keeping out of the way, in the outer light of the fire. Then Mary's chuckle came out of the darkness (I hadn't seen her until that moment, sitting there beyond the firelight): "Ha-ha, you white man."

I followed that bunch for several weeks. I never saw such a goddam lot of improbable people. Sukmit was the only acknowledged shaman, but he wasn't a leader. He was no chief, no *weheelu,* among them. We went around the brush. We would stop anywhere, evidently by common consent.

CONTINUED

JOAQUIN & GYPSY'S

JOAQUIN

'37 CHEVY

COLLAPSIBLE CLOTHESLINE

Some years ago Joaquin De La Luz traded his '48 Triumph motorcycle for this vintage Chevy flatbed, and with little money, much imagination, and found discards set about making one of the most unique homes ever to roll along America's roads. For the past five years, Joaquin, Gypsy and their three kids — Heather, Bear and Serena — have moved around the country and were last seen parked along California's Feather River. Following are some tips on mobile design, and living on the fringe prepared by the De La Luz family, and from the photos you can see it's true when Joaquin says: "I love trucks."

I have found more freedom in my building designs by not confining my ideas to a planned form. If you plan what you are going to build then you have to find materials that conform to that plan.

Every area has its own unique throw-aways.

If you build a house on wheels it is best to build the frame of good solid material, preferably all the same dimension.

Bending bolt shank works better than lock washers on house trucks (esp. metal framework). See sketch above.

Never use less than 5/16" nuts & bolts on house trucks. (When bolting studs together.)

Make truck level before starting to build.

Use a square: it pays off in the long run.

Notch every corner you can when building with wood.

Junk building material does not mean junk workmanship.

A good solid frame is most important *even* if you have to buy the wood.

I have found it much easier to *find* building materials to cover a building with than to build the framework with.

Build housetrucks with as low a center of gravity as possible. It is better to have steps over wheels than to have floor too high from chassis. *This is very important.* It means you can have more space inside and lower clearance from ground to roof outside. 4" makes a *big* difference.

Lower center of gravity also means much better handling on the road.

Heavy leather makes good hinges for cupboard doors.

A chainsaw can even cut round shapes like archs for truck roof.

I have found the chainsaw the most useful tool for my style of building. I used my chainsaw to build the entire framework for my house. It was built with green Douglas fir rough cut 1x4 that I bought new from a mill. Every board was dripping wet as it was winter in Oregon at the time. The chainsaw cut smoothly thru every board, an unequaled feat for sawing wet wood. A mini chainsaw is best for building with.

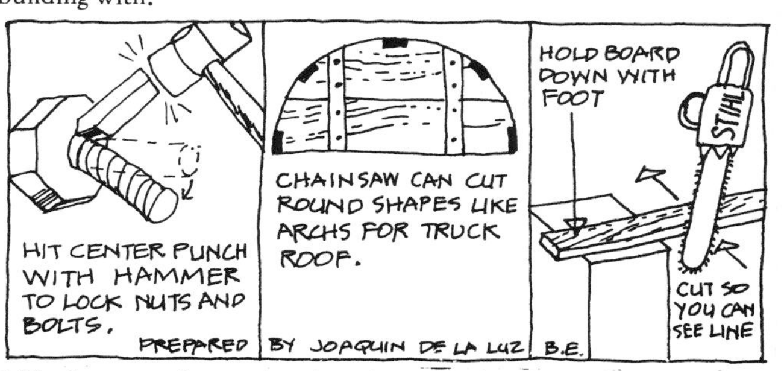
HIT CENTER PUNCH WITH HAMMER TO LOCK NUTS AND BOLTS.
CHAINSAW CAN CUT ROUND SHAPES LIKE ARCHS FOR TRUCK ROOF.
HOLD BOARD DOWN WITH FOOT
STIHL
CUT SO YOU CAN SEE LINE
PREPARED BY JOAQUIN DE LA LUZ B.E.

Utilize beer cans by cutting them into pleasing shapes for shingles — light weight — no rust.

Use bailing wire for "ground wire" cutting electrical installation costs in half.

After cutting out rear wall in cab of truck, join house with cab with an old inner-tube, cut & tacked to keep air tight. (Rubber allows for movement between house & truck.)

The waste of America is the richest in the world.

Building with cast-out wood you are sometimes faced with cracked pieces — cut different shapes out of whatever metal is available & tack over crack to strengthen.

Old produce crates make great spice shelves etc., many uses!

Old cans that have been discarded such as anti-freeze cans of 1 gal. size cans, can be cut and used for punched tin cupboard, making holes with a nail.

Third Printing Notice: Joaquin and Gypsy have found some land, moved into a house(truck was getting cramped with three kids) and Joaquin now has a 10-wheeled truck for hauling logs. Their housetruck(minus furnishings) is for sale. Plans for building the truck are also for sale for $4. Joaquin de la Luz, Rt. 3, Box 16A, Yreka, Calif. 96097.

CONTINUED

We would stop in the brush. Always there was a spring near by. I never knew where we were going. We were going somewhere. I didn't care at all. We were going somewhere, maybe. And if we were not going somewhere, we were not going, that's all. In the evening we would make a fire, several fires (there were several families of us). In the morning we moved again. I don't know where we were going. I don't think the Indians knew. We made quite a procession through the sagebrush, about six or seven of us, my car usually tagging at the end. I didn't know where we were going, nobody seemed to know where we were going, and then the night would settle on us, the fires would die down. The coyotes would begin barking from out in the brush. The Indians' dogs would howl back. Then everything would smolder back into the darkness.

Then one morning I was made to realize that there was something wrong about the white man's conception of the "taciturn" Indian. That happened the next day. We were going along the sagebrush, no road, just sagebrush, wind left, wind right, avoid this big clump, here's bad one, bump into the ditch...but there is no road at all anywhere, you are going through the brush, bumpety bump, all of us, six, seven, maybe eight cars, eight tin lizzies rattling through the sagebrush. Then, one morning, we had to stop. One of the tin lizzies was on the blink, and everybody got out to help. Then I witnessed something that amazed me. I had made up my mind that these men were straight out of the old Stone Age. I myself am not a mechanic; I hate machines; I am all thumbs; I don't understand machines; horses, yes; machines, no. And here I was watching these Stone Age men unscrew and rescrew and take things apart or out of the engine and spread them on a piece of canvas on the ground... but the amazing thing to me was their argumentation. It was perfectly logical. "...Can't be the ignition, look, I get a spark...I tell you, it's in the transmission...Now pull that lever..." Maybe I was over-impressed because the simplest machine smells of magic to me. Maybe I missed a lot of their argument because off and on they would lapse into Pit River. They called the battery *hadatsi* "heart"; a wheel is *pi'nine* (a hoop used in the old days for target practice); and so on and so forth. But certainly they made use of logic just as any white man would. Finally the engine, or whatever was wrong, was repaired. Then I overheard one young fellow say to another: "You know why this happened? Because he has been sleeping with his woman while she was menstruating! That's against the rules."

At last everything was fixed: the engine put together again, everything rescrewed...but the trek of the tin lizzies was not resumed. We just stayed there.

I don't know why we stayed there. We just stayed there, in the middle of nowhere, in the middle of the sagebrush. After that car which had broken down had been repaired,

CONTINUED

HOUSETRUCK

67

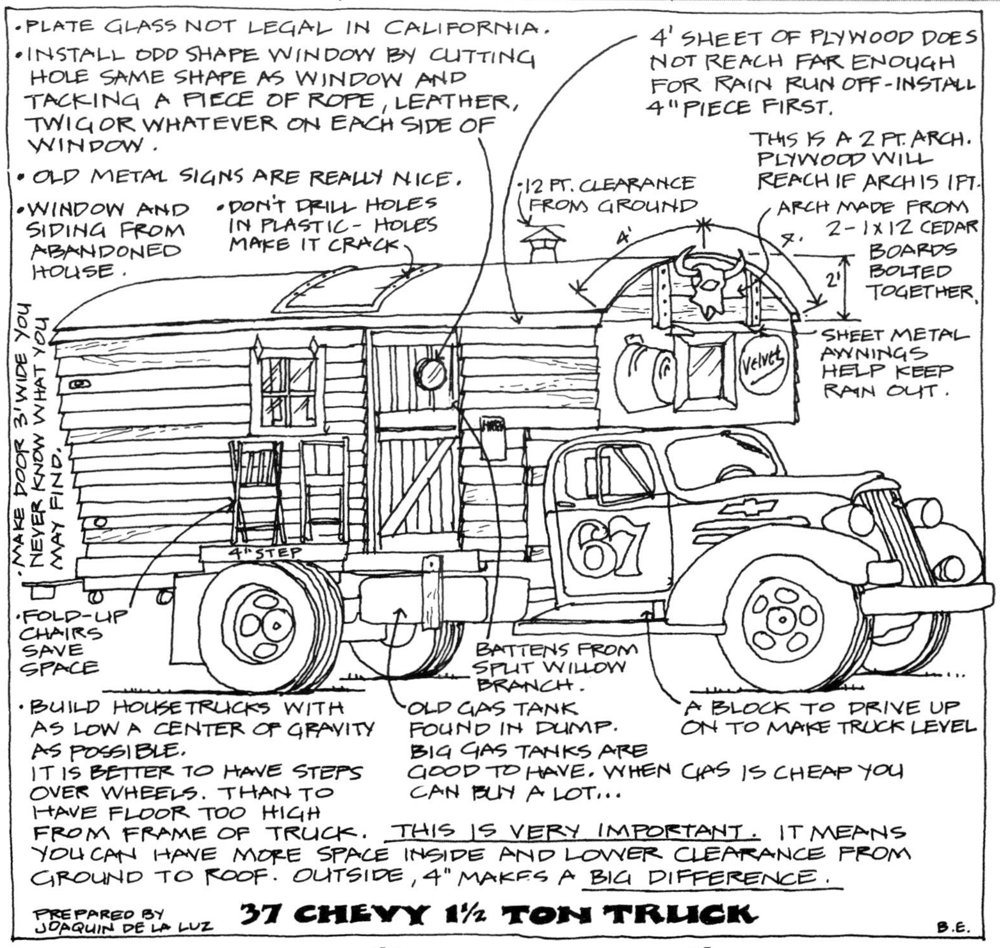
• PLATE GLASS NOT LEGAL IN CALIFORNIA.
• INSTALL ODD SHAPE WINDOW BY CUTTING HOLE SAME SHAPE AS WINDOW AND TACKING A PIECE OF ROPE, LEATHER, TWIG OR WHATEVER ON EACH SIDE OF WINDOW.
• OLD METAL SIGNS ARE REALLY NICE.
• WINDOW AND SIDING FROM ABANDONED HOUSE.
• DON'T DRILL HOLES IN PLASTIC- HOLES MAKE IT CRACK.
4' SHEET OF PLYWOOD DOES NOT REACH FAR ENOUGH FOR RAIN RUN OFF - INSTALL 4" PIECE FIRST.
THIS IS A 2 PT. ARCH. PLYWOOD WILL REACH IF ARCH IS 1 FT.
12 FT. CLEARANCE FROM GROUND
ARCH MADE FROM 2 - 1 X 12 CEDAR BOARDS BOLTED TOGETHER.
4'
2'
SHEET METAL AWNINGS HELP KEEP RAIN OUT.
MAKE DOOR 3' WIDE YOU NEVER KNOW WHAT YOU MAY FIND.
Velvet
67
4" STEP
• FOLD-UP CHAIRS SAVE SPACE
BATTENS FROM SPLIT WILLOW BRANCH.
OLD GAS TANK FOUND IN DUMP. BIG GAS TANKS ARE GOOD TO HAVE. WHEN GAS IS CHEAP YOU CAN BUY A LOT...
A BLOCK TO DRIVE UP ON TO MAKE TRUCK LEVEL
• BUILD HOUSETRUCKS WITH AS LOW A CENTER OF GRAVITY AS POSSIBLE. IT IS BETTER TO HAVE STEPS OVER WHEELS. THAN TO HAVE FLOOR TOO HIGH FROM FRAME OF TRUCK. THIS IS VERY IMPORTANT. IT MEANS YOU CAN HAVE MORE SPACE INSIDE AND LOWER CLEARANCE FROM GROUND TO ROOF. OUTSIDE, 4" MAKES A BIG DIFFERENCE.
PREPARED BY JOAQUIN DE LA LUZ
37 CHEVY 1½ TON TRUCK
B.E.

KITCHEN

I naturally expected that everybody would get back into their cars, and the procession be resumed. But no, nobody got back into the cars; everybody was drifting around, sitting here, sitting there, gossiping, yawning.

I asked a man: "Are we going on, or do we camp here?" He answered: "I dunno. I am not the chief. Ask that old man over there." I went to the old man over there. He said he was not the chief. Ask that fellow over there. That fellow over there was a middle-aged man. He said: "Hell, I am not the one to say, I am not a chief!" "Well, who is a chief here?" "I dunno. That old man over there, I guess. He is old enough to have the say. Go and ask him." That old man over there was the same old man over there, and he gave me the same treatment. He was no chief. Who said he was a chief? They could start when they liked, when they jolly well liked, he didn't care, he didn't even know where they were going, where the hell were they going, did they know where they were going, did I know where they were going??...He sat on the foot-board of one of the cars. He was squinting into the afternoon sun; it was late afternoon, by then. He was chewing tobacco and spitting the brown juice. He paid no more attention to me and went back to his reverie, squinting into the sun.

I noticed a woman had started a campfire. Very soon another one did likewise. So I went to my car and drove it next to Sukmit's. Old Mary was sitting on the ground in the shade of the car, weaving a rough basket of willow twigs. "Where is Sukmit?" "*Tsesuwi diimas'adi,* I don't know, went off in the brush some place, that boy is crazy, *yalu'tuusi,* always looking for *damaagomes;* you bring me firewood, white man, I cook." "All right."

The sun was going down. I heard two or three shots, off in the sagebrush. I made our fire. There were four or five other campfires. A man came by; he had several hares by the ears; he tossed one over to us. Mary drew it, threw the guts to my bitch, hacked it in four pieces, and stuck these on sticks to broil over the fire. No sight of Sukmit. We ate. Then Mary told me an old-time story. I spread my blankets on the ground. I rolled a cigarette and watched the stars. Some coyotes started a howling, not far off. My bitch stood up, all bristles, and she howled back (in answering coyotes, most dogs howl instead of barking). I went to sleep with my head full of old-time stories, tin lizzies, *damaagomes* mixed up with engines, coyotes and sagebrush.

I was awakened by the usual quarrel between Sukmit and his mother. The smell of coffee was in the air. We ate. I observed the camp. There were no preparations for starting. Everybody lolling around. Mary took up her basket and kept on weaving. Another woman went to the spring with a bucket (there was a spring, nearby; of course they must have known). A man was tinkering with his car.

CONTINUED

ELLIPTICAL CAMPER

The camper on the back of this ¾ ton flat-bed pickup is constructed of wood and polyurethane foam. The shape is a modification of an egg truncation, enlarged in the shoulder area for more room, but still maintaining the strength and aerodynamic advantages of this type of shape. The structure is basically a light wood latticework on which shingles are hung. Polyurethane foam was sprayed on the inside to seal, strengthen, and insulate it, giving the structure the advantages of lightness in weight and the adhesive properties of the foam and the shingles....

The total structure is quite strong, and weighs around 400 lbs. It has a floor of ½" plywood supported by 1x2's which are diagonally spaced at about 26" with interlocking notches. The oval edge of the floor is a laminated 1x4, attached to the flat-bed of the pickup by six bolts. The flat-bed and camper unit have been extended to nine feet, which proved a reasonable way to gain space without sacrificing roadability.

The framing was done like a boat, using ribs cut out of a sheet of 1¼"plywood. This method proved very accurate, but I feel lamination would be a more desirable way to do it. It would be cheaper and stronger, and would have a better interior appearance if nicely finished. The horizontal stringers are 1x2's resawn in half, cutting them on an angle is a good method to be sure the shingles will pull into the structure. The ribs are spaced 22" with the shingles being spaced at 12" at the widest part. Putting on the shingles is the time to do the precision work; until then the structure is very "giving," since the foam will help cover mistakes. I crisscrossed the outside of the structure, but inside, the stringers tightened with tension wires and turnbuckles, added the strength of triangulation until the foam was shot on. I am very pleased with the strength obtained from these, and it seems a cheap way to add strength....

Linseed oil rubbed on the shingles makes them more waterproof. I would suggest taping the inside seams of the shingles with Scotch Plastic Tape no. 483 to keep the ultraviolet rays from getting at the foam through the cracks, and allow the foam the added flexibility at those stress points....

The unit has very good roadability. There is very little of the instability caused by too much weight too high, it moves through the wind very well, and is not troubled by buffeting sidewinds. The worst possible shape to try to push through the air at highway speeds is a box. The main reason that most of the campers on the road are a box shape is that they are considered the most economical to construct. This method of construction will permit compound curves to be produced cheaply by the amateur home-builder. I paid $90 for the foam, $50 for the plexiglas (making my own frames and cutting my own plexiglas, and just paying for the blowing), and a hundred dollars for the wood and the hardware....

Terry B. Trenholm
Creswell, Ore.

GYPSY WAGON

Four years ago Betsy Craig heard of some people traveling by horse and wagon in England, near Stratford-on-Avon. Betsy located and joined them – a caravan of 3 wagons and 5 horses – and the group traveled about 10-12 miles a day. Inspired by the adventure, Betsy returned to the southwest U.S., got the National Geographic book *The Gypsies* for reference, and without previous building experience, put together her own gypsy wagon.

She traded a horse for an old logging wagon, then drew plans on graph paper. The flooring was cull lumber, the frame is $75 worth of new dry lumber notched and bolted together, and the roof skin is ½" plywood. The siding is ¾" plywood screwed in place, and front and back siding is tongue and groove pine siding she found. Windows are used scrap glass, the door is a used door cut in half to make a dutch door.

Inside I built in everything I wanted, a bed 4' off the floor with closet and storage space underneath, shelves and cabinets for a kitchen with a 3 burner stove set into the counter, a small work table in one corner, and corner seat which opens for more storage. A tiny pot belly stove makes it too warm easily. I insulated in some places with styrofoam but mostly with foam rubber on the ceiling and around bed and couch areas.

The plywood exterior is all covered with striped awning canvas – multi-colored and cemented to the plywood with linoleum cement. (I think some kind of polymer like Roplex would be better and plan to paint a coating on when the weather is nice – there was a problem of snow melting and moisture freezing again and cracking the canvas causing some seepage when it melted again.) The interior is covered with many different fabrics – silks, velvets, materials from many countries and rugs made extra thick with layers of rug matting as the floor gets cold in winter.

Recently we hooked the wagon up to the back of our pick-up and traveled about 10 miles. It was slow and bumpy in the wagon but everything held together and all went well. We haven't yet tried it with horses.

Betsy Craig
Redwing, Colo.

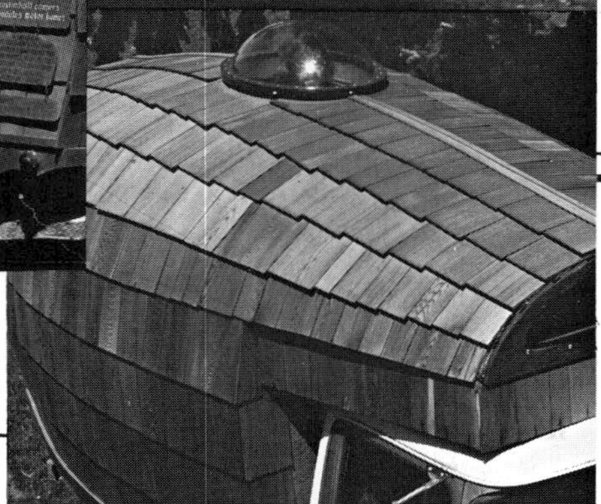

CAMPER

Tribal videographer's van is equipt with portable ½"-videotape studio for recording, editing & showing, life support & medicine kit. Alternate power systems from large capacity truck battery, built-in DC-AC inverter and 100' extension cord for standard line power. Portable motorcycle battery powers recorder & camera (or monitor) for up to 6 hours of remote recording (or playback). Full editing compliment of 2 recorders & 2 monitors consumes about 150 watts, approximately that of a bright light bulb.

This outfit is the base of a planetary cultural channel connecting diverse peoples along the Pacific Coast, operated by Tribe Of All Nations, 13 Columbus Ave., San Francisco, Calif. 94111. Loren Sears, videographer.

GYPSY WAGON

CONTINUED

Another one went off into the brush with his gun. That old man, the supposed "chief," was going around poking at things with a stick. He was almost blind. The morning was drawing on. I took out my notebook and started working on linguistics.

The days went by. Not so many days, but four, five days, maybe. I don't remember exactly. I was not taking notes. I was living. Sagebrush. Old-time stories, hares cooking over the fire, slow gossip, So-and-So is poisoning So-and-So, I don't believe it, yes he is, how do you know, well his paternal aunt belongs to the *hammaawi,* and they poisoned his *apau.* "That doesn't make him related!" "Who said related? I didn't say related. I said they poisoned him."... The days went by, four, five, six days.

Then it happened. It was midday, or near. I heard a man say, way off: "*S.huptsiidzima.*" If he had said *lhuptiidza toolol,* "Let's all go," it would have been different. But no, it was not in the imperative mode, it was in the indicative: "We are going, all of us, *toolol,* we are going, *s.huptsiidzima,* we are all going." He didn't say: LET'S ALL GO! No, he merely stated a fact: WE ARE ALL GOING.

It was like a whirlwind. I turned around. Women were throwing baskets into the tin lizzies. Then without any further warning or consultation one of the tin lizzies started off in a cloud of dust. Another was right on its heels. Then a third one, but this one had hardly started when someone yelled, "Hey! you are forgetting your baby!!" The car backed, a young woman jumped out and ran to a juniper tree where the baby was sleeping in the cradle-board swinging under a branch; she slung the cradle-board over her shoulder and ran back to the car, laughing and laughing; everybody was laughing; then the car started again.

Sukmit yelled at me: "*Lhupta,* let's go! For Christ's sake, are you going to stay here forever?!" I picked up my papers and ran for my own car....We go, we go, we wind in and out, all afternoon, all the cars more or less following each other, we skirted the town of Alturas, never stopped, we were going north, the sun went down, there was a moon, we kept going, somehow or other no car broke down, not even a flat tire, we had luck with us, and toward morning we stopped on those flats by Davis Creek....

END..

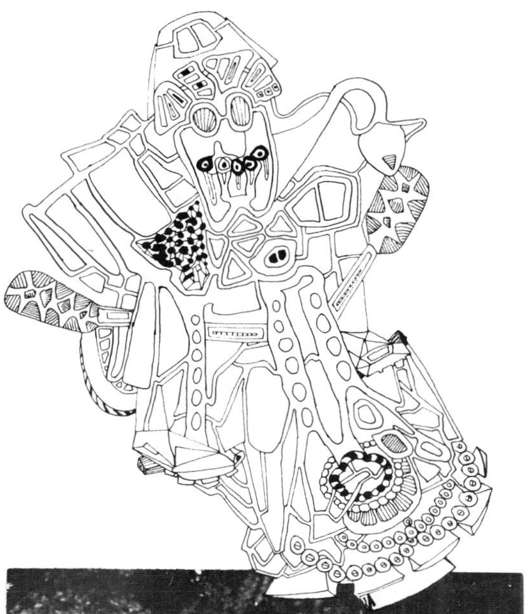

In Holland we saw this 50cc motorbike pulling a trailer, the rider and his two friends (on bikes behind him) off for a camping trip in the country. The trailer was made of lightweight tubes and bicycle tires, and carried a tent, food and camping gear. A European Winnebago.

WAGON, CAMBODIA,

YUCATAN

HOUSEBOATS

HOUSEBOAT, KASHMIR

THE JUNK

THATCH, KASHMIR

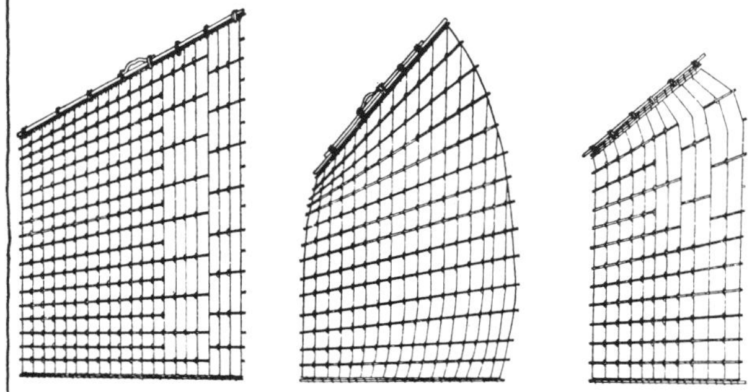

MIDDLE YANGTZE SIANG RIVER UPPER YANGTZE

TYPES OF CHINESE LUG SAILS

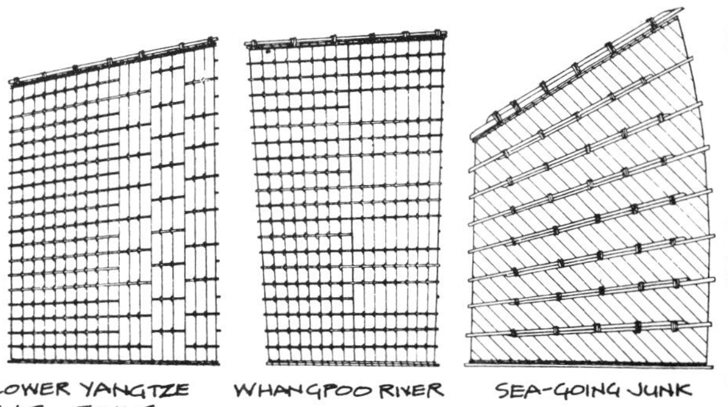

LOWER YANGTZE WHANGPOO RIVER SEA-GOING JUNK

The Canton water roads are, of course, well known. Here boats are tied up to the banks of the river and to each other in regular rows, 'packed like the scales of a fish,' as the junkmen say, and constitute nothing less than a floating city. Lanes are left at intervals of 20 or 30 boats to facilitate communication. This is very necessary, for the nautical 'commuter' returning from work may, as likely as not, find that his floating home has moved to another 'street.' No phase of life is unrepresented among this population. Kitchen boats supply hot food at a low rate. The barber calls in a small sampan which he rows himself, calling attention by ringing a bell. The river doctor also gives notice of his approach by beating a drum; and, when his medicines prove of no avail, there are floating mortuaries.

When he handles bamboo, the junkman's ingenuity finds its widest scope. He eats it in the shape of bamboo shoots, drinks out of it when it is made into a cup, sleeps on it when it is cut into bamboo shavings, which make excellent stuffings for mattresses. He uses it as a medicine and is finally carried to his grave by means of bamboo poles. Among a thousand and one nautical products for which it can be employed may be mentioned rope, thole-pins, masts, sails, net-floats, basket fishtraps, awnings, food baskets, beds, blinds, bottles, bridges, brooms, foot rules, food, lanterns, umbrellas, fans, brushes, buckets, chairs, chopsticks, combs, cooking gear, cups, drogues, dust pans, pens, nails, pillows, tobacco pipes, boat hooks, anchors, fishing nets, fishing rods, flagpoles, hats, ladders, ladles, lamps, musical instruments, mats, tubs, caulking material, scoops, shoes, stools, tables, tallies, tokens, torches, rat traps, flea traps, back-scratchers, walking-sticks, paper, joss sticks and finally rafts.

From *Sail and Sweep in China*

HOUSEBOATS, MEKONG RIVER, SO. VIETNAM

BOATS AS HOUSE, SAN DIEGO, CALIF.

HUGH'S ROOM

UP TO TALBOT'S ROOM

STEPS TO MAIN ROOM

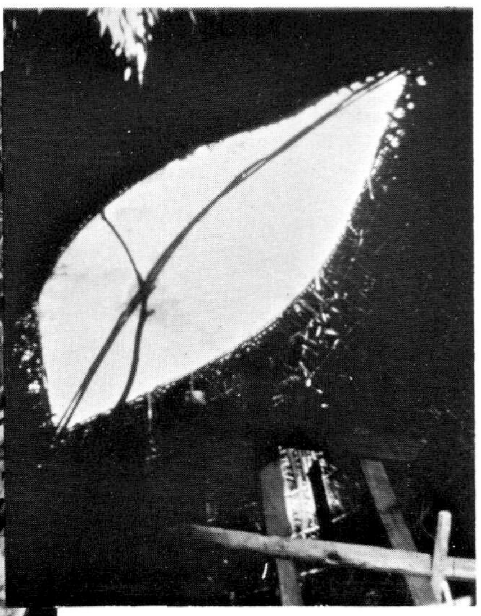

SKYLIGHT

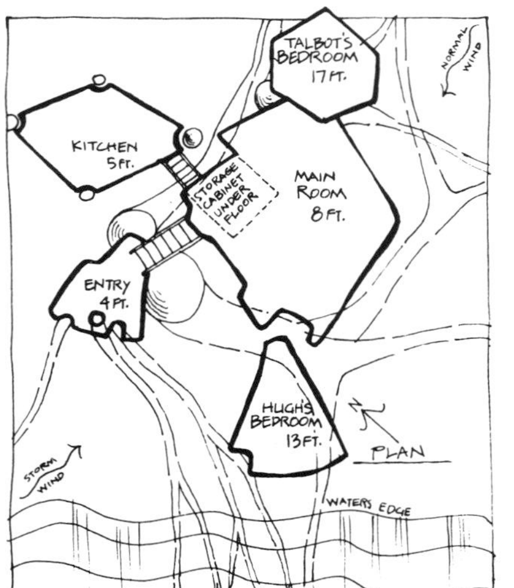

One of the nice things about thatch roofs is the life content —all the time lizards running around in it, & various insects, spiders, etc., couple of possums nest in it (once a big commotion when a 7 foot snake raided one of the nests — for a while it was raining baby possums in the bedroom. Snake ate 2 baby possums & left), & spot-breasted wrens roost in the eaves. Spot breasted wrens sing duets, one starts the song and another comes in with the singing and finishes it up....

The above is from a letter from Hugh Brown. Hugh and Jim Talbot built this five-level tree house on a Caribbean Beach near Trujillo, Honduras. They used local hardwood for flooring, thatch for roofs, and barbed wire as tensile framework for some of the floors and roofs.

...Between barbed wire and thorny palm thatch stems' effects, I once counted something like 220 small cuts on my right hand in various stages of healing; most minor, none serious...

Here is Hugh's story of tree-life in Honduras: watching birds, co-existing with insects, and what he calls "Solitary Life Sufficiently Removed From Anywhere Especially The United States."

•

After several months with a Rice degree and of being unable to come to concrete terms with the concept of gainful employment, I became aware that it was time to leave.

I had previously acquired a piece of land near a swampy area called El Mal Paso, near Trujillo, on the Caribbean Sea in Honduras, the obvious place to go. So in early April, 1970, after the rainy season on the Honduran North coast was over, I left. I was accompanied by two hundred pounds of baggage (too much) and Jim Talbot, who was convinced he would learn more architecture in Honduras than he would in the rest of the semester at Rice (and he got the Architecture Department to give him credit for it).

Soon after our arrival at the land, the house began to occur — a place to live by the beach. The design of the house evolved slowly, not so much from a master plan, certainly never from anything drawn on paper, but rather from a process in which each design decision directed the next through constant argument between Talbot and myself, which always produced better designs than what either of us had in mind.

What happened was a five-level tree house: from an entry porch at four feet, to an observation platform at eighteen, all built among five trees. The floors and framing were all of locally-cut mahogany, which is more resistant than pine to rot and termites. Since trees grow in irregular patterns and their movement with the wind had to be considered, we decided against having a rigid roof and instead made a tension grid of barbed wire which we thatched with palm leaves. This made for a free-form roof that we could adapt to cover each room in a way that helped define its space, and yet was still a single, fairly functional, continuous cover.

There were just minimal walls in only one of the rooms, all the rest with just roof, tree and floor, which reduced the difference between inside and outside and helped keep the house cool.

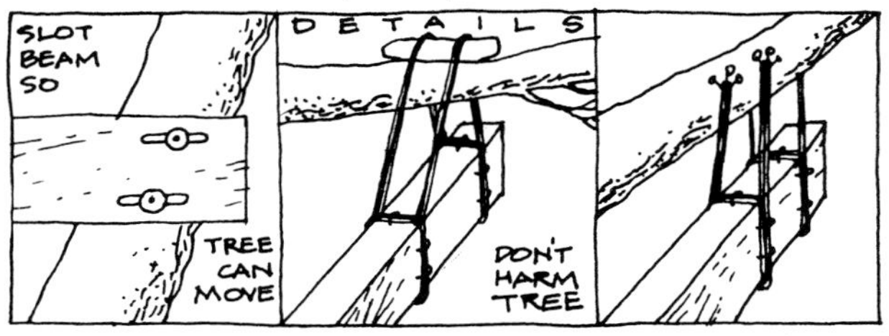

In late August, with most of the design work and about half the construction done, Talbot left to return to Rice for the fall semester, armed with dozens of photographs and drawings.

Furniture, as such, was kept to a minimum...I had no bed, but just slept on the bedroom floor, where I slept being defined by the presence of a voluminous gnatnet. For a pillow I used lumber scraps. Places to sit in the house were limbs, steps, or a stack of boards.

•

After a four-month introduction with Talbot began what I had come to Honduras to find: Solitary Life Sufficiently Removed from Anywhere, Especially the United States. I was, of course, not absolutely isolated from humanity; there were four thousand Hondurans living in Trujillo six miles south across the bay and another thousand in Puerto Castilla, three miles west, but my nearest neighbor lived two miles away.

I was alone as much as I wanted to be, and there was absolutely no one I had to answer to about anything at all. Naturally, I was the object of no small amount of curiosity to the surrounding population. My house was near a road, and often carloads (sometimes busloads) of people would get out and come trooping in to gawk at the crazy gringo heepee and his Tarzan-house.

I am not sure anyone there could really believe or understand why I was there, which was because I liked it, and that what I did was to live, think, and watch the birds. Such an answer to the question makes absolutely no sense in the context of Honduran society and culture. Sometimes when gawkers came to my house I would hide and listen to them:

"Meester Hugh! Meester Hugh!" (pause)
"He's not here."
"Very pretty construction of the house."
"Why does he live here, anyway?"
"He doesn't want to go to Vietnam."
"Why not?"
"Oh, it's some religion they've got."
"Hey, look, he's got food here. And people say he doesn't eat food, that he has some sort of pills he takes. I told you he ate food."
"There's Trujillo across the bay."
"Ow, there are mosquitoes here, let's go."

The role I adopted in the society was one of an outsider, and an unpredictable social element. People didn't know what to expect from me so they couldn't lay any expectations on me. This arrangement had various advantages, one of which was that, within fairly wide limits, I could do whatever I wanted to do.

There were, of course, unanticipated results of living thus. I became, contrary to my wishes and actions, the local bogyman. Since I was much bigger and much hairier than anyone in town, and lived alone in a tree off near the swamps, far from anyone else, the parents used me to discipline and generally terrify little children. I would be pointed to while the child in question was told, "You see him? Well, if he finds children who eat dirt (or beat up their little sisters, or throw beans on the floor, or whatever), he takes them away and EATS THEM UP."

There was no real way to shed this role without changing my whole trip, so I just did nothing to encourage it and tried not to corner children who were afraid. However, there were a lot that were not, and made the most of it, which was fun. There were a few hassles, but basically, the people and I, we had a good time, it was all right.

•

During the construction of my house, I had been watching and noting what birds I saw. As the house got closer to completion I got more and more into birding and it became my dominant physical occupation and a good point around which to organize my life.

The Trujillo area includes a wide variety of natural habitats, all of which are well supplied with birds, both resident tropical species and migrant and wintering species from the North.

In about two years of observation I noted 345 species there, or just over half the Honduran total. Every two or three days or so I would spend most of a day out birding. Sometimes I walked west on the nearly deserted North beach past mile after mile of sand dunes, coconut palms, and sea grape trees to the point, where there were always several hundred gulls, terns, pelicans, and shorebirds....

Often I went up on the mountain behind Trujillo, hiking up along the river valleys into the jungle. The river water is very drinkable

HOUSE

by Hugh Brown

THE WRITER

COROZO THATCH

INTERIOR, HUGH'S ROOM

TALBOT'S ROOM

ALASKA

TOP: SOMEWHERE IN INDIA. BELOW: CALIFORNIA.

and clear, and on their way down through the granite of four thousand foot high Cerro Calentura each of the four north-slope rivers has made a long series of very inviting bathing pools, overhung with the lush greenery....

Much of the time not spent birding was absorbed by the mechanics of daily life, which gradually adjusted themselves to the situation. Cooking reduced itself almost completely to making one large potful of food, usually every other day. Into the pot, about two-thirds full of water and hanging over the fire, went, in order, about a teaspoon of salt, a couple of tablespoons of sugar, a tablespoon of cooking oil, two cups of dried beans, three onions, cut up, some sweet peppers, cut up, two cups of dry rice, and about a pound and a half of white cheese, cut up. This was cooked together until the rice was done and then a canful of tomato paste was added.

Most of this I would eat slowly, drinking a lot of water to help with the starch, over the course of about three hours, while I read, got the birding notebooks caught up, wrote letters, or mended clothes. Often I left about an inch or two of food in the bottom of the pot to eat when I woke up before dawn the next day prior to going out birding. It tasted a little sharper in the morning, but I never got sick from it. This system worked out well in that it was necessary to leave for birding around dawn, no time to cook anything, and to have a lot of available energy for the day.

I never got into the fishing/farming/feeding-off-the-land trip. It might have happened had I been living with half a dozen others, but as it was, provisions were cheap enough in town so that it seemed best to stay in the money economy, since my cost of living was under a dollar a day anyhow. All my burnable trash went into the cooking fire. Fruit peels thrown into the bushes were removed quickly by the leaf-cutting ants. This left only empty tomato-paste cans, which I threw under the entry porch where they would not be stepped on and where the salt air rapidly converted them back into iron ore.

Bathing was done in the bay, or in town or on the mountain if I felt like having fresh water. Fresh water for drinking and cooking came from a dependable leak in the Puerto Castilla water pipeline, about a quarter mile away.

I tried as much as I could, to frequently examine the details and nature of the day to day routines of my life, trying to see what was really useful and necessary for living, and what was irrelevant social custom. Often this happened automatically. I remember when I realized that for days, perhaps weeks, I had been completely unaware whether or not my socks matched. I also stopped removing most of the insects from my food. I tried to separate what I liked from what I thought somebody thought I ought to like, which is not always easy, but is vital.

The presence of insects in the tropics is intense and where I lived was no exception. Living near a swamp as I did, there were many biting gnats and mosquitoes around. I gradually became more used to them in a variety of ways. I learned how to dress and organize my daily life to minimize their interference. Their presence was one reason I built my house in the trees – most bothersome mosquitoes are strongly ground-loving. On ground level there might be a couple of dozen mosquitoes after me, but a quick climb up the eight feet to the main room would shed them, at least during the day. While I was cooking on ground level, the smoke from my fire drove most of them away from my kitchen, and from the house above.

My body built up a tolerance for the itch fluid mosquitoes and gnats inject and often I was completely unaware of the bites. But perhaps most important is becoming emotionally immune, to stop feeling attacked when fed upon by insects, to be able to stay hang-loose about it all. I did not use insect repellent because it is expensive, ineffectual, mostly, and probably harmful when used on a long-term basis.

Every few weeks or months the army ants would pass through. The colonies consisted of about sixty thousand black, half-inch ants. Many carried larvae between their legs as they went, looking as if they were carrying grains of rice.

They would mass on the ground and begin their entry by sending long, exploratory, flowing columns of ants up the trees, and soon they all would be flowing across the floor, into the roof, everywhere. It was unnerving at first to see so many insects at once in an orderly, somewhat threatening invasion of territory, that I was unable to control, or even substantially affect – a bit hard on the ego. Besides, I could hear them marching (they make a low hissing sound).

The first time they came I got all uptight and attacked them, but later I thought it over and decided to just get out of the way and let them do their thing, which was to clean out the wasp nests, roaches (some of which were three inches long), and other small inhabitants. I once put out my dirty cooking pot for them, hoping they would eat it clean, but no such luck. I realized that the main body of them did not occupy all of my house at any one moment, so I would move around to where they were not.

My house was home for innumerable invertebrates, including two termite colonies, scorpions, blood sucking bugs, and small black weevils that continually fell out of the thatch in the bedroom.

It also housed various vertebrates besides myself. A possum of frequent reproduction made her nest in the thatch of the roof over my bed, and a small rat lived in the more permanent end of my clothesline. About half a dozen tree frogs shared the storage cabinet with a large tarantula, and little geckos and bark anoles constantly patrolled the treetrunks for insects.

Being extensively away from other people did a lot to my head. When one is primarily a social creature, there are hundreds of games, pretenses, images, fears, and fantasies having to do with human interrelationship that are always in the mind. They produce a lot of mental noise and tension, and prevent deeper and more generalized levels of consciousness from being heard.

Over a period of several months layer after layer of excess mental baggage dropped away. Sometimes it would happen without my being aware of it until much later, when I would realize I no longer felt at all about some relationship or aspect of society the way I had before. Progress would also be made through conscious thought processes. I did a good lot of thinking in my tree, thinking about just about everything that had touched my life, getting many perspectives and relationships I had failed to see before. Much time was spent without rationality and verbality, just letting memory's images flow, images from childhood, from Rice, from previous summers in Honduras, from imagination and dreams.

Occasionally some deeply repressed realization would surface. It generally would require a period of labor before birth into consciousness. All day I would not be able to concentrate on anything. I would go swimming and realize I didn't want to swim, fix food and not want to eat, start to read and be unable, begin to write with nothing to say.

After it got dark and I was under the gnatnet for the night, a long and intensely varied sequence of images and ideas would flow through my mind. I would momentarily stop on one, and suddenly I would realize one of those basic understandings about life that make its disparate parts fit together and harmonize, the sort of realization that puts everything else in a different place and makes one stand back in awe. The heightened tension thus broken, I would drift off to sleep amidst waves of euphoria and calm....

The constant presence of others in one's life tends to focus attention on a small area of the visual field directly in front, and on particular objects in the visual field. Alone, my attention tended to equalize over my entire visual field, and I became much more aware of what was happening in directions other than straight ahead.

The normal feeling of awareness of the world as a stage before the audience of oneself became shifted towards one of being immersed in a sea of sensed action. I became much more aware of what was in directions other than the seen, what was behind me, above me, below me – a different sense of space. Perhaps extensive birding reinforces this effect, but it seems to be primarily caused by solitude.

After being away for months from the constant noise of city life my hearing opened up to lesser sounds. Often at night I could hear the bass end of music playing in the bars in Puerto Castilla, three miles across the water, or the electric power station in Trujillo, six miles away. Sometimes on calm mornings I would awake to hear voices from fishermen in canoes far, far out in the bay.

The sound of the flight of birds was something I had never really noticed before, the difference between the sounds made by the wings of various birds, whether or not the individual wingbeats are distinguishable, what sort of pitch and rhythm the sound has, or peculiarities, like the squeaking of a vulture's wingbeats.

After one gets used to walking around in forests most short-range navigational decisions become almost automatic. It is as if one's feet think for themselves, deciding what to go around, what to go over or under. Main attention can be devoted to watching and listening for birds while one is walking along between trees, around rocks, over palm fronds, dodging spines and vines. The best route simply presents itself, without having to be thought out. Often the less one consciously, rationally thinks about one's path, the easier the going becomes....

I had come to Honduras knowing I would not spend the rest of my life there. Months before my departure, when I was still very actively involved in life in El Mal Paso, I knew when it would be, when I would feel the movement northward. When that time came, in September, 1972, I left my house on the Honduran beach and returned to the United States.

END...

TOWERS

LAKE VELDES, YUGOSLAVIA

ADOBE TOWER

POCITELJ, YUGOSLAVIA

MINARET, JAYCE, YUGO.

CALIFORNIA TOWERS

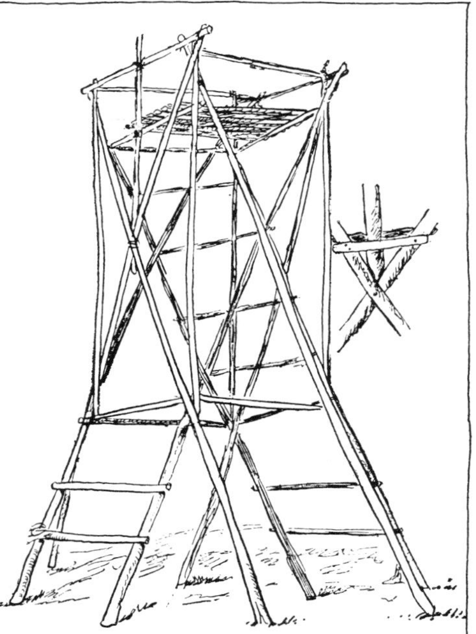

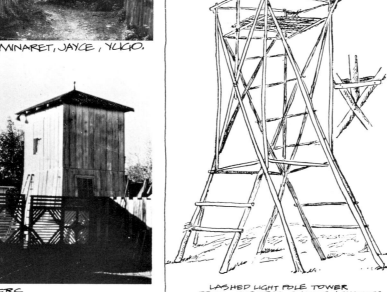

LASHED LIGHT POLE TOWER
FROM SHELTERS, SHACKS & SHANTIES

WATTS

Over a time of 33 years, Simon Rodia built what are now known as the Watts Towers in Los Angeles with no plans, no building permits, and no help from anyone else. Rodia, who came to this country from Italy at age 12 used salvaged pipe and steel, concrete, 70,000 sea shells and innumerable junk tiles and bottles. The tallest of the towers is over 60 feet. In 1954, work done, Rodia deeded his property and towers to a neighbor and moved away. In 1957, city officials declared the towers unsafe, but a committee to save the Towers prevailed upon the officials to conduct tests before proceeding with demolition. In October 1959, a heavy crane backed up and with electronic guages made a pull test of the tallest tower. In front of hundreds of shouting observers, the tower quivered slightly, but stood firm. One sea shell fell off.

BOLLINGEN

From *Memories, Dreams, and Reflections*
by Carl G. Jung

Gradually, through my scientific work, I was able to put my fantasies and the contents of the unconscious on a solid footing. Words and paper, however, did not seem real enough to me; something more was needed. I had to achieve a kind of representation in stone of my innermost thoughts and of the knowledge I had acquired. Or, to put it another way, I had to make a confession of faith in stone. That was the beginning of the *Tower,* the house which I built for myself at Bollingen.

It was settled from the start that I would build near the water. I had always been curiously drawn by the scenic charm of the upper lake of Zurich, and so in 1922 I bought some land in Bollingen. It is situated in the area of St. Meinrad and is old church land, having formerly belonged to the monastery of St. Gall.

At first I did not plan a proper house, but merely a kind of primitive one-story dwelling. It was to be a round structure with a hearth in the center and bunks along the walls. I more or less had in mind an African hut where the fire, ringed by a few stones, burns in the middle, and the whole life of the family revolves around this center. Primitive huts concretize an idea of wholeness, a familial wholeness in which all sorts of small domestic animals likewise participate. But I altered the plan even during the first stages of building, for I felt it was too primitive. I realized it would have to be a regular two-story house, not a mere hut

TOWERS

I no have anybody help me out. I was a poor man. Had to do a little at a time. Nobody helped me. I think if I hire a man he don't know what to do. A million times I don't know what to do myself. I never had a single helper. Some of the people say what was he doing... some of the people think I was crazy and some people said I was going to do something. I wanted to do something in the United States because I was raised here you understand? I wanted to do something for the United States because there are nice people in this country.

Simon Rodia

TOWER

crouched on the ground. So in 1923 the first round house was built, and when it was finished I saw that it had become a suitable dwelling tower.

The feeling of repose and renewal that I had in this tower was intense from the start. It represented for me the maternal hearth. But I became increasingly aware that it did not yet express everything that needed saying, that something was still lacking. And so, four years later, in 1927, the central structure was added, with a tower-like annex.

After some time had passed — again the interval was four years — I once more had a feeling of incompleteness. The building still seemed too primitive to me, and so in 1931 the tower-like annex was extended. I wanted a room in this tower where I could exist for myself alone. I had in mind what I had seen in Indian houses, in which there is usually an area — though it may be only a corner of a room separated off by a curtain — to which the inhabitants can withdraw. There they meditate for perhaps a quarter or half an hour, or do Yoga exercises. Such an area of retirement is essential in India, where people live crowded very close together.

In my retiring room I am by myself. I keep the key with me all the time; no one else is allowed in there except with my permission. In the course of the years I have done paintings on the walls, and so have expressed all those things which have carried me out of time into seclusion, out of the present into timelessness. Thus the second tower became for me a place of spiritual concentration....

SWEDEN

TOP, RUSSIA
BELOW, POLAND

POLAND

POCITELJ, YUGOSLAVIA

A termite tower in northern Kenya may house a colony of close to a million inhabitants. Constructed of earth and vegetation mixed with termite saliva and excreta, the thick walls become as hard as rock and efficiently control internal temperature and humidity. In some mounds, a network of channels near the surface provides air conditioning for inner chambers.

EASTERN EUROPE

The photos of the fantastic shelters and landscape of the Cappadocia region in Turkey are by Paul Oliver and Herbert A. Feuerlicht; each happened upon the area independently in 1972. Paul Oliver is editor of Shelter and Society, Shelter in Africa *and a book to be published in 1974,* Shelter, Sign and Symbol. *Herbert A. Feuerlicht is a sculptor and photographer and text below in italics was prepared by him. An excellent article by Roberta Feuerlicht,* The Cones of Cappadocia, *with photos by Mr. Feuerlicht appeared in the April 1973 issue of* Natural History *magazine.*

THE CONES OF

by Paul Oliver

About two hundred and fifty miles south-east of Ankara, Turkey in central Anatolia, there's a strange region of Cappadocia which lies around Nevsehir and Kayseri. It's a little like the eroded landscape of Bryce Canyon, but on a gigantic scale, where every pinnacle is riddled with holes and caves. These caverns are man-made, carved from the eroded rock of the pinnacles to make churches, monasteries, granaries, stables, houses and farms.[1] A unique, bizarre landscape, it has been burrowed into, hewn out, excavated and hollowed by generations of people for close on two thousand years.

It's the composition of the tufa rock that makes this possible. Tufa is compacted volcanic dust, which in this region settled from the eruptions of the volcano of Erciyes. The rock is easy to carve but when the new surfaces are exposed to the air they harden.[2] This allows the house-sculptors to work the material into durable homes that are literally a part of the landscape itself.

Kemal and Zehnep[3] are a typical, camera-shy, peasant couple who have carved out much of their own farm house. They appear to be in their sixties but are probably younger, the rigours of farming in this semi-desert, but often cold and wind-swept region, showing in their dark, seamed faces. They own a small holding of a few acres, keep a cow, goats and chickens, raise fruit, vegetables and grain. There's a little surplus, enough to enable them to buy some commodities in the towns like Urgup or Augula, and there's enough time left at the end of the working day for Kemal to peck out a little more of another room. Kemal owns two pinnacles as part of a divided patrimony inheritance; he lives in one and

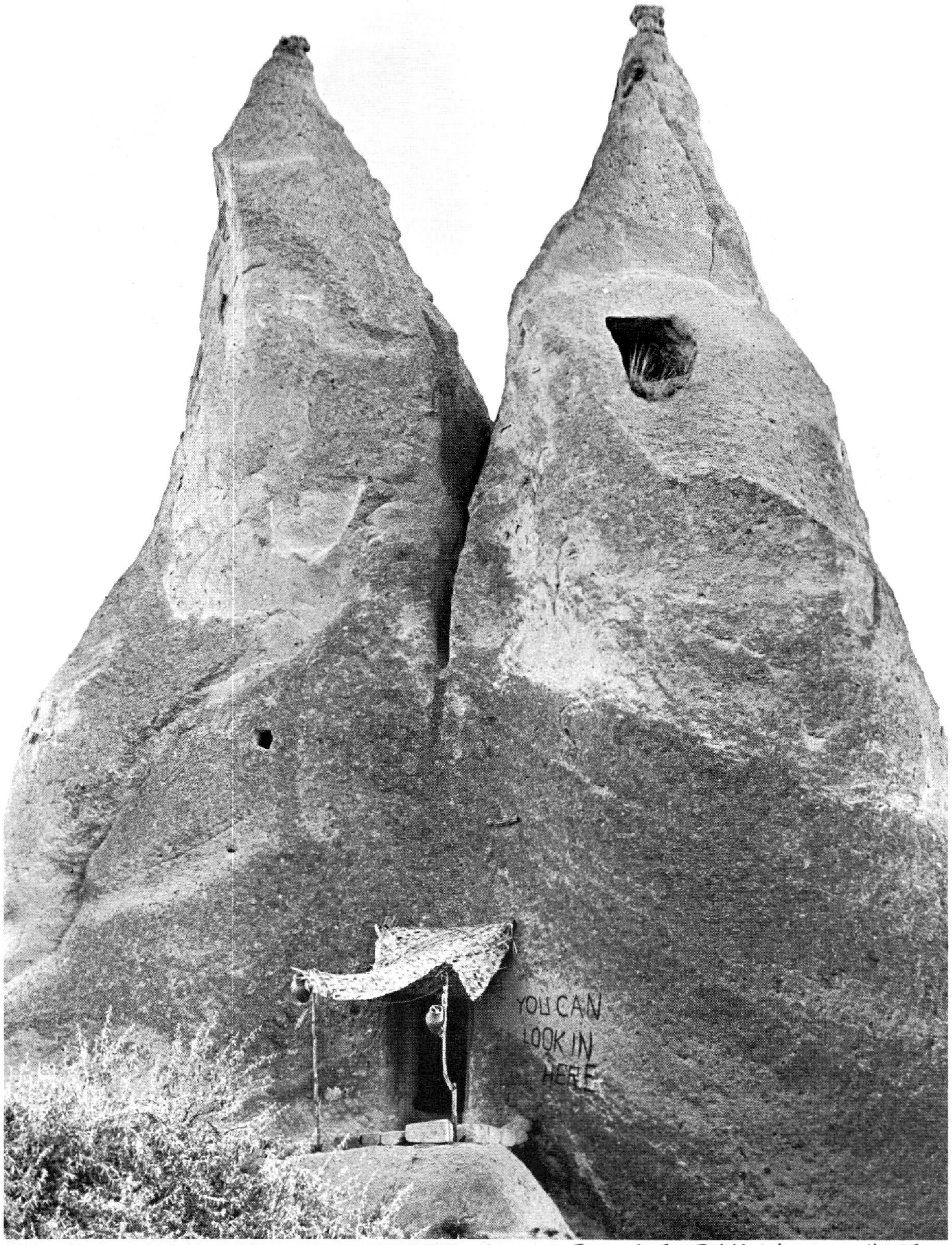

1

ROCK AND ROLL, SOFT DRINKS AVAILABLE HERE.

The people are quite friendly and hospitable. The tradition of hospitality to the wayfaring stranger has persisted through the ages, and extends through much of the Balkans, although the welcome of some of the younger generation of Western "back-packers" has begun to wear thin due to an abuse of this hospitality. The guest is never asked to leave; he is rather expected to understand what is appropriate.

4

ZEHNEP'S KITCHEN. WOOD STOVE, BUT NEW ELECTRICITY FOR LIGHTS.

3

2

INTERIOR OF ANCIENT JAIL, EACH CELL CARVED SEPARATELY FROM THE STONE.

AMPHORA AND PITCHERS LINE WALLS OF A STORAGE ROOM. 5,6,7,8→

THE BASE OF THE CONE IN THE CENTER PICTURE IS BEING ERODED BY SPRING FLOOD WATERS. DARK, HARD TOP STONE, SOFT MIDDLE LAYER, HARDER BOTTOM.

uses the other as a granary and pigeon-cote, the latter mainly for the fertilizer.

Extremely neat and orderly, he has tried to make his rock dwelling look more like a conventional house by painting the outside at ground level with white-wash. Inside, the walls of whitewashed tufa are dressed finely in the main reception-bedroom and are more crudely treated, left the natural tufa grey, in the kitchen and store. Floors are swept regularly but Zehnep is proudest of the bed-living room, which is hung with locally woven rugs. A low couch spans the room below the glazed windows and decorative woven

cushions and embroidered pillows are laid out neatly and symmetrically on it.

Carefully cutting away the tufa with an adze-like tool, and carrying in his mind where the steps should be, Kemel has recently hollowed out a new room with a staircase leading to a chamber above, which has taken him three years to complete in his spare time. This new upper room is used for weaving and also for sorting fruits but he has fitted a wrought iron balcony so that he and Zehnep can admire the view. In a niche on a precipitous part of the adjacent pinnacle he has placed a wall in front of a cave, which now forms his pigeon-cote. The nesting boxes are also cut into the tufa and eventually he will dismantle the wall to get at the *guano* – if the stone doesn't collapse first. When it does, he uses the tufa rubble to build a wall round his plot; nothing is wasted in this simple but timeless economy.

In recent years the discovery of the rock monasteries of Goreme has brought increasing numbers of tourists to the region. Kemal and Zehnep hope that some of the money tourists bring will come their way; one can only pray that it will bring them happiness, for their modest, rational way of life may not gain much from the dubious benefits of an economy based on tourism.

Notes

1. Further details of these carved churches and monasteries will be found in "Another Thick Wall Pattern – the Cappadocian Way of Making Spaces" in Paul Oliver, *Shelter, Sign and Symbol,* Barrie and Jenkins, 1974. The chapter is by Suha Ozkan and Selahattin Onur, to whom, and to Aydin German, my thanks for making my own trip to Cappadocia possible.
2. Ozkan and Onur, ibid. give details of the chemical process involved.
3. Actual names with-held at their request.

9

10

A typical household we visited in Ortahisar had a receiving room where guests are entertained. One or two glass front cabinets displayed the family heirlooms, special dishes and glassware and the like, and the rest of the wall space was lined with continuous seating areas, much like benches or day-beds, except that they were built up of clay brick or some similar material plastered smooth and whitewashed. Over all was an interlocking overlapping multilayer of rugs which lapped over to the floor. Several pillows were scattered against the walls to lean on, and the depth from front to rear was about a meter. Shoes are removed at the entrance and left just outside. The floor of the room is hardpacked earth and is also covered with a multitude of rugs all overlapping and several layers deep. A small low serving table was soon laden with fruit including apricots, watermelon, and very small pears, as well as goat's milk cheese in large pieces and flat loaves of a heavy bread somewhat orange in color and heavily laden with caraway. The bread, cheese, and fruit are fairly staple as first offerings throughout the Balkans, although sometimes sliced fresh vegetables such as tomatoes and cucumber will appear. Sometimes tea is served, sometimes just water but rarely, if ever, coffee, at least not in Turkey. Liquor was never served to us in a Moslem household.

CLOSE UP OF ENTRY TO APARTMENTS IN CONE SHOWN ABOVE. NOTE STAIR AT RIGHT. 11

Water is a real problem for these people, as the soil is quite fertile, but they have no plumbing, either for domestic or agricultural water supply. Water is carried in jugs by donkey, or on yokes across the backs of those too poor to own livestock. It is a precious commodity, which destroys and washes away the soil at spring flood and must be zealously sought and rationed in the drought of summer.

12

CAVE INTERIOR

Photos:
Herbert A. Feuerlicht: 1,2,3,7,8,9,11
Paul Oliver: 4,5,6,10,12

NEW YORK CITY

Ned and Micheline Cherry and their three children have lived in New York City for the past 10 years. Micheline works off and on as a nurse, Ned has worked on numerous architectural projects, received his architectural degree from Columbia in 1970, and his license in 1973. Ned has stayed in the city because he feels he can do something to help a grim situation. After years of working with both uptown and maverick architects, Ned put together some thoughts from his experiences; not a very hopeful picture:

Sitting here in the relative security of an 1890's old-law, rent-controlled, 5 story walkup tenament, in a reasonably civilized part of New York City, one wonders what kind of meaningful contribution one can make to a book entitled *Shelter,* dealing with indigenous, people-built, non-architect designed, alternative habitat. Being an architect, of sorts, dealing primarily in "community oriented architecture," i.e., child day care centers, housing rehabilitation for lower income tenants, community centers for the elderly, etc., makes it even more difficult in terms of dealing in alternatives which are not really optional to the majority of the people, as are artists' lofts in converted ex-commercial/manufacturing warehouse type buildings in Soho, but these are shelter for the privileged with options, options to split from the city at will, options to put down a lot of money for rent or lease and reconstruction of previously uninhabitable space. Lots of people are doing it, waiting for small textile merchants, dealers in recycled rags and the like, to fold up from city pressures, and then taking over huge (2500-5000 sq. ft.) spaces, doing some underground (like PVC plastic, thus far kept out of NYC by unionism) plumbing, a sleeping loft, etc. and voila, 5000 sq. ft. of space, equivalent to (5) 3-bedroom low-income apartments, occupied by artists who don't paint, sculptors who don't sculpt, perhaps an accompanying mate and the family dog, where once were employed 50 people providing life to the city now becoming the playground of the rich and the ghetto for the poor; the cycle is complete.

Another thing is go to the country, build a dome, take your much needed-in-the-city skill to the country and do carpentry for the rich who not only own all the brownstones in the city but all of the nice old farmhouses in the country. It's like driving a cab in the city, a would-be hip profession, yet one catering primarily to the wealthy and/or profligate or else ripping off a welfare mother hustling downtown for an appointment with the social worker and afraid to rely on public transportation which might mean a check cutoff. It's all about who are you serving with your energies. Now there are options like doing freebie carpentry where it's needed most and doing equal time in the suburbs for big money to make it all come out, like Robin Hood. Architects can do the same thing, but it kind of hurts designing duplexes for Bosie Cascade Board Members for bread while doing a preliminary zoning analysis, feasibility study and design for a group in Harlem that's been waiting for three years for a place to put their kids while they go to work, when you know that the day care center isn't going to be built anyway and you'd better believe that the duplex is going through even if the board member has to pay off the building inspector to get his illegal "picture window" put in, the one that was in *House Beautiful* last month, you know the one.....

Well there are some successful squatters in New York City for awhile anyway; it's called in Harlem, project "discovery" like when the white people came here a long time ago and "discovered" amerika although there were other people here already; sometimes it works with building sites too, when people are "discovering" all these neat building materials and taking them back to 127th St. to make a park because there aren't any within 20 blocks. Yeh it's called "Project Discovery". Anyway there are really a lot of nice places to live here in newyorkcity but you gotta have a lot of money unless you're lucky like me and live in a rent-controlled apartment — 5 rooms for $91 month right around the corner from Mayor Lindsay and 500 feet away from where a girl was raped and murdered last week, practically in full view of the mayor's police staff who are only concerned with keeping out mothers who want more day care centers. Anyway, when I move out, my apt. becomes automatically decontrolled and the sky's the limit on the next rent and since Mayor Lindsay lives right around the corner and this is a civilized neighborhood, the next rent could be $400/month — probably suitable for a $75,000/yr. doctor from the hospital on the corner except he probably wouldn't like a 5 story walkup even around the corner from Mayor Lindsay so they save it for 5 nurses at $80 apiece or maybe 10 stewardesses at $40 apiece but you can bet your ass that it won't go to anyone who can only afford $91/month which is about 75% of the people in funcity. We also have Mitchell-Lama middle income housing here starting at from $90 to $120 per room with the ceiling for qualifying for "middle-income" housing at around $54,000/yr. now. But there are lots of nice places to live in New York City, some even have lots of trees lining the streets but the rents are proportional to the height and diameter of the trees. A lot of the trees were planted by low income immigrants, from the south as well as from the rest of the world but as the trees began to grow so did the appetites of the landlords so they evicted all the low-income people which is legal if you provide 25% more units in the building at a minimum of 395 sq. ft./unit, and turned 5-room apartments for the working classes into 2-room studio apts. for the leisure classes and everybody is happy except for the people who planted the trees, watched them mature and now can only find similar rent housing in the Bronx or Coney Island where the trees have to be planted again. Again the complete cycle.

There are also some mildly successful communal efforts in the city but most of these people don't seem to want to stay long and lots of them move to the country, taking along their needed skills...Then there's subsidized housing which Tricky Dick has put a screeching halt to and a myriad of government programs to assist the disadvantaged like the Municipal Loan Program which ended in corruption, the Emergency Repair Program which ended in corruption, FHA assisted home ownership programs which ended in corruption, direct-lease, privately developed day care programs which ended in corruption, model cities, head-start, hot lunch, infill-housing, adventure playgrounds, vest-pocket parks, all of which are threatened with financial extinction while we get $350,000 playgrounds on 5th Avenue in Central Park, luxury housing built on top of public schools in "good neighborhoods" while ghettos get "found space" classrooms in converted bowling alleys, you know like indigenous architecture; Columbia University eliminating its night program in architecture and requiring a bachelor's degree for admission where once people with no bread nor previous educational opportunities could study architecture to come out to discover that the only way to make a living in the profession was to work for Skidmore, Owings & Merrill on hotels in South Africa with separate toilets for blacks and whites or maybe a Dow Chemical showroom on Madison Avenue because we're not building any low-cost housing for the next 18 months or any day care centers for people on welfare because we want them to work but we don't want to encourage them to have children because there are people at Stanford who say dumb people have dumb kids so let's not have any day care centers and if developers can't make money building subsidized housing we shouldn't build any so let's build more luxury housing and office buildings so everybody can make money and the welfare mothers can go to work and not have dumb kids. There are lots of nice places to live in urbania, but you've got to know when the trees were planted....

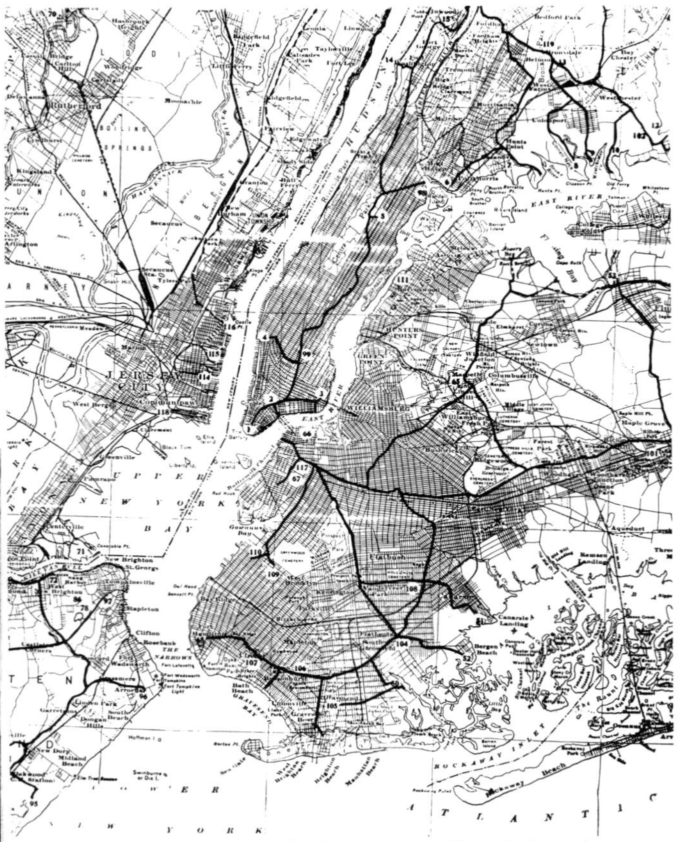

Indian Paths in New York City. Werpoes village indicated by number 2.

GUTTED HOUSE TRANSFORMED WITH SCAFFOLDING & HARDBOARD MEZZANINE FLOOR STRUCTURES. CASTLE ROAD, KENTISH TOWN.

300 SQUATTERS IN BLOCK IN SOMERS TOWN, LONDON. HOUSES BACK TO BACK – THE BRICK FENCES WERE PUSHED OVER – GARDENS & PATHS CREATED – IT'S NICE HAVING "JOINT ACCESS.

COUNTY COUNCIL GUTTED BUILDING. THIS TYPE WAS USED FOR TRANSFORMING WITH SCAFFOLDING. KENTISH TOWN.

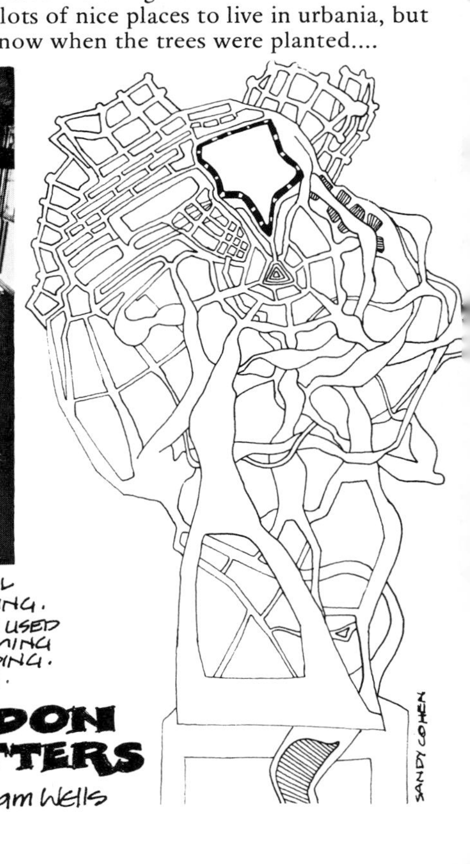

LONDON SQUATTERS

by Graham Wells

MEDIEVAL NYC

(The following is extracted from a treatise on architecture (as yet unpublished), File Under Architecture *by Herbert Muschamp.*

Jung was afraid to go to Rome. He tried. He got as far as Pompeii before he backed out. He feared the master pull of such an intense concentration of mundane energy, the dense layering of secular rituals which had shaped the culture from whose last decadent gasp he worked his whole life to escape. The air in Rome was poisoned, especially at night when the ghosts of history fluttered down like antique costumes, preventing the modern age from discovering itself.

Pitched halfway on the space/time scale between civilization and the universal city, New York City has no such restraints on its virility, and there the morality of poison passes from the mundane to the hysterically sublime. Ugliness, fatigue, drama, boredom, danger, noise and hopelessness combine in quantity sufficient to mesmerize, transmogrifying random quirks and flashes into the mercurial gestalt of the holy city, the place where the simple act of being is enough to sustain the progress of the hours. Nothing is believed, no one is saved. The dogshit dialogue, the many city forms of grit, the grace of garbage and the innocent squalor of ragbag mistakes take one ironically closer to Eden than all the attempts at beauty control, the drawing board versions of Paradise, could ever manage to do. The glass of curtain walls and shop windows reflects dimensions which cannot be duplicated by men except in the insect lifespans of fashion. The highest ideals of humanity are consumed with unparalleled relish, instantly digested, and thrown out with the trash. In the dormitory of danger, there is no fear of death.

One lives there and beholds no architecture; all is perennially plastic, intuitive, sacred beyond the sum of its secular parts. It was very trendy of Venturi to take his Yale design class to Las Vegas for a semester, but Las Vegas is just an X-ray of very old-fashioned architecture, just enlarged in neon. In New York the ether is so dense you can't even see the electric signs, much less the buildings behind them. It is always a shock to notice, as one does on rare split-second occasions, that New York has buildings, and that the buildings there are just like buildings anywhere else, no different from buildings in the great waste stretches of small Midwestern cities. For example: the same old street row architecture, the same standard details around windows and doorway, arbitrarily varied from block to block, from building to building. Usually your eyes bear witness to nothing architectural; building after building, all blending into a rising and setting miasma; the reality of a figurative climax at every red light, the flurry of a million separate moving parts hitting, missing, hitting and running, generating the white-heat conditions that make a mirage.

New York is the world's most medieval city; the narrow streets, buildings meeting at the top, squalor, paranoid nightwalkers, the skill required to steer yourself around the Scylla and Charybdis on every block, a whole life of drowning flashbacks encountered on every corner waiting for the light to change. Heading uptown or downtown in a cab, falling apart, lighting cigarettes, humming songs in a traffic jam, laughing at the amazement of being always two or three hours late, and in the midst of it you are always there, in New York, pretending not to wait, and sometimes X and Y get together and you just have to celebrate the fact that there's no design like the present.

Far from the semitropical drumbeat island where you alone possess all the energy you need to shape the day's events, this is a place where there are so many psychic covers placed layer upon layer between you and the planets that you have to buy houseloads of extra energy from Con Ed to forge the simple connections necessary to personal sanity. The Pan Am building provides a perpetual full-moon logo over Manhattan's lunar frenzy; the towers at night form a gigantic planetarium showing forth the fluorescent constellations of the New York process. People have gathered together here to turn out weekly the editions of the catalogue of software essential to the process of filling up time. That this appears to some to be a fundamentally immoral practice is irrelevant. Those who think that the emperor has no clothes are unfit for their station, and go off to become critics.

For in New York one learns that environments are epiphanies of human nature; one passes judgment on them at the risk of exposing one's failure to use them to advantage. Environments are practical jokes played on space; when the joke begins to backfire, it's time to begin again. The demolition of buildings, an activity of paramount importance to the working strength of New York, is highly valuable in providing visual evidence of the mortality of the environmental conventions we often take for granted. Here one day and gone the next, environments are shaped by perceptual patterns determined by the character of our motivations. As motivations change, so does perception and thus "the environment" is altered.

There is nothing at all new about environmental pollution apart from the fact that only recently have cultural motivations combined to create the formula by which pollution is perceived as an enriching aspect of conditioned space. Space contains both litter and litter basket. Culture created the town and the outskirts of town, and then obliterated the barriers as populations grew. Culture defines the awareness of room in terms of a room or several rooms and is free to shatter that convention whenever necessary. Space abides over the eternal process of its use, whether for just or unjust, moral or immoral purposes. None of them last. There are no implications for design....

Werpoes village enlarged. Broadway, Park Row, Bowery, Greenwich Ave. follow old Indian paths.

New York City produces over 30,000 tons of garbage per day. 20% of this consists of quality paper: canceled checks, rough drafts of Broadway hits, executive memos, IBM punch cards and so forth. On Mondays alone, seven million tons of the New York Times are picked up by NYC garbage men. In addition to this, New York's 600,000 dogs deposit 200,000 pounds of dogshit on city streets and sidewalks daily.

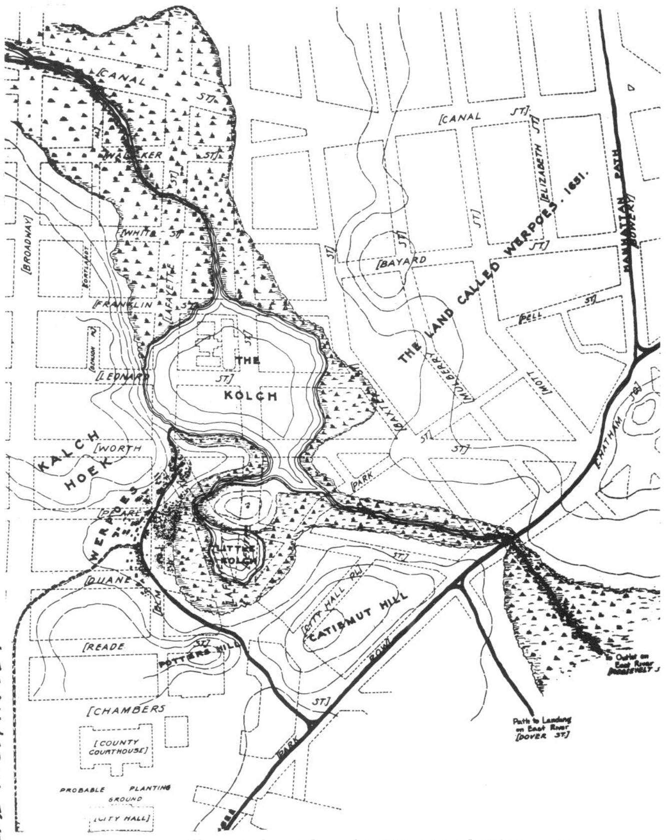

Werpoes village enlarged again. New York City maps: Indian Paths in the Great Metropolis *from Museum of the American Indian, NYC. Thanks to Peter Warshall.*

"LADY ROSE" MADE A HAND LOOM TO RECYCLE HER UNPICKED WOOL FROM OLD JUMPERS AND SOX. LATER, SHE & STEVE MADE AN UPRIGHT SHUTTLE LOOM FROM OLD PACKING CASE WOOD, MADE A RUG, THEN SAT ON THE RUG BY THE FIRE FED BY THE LOOM, TIME FOR CHANGE.

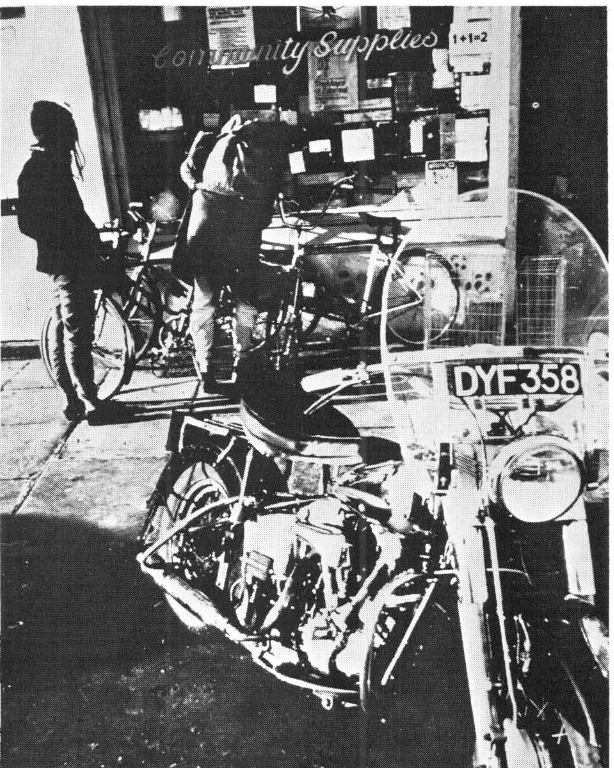

SQUATTER WHOLE FOOD, INFORMATION AND COMMUNICATION CENTRE. KENTISH TOWN.

BLOCK WITH 500 SQUATTERS, KENTISH TOWN. ALL HOUSES IN TRIANGLE EXCEPT 2 FOR MOST CORNER DWELLINGS. PLUS COMPLETE HOUSING ROWS ON EITHER SIDES OF TRIANGLE — BARING THE ODD PRIVATE BUSINESS. THE SCRAP YARD IN TRIANGLE HAS BEEN PROPOSED AS AN OPEN SPACE FOR RECREATION — EXPERIMENTAL STRUCTURES, FREE PEOPLES SPACE, ETC, ETC, ANYTHING.

RIGHT, SQUATTERS EVICTED, BUILDINGS SMASHED FOR OFFICES.

ADOBE BRICK AND PLASTERED APARTMENT BUILDING
JEDDAH, SAUDI ARABIA

BARRIOS

Why should governments try to do everything in their power to suppress squatter settlements when often they work so much better than the government housing that is supplied at much greater cost and effort? A most striking example of this absurdity occurred in Lima, Peru where at least one quarter of the population now live in squatter settlements, or *barriadas.* The way these *barriadas* were formed is instructive.

After the Second World War, they spread very quickly because of rapid urbanization. They were immediately regarded as social evils, as seedbeds of communism and as centres of crime and prostitution. In fact, as it turned out, the realities were quite different: the incidence of crime was much lower than in the urban slum and the sociopolitical views of the inhabitants were, ironically enough, conservative. But the government, social workers and architects were determined to regard them as social evils, and therefore they were made to fit this role at all costs. Naturally, the would-be squatters became activists in a literal sense. The only way they could gain rent-free land and leave the unendurable, urban slum was to form organized platoons and *invade* government-owned land during the night. Invasions had to be carefully organized with an avant-garde made up of lawyers to choose the site, lay-out men to draw the boundaries for streets and lots, and a woman known as the 'secretary of defence' whose precise role is somewhat obscure but whom one imagines acting both as a normal secretary and a buffer to the police. After an invasion (there have been more than a hundred around Lima) the police manage to counterattack and lose, or if they do happen to win and clear the site, it is only a matter of time before it is reoccupied and sooner or later recognized by the authorities. Then given a few years, a two-stage development sets in where the squatters set up more permanent abodes, built of cement rather than straw, and start their own form of sociopolitical organization. Soon they are organizing yearly elections of their own governors in a country where local democracy was unknown for more than sixty years. Inevitably there are drawbacks of an administrative kind: the squatters have trouble constructing large-scale facilities such as sewer systems and often have to depend on the central city for schooling. But the area in which they undeniably excel is in community spirit and popular initiative. There are endless examples of communal construction and communal services and, perhaps more important, self-help. In the *barriada*, the individual can construct and destruct his house according to need without government interference. For instance, if he needs a second story or a new room he can add one without restriction; if he needs a yard for raising chickens and guinea pigs he can chop down an old room. Thus the needs are satisfied according to individual priorities and not according to bureaucratic protocol: shelter comes before amenity; walls and roof before electricity and a bath. In a very real way, the *barriadas* are now proving that the best and most satisfying low-cost housing, is in fact the cheapest. Hence the Peruvian government has even had a change of heart and is now beginning to support their development and ask architects such as Aldo Van Eyck, Christopher Alexander and James Stirling for their suggestions on aiding the 'Young Towns' as they are now approvingly called.

The reason for the recent popularity of *barriadas* in architectural circles is that they dramatize in a non-romantic way the opportunities for determining one's own way of life and life-style. Many other examples of this individual initiative could be given in the West, from the 'self-build' groups in England to 'Drop-City' in America, but none of them go to the extremes of the *barriadas* because the city authorities are much more adept at fighting off invasion and keeping the poor in their place. As some activist has said, the only way to get out of a Western ghetto is to burn your way out. There is no chance of occupying new land and the poor can only rent tenements from public housing authorities or from slum lords.

Architecture 2000,
by Charles Jencks

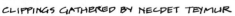

CLIPPINGS GATHERED BY NECDET TEYMUR

HUMAN CHAIN TRANSPORTS MATERIALS TO HILLTOP TO BUILD WATER RESERVOIR. BARRIO DWELLERS OF LIMA, PERU.

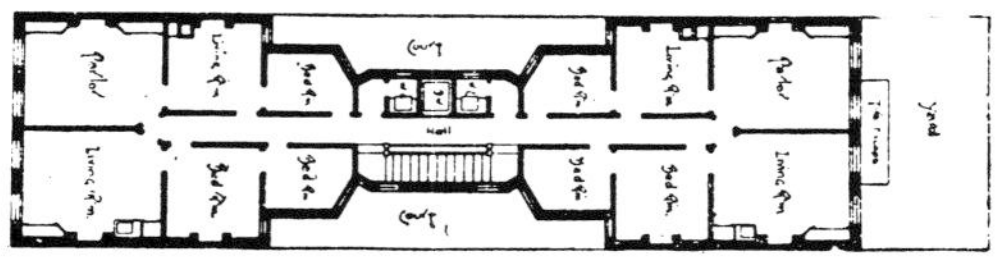

The medieval French believed that Gypsies came from the Central European fields of Bohemia. They called these Gypsies Bohemians, and when the age of romanticism touched French literature this name was exaltingly used to denote the penniless and carefree writers, poets, journalists, artists, actors, sculptors, and other members of that wide group which later the French and Russians so aptly labelled "intellectual proletariat."

Bohemian circles could be traced in America at least to Edgar Allen Poe and the "Pfaffians" who frequented Pfaff's restaurant at 633 Broadway. Later the circle included Walt Whitman, and at the turn of the century there were Maria's, the Bohemian Club in San Francisco, and the Bismark Cafe in Austin, Texas, where William Sydney Porter (O. Henry) published a magazine called *The Rolling Stone.*

The Greenwich Village area in New York was literally and figuratively on the border between the ghettoes to the south and east and the baronial stores and mansions to the north. Before 1910 the Washington Square environs were largely populated by social workers and political reformers. But in its days of magic the area had low rent, cheap food, and a quiet, self-contained village atmosphere. Artists began to pour in and spring up out of the square. Broken-down stables were renovated into studios, and galleries and workshops materialized everywhere. The low-rent flats and cheap restaurants filled with painters, reformers, writers, revolutionaries, sculptors, dilettantes, anarchists, actors, adventurers, poets, and dancers of all varieties, all of whom seemed to know each other.

Harvey Wasserman's History of the United States

It's ours. Right down to the last small hinge it
all depends for its existence
Only and utterly upon our sufferance.

Driving back I saw Chicago rising in its gases and I
knew again that never will the
Man be made to stand against this pitiless, unparalleled
monstrocity. It
Snuffles on the beach of its Great Lake like a
blind, red, rhinocerous.
It's already running us down.

You can't fix it. You can't make it go away.
I don't know what you're going to do about it,
But I know what I'm going to do about it. I'm just
going to walk away from it. Maybe
A small part of it will die if I'm not around
feeding it anymore.

Chicago Poem
Lew Welch

L.A.

Mention L.A., and most people think of smog, freeways, Hollywood, TV, Disneyland, housing tracts, barbeques, and swimming pools in about that order. These things aren't *products* such as food, steel or timber, but are concepts, *invisible*: transportation, mass communications, advertising, poisonous gases. They are, in their present form, all recent to human experience, as is L.A.; just 50 years ago, it was a village surrounded by orange groves.

L.A. has always attracted people out to make it, it's an open raw market for the new and untried. A few years ago, a friend ran out of money, so he moved his family back to L.A. He was a musician and writer, besides being a native of L.A. He felt he could make it there one way or another. He eventually did, setting up a small advertising agency, specializing in record album covers, and publicity campaigns.

People in L.A. are pre-occupied with *image.* One of his first jobs was at Capitol Records in Hollywood. Once after a meeting there, he had to give an executive a ride home in his small, old funky Renault. He was embarrassed, he simply couldn't afford a better car. The executive was uncomfortable, suspicious. He asked, "*Driving this old car, what are you trying to prove...*"

My friend was lucky, he got out of L.A. after making some money, but most people there just get caught in the big hamburger machine, any substance in their character is processed into style and image. They roar around the freeways 70 mph bumper to bumper, realizing that the smog makes the steak and swimming pool possible.

In this sense, L.A. is the harbinger of the future. There aren't many people around the earth that don't want the gadgets of America: cars, the tract house, color TV, if they can get them. A lethal kind of paradise has replaced the orange trees of Southern California. With luck, we'll run out of oil before terminal Los Angelization grips the earth.

Bob Easton

On highway 14 between Albuquerque and Santa Fe, N.M. is the abandoned coal mining town of New Madrid. 340 acres, a restaurant, liquor license, museum with 14 antique cars, a host of dilapidated buildings: sale price, $500,000. We photographed it in November 1972, and it reportedly sold to Ballantine Associates of Palo Alto, Calif. in August 1973.

TOKYO COMMUTER DIVES IN TRAIN WINDOW

BANANI

Banani, a cliff-debris village of the Dogon people of Timbuktu.

Up against the *Baniagara* cliffs of *Banani.* Note three children on flat roof and man in white above to left of them. In cliff crevices are built the *Tellem,* structures that contain woóden statuettes which Dogon say were left behind by former inhabitants of the land. "Those who were here before us" fled at the coming of the Dogon. But their mythical descendants are still considered rightful owners of the land.

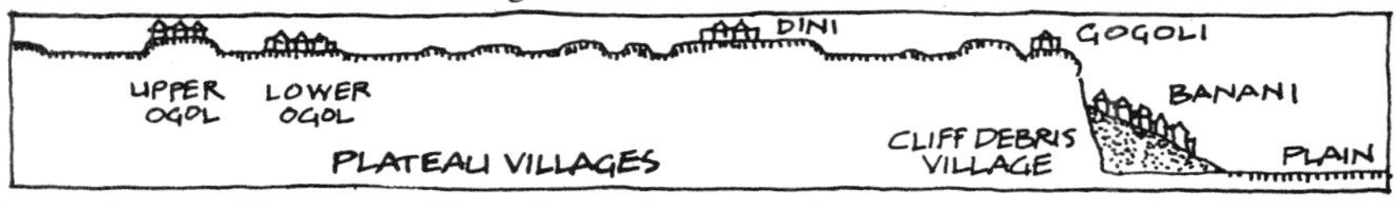

My house is my village, my village is my house. In his essay on the Dogon people, Fritz Morgenthaler tells of an experience with Dommo, of Andiumbolo where he learns that the Dogon definition of *house* is the people living there, not the structure, and why a Dogon house is never sold.

> *"...one day he took my hand and led me over high rocks up to Andiumbolo. Stopping at the entrance he said, 'This is my village.' He spat, took my hand again, and said, 'I want to show you my house.' "*

Dommo leads Morgenthaler on a circuitous route through the village for over an hour, visiting first the elders' council-place, then the chief's place, the priests', the family elder's. Finally, Morgenthaler turns to go home and Dommo asks, disappointed:

> *"Don't you want to see my house? I don't live where we just were — when my wife is working in the fields or is sick and can't cook, I eat in the big house. I'm home, here." So we walk into a yard near the entrance to the village. It is the place where, more than an hour ago, Dommo said he wanted to show me his house.*

Each house was Dommo's in a particular way, and their route led them back to where they started, when they entered the house Domo shares with his first wife and their children. It is his "own" in this particular respect, as are the other houses. As with houses, so with villages. For each individual, the village is "his" house, other villages are "his" village. More on the Dogon in *Meaning in Architecture.* See bibliography.

Looking down on Banani.

$40 A MONTH

CENTRAL SHAFT CUT THRU 4 FLOORS AND SKYLIGHT INSTALLED IN ROOF. HAND POWERED OAK CHAIR ELEVATOR.

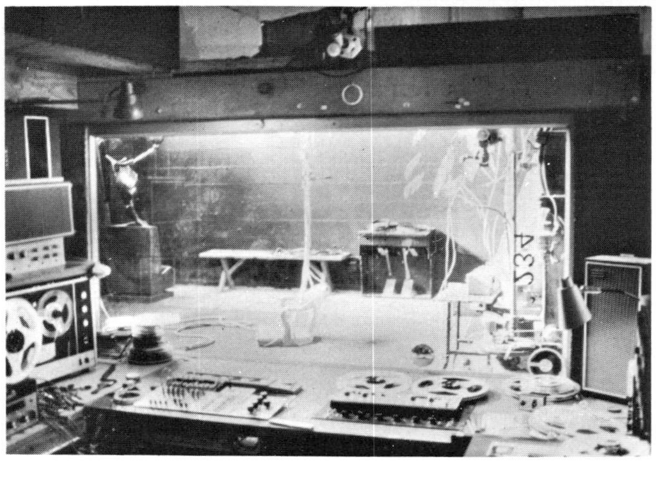

In an old 500 room condemned big city hotel, Roger Kent has created an imaginative living and working space, which includes a recording studio, hand powered oak chair elevator, numerous spaces connected by stairs and passage ways, and rooftop strawberries. Roger spent six months tearing out parts of the building, then using the salvaged wood to rebuild. There are now four other living spaces in the building, with about eight people, although at one time there were 50 inhabitants.

"It lasted about three months," said Roger. "50 rooms were occupied and the other 450 felt haunted. The building was open and the halls were like city streets, full of strangers—wierd. Finally everyone except two of us got evicted because there was only one toilet for 50 people."

From that time, the building has been kept locked and things are well-ordered, but occupants expect demolition any time, as the building is in an urban redevelopment area. For his 40-room space, Roger pays $40 a month.

$40 A DAY

The Hyatt Regency, on the San Francisco waterfront, is one of six bizarre futuristic hotels being built in America to cater to tourists and conventioneers. In the lobby is a rock and roll lounge with glowing (Clockwork) orange plastic tables, a huge geodesic-like sculpture, recorded bird calls, and transparent gondola elevators to carry you to the top, where there's a revolving bar that looks out on a dying and increasingly concretized San Francisco skyline. These hotels are designed and built as slick images for middle America: the city of the future, Playboy bunny cocktail waitresses, a controlled environment, an escape from the horrible conditions outside. As we sat in the top one day, amidst mirrors, foxy waitresses and tourists from Omaha we noticed that there were no ships in San Francisco docks. Then we realized that San Francisco is through as a port. Since Oakland automated its docks, cargo is unloaded there by machine. San Francisco didn't, and now must turn to tourism for income. *Disneyland by the Bay.* It all may get rearranged one of these days for the San Andreas fault has had no relief since the great quake and fire of 1906, and the Hyatt and other new highrises are built on filled land.

DIASPAR

The City and the Stars is Clarke's version of the pastoral and urban conflict in human history. Diaspar is a city of a future a billion years hence, but really, as in most such projections, as McLuhan has noted, it is a city that is only an idealized version of the present. Diaspar is a triumph of technology, closed in upon itself under a huge dome, its every function, from the moving sidewalks to the materialization devices that create one's every need, is taken care of by an enormous computer. In this perfect city man is not born of woman, but he emerges fully grown from the Hall of Creation to begin a life of thousands of years. After the fullness of millennia, man returns to his source; his atoms and his program are all undone for a time until, once again, he is reincarnated for another existence. But once in a million times or so, a man emerges from the Hall of Creation who is not a repeated version of those who lived before. Alvin, the hero of the novel, is just such a "unique."

Alvin shows his uniqueness by being uneasy in Zion. A vague unsettledness wears away at him, and while all the others seem to fit perfectly into their technological paradise, Alvin grows restless and wants to know the inconceivable: what is outside Diaspar?....

Because Diaspar is perfect, the founding scientists knew that its survival would depend on one calculated aberration of absolute perfection, and so they created the office of the Jester and gave him access to the central computer. Aided by the computer, the Jester could harass the perfect citizens in just the right degree of variation from absolute order that would ensure the continuation of the city. The Jester helps Alvin as part of just such a possible harassment, and, therefore, brings him to the central office where he can survey the entire city to look for a way out of the perfect sphere....

Clark's 1953 novel is uncanny in its prophecy for his billion-year future seems to be an imaginative description of our present. The United States is Diaspar and the land of the Hopi is Lys. We are the people closed in a culture of machines in which artificial intelligence, artificial organs, and extrauterine birth are promising to give "man" the future he has always dreamed of. The Hopis are the small spiritually and psychically advanced people watching us move toward apocalypse and remembering the prophecies. Some of the people of Diaspar are trying to find the way out, but as they leave the central computer behind them to return to the earth where they can grow simple food, build a culture of consciousness, and make a sacrament out of the sexuality of their own bodies, they are only moving into the suburbs of Diaspar where the technological pollution of earth is matched by the internal technological pollution of their minds. For them the only way out of Diaspar is through the door at the top of the head....

At the Edge of History

100 MORE YEARS

Michael, Patrick and Kerry Geraghty — three brothers from California — moved to England a few years ago, located an abandoned house in a small village, rented it, and began restoration. In about a year's time, they had salvaged a beautiful house out of the ruins, fit for another few hundred years of occupancy.

Looking back it seems like a long slightly nightmarish type dream. What we had to work with was a Queen Anne farm house, red brick, 4 up 4 down, foot and a half damp thick walls, mushrooms growing from the ceiling, ceilings on the floor, and generally what is known as dilapidated, and all rather dicey.

We got together on our work plan which consisted of a series of tangents upon tangents, leading off of tangents, leading off of tangents. Miraculously the first thing that went in was the vegetable garden. We burned huge amounts of rubbish so we could see what we were doing. By popular demand the rat families were requested to leave and we moved in.

Basing our headquarters in what is still known as 'the tower', with no electricity, no water, we launched intermittent surveillance forages and work parties, for a period of not less than 9 months and not more than a year.

Basically it was all here: oak rafters and door jambs, oak window frames in the front, red tile kitchen floor, leaded window panes, a fireplace in every room. Basically what we had to do was to uncover it and set it right: fix the tiles and gutters, plaster, paint and scrape, lay the stair carpet and put in roses, that sort of thing.

The most rewarding tasks were uncovering the 7 or 8 layers of paint over everything under the sun — green yellow brown yellow green yellow brown — 300 year old dark oak that had never seen the light of day. The best way we found was with a good sharp chisel, held straight out and pulled straight down. Very chippy but fast.

For the black floors we rented a belt sander and took shifts night and day, course medium fine. They came out a nice bright nut brown.

What else — we tried wallpapering; one strip on the ceiling which was so full of bumps we got hysterical, tore it down and painted everything magnolia white.

In the bathrooms we tried to keep the original fitments whenever possible: chain pull toilets, long deep bathtubs. In the kitchen we went practical and bought a stainless steel sink and built a stove from spare parts of a restaurant stove. And we built lots of counter space (the eternal cry) with a big long narrow oak table down the center. We've always wanted a fireplace in a kitchen so we built one using the old boiler flu, and we raised the level of the fire a couple of feet off the floor.

Everything started to look as if it had always been that way. That was our basic criteria — that it should look as if it had always been that way, and if it hadn't, it ought to have been.

Now all we have to do is build the tea house on the mound, the Japanese bath under the tower, the work shop in the cow-shed, the glass room, plant the fig tree, the apricot trees, the grape trellis. All in good time. The vegetable garden, picking the apples, mowing the lawn. Now the seasons start to go by.

Cherry tree by the garden gate
In half full bloom
Tea pot on the stove
Fixing up an old house in the spring

Michael Geraghty

CARPENTER GOTHIC

BUILDINGS ABOVE AND BELOW HAVE BEEN DEMOLISHED...

Phil Palmer is a photographer and writer from Petaluma, Calif. who has been photographing in the San Francisco Bay Area for the past 20 years. One of his many projects has been to photograph and record the finely crafted redwood houses of San Francisco built in the 1800's in preparation for a book to be called Carpenter Gothic.

"Carpenter Gothic" is what some San Franciscans call the style of their city's old wooden row houses from before the turn of the century. Others refer to it as Victorian....

Mills in California produced tremendous quantities of the moldings, scrolls and such to adorn the facades of these homes. The material used was primarily redwood. Many of the jigs used to shape the ornamental woodwork still exist; it's a wonder that some enterprising builder has not taken advantage of them.

In spite of "progress," hundreds of the old buildings still stand... Many, obviously, need refurbishing, if nothing more than a coat of paint, replacement of a sagging sill, a new roof or window. However, city building inspectors say that the interior of many of them are in dreadful shape with defective wiring, inadequate plumbing, rotted wood, etc.

But fortunately, many San Franciscans who have succumbed to the charm of these old homes have preserved them; in many cases their interiors have been remodeled to meet modern standards of comfort and building codes. And the Department of Housing and Urban Development has granted money through its Open Space Land Program, which, together with funds from local private and public agencies, have made it possible to designate some of them as landmarks.

Phil Palmer

This is an air house, a space ship & it's really been good to the 2 to 4 people who live here. It's warm in the winter (3 armloads of wood heats thru the coldest day & night) & always light, full of thriving plants. It leaks in perverse defiance of all efforts on our part, but we catch the drips & pour them on the plants. Our next project after the interior is finished is a greenhouse. We have over 2000 sq. ft. of floor space & about $1000 invested.

A little house, 8'x10' very near the 40' dome we live in, is my answer to having my own room (finally) after kids, communal scenes, old men, years of scrabbling around among people all the time. I find I'm something of a hermit actually. I like to have my own piece of space at least under control where the chaos of kids and coffee cups won't follow me. A space uncluttered by projects, obligations, dishes to be done.

Peggy

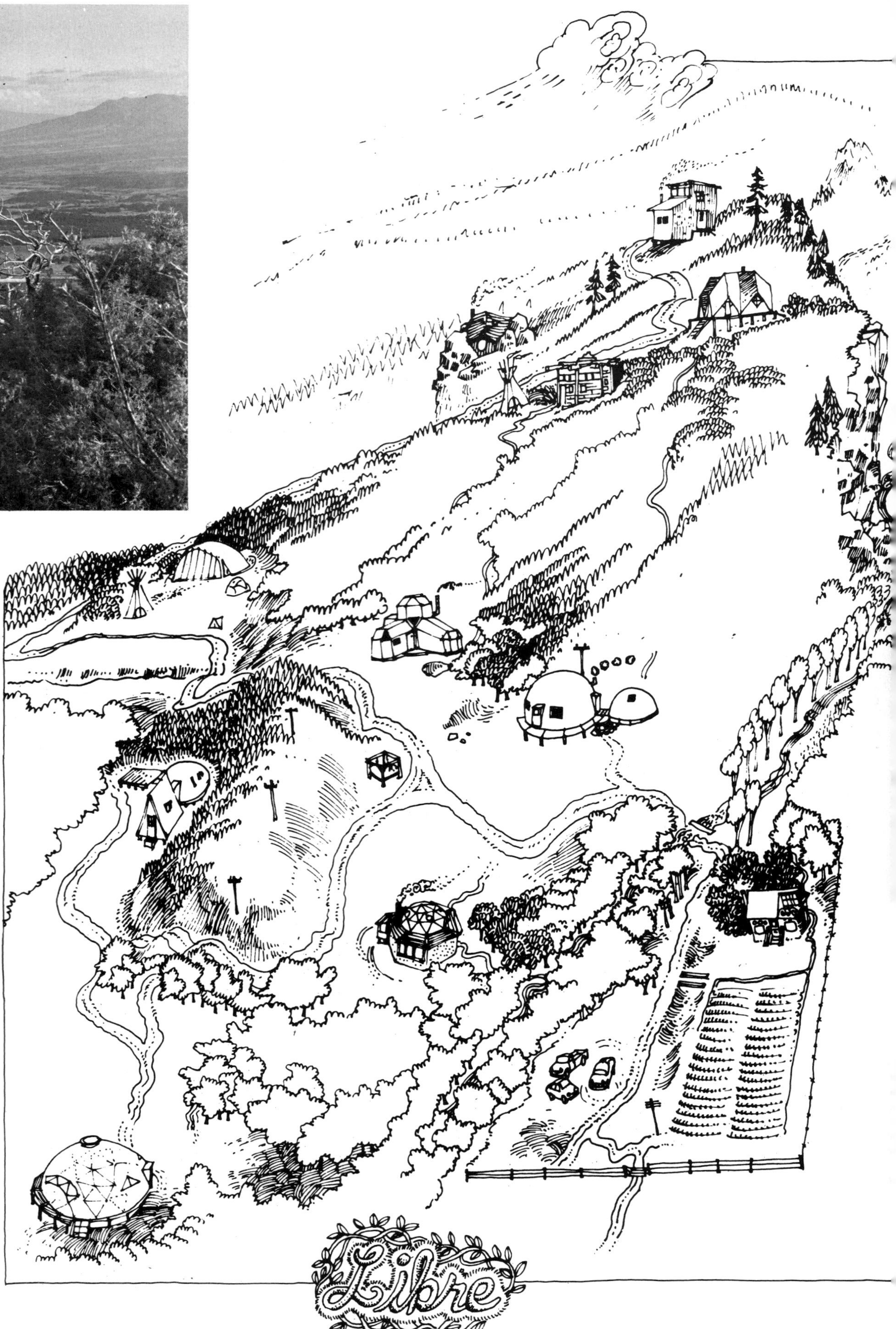

So now we've seen 5 years of seasons go on this old 12,800' mountain, mostly winter & 90 hectic days of summer. We live at 9000 feet; that's marginal they say.

For these 5 years we've thought, dreamed, built, torn down, rebuilt & maintained shelters — a real Milarepa trip — everything from superlight zomes to the heaviest rocks, adobes & logs. It goes on & on, but now there's not so much pressure. There are roofs & walls, at least, to contain the potbellys. Home, we call it, six children born here, one died.

When we were considering buying this 360 acres an expert told us, you don't want that place, it's nothing but a pile of rocks. It is indeed, an epic pile of rocks. Every bit of good soil we've got, we built, hauling shit & mulch. Every drop of water we get we use many times & with care. We're building shelters for water, our houses have all become rain catchers. It takes a long time to learn what you have to do to live in an ecosystem. This Cuerno Verde Mountain, Greenhorn, is our guru. A greenhorn is a berserk young buck with his antlers still in velvet, we are the greenhorns of Greenhorn.

I took this weird drug once called Parade. For three days & nights I hallucinated the evolution of the earth & all the creatures & things that live here. Far out. Now it seems we're living it out slow, that evolution. After shelter comes food — agriculture & animal husbandry. Life-learning what the old cliches *really* mean — chicken, a turkey, cowed, get your goat, the birds, the bees & the flowers.

It seems slow & hard most of the time, especially when it's all laid out for food stamps in the King Sooper 50 miles away. Why bother when the slow getting-it-together with the earth is such a hassle. Somehow the body knows, and trusting the organism comes with trusting the mountain. So slow though. All that shit we read in books was true. How to do it by people who have done it. But the body can't learn from books. The body learns slowly by doing. And it seems that each man & woman body has to learn all the things in the evolution parade by lurching down & up the road step by step. All in cycles. All in good time.

Libre came out of the explosion that started the decline of Drop City (I was there the other day, so weird wandering thru those deserted, ripped off structures, that so much love & agony & labor went into the building. There were ghosts behind every broken window & half-off-the-hinges door. Sad hippy ghost town with a huge pile of human shit deposited on the drainboard next to the sink. I guess we'll go down there & scrounge what we can) too many rats packed in too small a space. So we built our houses at Libre out of sight of one another. You can see far here, it expands the idea of space. We need more space & soon learn to give more space. Then comes the approach, slowly, in good time, there's no accounting for taste, begin the dance, contact, recognition. Your mirror — your mirror. So it goes, just folks beginning to learn how to live together on the earth, approaching slowly & from some space.

Roses are red, violets are blue.
We built our shelters, you can too.

Plant a seed & watch it grow. It never grows quite the way you thought it would or wanted it to, but it grows, & feeds you, & sometimes it even gets you high.

rabbit

We are united by our desire to make
Total Revolution against the mindless mask
of greed that passes as American Civilization
We banded together on this mountain-side
to gather our forces
and to catch a close look at
wild and creative Nature.
We draw our strength from the land.
With this strength we make shelters for our families...
We make every detail just so...
Just the way we want to live with it.
We care....
We care that our homes are places of joy and comfort...
We love every log and nail and smashed thumb....
We have no special skills.
No one is a trained architect or construction engineer.
We are a typical group of ding-bat hippies
who with great ingenuity, determination
and fair luck managed to scrounge an unbelievable amount
of cheap or free materials...
We love the tipi for its elegance and simplicity...
We love the spacious and airy domes....
We love the dense, earthy feel of log and adobe houses...
In fact,
We love anything that someone builds out of love.
In this quest for a little satisfying shelter,
we found that it is much easier and more rewarding
to pool our energy with like-minded individuals...
It was our hope to combine the efficient and ecologically sound
principles of communism
with the exciting and unpredictable forces of creative anarchy...
We wanted it both ways...
fierce individuality within a tight
communistic, tribal framework.
Our experience so far has proven that
the two ideologies are not necessarily exclusive...

volution of a star from a frame of a pyramid/symmetrical rowth from a center point sixteen feet above the platform ase/floor space is a thirty-two foot square divided in four qual areas — living room, kitchen, workshop, children's oom, two lofts for adult sleeping and meditation/space tilizes no center support/frame construction, ½ inch plywood kin, T-Lock shingles, 50 gal. drums at each point where roof rains to collect water. Constructed in four months by two en, one woman, two children, one cat/total cost $1,200... house form which echoes the mountain peaks.

A hobbit house built for virtually no money out of logs & adobe. Always cool in summer & warm in winter; it feels like it's been here for centuries. Trucks fixed, showers, fortunes told, shady deals.

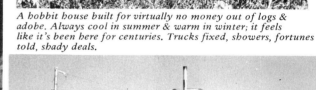

DOME ON LOWER ROOMS BUILT WITH STACKED RAILROAD TIES.

We love and fight excessively.
We chop wood, carry water, and stare into the sunset.
We live at the end of a boulevard of broken axles
and shattered differentials.
We laugh a whole lot when things are going bad...
And sometimes we cry when things are going well....
We like to share...
We naively expect others to do the same.
We try to take just what we need and leave the rest...
Isn't that the real meaning of ecology?
We get very high dealing with our building materials.
The magical spirits of logs and stone speak to us...
We carry them home and use them in a special way.
We frequent the junkyard.
Anything with soul qualifies...
We are not here to spend money...
We are trying to subvert the building industry in its
present form.
The effect of its dehumanizing and impersonal
structures is profound on the national psyche.
We are conditioned to believe that we don't have the
special abilities to do things for ourselves.
Call the contractor!
Call the auto mechanic!
Call the farmer!
Call the plumber!
Call the electrician!
Call the architect!
Do it yourself you Lazy Creeps!!!
Make it Real!

A HOUSE BUILT AROUND A LARGE BOULDER.

SAMOAN CHIEF'S HOUSE. FRAME IS WOVEN BAMBOO THATCHED WITH SUGAR CANE. FLOOR IS CRUSHED CORAL COVERED WITH GRASS MATS. PHOTO BY GEORG WEGENER 1903.

THE DOME

One of man's earliest shelters, before he had the metal tools to cut timber or hew stone, was the small woven dome. The framework was made of pliable branches or saplings, woven together, utilizing the strength inherent in a double-curved surface to span a useful space. It was then covered with leaves, thatch, or animal skins, whatever was locally available. The shape was:

> *...not just a utilitarian form of vaulting, which had originated for structural or environmental reasons in some one country, but was primarily a house concept, which had acquired in numerous cultures its shape and imaginative values (based) upon an ancestral shelter long before it was translated for ideological reasons into more permanent and monumental form by means of wood carpentry and masonry...*[1]

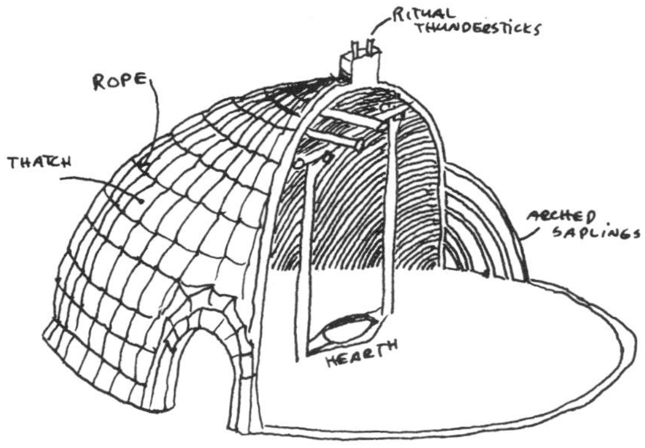

CONSTRUCTION OF ZULU INDLU, BUILT BY WOMEN.

The circular earth lodges with center posts probably evolved from these small woven domes, and were then expanded to rectilinear shape in the early days of agriculture. (See p. 21.)

Later as tools were developed in the changing societies and a greater variety of materials became available, two new types of hemispheric construction came about: the wooden dome, "...with an elaborate system of girders and ribs which was the result of experience acquired in shipbuilding...."[2], and the masonry dome of brick or cut stone which was used for a variety of purposes: shelter as with the *trullo* of Southern Italy, vaults for granaries, or as grand monumental structures.

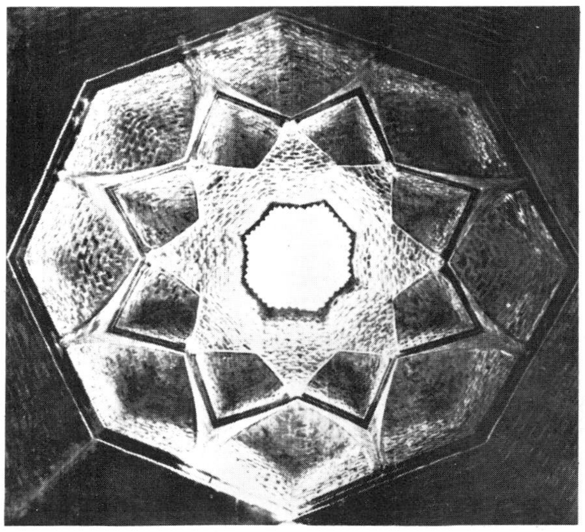

VAULT AT MASJID-E-JAMEH, ISFAHAN.

TWIGLOO, SEE P. 10.

With the invention of concrete by Roman engineers there was a fourth type of dome construction: the monolithic concrete dome, and immense vaults were cast over the baths and public structures of Imperial Rome.

STONE TRULLO DOME, LOOKING UP.

It was not until centuries later that the fifth and most recent type dome was invented: in Jena, Germany in 1922 Dr. Walter Bauersfeld built a subdivided icosahedral hemisphere, the structural framework of light steel bars: the first geodesic dome. (See pp. 110-111.) The frame was covered with a thin layer of concrete, based upon the thickness ratio of an egg shell to its diameter. This was the world's first thin shell concrete dome, the building technique later furthered in construction of large structures by Pier Luigi Nervi of Italy and Felix Candela of Mexico.

FIRST GEODESIC DOME, 1922. SEE PP. 110-111.

Of these basic types of dome construction, it is the latter — the *industrial* type — that is today most commonly associated with the word *dome.* These structures were made possible by the industrial revolution, power tools and new building materials; they are generally mathematically-derived, precise, and built of straight framing members (struts) connected at joints (hubs), or constructed of individual panels fastened at edges. (On page 13 of *Domebook 2* are described the basic types of industrial, or *polyhedral* domes.)

PEASE DOME.

BERLIN 1926. PLANETARIUM AM ZOO. DYCKERHOFF AND WIDMANN.

Some 30 years after the Jena dome and considerable European development of domes and thin shell construction, Buckminster Fuller patented the same subdivided icosahedron principle (in 1954) and built a variety of what he termed *geodesic domes* in the United States – for the military, at colleges as test structures, and the "Pease Domes": 39' diameter plywood hemispheres manufactured by the Pease Woodworking Co. of Hamilton, Ohio and by various other firms franchised by Fuller throughout the country. In his lectures throughout the world in the '50s and '60s Fuller popularized domes as a breakthrough in building technology, as the most efficient structures yet invented.

Fuller envisioned dome components being mass produced on assembly lines, yet his factory produced plywood domes never caught on as housing, perhaps due to leakage, or perhaps due to difficulties in subdividing interior space, installing doors, windows, counters, or adding on at a later date. The fact that the dome uses less materials in its frame construction is offset by these drawbacks, and by the fact that the structural shell of a home-sized building is only 20% of its final cost.

DOME INTERIOR.

DROP CITY.

Then in the late '60's, largely due to Fuller's inspiration, a phenomenon began in America: domes became associated with a new lifestyle, the subculture, doing "more with less", living in the round, and ecology. A group of architecture students and artists heard a Fuller lecture in Boulder, Colorado and soon afterwards founded Drop City on the outskirts of Trinidad, Colorado. They built two geodesics, then later two of the *zome* structures invented by Steve Baer. They wrote: "...corners constrict the mind. Domes break into new dimensions...."[3]

In 1968 (Haight Ashbury days) the media discovered that domes photograph well. They looked exciting, colorful, new, and in a wave of excitement outlaw builders began a spontaneous series of shoe-string experiments in new polyhedral design and untested materials. These were the days of the moon shot, the Moog synthesizer and computer graphics. As Fuller romanticized science and technology, the geodesic dome became a *metaphor* to builders for the space age, the age of transcendant science. Fuller talked of the tetrahedron as the "building block of the universe", and the skins of cells as geodesic nets. He implied that the lightest weight *transparent* dome was an image of structure in its purest manifestation and that you were somehow in touch with the universe in building a dome.

NEW MEXICO.

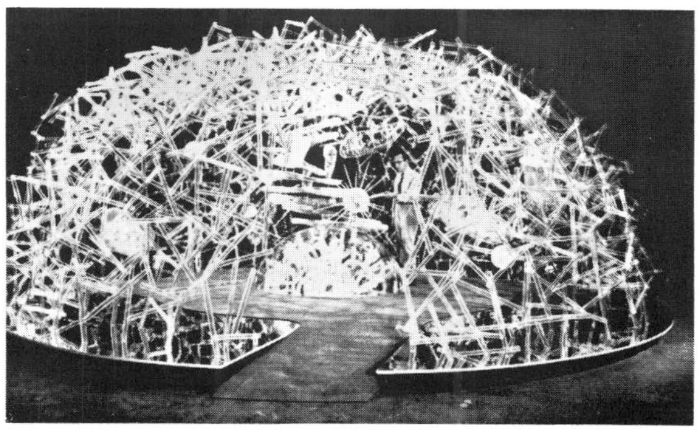

HUMAN CELL MODEL. MUSEUM OF SCIENCE, CHICAGO.

Domes began to appear in the southwest, where there were little or no building code restrictions. In contrast to Fuller's vision of assembly-line production, most of these individual domes were constructed by hand, and the young owners not only built, but lived in the structures they designed. The builders were naive, hopeful, usually impractical, always inspired. Materials ranged from chopped-out car tops caulked with tar to shiny aluminum sealed with space age silicone. Icosahedrons, triacontahedrons, triple fused rhombic-icosa-dodecahedra. Families of from 3 to 30 living in one room. Building inspectors presented with *fait accompli.*

Two competent commercial companies were making domes during this period: Bill Woods produced fiberglassed plywood Dyna Dome kits in Phoenix, and in 1969 Steve Baer founded Zomeworks in Albuquerque to produce zomes and solar heating devices. And in 1969-71 we built 17 experimental geodesic domes at Pacific High School in California's Santa Cruz mountains; this eventually led to *Domebook 2,* a result of our experiences and those of a network of builders established by *Domebook One* and the *Whole Earth Catalog.*

WISCONSIN.

CHRIS, JONATHAN, MARK, WAYNE. PACIFIC HIGH.

INTRODUCTION to DOMEBOOK 3

In 1971 as we were putting together *Domebook 2,* with a great deal of the evidence in, we began to feel uneasy. Enough time had now passed to begin evaluation of the experiments, and much of it was not working. We had built small wooden buildings before the domes and we now began to have more respect for our past work. We realized we had more to communicate about hand-built shelter than just domes. Thus *Shelter.*

BURNING FOAM DOME.

DOME AT PACIFIC HIGH SCHOOL.

Domebook 3 is within *Shelter* because it contains the best of recent dome information and because with a few years' time we can see more clearly the relation of domes to our earlier work. *Domebook 2* made domes look too easy, too much like a breakthrough solution, too exciting. Here we hope to correct that impression and to present our changing thoughts on "revolutionary" design, new materials, and the latest housing technology. We also hope to show how the polyhedral dome fits within the choices available to builders today.

Notes.
1. *The Dome, A Study in the History of Ideas,* E. Baldwin Smith, p. vii.
2. Above, p. 13.
3. *Shelter and Society,* Paul Oliver. P. 158.

THE WONDER OF JENA

The above photo is of the world's first geodesic dome, built on the roof of the Carl Zeiss optical works in Jena, (now East) Germany in 1922. It was also the world's first lightweight steel structural framework, and when covered with ferro cement became the first thin-shell concrete structure in history. What is even more remarkable about this dome is that it was almost incidental to a spectacular scientific and technical accomplishment: invention of the planetarium projector. The inventor of the projector and the dome was Dr. Walter Bauersfeld, chief designer at the Zeiss works. A brief history of the astronomical devices that led to these inventions follows; it is a story of the foremost breakthrough in astronomers' attempts to "...create the illusion of the mysterious, silent march of the worlds of nature."[1]

Copernicus

The Gottorp armillary sphere.

The Farnese Globe.[1]

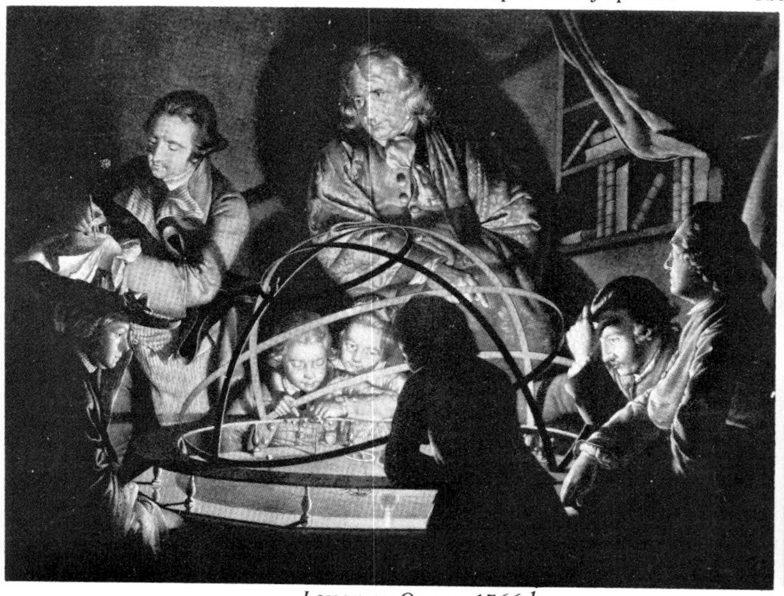

Lesson on Orrery, 1766.[1]

Walter Bauersfeld

A planetarium

Although the concept of the sky as a sphere may have occured as early as 2,000 B.C. in China, [2] it is recorded that in the 6th century B.C. the Greek Anaximander taught that the stars and planets pass not only above, but beneath the earth. He probably made a globe to demonstrate this, but it has never been found. Greece's first scientific astronomer, Eudoxus of Cnidos (about 400-355 B.C.) constructed the first known complete celestial globe, which became the model for future globes. In 73 B.C. in Italy, a white marble statue was discovered, depicting the god Atlas supporting a celestial sphere which is 26" in diameter. On the sphere are inscribed not only constellations, but circles representing the elliptic (a great circle) boundaries of the zodiac, and the major parallel circles.

Many globes and charts of the heavens appeared after (and before) the Farnese Globe; but the first real *instruments* of astronomy were the *armillary spheres,* which consisted of a framework of circular rings representing the various astronomical circles, and horizontal rings to indicate the horizon, equator, elliptic (path of the sun) and a vertical ring for the meridan. One such device, the Gottorp Armillary Sphere, built in 1653 by Andreas Busch was a marvel of craftsmanship and art, mechanized to show movement of the sun and with six silver angels representing the known planets. The part of the framework bearing the equator was made to rotate with respect to the zodiac at a rate corresponding to one revolution in 25,000 years, which is the rate of the precessional motion of the earth.

A remarkable device, also constructed in Germany by Busch in 1664, was the Gottorp Globe, a water-powered 3½ ton, 10 foot diameter sphere that rotated once every 24 hours. Inside it was a platform for 12 persons and on its interior was a map of the sky with gilded stars. All other globes to that time had shown the sky from the unnatural position of the observer on the outside of the celestial sphere.

Early in the 18th century a fine mechanical planetarium was built by John Rowley for Charles Boyle, fourth earl of Orrery. Called the "Orrery", similar instruments since have borne the same name. These devices incorporated the "new" concept of the solar system originally proposed by Copernicus, that the earth was round and revolved around the sun once a year.

In 1758 a large globe was built by Roger Long at Cambridge. It was 18 feet in diameter and accomodated 30 people. In 1911 Dr. Wallace Atwood, director of the Chicago Academy of Sciences designed and built a 15 foot diameter electrically driven globe that is still in use today.

•

The difficulty faced by astronomers at this point in history was in constructing a globe to accomodate a much larger audience. In 1913 the Carl Zeiss optical works of Germany undertook the problem of designing a huge sphere that would both hold a large number of people and show the motions of the planets as well as the stars. After much work, no satisfactory solution was found. Then in 1919, just after the end of World War I, Dr. Walter Bauersfeld of Zeiss:

> *...caught an entirely different idea: reversing the plan of a mechanically rotatable hollow sphere with illuminated images of the stars, he transferred the entire mechanism for the movements to a collection of projectors which would project luminous images of the stars on to a stationary white hemispherical dome of much larger dimensions than those originally conceived. Within the dome, the centre of which would be occupied by the projectors, all would be in darkness. By means of suitable mechanisms the projectors would be moved and guided so that their illuminated images of the heavenly bodies would conform on the dome to the motions which actually occur in nature....* [3]

For five years a large staff of scientists, engineers and mechanics worked with Bauersfeld at the huge Zeiss plant in Jena to design the projector and the projection dome.

> *The projection of the starry sky required a certain number of projectors, arranged in the center of the dome. Each projector should illuminate an area of the same size as the dome. If the vertices of an icosahedron are cut in such a way that the new surface consists of 12 pentagons and 20 hexagons the area within each is nearly of the same size. The projectors are arranged in the centers of the pentagons and hexagons and produce 32 starfields on the dome. (Actually only 31, since one area is used for the support.)...*[4]

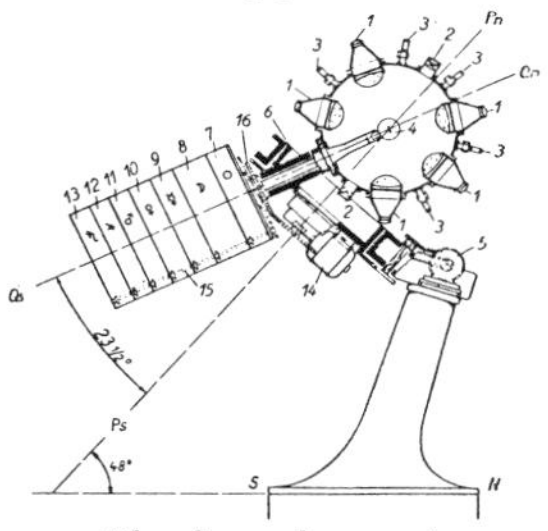

The first planetarium projector.

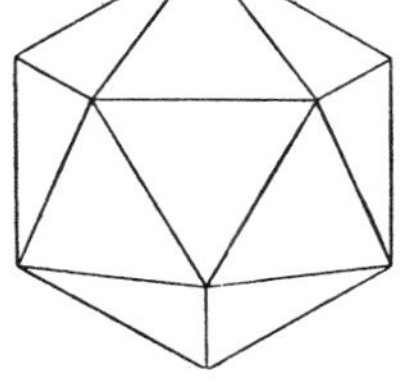

Icosahedron. Slice off vertexes to get...

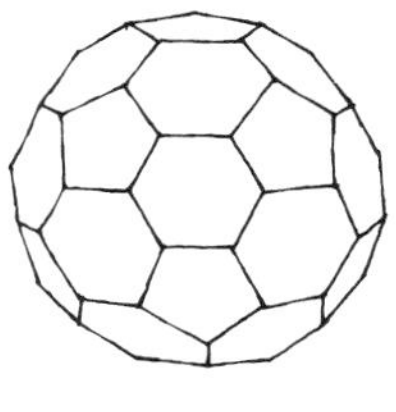

truncated icosahedron. 12 pentagons, 20 hexagons.

To test the projector Bauersfeld needed a hemispheric dome as a replica of the sky. It had to be lightweight, as it was to be placed on the roof of the Zeiss factory in Jena. He built a light iron rod framework, the design a highly sub-divided icosahedron, with great circle arcs. Thus both the dome frame and the projection pattern were derived from the icosahedron.

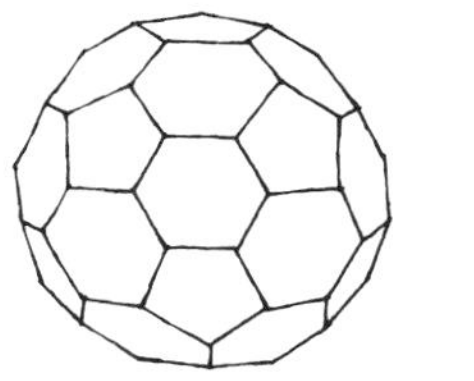

Projection pattern 12 pentagons, 20 hexagons.

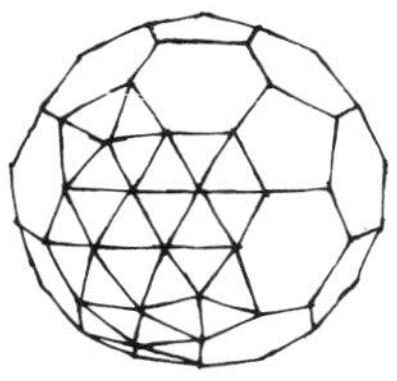

Connect centers of pents, hexes to 3-way grid to stabilize. Truncate to get dome.

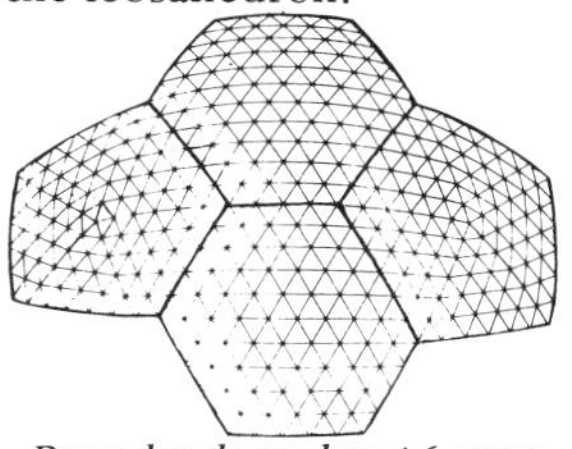

Rounder dome has 16 struts between one pent center to another. This produces great circle arcs.

Not until the complex skeleton (3,480 struts accurate in length to 2/1,000 of an inch) was complete did Bauersfeld seek professional construction advice.

> *We planned to cover it first with a fine network of thin wire in order to embed the whole construction in a layer of gypsum of about 1½" thickness. But gypsum did not appear advisable because it could not be waterproofed and so we inquired of an engineer of Dyckerhoff and Widmann, who were engaged with factory buildings of ferro concrete for the Zeiss Works, if he could not suggest a waterproof cement instead of the gypsum. He replied, "Yes, just recently we have tried a new method of sprinkling cement of viscous consistency by a hose similar to that of fire-fighters. If in the interior of your framework we fix to it a wooden shield of suitable spherical curvature, against which we sprinkle cement in thin layers one after another we can avoid the concrete running off the inclined surfaces. Within a few days the cement will be stiff, we take away the shield and you get a fine smooth surface in the interior of the dome which is to be sprinkled by a white colour to represent an ideal surface for the projection."*[5]

Basing their design on the thickness ratio of an egg shell to its diameter, Bauersfeld, and Mr. Franz Dischinger and Dr. Ulrich Finsterwalder of Dyckerhoff and Widmann then built the world's first lightweight thin shell concrete dome. Although the firm did not again use the icosahedral dome geometry, the invention was perfected in later structures and made possible clear spans of lighter weight than was previously possible.

In August 1923 the heavens were for the first time accurately reproduced in all their brilliance on the Jena rooftop dome. The stars and individual motions of the planets appeared on the dome's interior and the effect was so startling that even the men who designed the planetarium were astonished, as were early spectators. Newspapers referred to it as the "wonder of Jena".

As the planetarium began to be widely publicized, representatives from large cities in Germany asked Carl Zeiss to sell them planetaria of this kind. This caused the inventors to redesign the first projector which showed only the skies over Munich, to a model that could be used anywhere in the world. 25 of the latter were subsequently built and in May, 1930, one was opened in Chicago — America's first projection planetarium — the Adler Planetarium. The most recent development in planetaria is the new Space Theater in San Diego, a 76' 3/4-sphere dome with a computer controlled projector.

•

The "great circle" principle used in the Jena dome has been in use in the Orient for centuries to weave fish traps, hats and baskets. And the same principle is evident in a remarkable sculpture in China's Summer Palace of a lion holding what appears to be a five frequency geodesic sphere under its claw.

We made an error in *Domebook 2* in stating that Buckminster Fuller was the inventor of the geodesic dome. Fuller's contribution, rather than origination of the great circle principle, or its earliest structural utilization, is rather application of the word *geodesic* to this type of polyhedral building framework, and its popularization and commercialization in the United States.

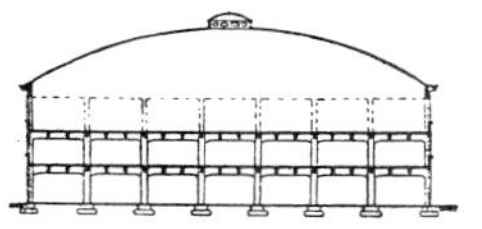

Schottkuppel, the second ferro cement dome built in Jena.

St. Peter's

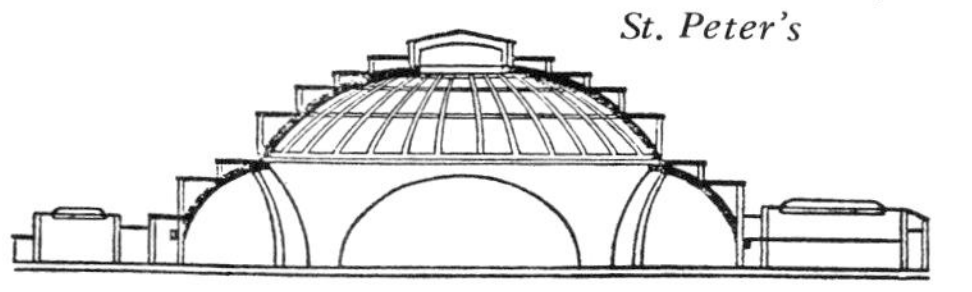

Jahrhunderthalle

Weight comparison of three large domes:

	diameter	weight
St. Peter's dome, Rome	*40 meters (130')*	*9842 tons*
Jahrhunderthalle, built 1911, Breslau, Germany	*65 meters (211')*	*1476 tons*
Schottkuppel, built 1924 in Jena, Germany	*40 meters (130')*	*325 tons*

During the past 30 years the shell concrete construction has expanded over the whole world. At the end of 1941 the firm of Dyckerhoff and Widmann estimated the total area which had been roofed by shell construction at more than 1,600,000 sq. yd....

Prof. Walter Bauersfeld, London 1957[5]

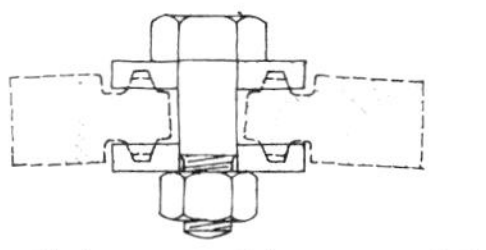

To make the light assemblage possible and for equal distribution of tension, the iron rods and discs of the locks (hubs) had to be made with extraordinary exactness, common otherwise only in an optical factory.[5]

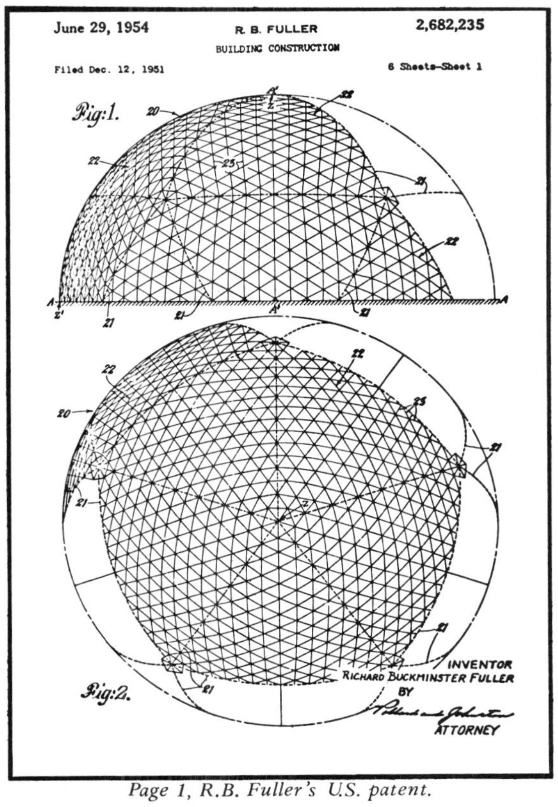

Page 1, R.B. Fuller's U.S. patent.

Homemade planetarium...

Stars are a very big thing in the Ozarks. So far we've been too poor to produce enough polution to louse up the sky. With a little know-how you can find directions, tell time and even the seasons by finding familiar stars on a clear night. One of the best ways of learning stars is to have access to an expensive gadget called a planetarium. This devise projects the night sky onto the inside of a dome. The dome is big enough to hold both the projector and the group of people who come to see the show. Most big cities have fantastic planetariums but lousy skys. We had great skys but no planetarium so we decided to build one.

A very small used projector was bought at a garage sale for $1.50. New it sells for about $30 from Spitz. We priced domes and discovered $80 would buy only a 4 foot diameter prefab fiberglass hemisphere. This was too little for too much so we decided to try to build one. The biggest room available had only a 9 ft. ceiling which limited us to a 12 ft. diameter dome. This would give room for about 10 people sitting on the floor. We wanted a smooth interior with as few joints as possible and so picked a hex-pent design rather than the usual triangles. Corrigated cardboard reinforced at the joints with lath gave us a skin stressed structure that is light, easily built and inexpensive. We figure we got by for about $15. It's not too fancy but it does work and if you squint way down and stretch your imagination a bit you can almost believe you're sitting in the middle of a Missouri prairie watching the oldest and one of the best shows on earth.

Dick Boyt
Neosho, Mo.

Part of letter to Shelter Publications from Dr. W. Degenhard of Carl Zeiss, Oberkochen, Germany, June 8, 1973:

Thank you very much for your letter of June 8. It is very difficult for us to find the proper answers to your questions. You have to consider that Carl Zeiss in West Germany had to be rebuilt from the scratch. We have no access to the archives in East Germany. All patents and recordings were taken away either by the American or the Russian armies when they occupied Jena in 1945. Dr. Bauersfeld was among the 126 people who were brought to West Germany by American troops in 1945.

...We were not able to find any patent which covered the subject of the planetarium or the dome. At that time it was the principle of our firm, that such basic findings should be made available to the whole scientific world, so it may well be that nothing was patented in Germany.

The first presentation of a projection planetarium was widely published indeed. Since some of the previous attempts were made in the US (Chicago, among other places), I am certain that interested people learned at that time, what was achieved by Zeiss in 1922....

Sincerely yours,
Dr. W. Degenhard

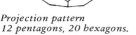

The Edward Longstreth Medal of the Franklin Institute in Philadelphia was awarded in 1938 to the firms Carl Zeiss and Dyckerhoff and Widmann for the shell-form construction of the planetarium domes.

great circle *n*: a circle formed on the surface of a sphere by the intersection of a plane that passes through the center of the sphere; *specif*: such a circle on the surface of the earth an arc of which constitutes the shortest distance between any two terrestrial points.

1geo·des·ic *adj* 1: GEODETIC 2: made of light straight structural elements largely in tension (a geodesic dome)

2geo·des·ic *n* the shortest line between two points on a mathematically derived surface

Great circle basket.

Lion sculpture at China's Summer Palace, on the outskirts of Peking, built about 1885 by the Empress Dowager. The sphere may be related to the fact that in the late 1500's, Portuguese missionary Mathew Ricci brought first knowledge to the Chinese that the earth was round and showed them celestial spheres.

Notes

1. Article *The Heavens in Our Hands,* George W. Bunton, Pacific Discovery, Dec. 1952.
2. Same as above.
3. *From the Arratus Globe to the Zeiss Planetarium,* Helmet Werner, Publ. Gustav Fischer, Stuttgart, 1957. (Available only from Carl Zeiss, N.Y.)
4. Letter to Shelter Publications from Dr. W. Degenhard, Carl Zeiss, June 19, 1973.
5. James Clayton Lecture: *Projection Planetarium and Shell Construction* at Institution of Mechanical Engineers, London, May 10, 1957 by Professor Walter Bauersfeld.

Technical references (Translations from the German by Ilka Hartmann, Bolinas, Calif.):

6. Fortschritte Im Bau Von Massivkuppeln, Der Bauingenieur, Berlin, Feb. 23, 1925.
7. Der Schalenbau, published on 90th anniversary of (and by) Dyckerhoff & Widmann, Munich, 1955.

Photos of Dyckerhoff & Widmann domes, courtesy Herr Schrimps, archives, D&W, Munich.

We are grateful to Charles Hagar, director of the Planetarium Institute, San Francisco State University for assistance. Mr. Hagar has just written a book (not yet published): *Planetarium: Window to the Universe.*

SMART BUT NOT WISE

FURTHER THOUGHTS ON DOMEBOOK 2, PLASTICS, AND WHITEMAN TECHNOLOGY.

BY LLOYD KAHN

He looked upon us as sophisticated children — smart but not wise.

Saxton T. Pope (said of Ishi)

This article was prepared a year after publication of Domebook 2, *reflecting then, as now, our changing views and evolution of thoughts on shelter.*

Metaphorically, our work on domes now appears to us to have been smart: *mathematics, computers, new materials, plastics. Yet reevaluation of our actual building experiments, publications, and feedback from others leads us to emphasize that there continue to be many unsolved problems with dome homes. Difficulties in making the curved shapes livable, short lives of modern materials, and as-yet-unsolved detail and weatherproofing problems.*

We now realize that there will be no wondrous new solution to housing, that our work, though perhaps smart, was by no means wise. *In the past year, we have discovered that there is far more to learn from wisdom of the past: from structures shaped by imagination, not mathematics, and built of materials appearing naturally on the earth, than from any further extension of whiteman technoplastic prowess.*

In May, 1972, about a year after we published *Domebook 2* I received an invitation to participate in a conference at MIT, *Responsive Housebuilding Technology.* Out of curiosity I decided to go, not thinking too much about the fact that I'd been invited as the editor of the *Domebook,* and that since that time I'd more or less given up on domes and was disillusioned with new materials and high technology as applied to building. I decided to bring along slides and videotapes of housebuilding in Northern California: shacks, driftwood buildings, interviews with real builders, and on video, the contrast between a crane dropping in a pre-fab and 25 men picking up and moving a small building: Machine vs. human energy.

So my son Peter and I took off for Cambridge. Our first helicopter ride, from Sausalito, smelly exhaust, a dreadful machine, to the S.F. airport. Then in a 747, five hours to cross the country! The huge jet was not 1/5th full, a terrible waste of fuel. When I went into the bathroom, the finely-built one piece aluminum washbasin and toilet stand gave me an insight into Buckminster Fuller's ideas of housing. Since Bucky has been constantly traveling now for many years, he spends an enormous amount of time in planes. He has always loved machines and metal (see the Phantom Captain chapter in *Nine Chains to the Moon)* and his fascination with air flight and aerospace technology lead him to dig aluminum efficiency such as the 747 in-flight bathroom. Bucky and many others (see Le Corbusier: *Towards a New Architecture)* think of houses as machines. Probably because machines were just beginning to demonstrate their remarkable clanking capabilities when Bucky and Le Corbu were at impressionable ages, their image is of houses being mass-produced, standardized, and now computerized. But I'm getting ahead of myself.

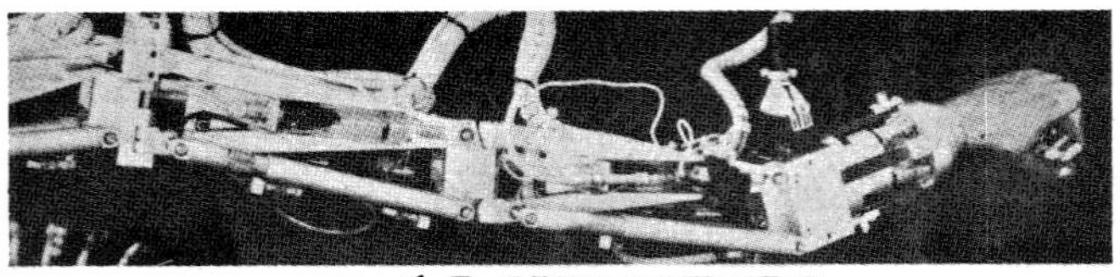

ROBOT ARCHITECT

The conference turned out to contain some ideas of architecture which made me gasp. Even though MIT has published some excellent books on native structures, the dominant theme (ironically) of Responsive Housebuilding Technology was computerized plastic flash. Right around the corner from the conference room there was a large computer being worked on by students, staff and others. It's in its own suite of rooms, with homey looking exposed wires running between machines, plexiglas panels so you can see the electronic wizardry, and rock and roll on the radio.

The computer is called "The Architecture Machine" and its creators seek to build an intelligent machine, one that they can have a dialogue with. Robot architect. It took me a few days to figure out what the machine could do, and was being trained to do, but here it is, and realize dear reader, that this is architecture at a leading American university, and that the project is well funded, and well respected:

Meet the robot architect and its functions (with code names):

SEEK is a mechanical device hooked into the computer that will pick up, stack and rearrange cubical blocks on command from the computer. In a museum exhibition two years ago, the machine, which can handle 300 cubes, and a colony of 60 hamsters were put together. The idea was to have the computer stack the blocks in a way the hamsters liked. The hamsters tended to knock over the blocks, running in and out (looking for their natural environment, but this was overlooked by the researchers) and SEEK was to figure out which way the hamsters liked the blocks stacked, and arrange them in that manner. Apparently what happened was the hamsters didn't like *any* way the machine stacked blocks, they didn't like the blocks, they didn't like being in the museum, and they just knocked blocks over. But the idea of it all, in the words of one of the computer team: "...If this idea was carried out in a peopled world, perhaps a giant SEEK could sense the behavior and actions of its people and provide a responsive, useful and friendly living space, better than what now exists..."

GREET is a doorway device of photocells which will recognize whoever passes through the doorway. Work is now in progress "testing the machine for ways of recognizing height, weight, stride, foot size, i.e. relatively constant characteristics." A series of photocells will sense the silhouette of passers-through the door and will compare it with a dictionary of well-known silhouettes and say "Hello, Richard," or whatever, as you pass through. The voice part of the computer is called SPEAKEASY.

HUNCH is a project whereby the computer will be able to understand sketches. In this way the architect can feed his rough sketches in to the machine and the scribblings will be made into perfect curves or angles and speed up the design process.

There are other things the machine can do, like a three-TV-screen unit which can display multi-images of the same scene from different points of view. But that is just a quick layman's view of it.

Now, also hanging around at MIT are pneumatic structure designers. Air buildings have been used at fairs, exhibitions, ice rinks, and now the technology is well enough along so that architects are able to construct them. Artists started out several years ago with polyethelene, and some designers made nice enough looking structures so that now plastic manufacturers, schools, etc. are interested. They appeal to the consumer-oriented U.S. public, as they are even newer than domes, and are flashier media architecture.

This computer/airbuilding/plastics thing that seemed to be on so many of these architects' minds jarred me, as it seemed roughly parallel with a logical extension of some assumptions I'd made 3-4 years earlier on the idea of housebuilding technology. The assumption, encouraged for a time in my mind by Bucky Fuller, was that we will have to depend upon *new* technologies, *new* materials, *new* designs to solve the housing crisis on an overpopulated earth.

Thus architects and designers look for the *breakthrough,* the previously undiscovered and hitherto-hidden quick and easy solution.

LOOKING BACK

Some scant background: Looking for new solutions to making family sized houses led me into building and helping others with a good number of geodesic domes. We were inspired, we had a vision, and we were in a hurry — we had people waiting for a roof over their heads. We tried every material we could get cheap enough: wood, plywood, cardboard, sheet metal, aluminum; fiberglas/Vectra cloth/polypropelene/all manner of horrid chemical caulks/vinyl/polyethelene/plexiglas/Lexan/ABS plastic/steel and on and on.

At this time I was intrigued with the space program, video, computer art, the Moog synthesizer — and I decided we would try any hi-tech application we could get our hands or Our work at Pacific High School, as described in *Domebook 2,* was exploring materials. We stuck to geodesic geometry as it was simple and gave us a rather neutral framework to work with in each case. Our main work, often missed by people thinking of the dome work in architectural terms, was in the realm of *materials.* With each material, the builders there tried to create as aesthetically pleasing a space as possible.

In all this work, we tried just about any plastic we could obtain. What I found out is that compared to the publicity by oil/chemical/plastic industry, plastics are going to have a very limited application in housing of the future.

While plastics have certain limited building applications (such as plastic sewer pipe, which an amateur can assemble) it is highly unlikely that the use of oil/chemical derived materials will ever be of significant use as structural or cladding construction, for these reasons:

PLASTICS HAVE SHORT LIVES

First, there are *practical* disadvantages to the use of plastics in building. They are extremely expensive compared to conventional building materials. This has caused me to thin that the cost of a material is roughly proportionate to the ecological damage done to the earth in removing and refining it. To find, for example, a plastic material that will resist sunlight without cracking is extremely difficult, or expensive, or both. There are virtually no plastics develope that are cheap and durable enough to cover buildings on an scale.

I recently went back to look over the 17 domes we built at Pacific High School, so these observations are based on experience: plastic foam gets easily damaged if not coated with something hard, and to coat it with something hard is expensive; it gets knicked and gouged very soon. It also turns an ugly oily brown color if not painted.

Polyurethane foam is said not to burn by foam salesmen, and it is true that it doesn't catch fire easily. But it is also true that once it *does* catch fire, it explodes like gasoline and releases poisonous cyanide gas. I've concluded that foam is strictly an insulation material, and even then to be avoided if possible due to cost, fire danger, pollution in its manufacture, and poison danger to the applicator.

We used vinyl for windows and in some cases to cover entire domes. After living and working with it for a few years I have become repelled by the material. It never loses its objectionable smell, it attracts and collects dust and although at first you think it is clear, after a while you realize that you are looking at trees and stars through a film of chemically-rearranged oil. Vinyl continually loses molecules from its plasticizer, which accounts for the film you see on auto windshields — from vinyl seat covers. In Viet Nam some GI's died from blood transfusions from vinyl bottles. This molecular migration probably also work subtly on your nervous system.

Fiberglass does a lot of things other plastics can't, but I don't like to work with it: it smells, has itchy glass fibers. Though it looks o.k. on surfboards, it is hard, shiny, unattractive to me as a building surface.

We had some spectacularly bad results trusting in caulks. Of course our 16-year old workmanship at Pacific High School was not that accurate, but even with super fitting, we were trusting too much in claims of manufacturers and salesmen.

After working with every possible kind of plastic clear or semi-clear window material, I've rediscovered glass. It is true that plexiglas doesn't break and is easier to cut, but it scratches easily and permanently, attracts dust and dirt, an just never has the sparkling clear, image-transmitting capabilities of glass.

Secondly, here are some personal *aesthetic* discoveries I've made in spending a few years around various plastic materials (I'd lived previously with more conventional materials such as wood, concrete, glass, brick, etc.)

I've found that the less molecular rearranging a material has undergone, the better it feels to be around. Wood, rock adobe as compared with polyurethane foam and polycarbonate resin windows.

...RU EARTHQUAKE, 1970. FOAM DOMES ARE SPROUTING ...AND-BUILT ADDITIONS.

OIL OR WOOD

...occurred to me lately that there is a profound difference ...tween the way wood and rock are produced, and the way ...stic foam and flexible vinyl windows are manufactured. ...nsider that a tree is rendered into "building material" by ... sun, with a beautiful arrangement of minerals, water, and ... into a good smelling, strong, durable building material. ...revever, trees look good as they grow, they help purify air, ...ovide shade, nuts to squirrels, and colors and textures on ... landscape. And *wood is the only building material we ...n regenerate.* On the other hand, most plastics are derived ... pumping non-renewable oil from the earth, burning/ ...fining/mixing it, with noxious fumes and poison in the ...ers and ocean, etc. Of course, saw mills and lumber ...mpanies rip stuff up with gasoline motors and saws, and ...oke fumes, but it strikes me that the entire process of ...od growing and cutting is preferable to the plastics ...oduction process. What *is* called for is tree-respecting ...rest management.

...wever, there are obviously many people who feel ...mfortable with items such as Tang, pink plastic hair ...rlers and the disposable dishes on airplanes. Discover ...ur ideals as you take your choice.

...end to feel uncomfortable around any oil-derived or ...ghly processed plastic material. Polyurethane foam seemed ... if it would be better than the others, but it, too, turns ...t to be ugly.

... addition to the *practical* and *aesthetic* disadvantages I've ...und in plastics there is the idea that one is dealing with ...ow, and the oil industry — that is the people Nixon works ...r.

...n still not afraid to use plastics, I just have a far more ...alistic picture of what they can do. It turns out, after ...veral years of varied experimentation that plastics ...n't stand the weather, or if they can they're ...tremely expensive.

...e foam builder tells us foam can be shredded up into ...ulch. Sure, I reply, it's a good mulch, but it stays in the ...il, and after you keep mulching with it, your soil ...comes more and more plastic and less and less dirt. ...etty soon you can raise plastic flowers!

...fter the MIT conference, Peter and I drove out to Cape ...od, spent Friday night in an old inn. It was a beautifully ...ilt 100 year old wood building with an elliptical spiral ...aircase said to have been built by an itinerant carpenter ...o built three such staircases on the cape. Next to the inn ...s a large barn which was being converted into an art ...llery. I had a drink with the owner in the inn's small bar ...d we started talking about buildings. I asked about the ...rn, and he said, "Do you want to see it tonight?" "Sure." ... walked into the large building in the darkness, and then ... switched on the lights. It was about the most dramatic ...y to see a beautiful old building, the sudden blaze of ...hts revealed a 100 year old mortise and tenon structure. ...ere were about four loft-levels, and at the top was a ...xagonal cupola. The inn's owner sensed something was ...ing on with me in the barn, so he went back to the inn, ...lling me to stay there as long as I liked. I climbed up all ...e ladders, up all the stairs, looking at the joinery (wooden ...gs.) Then up into the little cupola room which was above ...e roof line, smoked a joint, sat and looked out over miles ... countryside in moonlight. To the north, the water. ...tting there, 50 feet high, supported by hundred year old ...ooden structure, the futuristic plastic building notions ...emed strange indeed.

MACHINE BIRTH

...e pilgrims actually landed in Provincetown, before ...mouth. One of the first things they did, according to ...ks in Provincetown, was to steal the Indians' corn crop.

...wish I knew more history. Where did this western ...chnology start? Was it due to metals? Machines? ...ectricity? Resources? What started this thing that led to ...ath of American Indians, much wildlife and forest, ...ssive alteration of air, water and topography? What was ... spirit that invaded this continent, machined its way to ... Pacific Coast, then eventually got a stranglehold on ...ost of the planet?

JIB SAIL WINDMILL IN THE PLAIN OF LASSITHI, CRETE.

I sent an early draft of this writing to Bob Easton; here's part of his reply:

1. Science: got started by people studying the stars and biology for healing purposes. Certain principles of mechanics grew out of observing nature: stars, trees, animals. Leonardo. Newton. Etc.

2. The New World: Stories of fabulous riches in the East moved western man to explore and hoard — the development of consolidated power by the developing "nations" of Europe created this awareness of the Roman experience, of super abundance, super-power — lust for more riches, hoarding, super tribes competing for dominance by the ultimate in power display — the greatest accumulation of useless gems, gold. Ferdinand and Isabella. Henry VIII. The new world exploration breeds technology, better equipment to transport. Worship of material objects creates subsystem of technique necessary to masturbate this outrageous lust.

3. Slavery: The human slave was considered a machine by Romans, Greeks. European man in his exploitation of the New World riches could condone slavery abroad — possibly the church in its traditions dating back to Roman days would not allow slavery within Europe. The slave "machine" was profitable because it bred, needed cheap fuel, basically looked after itself, wasn't paid; is the "robot" of the futurists...

4. Slavery Ends: Outrage over conditions slaves are subjected to is voiced by humanists and artists of the 16th and 17th centuries — a new class — people who have moved thru the arrogance of accumulated objects into new levels of consciousness. These people bring tremendous pressure on the merchant/power/military class first in England, then the U.S., because they are of a higher class within the social hierarchy of the society...the children of the leaders (Dickens, Swift.) The pressure builds to end slavery — panic — the old order must change. The newly growing technical class is pressed by merchant leaders — possibly unconsciously — or perhaps independent innovators within the merchant class rise to meet the challenge — certainly within the circles of power and technique the fears were voiced. The biology-scientist becomes the gross engineer.

5. The Answer: Watt developes the artificial heart, the steam engine, and the others all follow: machines analogous to the rest of the body, including the greatest of all, electricity, the machine equivalent to the life force itself. The answer is the mechanical/electrical slave, the great source of wealth that western man created all by himself. No other culture developed this. China's war lords made gunpowder, etc., but is nothing compared to the incredible competitiveness of the fierce western white tribes. The new idea pioneered in America is: now everyman can have slaves — cars, labor saving devices, etc., plus the power high gotten off using power tools — the same high gotten off using slaves, basic to the small human ego, which is so susceptible to extending its range of influence and power.

6. However, the consumer-people of the western world are but children soon to be cast out of the warm cradle, because the monster slave has begun to die off: the young of today are instinctively cutting off its regeneration. The costs of using its services will soon begin a very rapid rise because of scarcity. The cost of gasoline, electricity, plastics will rise so they can only be bought by the industrialists to maintain their power. As the unions hoard the skilled jobs and knowledge, their power and wealth will die with them. As the medical professions develop more artificial drugs, the viruses will continue to grow more sophisticated to overcome those drugs and will kill off those who contact those germs/ viruses; since viruses only attack dead cells within the body, the ill-fed people/consumers will be susceptible to disease.

The next main stream culture will be made of the artists and humanists of today's subculture. Why? There may be no alternative. It appears now that the ultimate tool of the techno-fantasy people, the computer, says to turn itself off. (See World Dynamics, *by Jay W. Forrester, Wright-Allen Press, 1971.)*

Why not listen to Bernard Maybeck who wrote:

"The artist suspects it is not the object nor the likeness of the object he is working for, but a particle of life behind the visible. Here he comes face to face with the real things of life; no assistance can be given him; *he cannot hire a boy in gold buttons* to open the door to the Muse (our italics), nor a clerk or accountant to do the drudgery. He is alone with his problem and drifts away from superficial portrayals. After this he strives to find the spiritual meaning of things..."

Above quote from booklet: The Palace of Fine Arts and Lagoon, *by Bernard Maybeck, Paul Elder, 1915; quoted in* Five California Architects *by Esther McCoy, Reinhold Publishing, 1960.*

INVENTOR JOSEPH SALIM PERESS STANDS BESIDE HIS ARMOURED DIVING SUIT (DEPTH 1000 FT)

MAN ?

Now back to MIT. The computer people at MIT and the air building people have collaborated in various architectural visions. Example: an air building controlled by computer which recognizes people when they come in; and when say 60 people get into the building, the computer unrolls and blows up another plastic section to accommodate more people. The occupants have control over windows, for example — they can make windows appear or disappear. Computer allows occupants to change shape of building at will. "Hal, will you set the table for eight tonight?"

Another idea that's been around for a while, that came up at MIT: architect draws on cathode tube with magnetic pencil; design for a foam house is fed into computer. Computer operates a foam truck with barrels of foam, boom, and extruding device. The truck boom manipulates around, extruding walls of the house. The house is built with no human hands touching it.

Wait! at this point, the last day of the conference, I started yelling. (Sym van der Ryn had been arguing with them earlier.)

"This is an architectural conference, there are no *people* here, just *professionals* playing academic futuristic games. No women, kids, men here to react to your ideas, academic insularity. Moreover, you designers, especially the ones with artistic abilities, are making plastics and a totally impractical and weird shelter outlook appear seductively appealing to those folks who are always looking for something new and flashy. Spacy air buildings are deceptive, that's all. No one is ever going to really live that way, but it's good media. The same thing I learned with domes, they photograph well."

The planet needs non-polluting energy sources. Solar heat, wind electricity, methane from compost. Revive waterwheels; sawmills in New Hampshire were driven by water power. Put 2/3rds of the staff at MIT on developing clean(er) burning motor vehicles! Create a mind bank with the Architecture Machine and come up with a solution to internal combustion before the Chinese have two cars per family! If successful, you will be national heroes upon graduation, and receive free non-polluting cars the rest of your natural lives.

Architects, use your skills and desirable positions to assist in current housing problems. Help people! You don't have to find a gigantic new solution to housing. The answer may be in our *hands.* Whisk Whisk Whisk, the sound of 100,000 Chinese brooms sweeping snow off Peking streets. No snowplows. The shit of Peking collected and used for fertilizer. No sewage problem.

MIT, architecture schools, have you ever considered that in some cases, designs get about as good as they're going to get, and then don't improve for millions of years. Look at your hand! Is there a need to redesign it? Have architects, builders ever considered that our grandparents, but more especially the Indians, built far more sensibly than today's building industry? And that maybe looking for new structures and new materials isn't that important right now? That you can't think about building, or design unless you consider the life style? And that the extravagant use of resources in the U.S. now can't last, and is in fact maintained at the expense of subjugated, bombed, exploited third world people everywhere?

I was particularly disturbed by the vision of the architect sitting at the cathode tube, drawing his design into the computer, the computer causing the foam truck to build the house. The ultimate in laziness, machine worship. Machine can do anything better than man if we develop

machine enough, is the premise. Wrong! It's going to look like shit — guaranteed — it's going to cost too much, it's going to be ecologically unsound, it will only produce environments that machines or machine-like people will want to inhabit.

John Ryckman of Montreal sent us a photo of a Thai man weaving a rainproof head shield with the following comment:

He never heard of "great circle theory" — doesn't know geodesics from A,B,C, — and thinks Buckminster Fuller is nothing but a smooth-talking evil spirit.

So, there's a lot of trickery and hype afoot. I ran into a good deal of it and wish to pass along my disillusionments for the edification of those who won't therefore have to go through the same trial and error (much error!) process.

Buckminster Fuller's description of man (from chapter, The Phantom Captain, *Nine Chains to the Moon*):

Man?

A self balancing 28-jointed adapter-base biped; an electrochemical reduction-plant, integral with segregated stowages of special energy extracts in storage batteries, for subsequent actuation of thousands of hydraulic and pneumatic pumps, with motors attached; 62,000 miles of capillaries; millions of warning signals, railroad and conveyor systems; crushers and cranes...

HAND-OWNER-SELF BUILT

Here is a quick summary of some things I've learned about shelter:

1. Use of human hands is essential, at least in single-house structures. Human energy is produced in a clean manner, compared to oil-burning machines. We are writing for people who want to use hands to build.

2. It took me a long time to realize the formula:

Economy/Beauty/Durability: Time

You've got to take *time* to make a good shelter. Manual human energy. For example, used lumber looks better than new lumber, but you've got to pull the nails, clean it, work with its irregularities. A rock wall takes far more time to build than a sprayed foam wall.

3. The best materials are those that come from close by, with the least processing possible. Wood is good in damp climates, which is where trees grow. In the desert where it is hot and you need good insulation there is no wood, but plenty of dirt, adobe. Thatch can be obtained in many places, and the only processing required is cutting it.

4. Plastics and computers are far overrated in their possible applications to housing.

5. There is a huge amount of information on building that has almost been lost. We'll publish what we can, not out of nostalgia but because many of the 100 year old ways of building are more sensible *right now*. There are 80 year olds who remember how to build, and there are little-known books which we'll be consulting in transmission of hand-owner-self-built shelter information.

Before I left home, Peter Warshall told me to be sure to see the Peabody Museum of the American Indian at Harvard. So the first day of the conference, and twice thereafter that week, we went over to Harvard, and I was truly staggered. Seeing these things in real life rather than pictures — so unbelievably beautiful! Since I like to work with my hands, I usually look at the way objects are made. Chumash baskets! All hunting, religious, cooking implements are incredibly crafted, fashioned and ornamented by men and women in touch with the earth and its streams and breezes. Ingenious shelters! At the museum someone has made fine models of Indian villages with cutaways showing how their structures were built. There are even miniature baskets in the model settlements.

Walking amidst magnificence of Indian craftsmen with MIT dimly in mind, I realized that there may not be any wondrous new solution to housing at all. That there is far more to learn from wisdom of the past and from materials appearing naturally on the earth, than from any further extension of whiteman technoplastic prowess.

Relics of the past (Indians)
vs
Visions of the future (MIT).
No contest.
We've been losing ground.

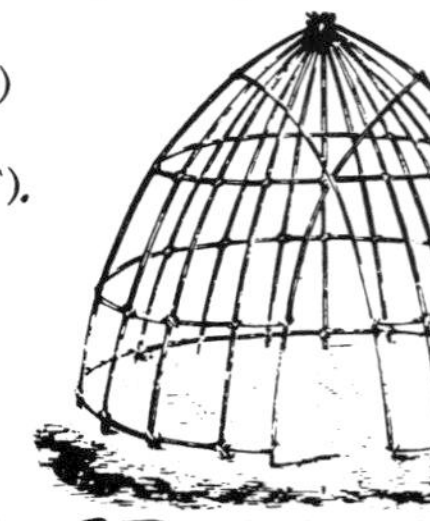

RESPONSES TO SMART BUT NOT WISE...

Ed Allen, who organized the conference wrote:

You have to understand that we are into a very broad range of possible ways to put the individual in control of his environment, from the sleek and mechanical to the folk-technological. We are not into closing off possibilities before they are adequately explored. We are keeping our stuff to ourselves and a limited audience of fellow researchers until it proves worthy; we are not into trying to popularize any half-tried technology, then later having to decide it was smart but not so wise. What you wrote was fair enough; what you didn't write leaves us with an image that is inaccurate at best. We're all human. We even smile sometimes, often at ourselves...

And in a later letter:

Smart But Not Wise *has had a wide reading here. The usual reaction is that you are either not wrong, or mostly right, but that you made a rather biased selection of MIT projects to discuss...*

Ed went on to point out that I neglected to mention that there were some other low-technology construction methods, including a soil-cement system he developed based on an African technique at the conference.

True. However, the main thrust of the conference was decidedly futuristic, as are many other such college gatherings. (See p. 117 on the Cal State L.A. *Shelter for Mankind* conference.) Colleges and universities clearly have a close relationship with corporate-backed technocracy and are encouraged to find further applications for new industrial products. Even though MIT does publish some excellent books (including Ed Allen's *Stone Shelters*) one can't help but feel corporate vibrations in looking at its monumental, fortress-like architecture and in walking down its gray corridors. Stewart Brand told me I wasn't the first to sense this and to read the chapter, *Getting Back to Things at MIT* in William Irwin Thompson's *At The Edge of History*, which describes a symposium at the Institute's new Center For Advanced Visual Studies (read art):

Another engineer from Bell Telephone Laboratories described a more up-to-date means by which technology could help the artist. He showed that we now possess the skill to place probes on a man's head and induce pleasant sensations in response to, say, the image of a woman on a screen. At last technological man could dispense with woman altogether to have an electronic orgasm in a geodesic dome, and while the sun was being obscured by the SST haze in the upper atmosphere and the earth was losing solar radiation, man could crawl back into the cave where industrial shamans could throw latrine images on the walls.

The guilty technologists had destroyed the environment; now their servile apologists, the technocratic artists, were trying to make up for it in an environmental art. One could almost hear the bureaucratic apologists discussing pollution at another future symposium at MIT: "Although the spin-off from our technology has caused undesired consequences through inadequately funded research and development *for habitational systems contiguous to the productive sector, we can be grateful that the forward thrust of that technology has given us the means to stay on top of the problem with relative ease. The artifical life-support systems that were designed for our astronauts in the sixties have given us the knowledge to create whole new environments in civilized containers of the geodesic variety. Surely, if we can keep man alive on the moon, or in the bottom of the sea, keeping him alive in Cambridge should prove no real problem.*

COLLAPSED TOWER, GATEWAY TO STANFORD UNIVERSITY. THE CALIFORNIA EARTHQUAKE OF APRIL 18, 1906. FROM REPORT OF THE STATE EARTHQUAKE INVESTIGATION COMMISSION, VOLUME I, PART I.

..."THE LEAST ENERGY."

...It certainly goes considerably beyond my expectations — for while I wasn't myself too inclined to trust the plastics and the adhesives, having been in on some of the early commercial trials some years ago — nonetheless, I am a bit surprised to find you publically admitting to similar conclusions after you went so far with your own experiments, and their publicity. More than that, though some might look upon it as simply a rationale, I'm very heartened to find that you are anxious to explore the whys and wherefores of the older building methods, and of low energy systems particularly....

Yet while I studied with Bucky, listened to his lectures four straight hours day after day at Black Mtn. as well as at the ID, I have come to question several of his fundamental premises. And to question them, I hope, in terms of his own, not my, logic. For instance, the dictum that you "cover the maximum area with the least material" should better read it seems to me, "the least energy." For in the interest of saving resources — human energy and all forms of impounded energy such as fuel and materials — the basic criteria should certainly be energy. This lets out a lot of modern materials that consume great quantities of fuel in their extraction, and processing, and fabrication if we are at all honest about it. It should also make us suspicious of technologies that require men to be so specialized that their training must be paid for by the community not only in high wages and long years of education, but also by the curious fact that they are seldom able or willing to perform other useful tasks — unlike settlers, farmers or modern owner-builders.

Nonetheless, some of the older technologies — like the brick walls of Iranian domes, and the pottery of China, Korea, and Greece resulted in the complete deforestation of these countries. Even glass manufacturing was almost immediately banned in England when Henry VIII, I believe, saw that it conflicted with the great demands for oak in shipbuilding and iron smelting. (And it would be very naive to believe that all the Indians, or other so-called primitives, did any better all the time. Archeological evidence demonstrates that the Hopi, for instance, considerably damaged their surroundings that the corn fields were moved to the lower irrigated flats before the Spaniards arrived with their herds of sheep. It has to be presumed that the Hopi cut down so much timber for agricultural clearings and for fuel that they changed the delicate balance in the humidity a few degrees and rendered their highland fields, their more easily cultivated unirrigated fields useless.)

And while it is just as well to remember that man does not live at the climax stage of vegetation, that we necessarily impose our own order for the species' maximum benefit or carrying capacity, it is not romantic or impractical to consider rational alternatives to ancient or modern forms of cultivation and building. I have soil texts indicating that the Chinese knew a good deal about the proper handling of particular and difficult soils by 100 B.C. and even earlier. Some ancient people have, on the whole, done remarkably well in their conservation practices....

Myself, I see much of our aesthetics coming from unconscious, or barely conscious intuitions anyway. And I'm not much on suggesting to others what's what. Bucky remarked that no one, even he, was right about anything much more than 50% of the time....

All the best,
Art Boericke

...WELL-TRODDEN PATHS.

... if we live long enough, one eventually comes to the personal appreciation;

"Beware the prophet or seer who claims that he is the proprietor and the fountainhead.

There is nothing 'new' ... under the sun. The only 'newness' of anything ... is but the surprise of our 'conscious-awareness' ... meeting an old and ancient thing for the first time. 'Newly-new' ... but only to our personal, subjective-awareness."

"NEW ... "ORIGINAL" ... but only as a subjective value-judging evaluation ... from the position of our previous ignorance.

Thus, the path of evolution ... and thus, by our ability to see and understand this, do we measure our individual fitness for survival....

Received your *Smart But Not Wise*. A...declaration of your fitness for survival ... in that it reveals your ability to have finally raised your eyes from the most very limited horizons of Bucky's navel ... to the sweeping horizons of man....

There is nothing 'new' ... ALL PATHS ... are old, ancient, very well-trodden paths.

John Ryckman
Montreal

:hanges in the quality of life as machines take over. Remember how a tomato used to taste before mechanized griculture?

t takes at least $100,000 to start as a farmer these days. Business men have replaced farmers and we now see the esult of mechanized farming: ruined topsoil, poisoned ivers and tasteless and expensive produce. If you rtificially stimulate celery with chemicals to grow too ast, it loses its natural defenses against predators, which nust then be sprayed with poison. The same type of short erm gain, long term loss will happen with mechanized ousing. We must find the ecologically-sustaining balance etween human and machine energy in both agriculture nd housing, and discover what different materials actually ost the earth.

Government and big business would like to automate ousing because it eliminates workers, provides higher orporate profits, consolidates wealth in corporate pockets.

One percent of the shareholders own 75% of the stock and 85% of the bonds in American corporations....

S.F. Chronicle, 7/8/73

Salaries of a few corporate executives, 1971:
Chairman, Standard Oil of New Jersey: $485,000.
Ralph Warner, Jr., Chairman of Mobil Oil: $410,000.
Chairman of IBM: $394,331.
J.M. Roche, Chairman of General Motors: $822,000.
Harold Geneen, ITT: $814,000.

Housing and Urban Development (HUD) sponsored *Operation Breakthrough* a few years ago, a contest for the est mass-produced housing. George Romney, who onsolidated four Detroit auto manufacturers into American Motors was hired by the government to attempt the same hing in housing. Such consolidation means higher corporate rofit, greater centralized control, mass-production, all ouses looking alike. Build houses like Fords, everything tandardized. All walls, windows, bathrooms, heating systems he same. Eliminate the craftsman, hire low-paid factory vorkers to pull the levers that stamp out the components. terile houses because the machines that make them, unlike uman hands, can only make repetitive motions. Sterile eople inside, isolated from sun, wind, rain and dirt. Life nslaving mortgages to pay off the corporate chain that roduced the houses.

George Romney: "The on-site builder will be around for a few more years..."

Machines, unlike humans, need be paid only with electricity, ubrication and spare parts.

The assembly line produces identical objects. Should homes e identical like cars?

Abundance creates a lust for greater abundance. Like the lust for innovation. The white man innovating technician trip. There are all these choices here because America is so rich. But this book is timely because we're approaching the time now when we're not going to have that abundance. The individual isn't much longer going to be able to choose between shakes and honeycomb aluminum. He's going to have to build with exactly what he finds, maybe dirt or salvage. It's starting to happen; maybe that's why we're back in the 50's in entertainment and so many styles now. Maybe the same will happen in design, we can't afford all the plastic. The '50 Ford.

We rented a car in England. 50 miles per gallon. It was adequate. England had abundance but lost her colonies. Now things are tight. Our friends in London have a well-insulated hot water heater that you turn on a half hour before bathing. Then you leave it on another half hour, and the water will still be hot the next morning. When you see how efficient the English are, and they're a rich country, you realize how far out of line America is. Things are going to get tighter and tighter. Americans are going to have to forget their indulgent conceptualization and start acting from their experiences.

TECHNOLOGY REVIEW

We're not against technology, mechanization, innovation, or plastics. We're against their misuse.

Technology is now warring openly against the crafts, and science covertly against poetry. The original meaning of these terms has long been forgotten....

"Technology" is a Greek compound noun originally meaning "the topic of craftsmanship", but now meaning "the application of mechanics to manufacture,"...

Robert Graves, Dec.2, 1971 *New Scientist*

Industry should produce raw materials, practical objects suited for mass production, and tools. There should be less centralized corporate manufacture of elaborate finished products that make the individual a consumer, therefore increasingly an invalid. Resources are abused and exploited in the production of useless consumer objects.

Individuals should be able to use the mass-produced products of industry to make goods regionally, to create local vernacular cultures. Raw materials could be used by people to provide those services and goods that are by nature local: food, shelter, utensils and craft objects. Industry, instead of mass producing food, houses and electric toothbrushes should do what local technology cannot: autos (but durable ones), tools, machinery, and basic materials.

Irresponsible corporate production	Responsive, responsible technology
disposable cars	*durable cars & parts*
mobile homes, plastic prefabs	*raw materials, plastic plumbing pipe, glass*
Aerosol can shaving lather	*shaving soap and brush*
747, Lear personal jet, SST	*DC3, 707*
centralized communications	*local tv*
corporate mechanized agriculture, mono-culture, chemicals, insecticides, high transport costs	*local, family crop farming, crop rotation, return of nutrients to soils, local distribution*
plastic clothes	*cotton, wool clothes*
prepared foods, McDonalds	*your own garden home-baked and prepared foods*
electric dryer	*clothesline—wind*

What is critical, especially in these times, is to determine the point at which *hands* enter the production process, and to define which parts of the process are suited for mechanization and which are more suited for the hand crafts. This is important not only for human, but for practical and economic considerations. For we are increasingly surrounded by dehumanizing, poorly crafted, highly expensive products of corporate-backed mechanical technology. And ironically, what may lie at the end of the world's most highly developed industrial society, is a picking up of tools, rediscovery of inherent individual capabilities, and getting back to work.

In England during the industrial revolution, art critic John Ruskin "...viewed with horror the appalling discrepancy between the promise of the machine and the actual squalor, suffering, and ugliness of victorian capitalism. He urged a return to hand craftsmanship as an antidote to the corroding effect of the machine. 'No machine yet contrived, or hereafter contrivable,' he wrote, 'will ever equal the fine machinery of the human fingers.' "

Greg Howell points out that in the development of airplanes there was a rapid growth and development period in the 40's and 50's when small investment returned great advances, but there came a time when ratio of advance to investment leveled off. Greg thinks the 707 was the last "good" investment, that the 747 probably cost as much to develop as a good proportion of what was spent in total, up to that time, and that development generally seems to follow this pattern.

Penthouse: *We're being deprived, then, of useful and labour-saving inventions?*

Burroughs: *Oh, my God, yes. I was brought up to believe in the tradition that the good product would always find a market. But it's not in industry's interest to put out a pair of socks that won't wear out. The original nylon socks would not wear out. I wore one pair through South America, walking through jungles and water. Now they wear out in a week just like any other socks. It's not to manufacturers' advantage to produce good products. It's not to their advantage to produce houses that won't wear out or anything else that won't wear out. Industry and government don't want these products because they would disrupt a very creaky social system. In other words, scientists are producing new inventions much quicker than the social system can possibly absorb them without great disruption.*

Penthouse *interview with William Burroughs.*

Amos Rapaport points out in *House Form and Culture* that excavated housing and settlement forms, of people long since gone, are often still usable, suggesting that: *...certain aspects of behaviour and ways of life are constant, or change very slowly, and that replacement of old forms is often due to the prestige value of novelty rather than lack of utility or even unsatisfactory relation to the way of life. Similarly, of course, acceptance of old forms may also be due to the prestige value of old things rather than any real or continued validity or utility of the forms....It seems clear that man has changed little in body and physiology since his beginnings. If man does, in fact have certain inborn rhythms, biological needs and responses which are unchanging...the built environment of the past may still be valid....*

We don't need new shapes in housing any more than we need newly shaped tomatoes (growers are trying to develop square tomatoes for easier picking and packing.) We don't need new foods. You can't improve on pure water. Are there really better building materials than earth, stone and wood?

ALBERT SPEER'S DRAMATIC LIGHTING AT THE NUREMBERG PARTY RALLY, CREATING WHAT SIR NEVILLE HENDERSON CALLED "A CATHEDRAL OF ICE" FROM *INSIDE THE THIRD REICH*

As the top representative of a technocracy which had without compunction used all its know-how in an assault on humanity, I tried not only to confess but also to understand what had happened. In my final speech I said:

Hitler's dictatorship was the first dictatorship of an industrial state in this age of modern technology, a dictatorship which employed to perfection the instruments of technology to dominate its own people....By means of such instruments of technology as the radio and public-address systems, eighty million persons could be made subject to the will of one individual. Telephone, teletype, and radio made it possible to transmit the commands of the highest levels directly to the lowest organs where because of their high authority they were executed uncritically. Thus many offices and squads received their evil commands in this direct manner. The instruments of technology made it possible to maintain a close watch over all citizens and to keep criminal operations shrouded in a high degree of secrecy. To the outsider this state apparatus may look like the seemingly wild tangle of cables in a telephone exchange; but like such an exchange it could be directed by a single will. Dictatorships of the past needed assistants of high quality in the lower ranks of the leadership also — men who could think and act independently. The authoritarian system in the age of technology can do without such men. The means of communication alone enable it to mechanize the work of the lower leadership. Thus the type of uncritical receiver of orders is created....

"The nightmare shared by many people," I said, "that some day the nations of the world may be dominated by technology — that nightmare was very nearly made a reality under Hitler's authoritarian system. Every country in the world today faces the danger of being terrorized by technology; but in a modern dictatorship this seems to me to be unavoidable. Therefore, the more technological the world becomes, the more essential will be the demand for individual freedom and the self-awareness of the individual human being as a counterpoise to technology."

From Inside the Third Reich *by Albert Speer, architect, Reich Minister of Armaments and War Production, prisoner.*

Eva Braun, Hitler's mistress, had herself filmed around their mountain hideout for years. Hitler had promised he'd make her a star in Hollywood when he took over the world...

CBS Evening News, 6/73

Walt Disney made war propaganda films during WW II. An Air Force general had written a book advocating a long-range bombing war against Germany, but no one was reading it. The general convinced Disney to do a film on it; Donald Duck starred. FDR saw the film and was convinced. Soon after, at a meeting in Canada, FDR was trying to persuade Churchill to follow the general's strategy and was getting nowhere...but a call to Hollywood put the film on a special plane to the meeting. Donald Duck convinced Churchill, as the people of Dresden remember....

From interview with film reviewer, KSAN radio, San Francisco, 7/73.

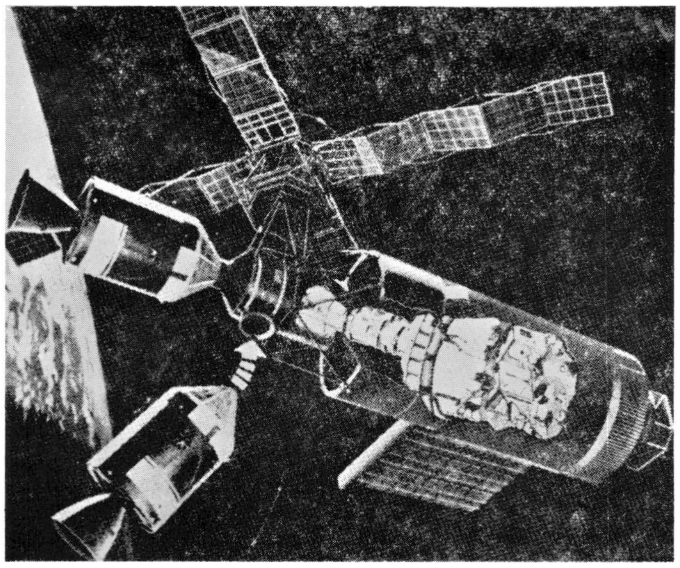

As the Skylab 2 astronauts prepared for 59 days in space, scientists said yesterday the Skylab I crew's 28-day mission showed the human body apparently adapts to weightlessness by shedding unneeded muscle tissue, calcium and red blood cells.

At a Cape Kennedy news conference, medical scientists said the decrease in muscle tissue and calcium shown by the three Skylab I astronauts was similar to losses exhibited by people confined to bed for long periods...

Dr. Michael D. Whittle, a Royal Air Force medical officer working with NASA, said the Skylab I astronauts experienced the greatest weight loss in their legs.

"We now think this weight loss is a pure and simple adaptation to weightlessness," he said.

The calcium loss seemed to be a result of decreased stress on the leg bones, Whittle said, adding that the loss was only about 1 percent of the total amount of calcium in the body.

"The loss would have to continue for a year before we would have to worry about a man breaking bones, but it may be a problem when we get to Mars," Whittle said.

Associated Press 8/73

We got to the moon and found there was really not much there. Dull, grey, lifeless. No monoliths. Scientists thought they detected water in orbit around the moon, it turned out to be jettisoned astronaut urine.

"Snoopy to Apollo, do you read me, over..."

ARTHUR CLARKE

L.F.: *You've written that machines will eventually supersede human beings.*

Clarke: *I suspect that's true, yes....I've concluded that we will have intelligent machines, and I think they will be superior to us in intellectual abilities — of all kinds. I suspect that our destiny is to act as a bridge between non-thinking, inorganic matter and our successors. I don't see how you can get from anything like a lifeless planet to anything like a super IBM machine without some intermediate stage rather like us.*

Interview with Arthur Clarke by Linda Ferguson, *Pacific Sun*, 4/12/73

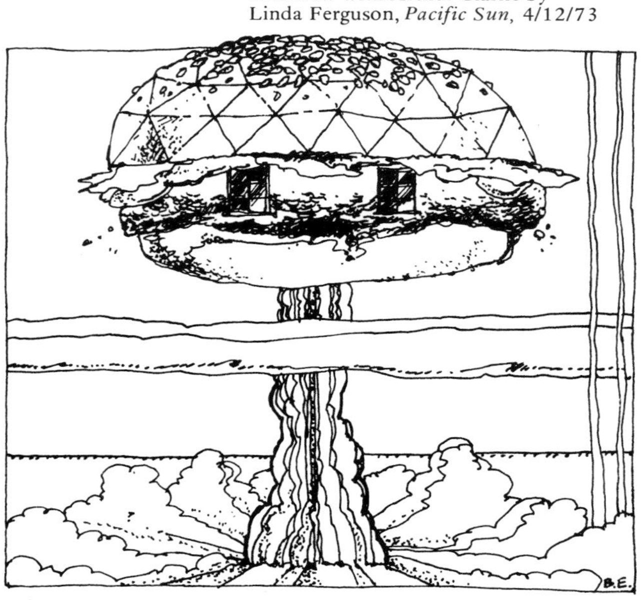

There is grave danger that stolen nuclear weapons may fall into the hands of terrorists, Mafia gangsters, black marketeers or perhaps some lone psychopath with megalomaniac visions of an atomic hijack....

Given Yankee ingenuity at building everything from rockets to steam engines to heroin laboratories in garages, says the report, "we would guess there are at least many thousands of persons who could make a fission explosive device if they wanted to and possessed the requisite nuclear material." Construction would take only "a few weeks."

"The only equipment required beyond what is commonly available for home workshops use would be ... ordinary Geiger counter (and) protective hoods," states the Ford Foundation document.

The nuclear materials needed for the weapons are poorly safeguarded and will become increasingly available as plutonium breeding plants are constructed to replace conventional generators.

"By 1980, tens of thousands of kilograms of nuclear weapon materials will be present in the U.S. nuclear power industry," warns a report. Not too many grams are needed to make a bomb that could kill hundreds of thousands.

The device itself, says the report, might be constructed by a single corrupt scientist, a band of outlaw engineers or a nuclear expert held captive by the Mafia or by terrorists....

Jack Anderson 8/73

The global village is beginning to resemble Los Angeles. The tribes of the world are becoming "Americanized", i.e., fragmented, specialized, superficial.

LET'S EAT

Hannon calculated that the McDonald's chain of 1750 restaurants used up the energy equivalent of 12.7 million tons of coal last year.

"That's enough energy to keep the cities of Pittsburgh, Boston, Washington and San Francisco supplied with electric power for the entire year," he said.

For every McDonald's customer, the energy equivalent of 2.1 pounds of coal was consumed in 1971, he figures and 2.4 ounces of packaging...

It took the sustained yield of 21 square feet of forest to provide the paper containers (and the packages they came in) carried out by each McDonald's customer.

And the energy content of the food on the McDonald's menu is just one-tenth the energy expended to get it from farm to consumer. The reverse is true in a primitive society of subsistence farmers in New Guinea. They get 16 times more energy from the vegetables they raise than the human energy used in farming and cooking them, Hannon said.

Their farm produce goes almost directly into their mouths, without ever being machine-stamped into a 1.6 ounce hamburger patty 3.875 inches wide (when raw and less when cooked), without ever being quick-frozen, shipped over great distances, grilled and then neatly packaged in a paper box. Only in American and other affluent countries does food take such a long, energy-wasting route to the consumer.

S.F. Examiner, 11/12/72

Douglas, Isle of Man

At least 37 people, many of them children, were feared dead today after fire flashed through a packed multi-story entertainment center built of plastic on this British vacation island...

Eyewitnesses said there was a loud explosion and a sheet of flame swept through the $5 million summer entertainment center, a cylindrical building constructed of acrylic panels just two years ago. It is one of the major attractions on this island lying between England and Ireland.

About 4000 people were dancing in the discotheque or in the amusement arcade or lounging in the sauna baths and the sundome.

People fled in terror, some with their clothes on fire as the heat grew more intense....

Reuters

PLASTIC

"Architectural styles will be more appealing, more individualized, as builders and developers learn to work with a greater variety of preformed concrete, plastic, metal and glass products," declares Jackson W. Goss, president of Investors Mortgage Insurance Co.

From *S.F. Examiner*, 3/18/73.

...IN LOS ANGELES

The buyers at a new $100 million waterfront housing complex here called Marina Pacifica won't need to pick out furniture before moving into their new homes. They won't have to choose dishes, place mats, silverware, ashtrays, flower arrangements or paintings for the walls either.

The staff of the development will select all of the furnishings for the buyers, producing what the complex's decorators call an "instant home," and what some outsiders may call "instant tastes."

The average price of the 1500 homes in the condominium complex is $55,000....

"We'll feed all the information about each individual unit into a computer," said Bea Cuthbertson, who is director of the project's interior decorating program. "In the event that Mr. Jones and Mr. Smith are next door neighbors and they have carbon-copy interiors, we'll go to one of them and suggest some changes.

"The owners won't have to do anything. They'll just turn the key and everything will be there. All they need to bring is a toothbrush and their clothing"...

Norman Danoff, the development's marketing director, said many buyers like the idea of breaking completely with the past.

If a buyer doesn't want something in the package, can he reject it? "No," Danoff said, "It's all ordered by computer from a central warehouse. People ask me what to do if they don't like a painting. I tell them we have to deliver it and that they should live with it for a while and maybe they will learn to like it...

New York Times. Reprinted S.F. Chronicle, 8/5/73.

echno: art, craft
gy: doctrine, theory
chnical: from Greek *technikos:* of art, skillfull/*tekton:*
rpenter, builder

Help;

Man I can't get it together. Maybe my head thick and the fact that I'm not to great on math, but I hung. In *Domebook 2* on page 24 you have a 3v 5/8 "alt" dome diagram, fine. But im dum or maybe it's just that I can't aford to take a chance, but you just show a on shot view and the order that the struts take on that shot, *but what about the rest.* I can't get it together in my mind for some reason. Can you get me a *breakdown on the rest.* Man, if you can you'll save my ass! I've only got one shot at it and its got to be right on!

As far as I know I'm the only guy around this area thats going to this and if luck holds out I'll really be making it. Your book is really together and I'm learning — all I need is for you to shove a little in my mind.

Do need the help.

Peace,
Rail
Dallas, Texas

Even though they were 600 miles from the nearest major civilization (Hawaii) and far off major shipping lanes, they recorded 53 manmade objects in 8.2 hours of viewing. More than half were plastic. They go on to compute that there are between 5 million and 35 million plastic bottles adrift in the North Pacific.

Science News 2/10/73

DOME MYTH — DOME TRUTH

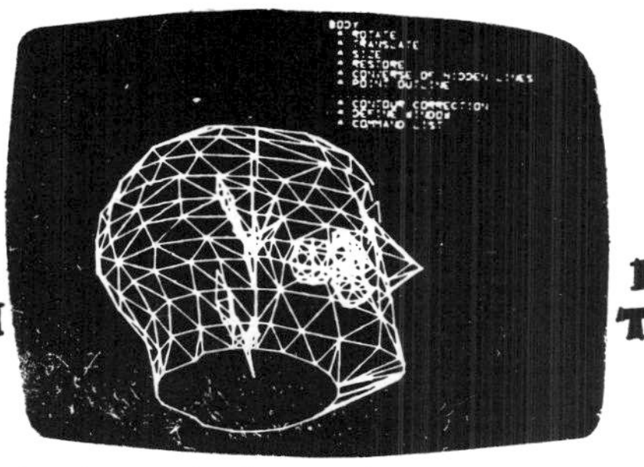

1. Our heads, the earth, drops of water are spherical, and thus our houses should be hemispheres.

Throughout history the dome shape has been used either in an economy of scarcity or as large monumental structures. Most cultures have moved away from the circular housing shape as soon as they were able to.

2. The cube is inefficient as the basis for structure.

It may well turn out to be that — given today's materials and tools — the cube is the most *efficient, considering length of the building's life, ease of expansion, and adaptability of less-processed materials.*

3. Crystals as a basis for building shapes.

Crystals are solid and small; buildings are large and must be hollow.

4. Principles of growth in nature applied to building.

The important thing to understand about the nautilus shell is that it is made in the sea, of the sea and designed for life in water.

Pure curves resulting from algebraic equations are never completely satisfactory in design. They all seem to lack life, while curves drawn by a human hand possess a live kind of beauty. This may be due to the fact that a mathematical curve represents only a particular function, very limited in scope, while life spells universality and constant change....

From *What is Design* by Paul Grillo and Paul Theobald, Chi. 1960. (I think out of print)

DOME HOUSES

We never intended the domes in *Domebook 2* to be middle class housing. They cost about $1,000 each, were temporary, but people who read the book then applied their own conceptions to the idea often tried to build real houses of them, and many suffered subsequent disillusionment.

A dome is o.k. as a room, but not as a house.

90°

What's good about 90° walls: they don't catch dust, rain doesn't sit on them, easy to add to; gravity, not tension, holds them in place. It's easy to build in counters, shelves, arrange furniture, bathtubs, beds. *We* are 90° to the earth.

Not important how much a building weighs. It *is* important how much a bird weighs, but a building doesn't have to move or fly. Certainly an adobe house weighs tremendously. So? Would it be better if it were built of plastic of 1/100th the weight and 10,000 times as polluting in its manufacture?

OTHER MISCONCEPTIONS

Spaceship Earth: Putting these two words together is the work of those who seek to *re-engineer* the planet.

Calling Earth a spaceship is like stepping out into a clear night in New Mexico and saying "Wow, it looks just like the planetarium."

World Game: Houston Control for Spaceship Earth. Centralization.

More with less: we should do *less* with less.

Large domes covering cities: look at stars through plastic? Never feel rain or wind?

Portable housing: bad idea. Once you build a floor (there are no portable floors), install plumbing and wiring and road or paths, it's extremely inefficient to move the building.

Energy slaves. Slaves?

Reform the environment not man. Shouldn't that be the other way around?

Despite all the hype, moon landing was dull. The only thing up there worth looking at was the earth.

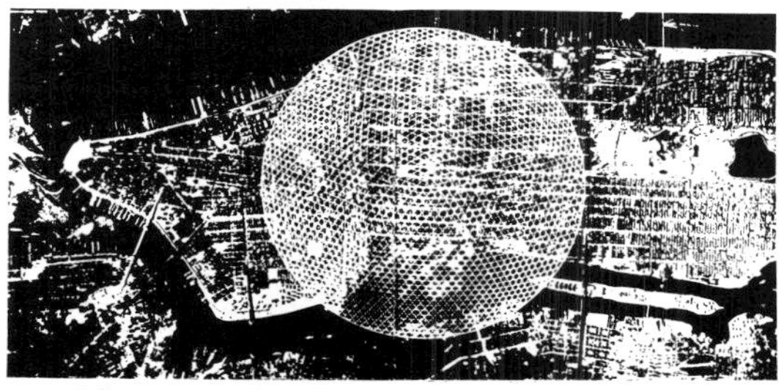

...THIS MAN POBLERI

The following extract was printed in Architectural Design *magazine, from a letter written to the Migrant Housing Task Force on a conference held in Los Angeles in Sept., 1972.*

The conference was called "Shelter for Mankind", held on the Cal State campus, just across the freeway from the Barrio, the largest concentration of Spanish speaking people in the U.S., where conditions are as you might guess barely tolerable: 6 freeways within a half mile, high concentration of poisons in the air, etc. There were over 40 speakers at the conference, mostly dome designers, air building dilletantes, mathematicians, psychologists, and the conference themes were preponderantly futuristic, with domes and plastics as central themes.

I had been asked to cover the conference by *Clear Creek* magazine, and went with serious misgivings about the conference staff using domes, plastics, and Soleri's and Fuller's ideas as solutions to "mankind's" shelter problems.

I was also a lecturer at the conference because of my involvement with the Domebook. I was paid $300, all expenses, and put up in the Biltmore. Emerging from the hotel at 9:00 Friday morning, there was no sky; a smog alert day. We drove up to the campus in the grey choking air. Foremost in my mind were the people who lived in that air full time, so I was quite interested to attend the first presentation that morning by the Barrio Planners who were as far as I could see the only group of *people* making a presentation amidst the academic professionals. There were very few in attendance at the Barrio Planners' meeting — too bad because they represented the human beings who the designers and future thinkers purport to be helping.

A short energetic man named Baez came to the mike: "The architects, the designers at this conference can do nothing. You cannot do better than our people have done for the past 10,000 years. We are children of the sun. We are an expressive people, we use our hands to talk unlike stiff, uptight Anglos. The outdoor patio, barbecue are ancient ideas of our people. There are two elements to deal with: sun and space. We want *air* and *light.* Forget about these domes and plastics."

Next a big man with a black eye patch named Trajo came to the front with some ideas about Soleri's arco-cities: "What are these anthill ideas of this man Pobleri? One man cannot design for the masses. I want a home, not a hole in a concrete monster. I want a yard, a lawn, a patio. To me it's a home, not a house. What do we care about domes? We want decent housing."

Mr. Baez returned: "You must find out from people their lifestyle. Spanish Americans are an outdoor people. You can live outdoors in Southern California. High rises and anthills don't give us space and sun. Domes don't give us easy access to outdoors...."

The rest of the conference went on its predetermined predictable course: domes, new technology, plastics, etc. At night Soleri presented a slick slide show of his cities to the sound of Gregorian chants.

The absurdity of the controlling interests of the country; a school and its patrons, industry sponsoring a conference on sheltering mankind in the heart of L.A. and inferring that we must depend on new technologies, new materials, new one-man design for shelter. No one listened to the Barrio Planners, who knew more about housing than any of the conference speakers.

Lloyd Kahn

BURSTING THE BUBBLE

enter Antioch's "Pneumatic Campus" at Columbia, Md.'s new wn is like crawling into the belly of a stranded whale. And the ng, I am afraid, has as much of a future.

e giant bubble which the students of the Antioch College branch Columbia have blown up for themselves — with the help of the S. Office of Education, the Ford Foundation's Educational cilities Laboratory, several air structure engineers and the odyear Tire and Rubber Co. — is, to be sure, an impressive ucture.

covers 180 feet square of a pleasant, rolling meadow with a ite-and-yellow striped polyvinyl skin that bulges up to 25 feet cause two small electric fans keep the air pressure a little higher ide than out. On the outside, the structure looks like an enormous pillow.

e skin is translucent, so the inside of the belly is light enough grass to grow. But the light and the atmosphere inside the stic enclosure are eerie and a bit oppressive. The space is vaguely dscaped with brick walls and furnished haphazardly as yet, with ffolding and sundry small structures like ski huts and geodesic ni-domes in search of a purpose.

good many idealistic people, particularly young architects, ntinue to be intrigued with Utopian visions of a future city all der one megastructure roof.

ucky Fuller would put a whole city under one geodesic dome. leri would stack a whole city into one mile-high, Babylonian per-structure out in the desert.

chitecture books and classrooms are full of these notions, so u can't blame Antioch for wanting to get with it. If the city of future is to be all under one catching tent, it certainly seemed ical to build one for an entire campus. The catch-all was to ter a new, more flexible education, conserve energy and the vironment and save money to boot.

t it doesn't.

r does the megastructure save energy in itself. It still must be at night and it certainly must be air-conditioned during the day, ich is not necessary in a conventional structure designed to ch the breeze. One irony of the Antioch bubble was that it got hot during the recent International Conference on Air Structures, ed to celebrate its completion, that the meetings had to be held a nearby motel....

e Antioch structure, too, is conceived as "a nomadic campus." was designed so it can easily be deflated and pumped up sewhere in Columbia or across the country, if that need arose."

at need?...

e best practical purpose of the Antioch bubble, however, may ll be to help deflate the all-purpose and no-purpose megastructure a. It is easy to get a structural high. But it seems awfully hard to campus and city planning down to earth where the people are.

Article by Wolf Von Eckardt, Washington Post, *reprinted in* S.F. Chronicle, 7/15/73

DOXIADIS CONFESSION

The famous global city planner Constantinos Doxiadis published a remarkable document last year. He entitled it "Confessions of a Criminal."

His first crime, he wrote, was that he had advocated the construction of high-rise buildings. They work against nature, he wrote, "by spoiling the scale of the landscape." They work against man himself. They work against society because they do not help the units of social importance — the family and the neighborhood — to function as it naturally must.

Doxiadis concluded that highrise buildings benefit only the few against the interests of the many.

DYNAMITE

Half a ton of dynamite was used to blow up a 14 story apartment house for low-income families in St. Louis, summer 1972. Less than 20 years old, the Pruitt-Igoe complex had precipitated uncontrollable dispair, vandalism and violence among its inhabitants. It had received an honor award for excellence from the American Institute of Architects.

Extract from speech prepared by Washington Post Architectural Critic, Wolf von Eckhardt for the Northern Calif. AIA meeting in 1973.

APARTMENTS BLOWN UP IN ST. LOUIS

DROP CITY REVISITED

Drop City was the first American hippie dome community, built on the outskirts of Trinidad, Colo. out of acid visions, idealism and chopped-out car top dome panels. Stop-off point for hundreds of hitchhikers on their way to Haight-Ashbury mecca in 1967, Drop City has in less than ten years become a dome ghost town. Best account of the place is *Drop City*, by Peter Rabbit, Olympia Press, 1971, and the article by Bill Voyd in *Shelter and Society*.

I was there the other day, so weird wandering thru those deserted, ripped off structures, that so much love & agony & labor went into the building. There were ghosts behind every broken window & half-off-the-hinges door. Sad hippy ghost town with a huge pile of human shit deposited on the drainboard next to the sink. Too many rats packed in too small a space. I guess we'll go down there & scrounge what we can.

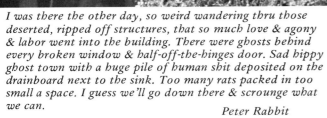

Peter Rabbit

PACIFIC HIGH SCHOOL REVISITED

Hot dry dummer 1969. An experimental high school in the Santa Cruz mountains with assets of 40 acres and under $100 in the bank. Bringing students to school in buses isn't working: trouble with drivers, insurance, schedules.

Suddenly someone, either Mark (director) or Michael (ex horse-trader now treasurer) gets an idea: we'll become a boarding school! We'll use room and board money to build with.

In August we start accepting students. Ah yes, Mrs. Peckinpah, we're building domes *for the students.*

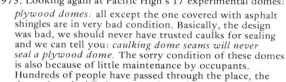

1973. Looking again at Pacific High's 17 experimental domes:

plywood domes: all except the one covered with asphalt shingles are in very bad condition. Basically, the design was bad, we should never have trusted caulks for sealing and we can tell you: *caulking dome seams will never seal a plywood dome.* The sorry condition of these domes is also because of little maintenance by occupants. Hundreds of people have passed through the place, the domes have had little care. To keep the water out all the plywood domes will have to be covered with asphalt shingles.

aluminum domes: they don't appear to have changed much, just suffering from lack of care.

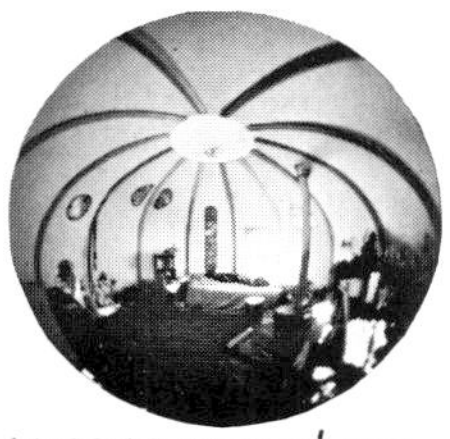

INTERIOR, MARTIN'S POD

Martin's pod: the shingles are weathering, turning grey, it's the best looking dome there.

foam domes: the little elliptical dome where Peter shot foam over burlap, then painted the outside with Elastron looks quite good. The inside still has the good feeling burlap texture. The other where we shot self-skinning foam on the outside is an unattractive rusty color inside and the outside skin is deteriorating. So much for self-skinning foam. Ananda had even worse problems with it

vinyl pillow domes: no sign of the vinyl deteriorating yet, although it's become dusty and dirty looking, which probably won't wash off. They look cold, impersonal, still smell strongly plastic.

overall: the only domes that look right to me in the woods are Martin's pod and the asphalt shingled dome. Most of the others are glaring, or depressing in decay. Plastic doesn't age with grace.

ALUMINUM DOME

VINYL PILLOW DOME

PLYWOOD DOME

Naive, hopeful, inspired, propelled with the energy of necessity we built 10 domes that first year, seven the next. We tried all the design ideas we could think of, all the new materials we could afford. Domebook 2 *came from those experiences. After the book we left the school and with the perspective of distance and time, began to evaluate the experiments.* Smart But Not Wise *(See p. 112) followed, and reflected our changing thoughts.*

In 1973 I went to photograph the domes at Pacific High School and as I walked through the woods looking at these sad, neglected buildings, I realized what we should have done:

For immediate space, build tents, using untreated natural canvas (See pp. 11-14). Or surplus army tents.

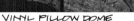

Then using tents for temporary shelter, build several different size platforms. Some could be round, most should be rectilinear. Then 8' *vertical* walls, with different simple roofs (See pp. 40-45). Then take the canvas used for tents and use it to cover roofs. Insulation could have been the same as with domes, i.e. pop-in foam panels (in this case between rafters, not struts, and no need to cut into triangles.) With canvas roof you could pull out sections of insulation to have more sky light.

The idea is to start with a building that can *grow,* that can be added to and changed by different occupants. The canvas, with care, might last 3-4 years. It could then be replaced with 1" sheathing and roll roofing, later with shakes (there were fallen redwoods nearby). Someone could add a fireplace. Rather than beginning with an abstract mathematical concept, we should have allowed the site, available materials, occupants' skills and needs determine the design. This would have allowed more direct participation, ingenuity. Kids could have used hammers instead of caulk guns, put in doors and windows easily, added on where needed. We could have taken our truck to San Francisco to scrounge doors and windows rather than to U.S. Plywood and Transparent Products. Finally, this plan would have cost considerably less because although domes use less materials, these materials are twice as expensive: clear grain struts, plywood. Cost comparison between one of our domes and a gable frame building of same floor space, redwood walls, roll roofing, 1973 prices: Pacific Dome: $929; gable frame structure: $643. And these buildings would have lasted for 75 years, instead of 10.

I have been involved in Dome construction both wood & fiberglass & foam for over 3 years now. During this time however I have discovered as you have that the house as we know it must have evolved from a much simpler approach using more basic materials & of course human labor & ingenuity. I am currently trying to compile a paper describing the many & various ways & materials with which a home can be built; i.e. bottles & cement (being the most obvious). In short, I want to hear more about the "organic house."

Any information you can make available concerning ideas either new or old using adobe, straw or recycled materials from 20th century production will be greatly appreciated. Maybe I can return the favor in the future.

I am about to receive my architectural license from the state of Texas but I already realize that the "owner built home" may be the final answer as it was in the beginning....

Allen Seale
Buda, Texas

Dear Lloyd:

It was good to hear from you. I've got both the Domebooks at home...and I've spent some considerable amounts of time with both of them. My personal preference for houses is wood — I'm building a log house in a hexagonal style at the moment — but, Jesus, it's getting so a person goes through some pretty fancy guilt trips every time he cuts down a tree anymore. Your method of using cull and driftwood is one answer. Around here we don't have much of that, though. All these mountains were logged pretty heavily 30 years ago, and most of the new growth is healthy, solid and fine. I have to get my logs from thinning the poplars on my land.

Anyway, I looked with real interest at the Domebooks because they seemed to present a genuinely viable alternative to forest rape. But I can understand your feelings about them now, and I'll be really interested to see the second book on shelter that you mention....

Eliot Wigginton
Foxfire
Rabun Gap, Ga.

SEALING WOOD DOMES

DOME AT PACIFIC HIGH SCHOOL COVERED WITH ASPHALT SHINGLES.

From many years experience, feedback from literally hundreds of dome builders and owners, we have concluded that there are only two ways to waterproof wooden domes: asphalt shingles, and fiberglass.

1. *Asphalt shingles.*

An entire building covered with asphalt doesn't look too appealing, but it *is* a way to keep out water and the cheapest solution. In the 20' dome we shingled at Pacific High School, we put tar paper underneath the shingles for double protection. It's probably best to use the self-sealing shingles at the top where pitch is not steep so they lock together, and wind is less likely to blow them off or rain get underneath. It's the solution when all others fail, hardly a space age technique. Buckminster Fuller's own dome home in Carbondale, Illinois was finally covered with grey asphalt shingles.

2. *Fiberglass.*

This is an *untested idea:* to seal the typical 39' diameter commercially-sold geodesic domes: nail 2x2's over seams. Then prepare pre-glassed plywood panels as follows:

—calculate the slightly larger radius, as you are expanding the original dome's diameter. See *Domebook 2* for explanation of chord factors. Calculate carefully. These triangles will probably be about ½" longer along each edge. Before cutting all triangles (see *Domebook 2* for good cutting methods) cut five for a pentagon, six for a hexagon and tack them up to see if they are accurate: seams should come very close.

—use 5/16" white fir or spruce, grade BC or CD, with knots plugged and touch sanded, with exterior glue. Other plywoods have sap or secondary material that come out over a period of time and are not good for fiberglass.

—try to find a local fiberglass or boat shop to glass panels for you. Costs should be 30-50 cents per sq. ft. Panels are glassed with chopped fiber and a good wax-free resin like Pittsburgh Plate Glass 58502. A typical panel on the 39' dome should get about a gallon of resin. *Stick to the same resin on seams.*

Next nail on the pre-glassed panels with 6 or 8 penny cement-coated nails.

Caulking: seams precaulked with same resin, mixed with talc (Fabreen C-400, made by United Sierra) or equivalent. Mix resin catalyst and talc until it's a consistency of warm butter. Apply with rubber squeegee, filling crack. Do this carefully, then go back and sand any bumps or irregularities until it's perfectly smooth.

Taping: Get a 4" paint roller (saw a 9" one in half), fill a Chlorox bottle with top cut off with same resin and catalyst, apply 4" strip of 1½ oz. matting over seam. Dip roller in resin, roll it onto seam, then lay the glass on it. Cross over every vertex at least twice. Put more resin on the tape until the glass disappears, let cure, then it's ready for paint.

Painting: Use acrylic house paint; because it's rubbery, it will seal up any pin holes. A good paint is Dunn Edwards Z 2456.

Using the same glassing system, you can build up hoods over doors and windows.

Shake Domes?

Shakes are a bad dome covering, as I can tell you from personal experience. I got my shakes free off the beach so used them but I wouldn't again, for these reasons:

—it's wasteful of shakes, as you have to overlap closely and as you get up towards the top, you must start cutting shakes into pie shapes to make the tightening circle. Shakes are linear, go best on linear planes.

—up around the top of domes, pitch is not steep enough, wind can blow them off, water can be driven underneath.

—putting them on is a nightmare with the literally hundreds of angle changes. Each edge where dome changes angle, and especially vertices, are extremely difficult.

—I'm not sure, but think shakes should have breathing space so they don't rot.

All in all, they look better than any other dome covering, but are a short term, difficult, and expensive solution.

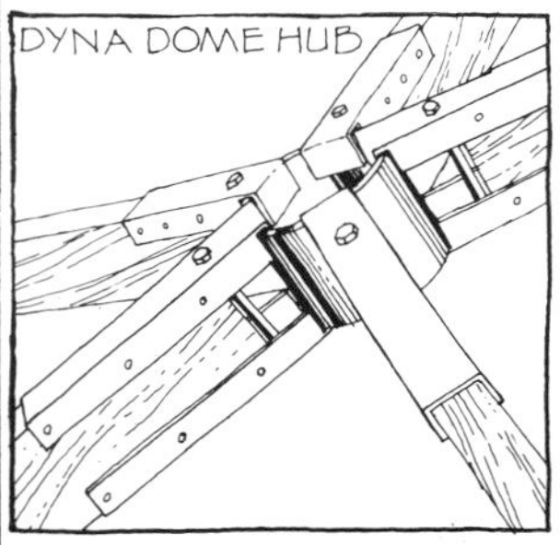

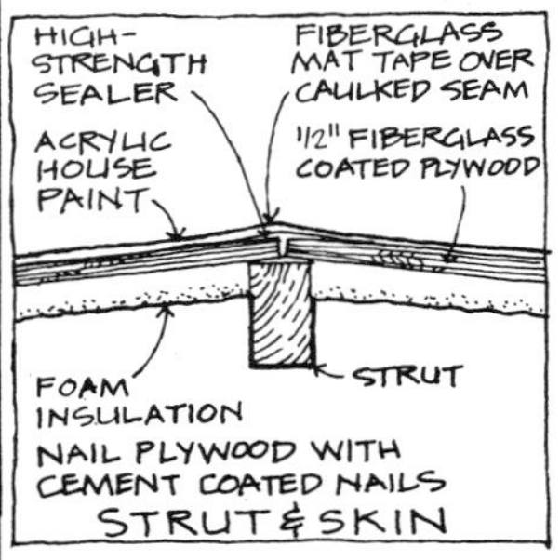

BILL WOODS

BILL WOODS

BILL'S SHOP

Bill Woods probably knows more about triangulated dome home building than anyone else in history. His firm, Dyna Domes, has sold over 500 dome connector kits with blueprints, and Bill has personally supervised construction of more than 100 domes throughout the country. His domes are based on the octahedron, which truncates at the hemisphere, and involve a patented connector kit, 2x4 framework, pre-fiberglassed panels, and fiberglassed seams. Bill is an experienced builder, sheet metal worker, and inventor. He is versatile with all kinds of building materials, including sheet metal, foam and fiberglass, and has built a 40' sphere on a pedestal, and a spectacularly lightweight 80' dome with 2x4 framework.

Bill and I have known each other for some years (I once worked a week with him in his Phoenix shop), and he knows of my subsequent personal disillusionment with most domes. Since he's the only domebuilder with years of experience I know and trust, I called him one night to ask his opinions of the future of dome homes, and his opinions of *Domebook 2:*

Bill: Well, I've got mixed emotions about the book. I'll be quite frank about it; in some respects it has helped people. Let me put it to you in the way I hear it, which you wouldn't hear: "The thing is fine but I can't build a dome out of it." Basically what they're saying is they can't cope with jumping from one page to another. (*Refers to the fact that we had frame instructions on one page, sealing on another, windows on another, etc.*) I understood that your original intention was to provide the chord factors, which seemed to be the hang-up with everybody. Then from that (you thought) there'd be a jillion ideas spring forth and the country was going to be overcome by it. Well, this is not the way evolution takes place. In some respects I think it has helped, but in others, it just got a lot of people frustrated. There's a lot of frustrated energy...The die-hard domer, shall we call him, has taken the book and it has given him a tool to work with. It would have been better if you had...put it into language like a set of blue prints to where if they're going to use asphalt shingles you spell it out in there to where they go down and buy 135 lbs. blue tab asphalt shingles... Tell them what kind of lumber they should buy. White pine or Douglas Fir. Whatever is in their area. Buy so many 2x4's this long. In other words, step by step for one dome all the way through. Then there's a possibility. The dome has helped a lot of people, I feel it has. I've got an awful lot of satisfied customers.

Lloyd: I know you do. But you know I don't want to do those blueprints. My idea has always been to relate to people our experiences, our mistakes, and let them take it from there. I'm sure you saw in Domebook 2 *that I wasn't any longer thinking domes were a solution to the shelter problem. What I think is the success of your operation Bill, is you yourself. You're an extraordinary builder, you work with people, get right out and help them, you've got a system down, you work hard, you keep it all together. And I think that's why your domes work. I don't see anyone else doing that.*

B: There's quite a bit of truth in that. However, I've taken the connector kits and sent them out with the blue prints and some people have actually ended up with a very beautiful product. They're just more than happy.

L: Well, I still don't see anyone else providing home-size domes successfully.

B: Well, there are a lot of rip-off artists.

L: There are a lot of new dome builders in the country right now. Maybe one or two of these guys is building a decent dome. But most of them are going to leak or fall apart. Look at those commercial plywood domes, you should see the letters we get from people. "Help, it's raining inside...." And they've spent their life's savings on it.

B: Well, here's the situation: It works for me, and I'm no genius. I can take one system and it will work. The way I profess it is this is an alternate method of enclosing space. It works for me and it works for some of my customers. All domes are not bad. There are some good ones and they are doing the job. The domes are standing up better than some of these 35 year mortgage tract houses.

L: Well...you said something about an evolution. Do you think a change is necessary utilizing whatever the efficiencies of a dome are? Do you think that's got to happen?

B: It has to happen. I'll tell you why. It is taking place right now. The reason I can see is because I'm in the middle of it, I'm not reaching Joe Square. Right now, 90% of my customers are creative. They are artists, teachers, draftsmen, the church. These people are usually about ten to fifteen years ahead of their time. And evolution has to take place.

L: Well, the revolution I'm looking for is to see people in this country get back to work. And use local materials, like adobe which doesn't hurt the earth. I'm talking to people who're going to build themselves, I'm telling them don't try a dome. Now if they want to buy one from you, that's different, I'm just saying it's basically too hard for an owner-builder.

B: I'm pretty much anti-credit. I don't know if it has ever occured to you, but I never provide floor plans because it makes people do things for themselves. That way it becomes part of them.

L: That's a good aspect in making a sort of neutral space.

B: I provide them with the tools to work with. When you go out to plow a field, you're not going to plow it with a hoe, you're going to plow it with a plow. I provide them with the tools to build the dome. They have to build it. You provided them in the domebook, I think with a hoe to plow a field.

L: Yeah. (Laughter.)

B: The point that I'm trying to make here Lloyd, and I hope we remain friends over this.

L: No, that's right on.

B: The part that I'm getting at is no matter what a guy does he should look at how it affects him and how it affects his surroundings. But the prime thing that he's got into it and what happens to us here in the United States is along the lines that you were talking about a minute ago is the idea of having Joe do it. But if a person puts his own house together, he's going to take pride in it. And he can build it with his own money. He can bypass all your fat cats, government agencies. If we get the money changers out of our lives, then our dollars aren't going to go down 10% every six months. As you say, we've got to go back to work. But also to utilize what we've got, we've got to utilize it better, shall we say.

L: I agree.

B: Like a conventional house, 13 to 15 thousand board feet of lumber to build it. This is one of the reasons why I'm still so strong on domes. The amount of material that goes into it is 2 or 3 thousand feet at the maximum. If we can gear our whole economy to this percentage our forests can produce enough to build out of.

L: I agree with you and all that, but...

B: The dome can keep the rain off of you, if it's put together right and provide us with warmth and comfort if we want it. If we're willing to put the effort forward and do it right the first time. Be sure of your technique. It is the technique that's the most important thing whether it's a square house, a round house or anything. Pass this knowledge along to people so you can save on pitfalls....

L: Maybe we gave people too much *information in our book. If we'd left a bigger gap, then it would've taken more brains to fill it in and there wouldn't be the messes you see around today.*

B: I know as I'm sitting here talking to you that unless we change our ways in the building industry and thinking in this country, we're doomed. I can see it and every day you read the newspapers it's there before your very eyes telling you. All you've got to do is look at it. Things have got to change. And if they don't change then we've really got problems. I think the thing that will help them more than anything is to have a depression and inflation.

L: The thing that disturbs me is that Domebook 2 *makes people think that domes are really more than they are. It got people excited. People that had never done anything before. Somehow domes got a hip image. We didn't figure all that was going to happen, it just came about. I've got to be honest about what I am feeling right now and at the same time I want you to appear in the book and say what you think. I can't write anything about dome in this book unless it's what I think.*

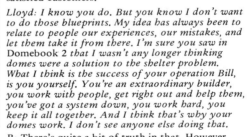

B: Well, that's fine, as long as when you put it in there it's your opinion, it's not the gosp...not that this is the way everything is. There could be

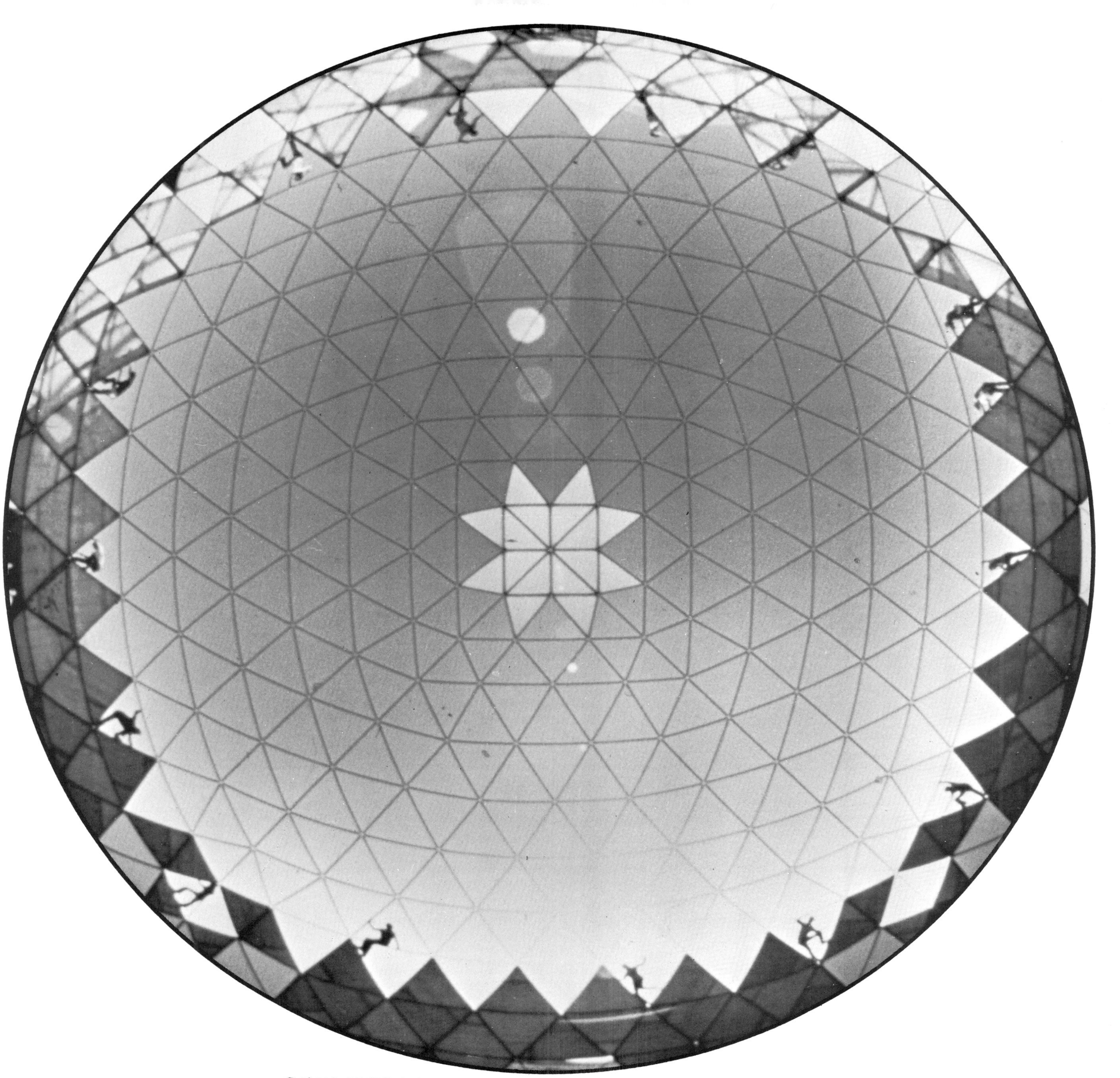

80' DIAMETER DYNA DOME CHURCH BEING BUILT BY PARISHIONERS. 2×4 STRUTS. PLYWOOD SKIN. PHOTO: JOHN DOMINIS.

something that does work that you may be aware of and that you may not be aware of. I know the dome and I know all kinds of construction. I can do a square house, a round house, you name it, I can do it.

L: We're wondering what domes we're going to show in the book. In other words, the only thing I can think of to do is there's a nice little tent dome that's being made in San Francisco. You're making domes. But other than that....

B: Could I make a suggestion? You're going to be showing all different kinds of shelter...things that are working for people. Why don't you show two or three domes that are working for people. Like here's a dome and it's covered with asphalt shingles and it's water proof and it's working.

L: Yeah. We're going to do that. Another thing; what about all the bucksters that are in the dome business right now. We may have made that happen, because we made domes look good.

B: Well, I wish I knew what to do about them.

L: I don't think that they're all crooks.

B: 95% of them are. One of the worst things about living in a dome is that people feel they have a god-given right to come look in on you.

Later, talking about some new domes that Bill has built:

L: Well, the octahedron turns out to be better than the icosahedron which we used.

B: Yeah. The icosahedron is no good for domes. When all avenues have been explored and gets the person nowhere they call Bill to get them out of a jam.

L: We used the icosahedron because of aesthetics.

B: Yeah. Well the point that I'm getting at is I'm strictly a man of mechanics. Make it work the best that it will work. I'm not a purist on anything. If I can move a board and save a sheet of plywood, then I'll move a board and save the sheet of plywood.

L: Look, Bill, what are you doing now? What is your operation now?

B: Well, I'm pushing the connector kits for people who are interested in building their own thing. I want to sell the connector kits and blueprints together, only together.

L: Well, what do they do with the skin? How do they get it fiberglassed?

B: We tell them to call a local boat company and have the glass sprayed on there. Quite a few have gone this route.

L: How much does that cost them to job it out.

B: Between 30 and 50 cents per square foot. My shell kit, I've changed that since the first of the year. I sell it for $3 a square foot of floor space delivered and I go out on each job. They're required to have three people on the job and I help them put it up. They don't need to know anything, they just shouldn't be afraid to climb. That's any place in the U.S. but doesn't include the floor. It includes everything to put the shell together with. The struts, fiberglassed plywood, nails, putty. No doors or windows, they're extra. It doesn't include insulation, but it includes one layer of panelling inside. With that system, I guarantee they can get a building permit. I can beat any building code in the United States if I have the time.

L: What size dome do you sell for $3 a sq. foot.

B: Any size up to 80 feet.

L: Didn't you build an 80 foot dome with 2x4's?

B: Yeah. Two 80 foot domes with 2x4's.

L: Have they done o.k. in the winter?

B: Yeah. These domes are in Lake George, New York.

L: How many triangles in an 80-footer?

B: Five hundred and fifty-two.

L: Well, how many triangles are in your 70 foot dome in Arizona?

B: 140.

L: On the 80 foot dome, how long are the struts?

B: About 6 or 7 feet. I put up the 80 footer from a cherry picker in 2½ days, one man. It took up nine days to get that dome completely built and water proofed.

At this point I asked Bill about dome leakage. Most of the bolt-together-triangle geodesic domes sold in the 50's and 60's leak, to great consternation of owners who were not told anything of leakage problems when they made the purchase. We've received many letters and calls as has Bill, from people living in these domes, which had a variety of tapes and caulks for seam sealants. There is far more expansion-contraction with the bolt-together system than with a single strut system such as Dyna Domes, or our plywood domes at Pacific High School. Bill's reply was that he can think of three solutions: first, simplest is asphalt shingles: they do work, aesthetic considerations aside. Secondly, somehow get all paint and caulk, etc. off and fiberglass as with Dyna Domes. However, this is a tremendous job, getting it down to bare wood. Third, and this is an untested idea: nail 2x2's over the seams, cover with ¼ or 5/16" pre-fiberglassed plywood, then caulk and tape seams with fiberglass. See fiberglass details, p. 120. Back to our conversation:

L: Have you seen the latest National Geographic? *There's a tribe in Bangladesh that makes beautiful buildings out of bamboo, these people working so well with just what they have around them....*

B: Well, we have become so involved in our great society and our great Cadillacs, automatic dishwashers, we've missed what the world is and what it is all about.

L: Yeah, right.

B: We're given soil and hands and minds. We're using our hands to turn on the TV and our minds to see how quickly we can destroy ourselves. (Laughter.) Every day I look at this stupid lunacy that's happening. I keep my own little island and my own sanity here. It's really weird, too...there is one good ray of hope in the whole thing.

L: Yeah?

B: We're running out of gasoline.

L: That's a great note to close on. You know L.A. is going to go on gas rationing.

B: I cannot think of a more favorable thing to happen than that we run out of gasoline. Then we have to maybe...change.

4V ICOSAHEDRON

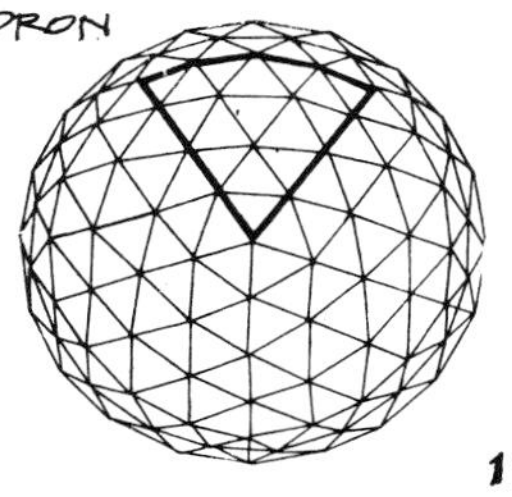

4V OCTAHEDRON

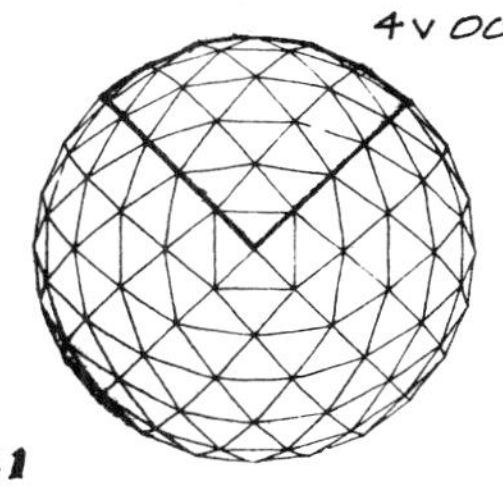

FERRO CEMENT BRONTOSAURUS
CURIO SHOP ON HIGHWAY BETWEEN L.A. AND PALM SPRINGS

FERRO CEMENT

Ferro cement and plastic foam are two methods of building free-form or double-curved shells. Ferro cement is a difficult, time consuming technique that produces a strong, permanent structure. Plastic foam is expensive, requires professional application, and is prone to a peculiar type of fire hazard: if temperatures near the foam get high enough: it can explode in flame like gasoline, and emit poisonous gasses.

The following instructions are reprinted from Peter Calthorpe's article on the mixture used on the ferro cement dome at Pacific High School. There are four pages of ferro cement information in *Domebook 2:*

Each person you talk to will give you a different formula for the mix; we used the simplest, funkiest method mentioned. Used Portland no. 2 cement in a one to two ratio with Olympia no. 1 even graded sand. We got some extra bags to measure the sand in, one bag of cement to two bags of sand; the cement to sand ratio is not that critical, we must have varied at times down to one and a half bags of sand — the critical ratio is water to cement. Even graded sand means that the sand has all different sizes of granules which allow the mix to be denser because the sand can close pack and leave less air spaces. About ten pounds of pozzolan, a kind of powder, per bag of cement is used to make the mix even denser. Three and a half gallons of water seemed like the minimum and we often added more if it felt too dry and crumbley. I tried baking the sand to check out the amount of water it contained, but my scale was too inaccurate; didn't use a sump test either just took a handful, pressed it into the mesh, and if it looked ok we'd use it.

The mixing went like this: water, cement, pozzolan, then sand. Let it mix for about five minutes to get all the water evenly distributed. It will seem too dry at first but after a while you'll get a sense of how it should be. The most important qualities about the mix is that it be dry and dense. If it is too wet, surface cracks will develop when it cures. I was worried about the mix but there was no need, you really can't go wrong. This dome was actually a test in how funky one could be with ferrocement and the conclusion is, very.

I was told that it took about ten pounds of mix per square ft. of surface area, so with a twenty ft. hemisphere which is 600 sq. ft. we figured 2000 lb. (20 bags) cement and 4000 lb. of sand. Also told one would need about one and half times as much mix as was to end up on the structure, so that hyped the order up to 30 bags of cement, which is just what it took. The cost of the mix was about fifty dollars.

This is a two-level ferro cement building by Ronnie and William Feldman of Cazadero, Calif. The downstairs room is 16'x25', the upper 12'x14'. A wood framework was built of 2x4's, plywood and redwood lath, then six layers of Japanese aviary wire with tempered steel rods in between the third and fourth layers. The builders say it was a very slow process (six months), that indispensable tools were an aviary wire twister and 30" bolt cutters, that they made a mistake on painting the cement with a sealer before thinking of insulation. This made it impossible to plaster or foam either inside or outside (won't bond to sealer). They put redwood mulch on the outside for insulation, but have a problem of condensation, as they wrote:

> *...it would be cold outside and warm inside with a fire going, hence water formed on the walls and ceiling and kept the inside damp...you certainly learn what your shelter should be once you have built one!*

BURLAP

Bernard Maybeck, looking for a cheap fireproof material tried coating burlap sacks with a foamy concrete called "bubble stone" It was a method of adding chemicals with cement to make a concrete so light that a mass the size of a bale of hay could be lifted by one man. The water and chemicals were mixed in an old washing machine, then Maybeck folded cement and sand mortar (without aggregate) into the froth "like adding sugar to whipped cream". Wet burlap sacks were dipped into the mixture and came out with about an inch of foamy concrete adhering to them. They were then nailed to the studs and sheathing, and in another case were hung from wires to make a wall.

From *Five California Architects*

LATEX

...Making building components by dipping fabric in concrete is becoming very exciting...One piece of wisdom we know for sure: Don't try to dip new burlap from the drygoods store. It won't wo You have to use old sacks that have no sizing on them. Concreted burlap is useful, but not enormously strong, because the burlap is too stretchy and allows the concrete to crack...

Dow Latex 460 is a concrete additive which replaces part of the mixing water. It imparts several interesting properties to the concrete:

1. The concrete will adhere tightly to almost anything — cured concrete, stone, foam plastics, even wood and steel. (Also hair, skin, clothing, tools. Wash it off before the latex dries, which take only a few minutes after the concrete is in place.)
2. The concrete needs no wetting during curing. The latex traps the water inside, ensuring a good cure.
3. Shrinkage and shrinkage cracking are drastically reduced.
4. Weather resistance is greatly improved.

These properties make latex concrete an attractive material for patching concrete surfaces, stuccoing over various materials, extra strong mortar that won't leak or let go of the brick, and so on. Do uses it to make concrete thin-shell domes which need no roofing. It is too expensive to use in large masses of concrete.

To prevent frothing during mixing, the latex is pre-mixed with a small quantity of Dow-Corning Anti-Foam B.

The mix:
80 lbs. sand
27 lbs. cement
8 lbs. Dow Latex 460

Write Dow, Designed Products Dept., Midland, Mich., 48640.

Ed Allen

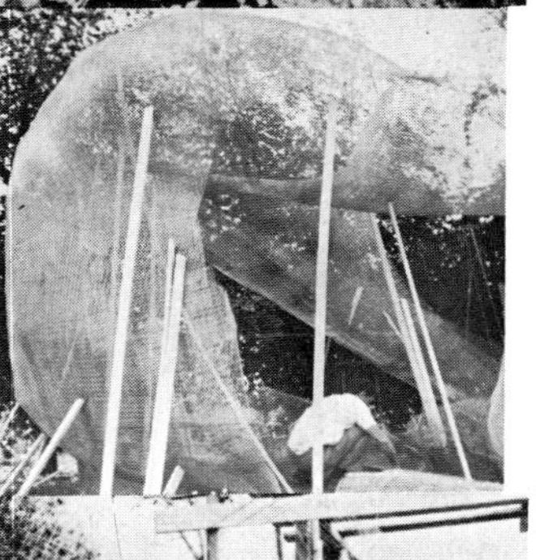

This free form 1500 square foot structure was designed by Vittorio Giorgini of NYC, and built for about $8,000 using much unskilled labor.

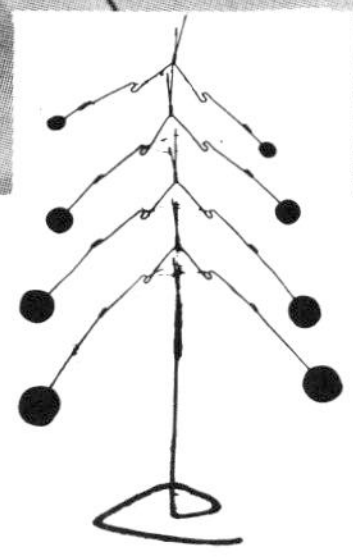

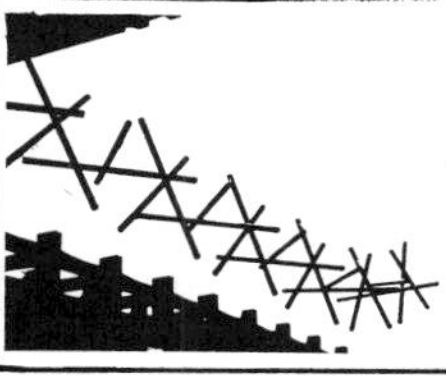

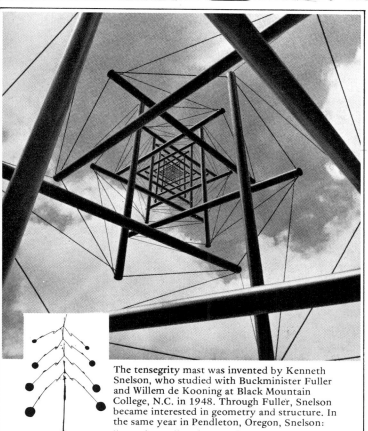

The tensegrity mast was invented by Kenneth Snelson, who studied with Buckminister Fuller and Willem de Kooning at Black Mountain College, N.C. in 1948. Through Fuller, Snelson became interested in geometry and structure. In the same year in Pendleton, Oregon, Snelson:

"...made small sculptures that moved: wire, wood, string, clay. Began with clay-weighted balancing toy in stacked series. Each hinged element supported those above. All moved like a spinal column. I replaced the wire hinges with thread slings. The sculpture still moved but everything seemed to float. The next change was to remove the balancing weights and to tie the floating elements to one another, eliminating movement. Suddenly I had made a closed system of solid objects which supported one another only through tension lines; a new kind of tension structure, in the same class as a kite, the balloon and wire-spoke bicycle wheel. For me, that small moment of discovery was especially pure and beautiful...

1949 – returned to Black Mountain College. Showed my new invention to Fuller who was surprised, fascinated; even published it in *Architectural Forum* in 1951. In 1955 he began to call it *Tensegrity*."*

**Kunstverein Hannover*, Kenneth Snelson. Books on Snelson's work pub. 1971. Extract from his biography.

TAO FOAM

Polyurethane foam is the best insulating material known. However, it is also expensive and prone to a peculiar type of fire hazard: if temperatures near the foam get high enough, it can explode in flame like gasoline and will emit poisonous gasses.

Thus foam must be protected on the interior with some type of highly fire resistant material, such as sheet rock or plaster. "Fire-proof" paint is not enough.

Below are parts of a letter from Charles Harker, who built this foam structure:

The photos show the house before application inside and outside of wire mesh and cement plaster. In actuality it is not a foam house, but cement with urethane foam core. We have advocated cement plaster to reduce fire hazard for 2½ years. The urethane industry was making false claims about fire retardance to increase sales. This year at Synfoam II – a conference for urethane foam builders at Blacksburg, Va. – John Degenkolb, U.S. Fire Marshall delivered a scathing paper[1] which demanded that the industry come up with realistic claims about their material or face the possibility of total prohibition....

No matter what any urethane foam dealer tells you, all urethane burns. This includes the so-called class A urethane with flamespread ratings of 25. The danger lies in interior and overhead surfaces. As heat builds up inside, the insulating qualities of the foam work against you. The material gives off gases that are flammable themselves. In our own experiments, we sprayed up a 6 foot dome with class A urethane, set a torch against the side. It started slowly, burned through to the inside and filled the interior with flame. The flames burst out at the top of the dome and burned back down the sides. The entire dome was consumed in 6-8 minutes.

It sounds dangerous and it is. But it is no more dangerous than any other type of construction if interior surfaces are covered with ½" or more of bonded plaster (Acrylic bonding admixtures are acceptable in meeting newly established government standards.) The interior surfaces are the problem. Exterior surface fires will burn out quickly once the heat source is removed. Manufacturers' claims are perfectly justified in the case of fire on exterior surfaces. We plastered our exterior anyway because it seemed like the only system which in one coat gave a surface which incorporated structural skin, ultraviolet protection for foam, an acceptable surface texture and color. It makes it seem "permanent." Ask us in a couple of years if we would do it again.

Charles Harker

Note:

1. Copies of Degenkolb's paper available from Synfoam Institute, P.O. Box 12027, Raleigh, N.C. 27605, and other information from Tao Design Institute, 1201 Bee Cave Rd., Austin, Texas.

RAW STUFF

Bill Bennett

Enclosed are photos and drawings of a ferro cement dome I built in central Pennsylvania during the spring and summer of 1972. I call it *Raw Stuff* after a quote by David Smith, the sculptor, who described art as being raw stuff. It was built as a sculpture and as living quarters for myself. The dome is small but so are my needs and my budget. At the base it is 14' in diameter and is 8' tall at the center. The doors and windows in the dome are from a car and extended about one foot from the surface of the shell. This makes the walls seem very thick from the inside. The outside surface has a knobby and less dome-like appearance because of this. Now to the construction.

The base is a 4" reinforced concrete slab on top of 4" of styrofoam and a layer of polyethelene film. Because of the way the foundation was dug, the front 3' of the dome rests on fill. There are 3 reinforced footers that go into solid ground in this area. 3/8" rod was used for reinforcing throughout the dome because I was able to get a great deal of it for free. It was very rusty but I was told that the cement would grip better because of this. A pipe with about a 2" hole was cast into the center of the slab. This would be used during the construction of the shell. The 3/8" rod radiated from this pipe and surfaced at what was to be the base of the walls. A 4" section of 3/8" pipe was welded to the end of the rod so the wall reinforcing could slip right in and be connected to the slab.

I put a perforated plastic drain around the perimeter of the slab to carry off the water coming down the hill. A form for the slab was made from masonite and wood stakes.

I wanted to use readymix for the slab but there was a problem. The site was about 12' higher than the road and I couldn't get the cement pumped uphill. A wheel barrow path was out because I didn't want to rip up any more of the hillside than I had already done. A bucket brigade seemed to be the best solution. I rounded up all the people I knew and some I didn't and we (about 25 people) moved 2½ yards of concrete in about 20 minutes. It was beautiful. The spirit of Amish barn building was present. Many thanks to all.

Three days later, I started on the framework for the shell. The first thing up was the cupola-like hub on a pipe that fitted into the pipe case into the slab. Welded around the lower rings of this cupola were 4" sections of 3/8" pipe corresponding to the 3/8" pipe coming out of the perimeter of the slab. 3/8" rod was fitted between the hub and the slab. They looked like the longitudinal lines on a globe. At this time, the door and window frames were wired in place. These were cut from a car donated by a friend named Raz. A sharp cold chisel and a 3 lb. hammer were used to de-door and de-window the car. Cutting the car up was easy. Makes you wonder when you get behind the wheel. I welded some 3/8" rods to these car parts and wired them along with 2 vent pipes and 2 stove pipe holes to the wire frame. Next, I wired a series of horizontals to the existing structure. The wires that crossed in front of the windows would be cut after the cement was plastered. Extra rods were wired around the doors and windows to compensate for this. The frame was very strong at this point. It had an aesthetically nice feel also. Like a big bird cage. There was an inside outside interplay at work also. You could enjoy the immensity of the entire woods and feel the intimacy of a small cozy space at the same time. Very enjoyable feeling. *Continued at left*

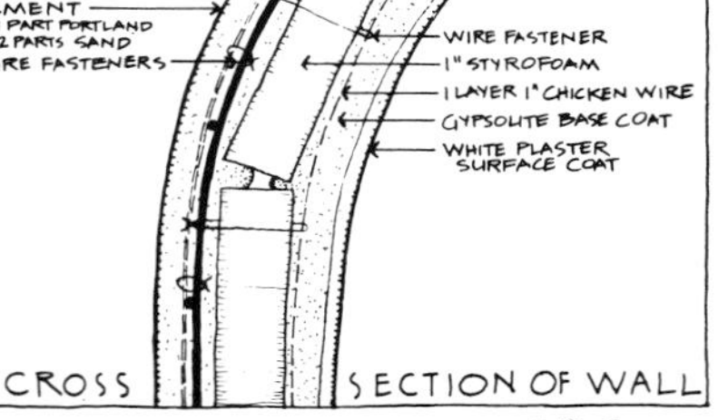

CROSS SECTION OF WALL

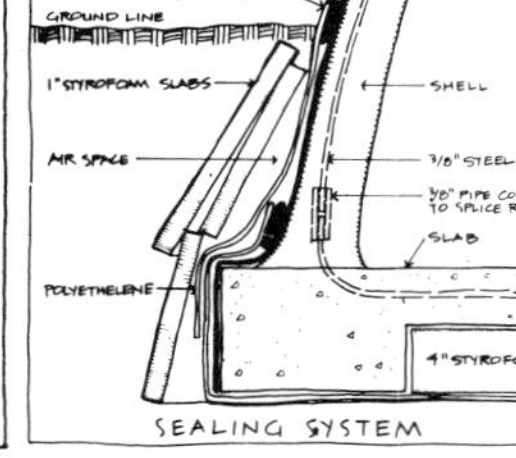

SEALING SYSTEM

At this time I built a scaffolding that radiated from above the structure on the same pipe that supported the cupola.

Wiring the mesh came next. One layer of diamond mesh plaster screen was wired to the outside of the 3/8" rod frame. I insulated the shell at this time with 1" styrofoam sheet fitted against the rod frame from the inside. The styrofoam was held in place by a layer of 1" chicken wire on the inside surface of the styrofoam. The chicken wire was wired to the plaster diamond mesh sandwiching the styrofoam in place. When all the mesh was tightly wired, a base or brown coat of gypsolite plaster was applied over the chicken wire on the inside. This was to act as a backup for the cement which would be pushed against it from the outside. This substitutes the overhead plastering of cement with overhead gypsolite plastering which is much easier.

Applying the cement was next. I used air entraining portland cement mixed 1 to 2 with even graded sand. The cement was mixed by hand in a mortar box and applied by pushing it through the mesh with the fingers and sealing the surface with a trowel. The cement plastering was done over about 5 days. A cement sand slush was used to join the wet cement to the set cement. Once hard, the cement shell was covered with wet dropcloths and kept wet for about 2 weeks.

When the base coat was dry, the finish coat of plaster was applied. I used quick set plaster and easy soak lime mixed about 50-50. Retarder was added to extend the working time. A slightly textured surface was achieved because of square tools and curved surfaces.

The final step was to weather proof the dome. It was fairly tig already and came through Hurricane Agnes with very little wa seeping in. The first step in the sealing process was to cement polyethelene extending from under the slab to the shell with a thick black pastey substance called plastic roof cement. Two c of foundation sealer were applied to all surfaces which would under ground. I then cemented left over polyethelene over the sealer and piled styrofoam insulation scraps on top of this. I w a little paranoid about having a cold leaky dome. The shell of dome above ground level was sealed with U.G.L. ready mix masonry sealer.

As final homey touches, I laid some bricks in an area where I placed the stove. And put in a wood floor made of no. 3 grade pine 1" x 12". I nailed these to 2 x 4" joists which were nailed to the slab with masonry nails.

One thing about this dome which I feel is significant is that 90 of the construction was done by one person. Ferro cement is a very heavy construction material but the ingredients and units construction can be broken into small pieces so that one perso can handle it. Great for hermit do-it-yourselfers like myself.

The car doors as entrances have worked out well. It is a very inexpensive way to get a complex machine. In one unit you ge weatherproof door, 3 operable windows and a lock. Quite a bargain.

I'd like to thank Niel and Connie for the land. Rit, Rich, Rog Raz were a big help. And thanks to the rest of the hummers o bucket brigade. Thanks also for *Domebook 2*. It really pushed and pulled.

I'm not living there now. I hope to get back soon....

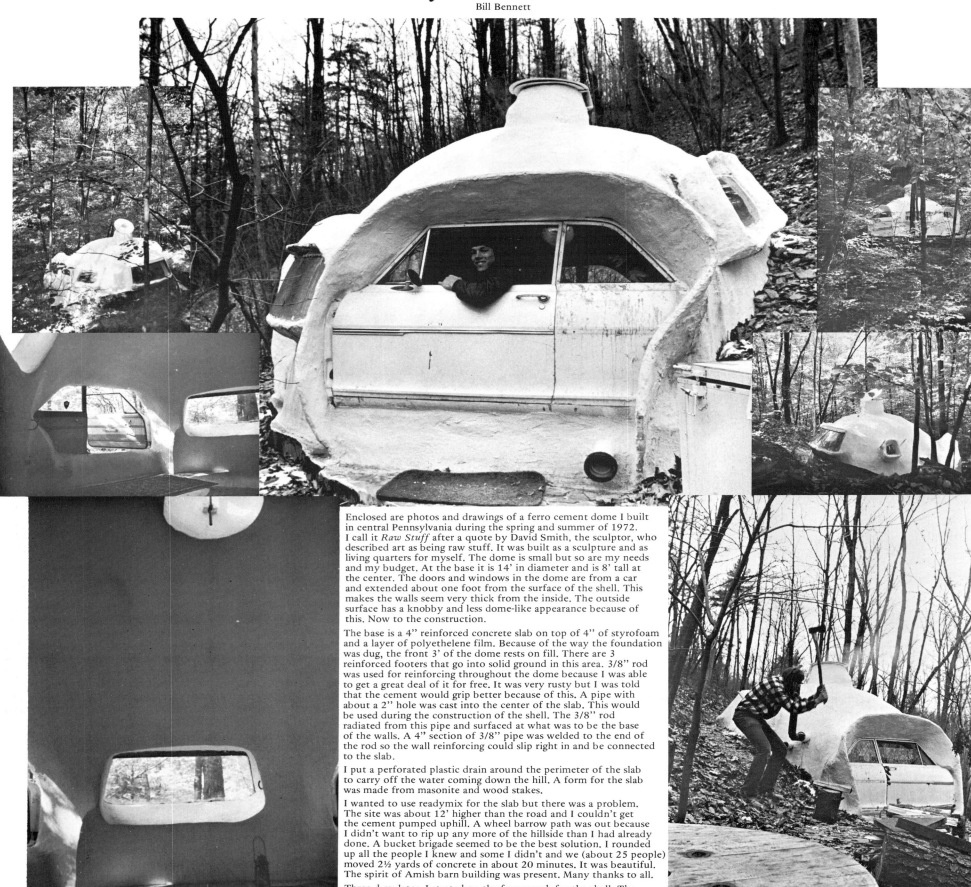

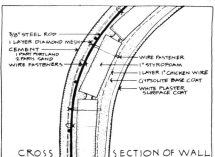

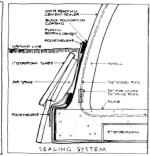

The House of the Century is what designers Doug Michels, Chip Lord, and Richard Jost call this ferro cement weekend house they built for Marilyn and Alvin Lubetkin on a small private lake near Houston, Texas. Doug, Chip and Richard were friends of the owners, and no restrictions were placed on design or construction. Said Marilyn Lubetkin: "I just knew we would get their very best...It was a trust-a total trust, and I just *knew* that it was going to be great from the beginning."

The following construction details are reprinted from the 6/5/73 *Progressive Architecture:*

ANT FARM ANT FARM

Beginning with a three foot three dimensional grid interval, ½-in. pipe was hand bent to form the compound curves. Held in place by wood shoring that later became the flooring, the pipe was the base for a layer of 3/8-in. steel reinforcing rods 6 in. apart. Four layers of chicken wire were then secured to both sides of the rods. Two reinforced concrete columns were placed to give support to the tower and its two floor levels, and reinforced concrete arches added extra strength to the tower-to-wing intersections. Specially designed door and window frames were installed. Three coats of high early strength portland cement, hand applied to the mesh by a Houston plastering crew, were moist cured for seven days. Battens at the locations of the pipe contours secure four inches of foam insulation. The entire inside surface was finally covered with upholstery, pleated at batten lines.

Pipes and ducts were run in the space above the slab and below the wood floor, which is laminated in butcher-block fashion out of 2x4's. Even the floor is sculptured, sloping down around the fireplace and stopping short of the building's perimeter. In the kitchen area, the laminations grow up out of the floor to form a sink which has been given a clear coating similar to surfboard finishes. As the floor nears the shower in the bathroom space, it again slopes off to form the tub. The toilet is standard, but there is a fiberglass structure incorporating the sink, plumbing pipes and a shower head. Between the bath and living areas, the furnace stands exposed, with a duct and the chimney for the adjacent fireplace rising up into the tower. A curved ladder provides access to the children's loft and the master sleeping loft in the tower....

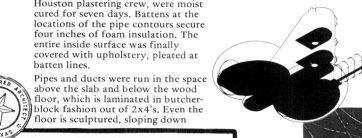

Jim Murphy

REGISTERED ARCHITECT · STATE OF TEXAS

THE HOUSE OF THE CENTURY

1972 2072

A Ferro-Cement Residence For Marilyn & Alvin Lubetkin

Designed & Built By	Architects: ANT FARM San Francisco, California
RICHARD JOST CHARLES LORD, JR. DOUG MICHELS	Contractor: NATIONWIDE BUILDERS Houston, Texas

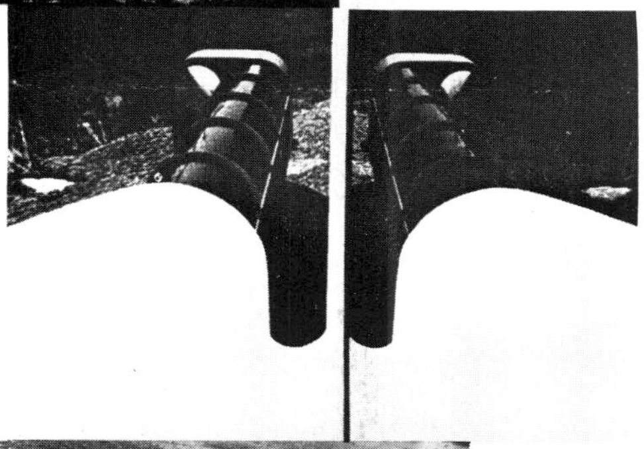

ENTRY

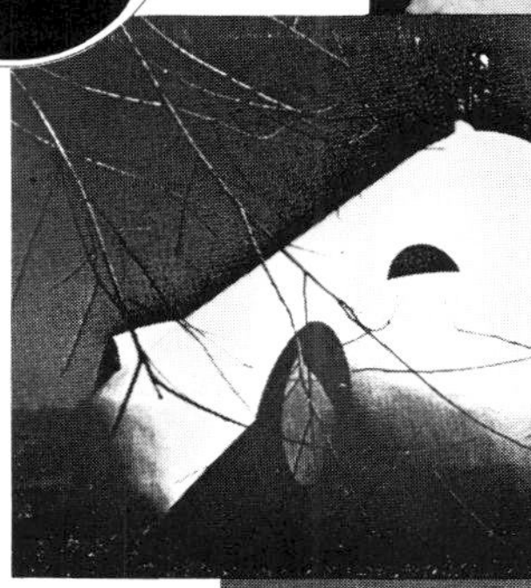

BATH

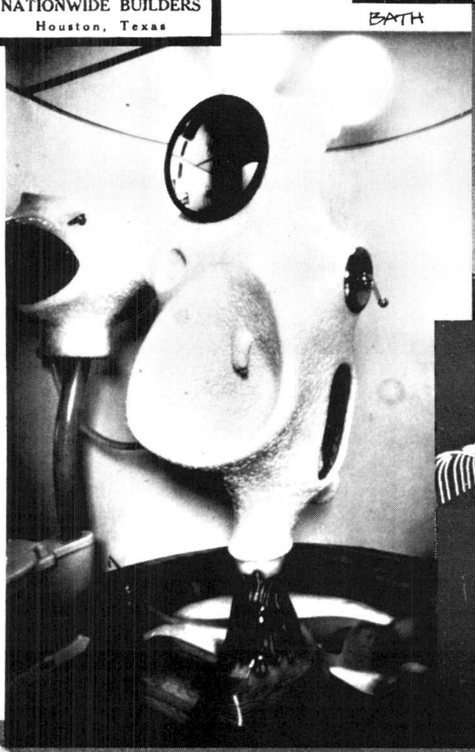

LIVING ROOM

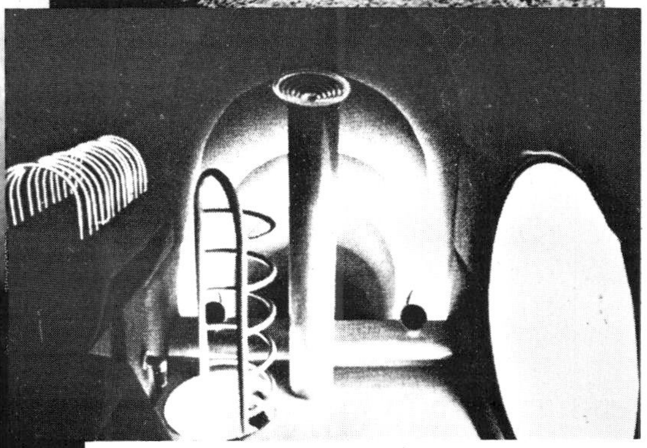

BEDROOM

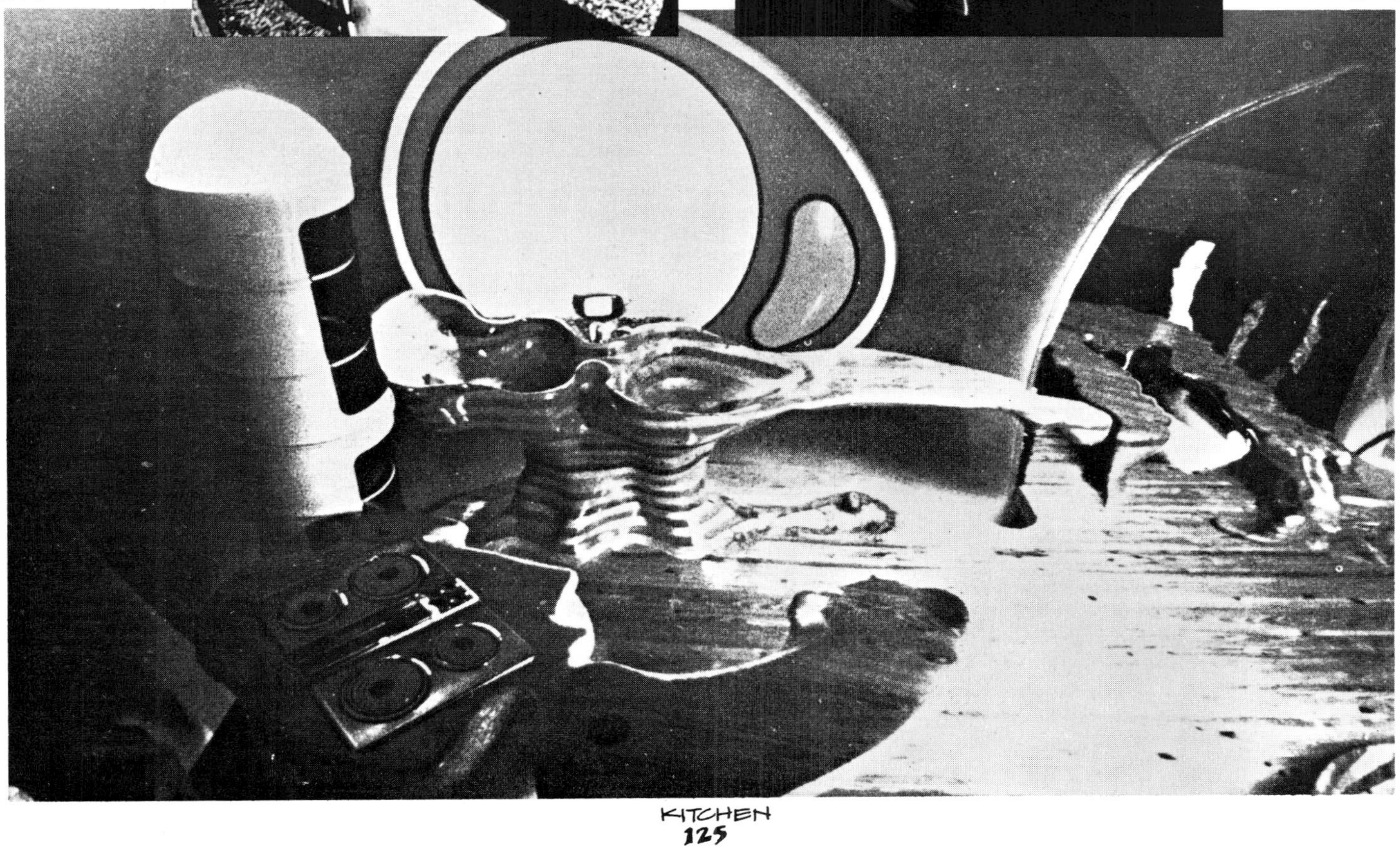

KITCHEN

DIVINE PROPORTION

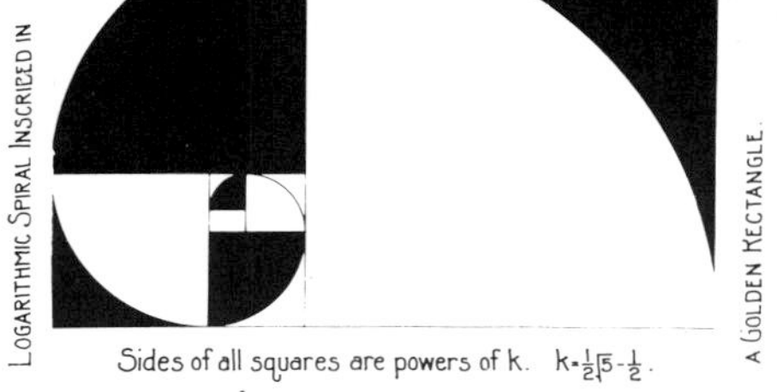

Sides of all squares are powers of k. $k = \frac{1}{2}\sqrt{5} - \frac{1}{2}$.

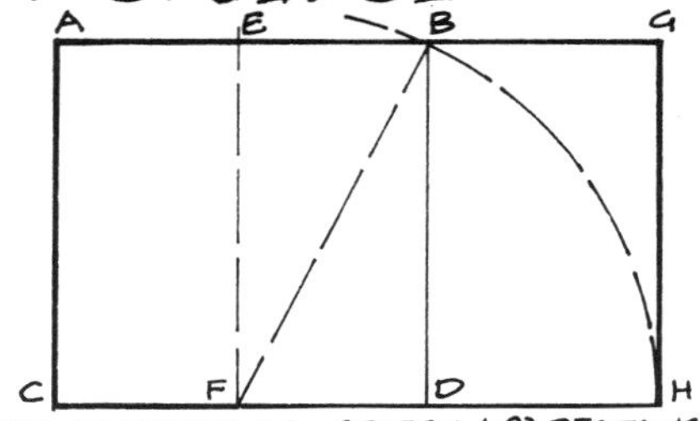

TO CONSTRUCT A GOLDEN (φ) RECTANGLE DRAW SQUARE ABCD, BISECT WITH EF. SET COMPASS AT DISTANCE FB. DRAW ARC BH. EXTEND LINES AB, CD. DRAW GH AT 90° TO CH.

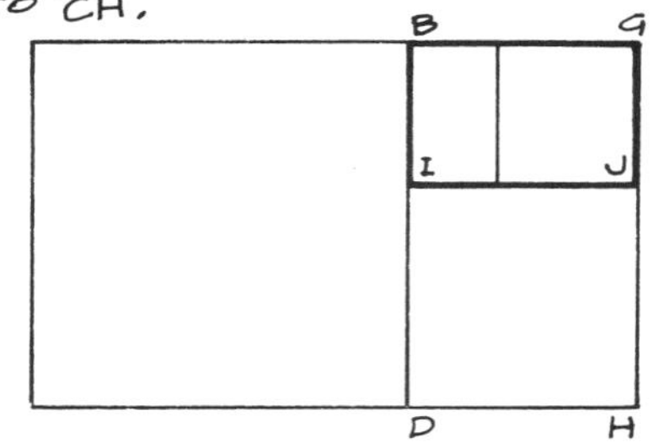

...NOW IF YOU DRAW A SQUARE IN BGDH, YOU CREATE ANOTHER GOLDEN RECTANGLE BGIJ, ETC...

The *Golden Section,* or *Divine Proportion* said by Kepler to be "one of the two jewels of geometry," and considered by Plato (in *Timaeus*) to be the key to the physics of the cosmos. This mathematical relationship appears repeatedly in growth patterns in nature, and has fascinated mathematicians and artists for thousands of years. The Golden Section, or (φ) as designated by the Greeks, is obtained by dividing a line

A ——— B

AT A POINT C

A ——— C ——— B

IN SUCH A WAY THAT THE WHOLE LINE

A ——— B

IS LONGER THAN THE FIRST PART

A ——— C

IN THE SAME PROPORTION AS THE FIRST PART

A ——— C

IS LONGER THAN THE REMAINDER

C ——— B

THIS MEANS THAT $\frac{AB}{AC} = \frac{AC}{CB} = 1.618$

THEN, TO DRAW SPIRAL...

How to draw a golden rectangle and a logarithmic spiral: draw circle with compass. With protractor mark off 72°/144°/216°/288° on circle, connect 5 points to make pentagon. Extend edges of pentagon to make 5-pointed star. ABO=golden triangle. Set compass at distance A-B. Mark off same distance on O-A to get P. Draw PB. APB= golden triangle. Now do same thing with APB: set compass on AP, mark Y on line BP. Keep doing this, making golden triangles, each smaller. *To draw logarithmic spiral:* Set compass at point P. Measure to O. Connect O-B; next set compass at point Y, measure to B, connect A-B; set compass at Z, measure to A, connect A-P. Etc. The *spira mirabilis,* or the equiangular spiral can also be generated from a rectangle. See bibliography for books with more information.

AOB is a *gnomon* To ABP. ABP is a gnomon to YAP, etc. A seashell grows in size but retains its shape. The sunflower has equiangular spirals.

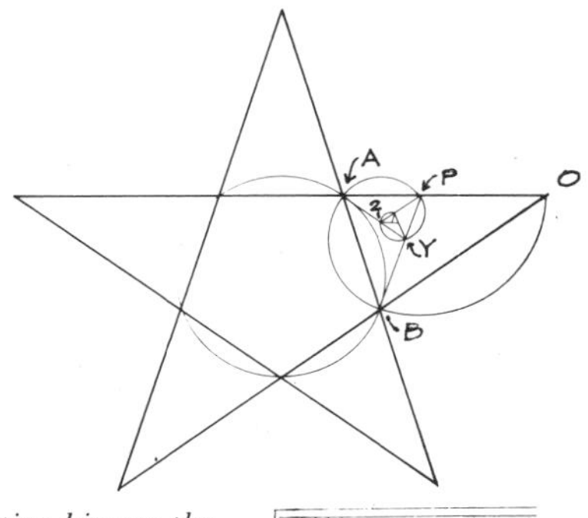

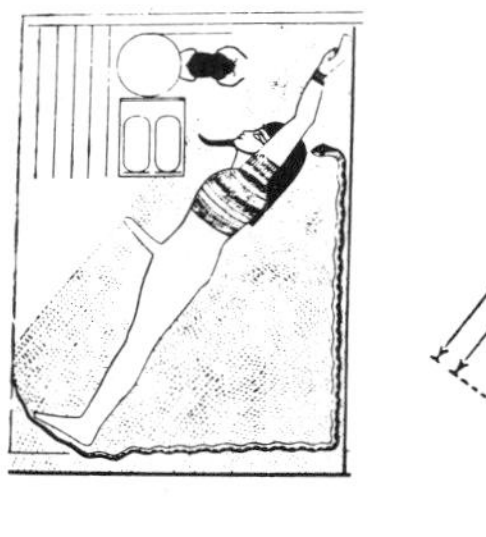

Egyptian king as the hypothenuse of a sacred 3-4-5 triangle formed by a snake. Schwaller de Lubicz shows the king as split into a φ^2+1 proportion by the phallus. The king's raised arm gives a 6/5, or 1.2 x φ proportion, which is exactly 3.1416, or π.

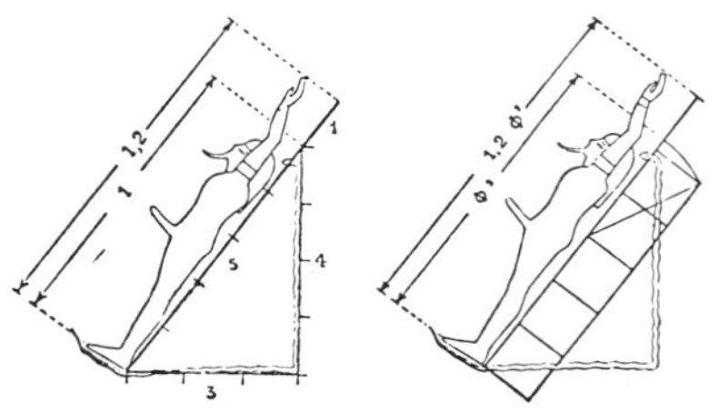

From *Secrets of The Great Pyramid*

OUR SHCOOL HAS DOMEBOOK TO. THE CLASS LOVES TO MAKE DOMES AND THINGS.

NO BURNT ENDS
CHARLES SELLERS
5142 46
SEATTLE WASH

CRIMPED WIRE

The most elegant model kits we've seen are the Heliwire crimped wire connectors invented by John Fieldhouse.

The connectors are broken from a prepared helix of coated mild steel wire. Several loops, usually three, fit together inside a tube to make each arm of the hub. The spring of the loops and the resistance of the tube wall hold the structure firmly together. Tubes and connectors can be altered and used repeatedly and no tools are required.

Heliwire is very suitable for both molecular and geodesic structures since any angular situation can be represented.

For information and prices: John Fieldhouse, 50 Swithland Lane, Leicester, LE7, 7SE, England.

Recommended straws:

Squash-proof Carnival king size plastic straws in packages of 100. From: National Soda Straw Co., 2323 S. Halsted St., Chicago, Ill. 60608.

FASTENERS

Re: Plastic straw model sphere hubs (pp. 6 & 7 *Domebook 2*). Try using Scovill round head brass (paper) fasteners (size no. 1 @$.68/100) and washers (size no. 1 @$.36/100) per Gilbert's catalog (590 Sutter St., San Francisco, Cal.) p. 174.

Also, to dodge candle and needle bit, try a Gem ticket punch with 1/8" round hole @$.65 (Gilbert's catalog, p. 328).

Best method is to plan construction so that all five or six strut ends are placed on fastener, then washer. Bend prongs, give light rap or two with hammer.

I don't think strut models get any neater/cheaper than this.

R. P. Scheld
Martinez, Ca.

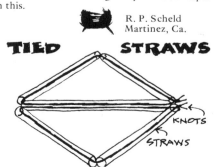

TIED STRAWS

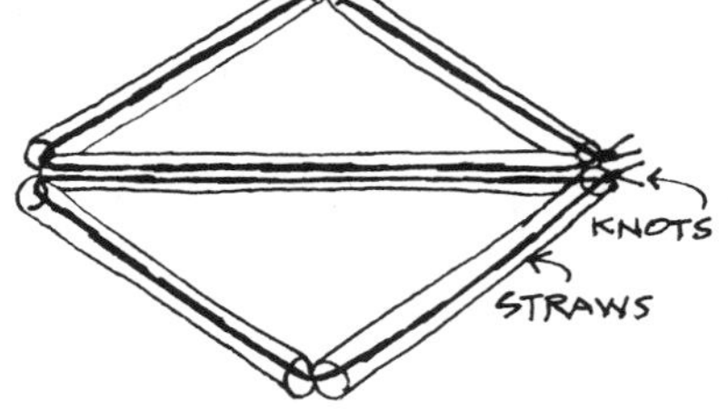

I wanted to build a model...I was able to construct the 3v sphere by using the straws and running monofilament (fishing line) thru the straws and tying triangles together. It goes fairly rapidly and is probably equally as inexpensive as welding. The structure lacks the rigidity of solid connectors but is nonetheless quite strong. Also the flex that is apparent is useful in that it enables you to bend and distort the structure to observe various, perhaps useful alternates.

M. L. Vermillion
Woodland, Cal.

CHORD FACTORS

A chord factor is a pure number which, when multiplied by a radius, gives a strut length. See *Domebook 2* for details.

SUPER ELLIPSE

Enclosed is a suggestion for a "superelliptical dome". It is a 4 frequency icosahedron, class 1, expanded onto the "superellipse".

$$\frac{x^{2.5}}{(1.25)^{2.5}} + y^{2.5} + z^{2.5} = 1$$

The superellipse is the brainchild of Piet Hein (Grooks). An article in the September 1963 (or '65) issue of *Scientific American* magazine gives details.

The superelliptical dome gives more useable floor space for a given dome size, I think, than does the elliptical dome.

Bill Wild
Vail, Co.

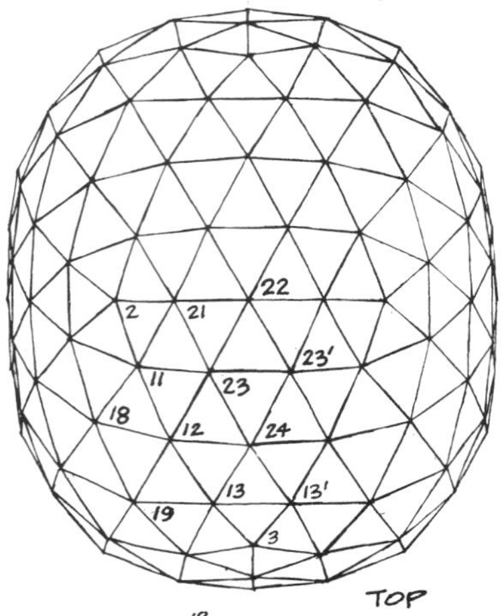

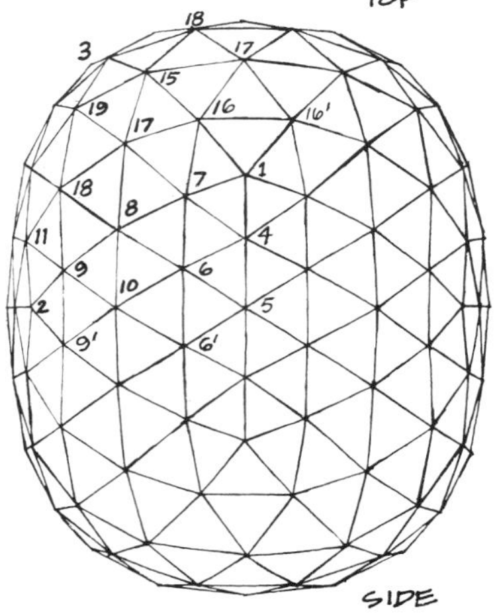

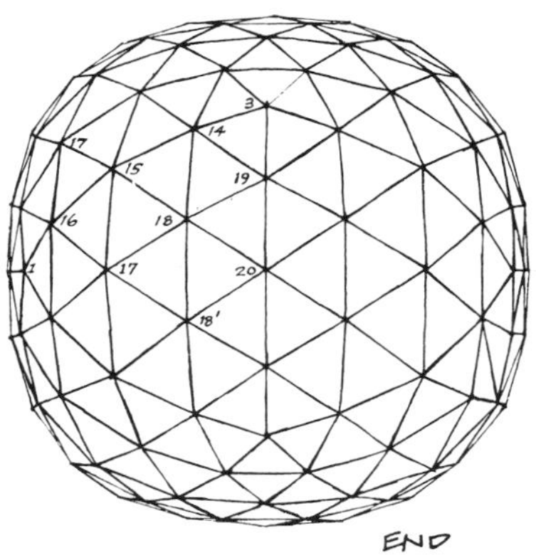

Strut	Chord Factor	Qty for Full Sphere	Strut	Chord Factor	Qty
4-5	.3055	4	18-19	.3949	8
1-4	.2791	4	19-14	.3372	8
5-6	.3127	8	10-9	.3223	8
4-6	.2958	8	9-11	.3466	8
6-6	.3382	4	11-18	.3330	8
7-1	.2689	8	12-18	.3636	8
8-7	.3225	8	12-19	.3576	8
9-8	.3297	8	19-13	.3394	8
2-9	.2813	8	14-13	.3300	8
9-9	.3302	4	16-16	.3854	4
8-10	.3460	8	16-17	.3836	8
6-10	.3385	8	15-17	.4158	8
6-8	.3386	8	18-17	.4029	8
6-7	.3217	8	15-18	.3952	8
4-7	.3150	8	14-18	.3805	8
11-2	.3083	8	14-19	.3564	8
12-11	.3507	8	3-19	.3177	4
13-12	.3355	8	19-20	.3753	4
3-13	.2732	4	19-18	.3442	8
3-14	.2879	8	18-18	.4226	4
14-15	.3614	8	18-20	.3719	8
15-16	.3849	8	2-21	.2658	4
16-1	.3299	8	21-22	.3034	4
7-16	.3611	8	11-21	.3366	8
7-17	.3780	8	21-23	.3396	8
16-17	.3324	8	11-23	.3042	8
8-17	.3909	8	12-23	.3527	8
8-18	.3841	8	12-24	.3242	8
17-18	.3690	8	13-24	.3252	8
17-19	.4016	8	24-23	.3629	8
17-15	.3816	8	22-23	.3503	8
19-15	.3594	8	23-23	.3262	4
9-18	.3677	8	13-13	.3108	4

3v TRUNCATABLE

Dome Cookbook of Geodesic Geometry, by David Kruschke shows actual derivation of chord factors and angles, and "...unlike *Domebook 2,* the chord factor results here are in close agreement with those of Buckminister Fuller's...." Here is David's 3-frequency truncatable dome, which sits flat on the ground, unlike the *Domebook 2* 3 freq domes:

The book is available for $1.50 from David R. Kruschke, 2135 West Juneau Ave., Milwaukee, Wisc. 53233.

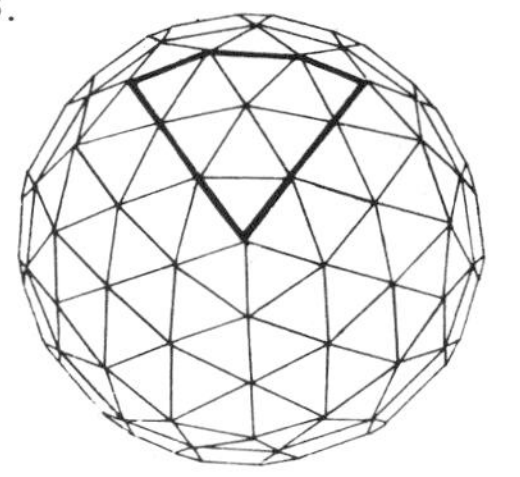

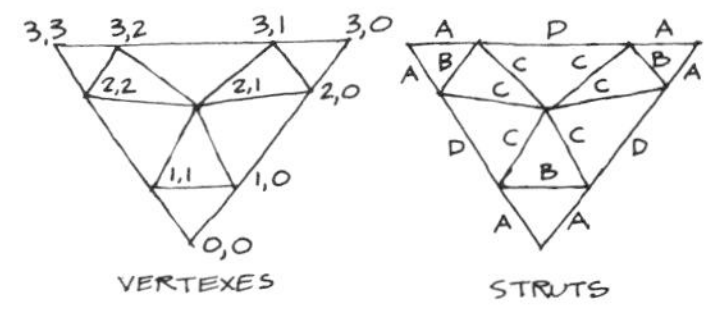

V,E,F(L) means the number of Vertices, Edges and Faces in one icosahedron face.

V,E,F(G) means the number of Vertices, Edges and Faces in the entire sphere.

3 Frequency Icosahedron, Truncatable
V(L) = 10 E(L) = 18 F(L) = 9
V(G) = 92 E(G) = 270 F(G) = 180

Length A	3,1	3,0		.3297
Axial	0,0	3,0	3,1	80.51°
Face*	3,1	2,0	3,0	54½°
Dihedral	3,1	3,0		166.34°
Length B	1,1	1,0		.3822
Axial	0,0	1,0	1,1	78.98°
Face*	1,1	0,0	1,0	71°
Face*	1,1	2,1	1,0	54°
Dihedral	1,1	1,0		167.97°
Length C	2,1	2,0		.4214
Axial	0,0	2,0	2,1	77.83°
Face*	2,1	1,0	2,0	58½°
Face*	2,1	3,1	2,0	63°
Length D	3,2	3,1		.4410
Axial	0,0	3,1	3,2	77.26°
Face*	3,2	2,1	3,1	63°

Base radius of cut off plane	.9822
Height of 3/8 cut off	.8124
Height of 5/8 cut off	1.1875

* Denotes angles that were obtained graphically to the nearest half degree.

OCTA 21

Hugh Kenner discovered what is perhaps the simplest geodesic structure, what he calls the *Octa 21.* The roof is diamond shaped, floor plan is a slightly elongated hexagon, with front and rear walls nearly vertical. You can make a structure that is 7' high, floor about 14' by 17½' using seven 10', fourteen 8' 2x4's. Hugh suggests first making a paper model. Specifications below:

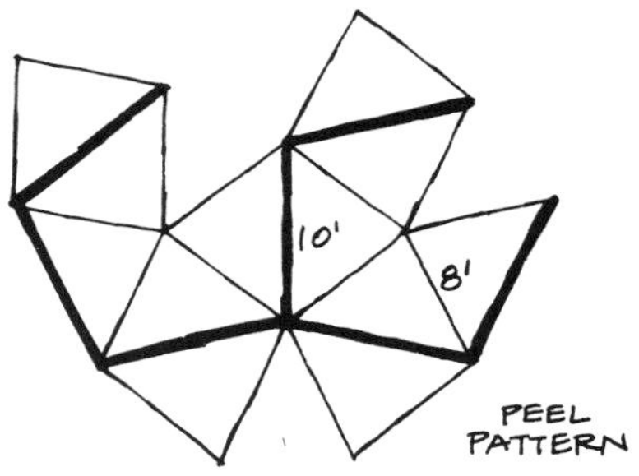

Chord Factors	For 14x17½ 2x4 frame	For paper model
Heavy lines 11.5	7-10' long	Heavy lines 5" long
Light lines 9.2	14-8' long	Light lines 4" long

About the vertices. There are only four up aloft: two 6-way, where short and long members meet alternately, two 4-way, all shorts. Around the base you have four 4's (long-short-long-short) and two 3's (short-short-short). That's not very many, and you could afford to take more time over them than would be sensible with the multi-vertexed structures described in the Domebooks. You could even do fancy carpentry and bevel the ends together neatly. Or use a dome-builder's hub system (Domebooks describe several).

It would be a good idea to subdivide the triangles, to keep those long members from trembling. Also to make it easier to nail on walls. Do walls, windows, doors, any way you like. Don't miss the possibilities of skylights.

But before you do anything, make a paper model. The diagram shows the idea. Lay it out carefully with compasses, heavy lines 5", light ones 4" (or 10 and 8, if you've big enough paper). Cut out. Fold to bring gaps together. Tape. Inspect.

Hugh Kenner

LOG DOME

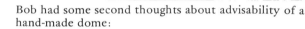

Bob Lander built a log-frame dome covered with shakes on an island in British Columbia. Large cedar snags left by loggers were milled into lumber for joists and decking, young fir trees were used for struts, and the connector was the plyhub from *Domebook 2,* with wood dowels. Bob designed the dome for a 60 lb. per sq. ft. snow load, said the dowel was stronger than a bolt, "...with a 1" dowel and 5/8" plywood, the dowel starts to cut through the plywood at about two tons pressure. A bolt would be smaller and cut sooner."

sing an Alaskan Mill: I felled large cedar snags (dead some thirty ears, but still standing), and milled them to lumber with the small laska Mill on an old Stihl Lightning. The Alaska Mill attachment ost $100, as did the saw I got to use it on.

found that to make 2x6's, 2x8's, etc., it was most economical of oth labor and material to slab the log into 2" thick slices of necessarily) varying widths, which are later ripped into the esired widths by snapping chalk lines on them and walking down he slab with a smaller hand-held chain saw held low so that the ar is almost parallel to the board being ripped. This is the lternative to squaring up the log before you begin. You'll need to ave another saw on hand anyway as you can't be taking the mill part every time you have to cut a branch away. Incidentally, this peration works as well in fir or, I assume, anything else.

cut about seven thousand board feet for the deck, which turned ut to be an octagon some 44' across, five feet off the ground. heoretically, one could easily make this much lumber in 2 months nd keep up with the chores as well, but don't count on it. The est I ever did was 350 board feet in a day.

JANE & BOB LANDER

PLYHUB WITH 1" DOWELS

Bob had some second thoughts about advisability of a hand-made dome:

It doesn't make sense to hand-fit a dome. Hand-fitting breaks down wherever there is a line, a place where things must meet, fit. In domes there are too many. I don't mean the skeleton. That's fun. I mean the skin, the doors, windows, porch, any other attachments.

A Volkswagen goes together because all the parts are stamped out by machine. Domes lend themselves to this. For the amount & sizes of material involved, they are the most practical structure. All short & light-weight members.

Mass produced domes. Land developments. Vulgarized. But that's all right. Domes have out-lived their usefulness as an art form. They're too restrictive anyway. If you're gonna hand-make a house, it might as well be free-form.

DOME ADOBE BY JERRY BURGLAND
PLACITAS, NEW MEXICO

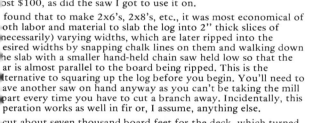

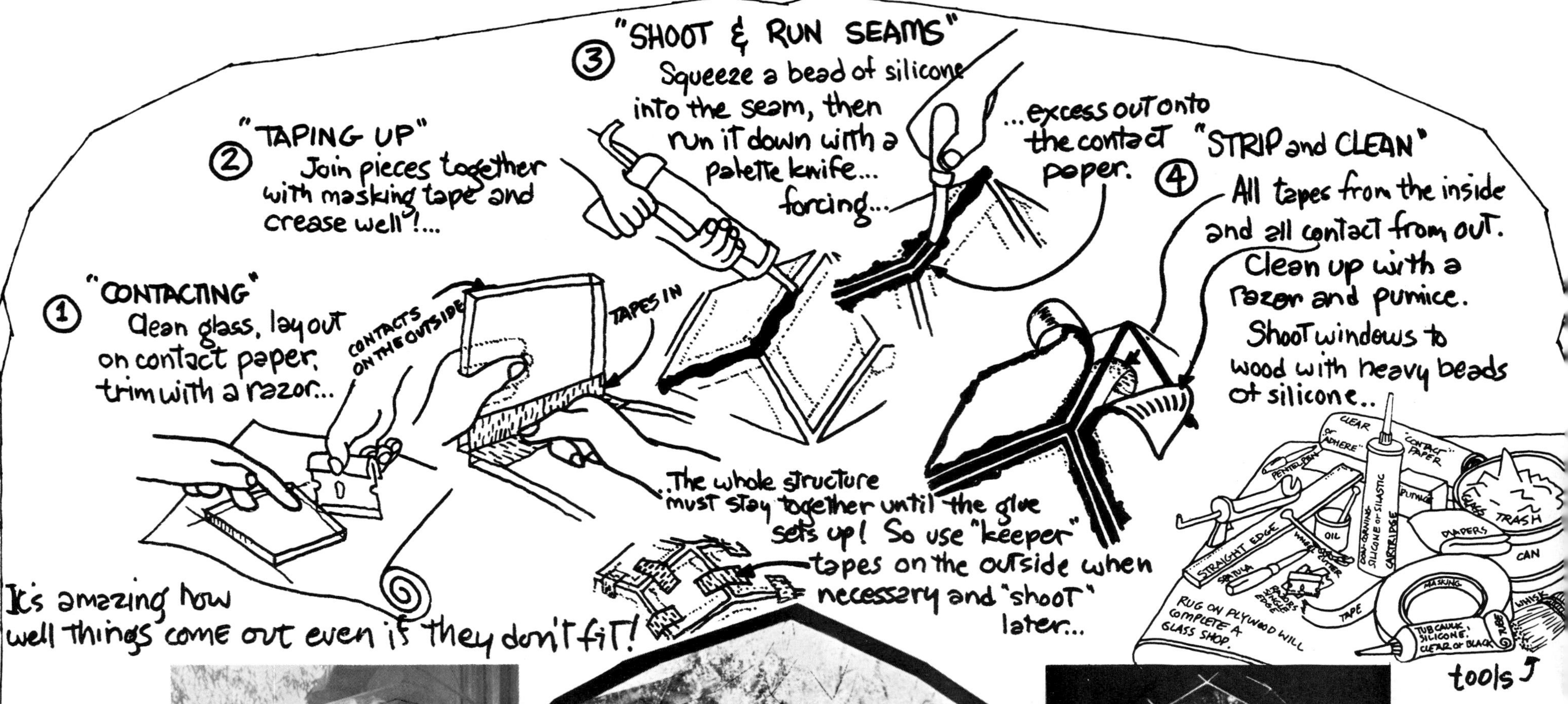

③ "SHOOT & RUN SEAMS"
Squeeze a bead of silicone into the seam, then run it down with a palette knife... forcing...
...excess out onto the contact paper.
② "TAPING UP"
Join pieces together with masking tape and crease well!...
④ "STRIP and CLEAN"
All tapes from the inside and all contact from out. Clean up with a razor and pumice. Shoot windows to wood with heavy beads of silicone..
① "CONTACTING"
Clean glass, lay out on contact paper, trim with a razor...
CONTACTS ON THE OUTSIDE
TAPES IN
...The whole structure must stay together until the glue sets up! So use "keeper" tapes on the outside when necessary and "shoot" later...
RUG ON PLYWOOD WILL COMPLETE A GLASS SHOP.
STRAIGHT EDGE
OIL
TAPE
CAN
GLASS TRASH
tools
It's amazing how well things come out even if they don't fit!

8 SIDED COLUMNAR SPIRAL ZOME

CRYSTAL WINDOWS

Our big windows been up for four years now and looks good for at least 10 yrs. Silicone is hi tech but very economical in use for glass joining... The scale of structure you can build seems unlimited. Single shell domes up to 10 feet in diameter would be fine. You'd have to build simple glass trussing into the larger span shelters like the garden houses we'll be building.

Silicone will work on plexi too, but it's a drag compared to; optical pure, easy to clean, a snap to cut! glass...

Windows to fit any shape! Crystals and the basic solids, shells and flowers all provide ample inspiration for regular structures, but even free forming will get you right into it! We've built right up from frames, and marking the next piece with pentel pens, exactly to fit!

Use a bead inside and out when sealing windows to wood. Bubble out from your home! Pickup on a good window trip!

Kim Hick

JEFFERSON STARSHIP

Built for Paul Kantner & Grace Slick by Roy Buckman

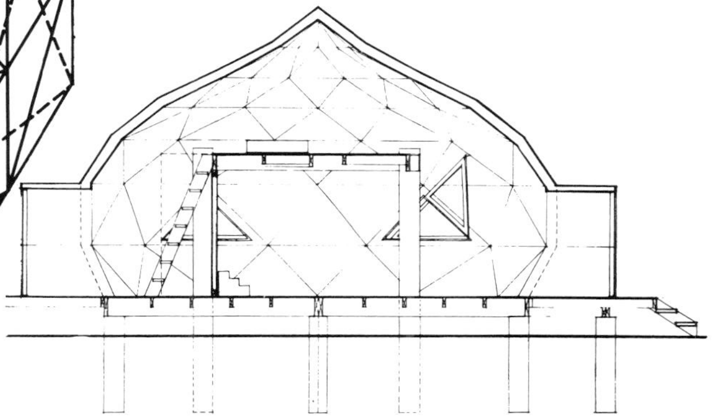

tside Dome Diameter at Floor = 21' – 4"
tside Dome Diameter at 2' – 0" above floor = 24' – 0"
side Dome Height above floor = 14' – 6"
te: 'Outside cut' dimension assumes use of 4" hubs,
d refers to outside surface of dome.

TRUT DIMENSIONS AND ANGLES:

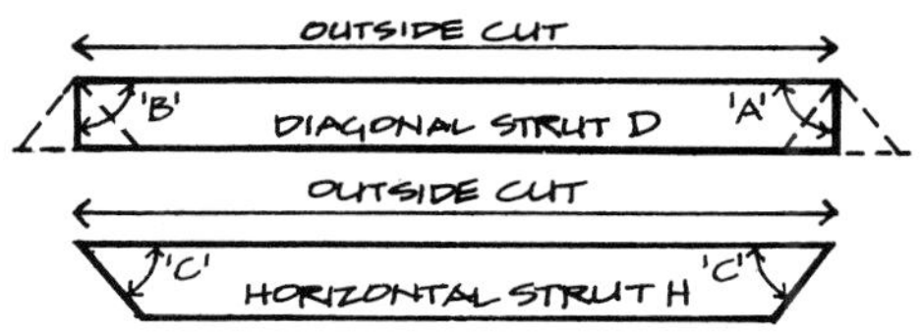

IAGONAL STRUTS (approx. 600 lin. ft. 2"x4")

trut ame	No. to cut	Outside cut	Angle 'A'	Angle 'B'	Chord Factor (center to center)
D_0	12	32.0"	$53^{\circ}22'$	$96^{\circ}18'$	2.40
D_1	24	35.64"	$96^{\circ}30'$	$91^{\circ}29'$	2.64
D_2	24	35.64"	$89^{\circ}39'$	$77^{\circ}9'$	2.64
D_3	24	44.99"	$77^{\circ}35'$	$75^{\circ}48'$	3.27
D_4	24	44.99"	$75^{\circ}48'$	$77^{\circ}19'$	3.27
D_5	24	48.77"	$78^{\circ}2'$	$71^{\circ}1'$	3.25
D_6	24	48.77"	$70^{\circ}45'$	$82^{\circ}36'$	3.52

ORIZONTAL STRUTS (approx. 360 lin. ft. 2"x4")

trut ame	No. to cut	Outside cut	End Angle 'C'	Chord Factor (center to center)
H_1	12	11.01"	$82^{\circ}30'$	1.00
H_2	12	29.19"	$84^{\circ}26'$	2.21
H_3	12	45.69"	$81^{\circ}49'$	3.31
H_4	12	62.45"	$78^{\circ}54'$	4.42
H_5	12	70.54"	$75^{\circ}40'$	4.97
H_6	12	73.17"	$75^{\circ}23'$	5.14
H_7	12	62.26"	$78^{\circ}7'$	4.42

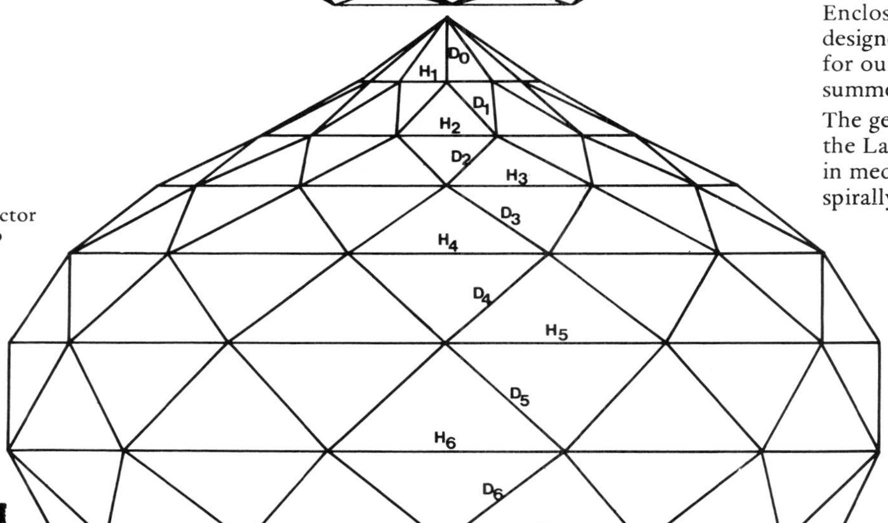

Enclosed are some photographs of a meditation dome designed and built by the members of Dhyana Mandiram for our Guru, Swami Rama, at our ashram center this summer.

The geometry was triggered by inspiration from Fuller, the Lama domes, the patterns of Col. Beard, and a feeling in meditation of a space emanating concentrically and spirally as waves of vibration from a bindu (point of light).

Namaste.

Dennis R. Holloway
Hindu Meditation Temple
Minneapolis, Minnesota

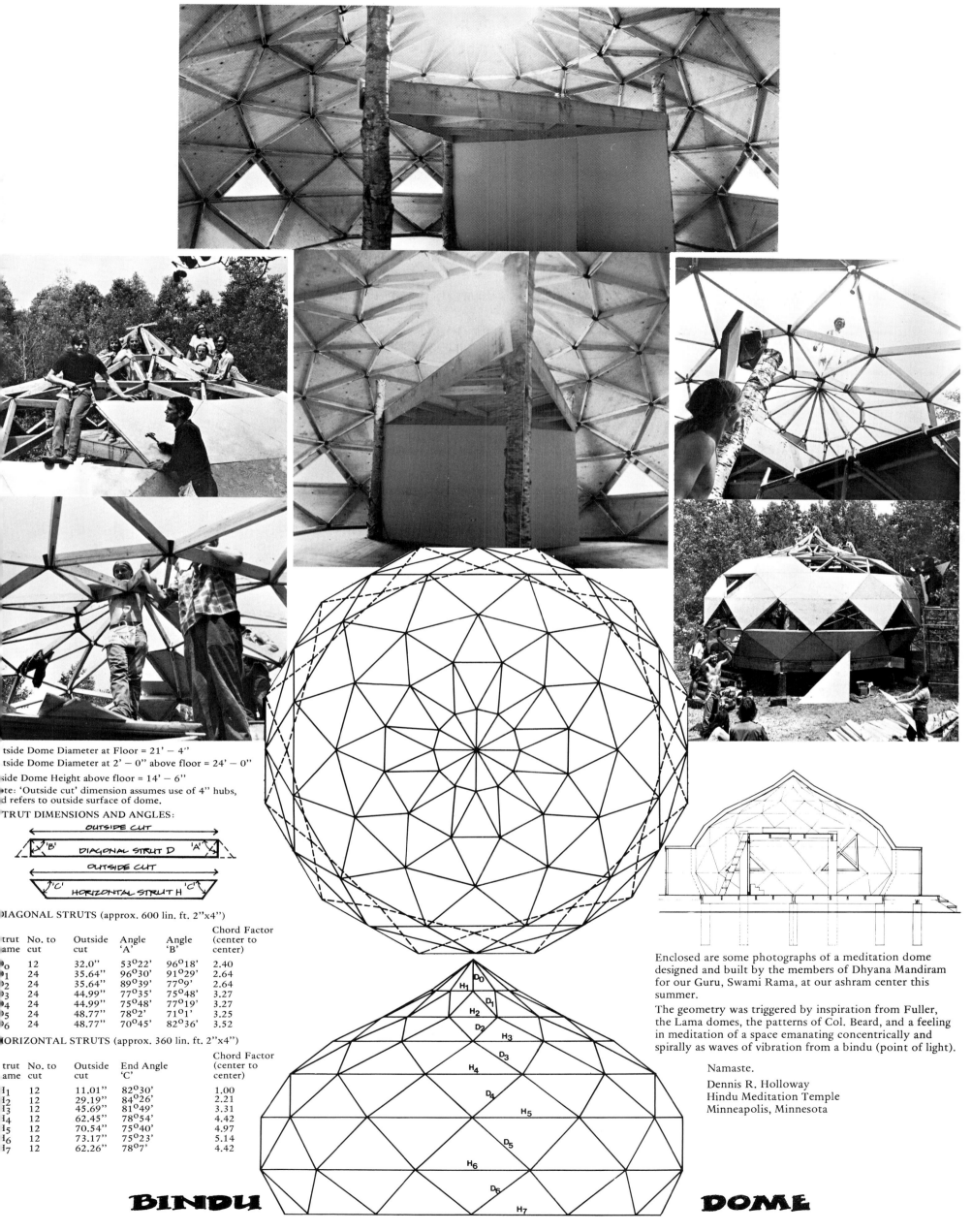

BINDU DOME

The Sufis tell a story about a metalsmith who was unjustly thrown into jail. He pleaded with his captors, and they finally allowed him to receive a rug woven by his wife.

Day after day the man said his prayers on the rug, prostrating himself in the direction of Mecca. After a time, he said to his jailers:

"I am poor, and I have no chance in life any more. You youselves are paid like slaves. But I happen to be a metalworker. If you bring me some tin, and some tools, I can build some small trinkets which you can sell in the marketplace. In this way we may both benefit."

The guards agreed, and pretty soon both the tinsmith and the jailers were making a nice profit. They used the extra money to buy food and luxuries for themselves.

But one day, when the guards went to the cell in their usual way, they found the door open. The man was gone.

Many years later, the man's innocence was established. He happened to run into one of the men who had imprisoned him. This man was burning with curiosity, and he asked the metalworker how he had managed to escape — what magic he had used.

The tinsmith answered: "It is a matter of design, and design within design. My wife had found the man who designed the jail locks. She wormed the design out of him. Since she is a weaver, she skillfully wove the design into the carpet, at the very spot my head touched five times a day when I was praying.

"You know that I am a metalworker, and to me this design looked like the inside of a lock. So I designed the plan of the trinkets to allow me to store up the necessary material to make a key — and I escaped!"

THE PASSERELLE OF THE WEAVERS

"Allah willing, we'll end up with the Alhambra Palace!"

Imagine a narrow North African passerelle, salmon-red walls and tall green palms swimming in the blue-blackness of twilight. We pass a peaked archway, which is a door into ancient magic. Inside, by the light of silver gas lamps, dozens of small children are assiduously holding slender black threads looped around their tiny fingertips. The threads stretch in an intricate pattern from one cousin to another brother. In one corner a wrinkled old grandmother works at a spinning wheel, and a small girl is waiting to take her freshly spun yarn to the dye market.

The men are performing an ancient dance, the dance of the rug makers, by gliding in and out among the little children while knotting the warp threads. They move according to a cadence which is sung by the women, who are sitting around the walls paying out measured lengths of colored wool, in a ritual drawn up unpolluted from the deep well of time.

Each district, each family has its own special song, and this gives each rug its unique design. With each change in rhythm comes a change in color; a new harmony makes a new pattern.

There is a legend still very strong in the Middle East and North Africa that these rug designs were all, at one time, carefully constructed by the members of a learned society for the purpose of preserving certain fragments of esoteric knowledge.

The secret of the rug is hidden in the music. I remember one rug shop in Tangier where 16 blind girls worked at wooden looms, while a lively old woman sang to them and they hummed along with their fingers. I came in with a professional storyteller who regaled them with tales of Malta. The 16 girls all laughed in unison, like a moonlight minuet of mountain bells.

As we leave the family rug shop and continue down the passerelle we come upon another lighted archway. We look in and see an old man with young eyes, bent over pages of calculations and peculiar diagrams. As we look in wonder, he beckons us in.

He is not a scribe, it turns out, and not a scholar. He is a "designer" and is busy with the plans for the grillwork on a certain archway that will adorn the mosque.

We inquire about the designs.

"You know," he says finally, "that it is forbidden for a servant of Allah to make images."

We nod.

He takes up a drawing which looks like a very ornate stencil of leaves and flowers.

"This," he says dramatically, "is the first *sura* from the Koran, in which Allah commands the Prophet: 'Read!' Mohammed protests that he does not know how to read. But Allah commands him again: 'Read!' "

It dawns on us that he is telling the literal truth: That leafy pattern is actually an exquisitely ornate Arabic calligraphy. Those "gaudy" designs on the walls of mosques are actually lifelines from the inner nucleus of the Moslem world: This is the Koran, which even illiterates know by heart.

"The Koran is a code, a mystery," he says in a voice full of emotion. "To solve it is to become enlightened." He takes us over to his table littered with number calculations, geometric designs and the ropework patterns of Islamic art.

Cont'd right hand column

DIAGRAM OF ORNAMENT OF THE OCTAGONAL DOME OF THE ALHAMBRA.

DECODING ARABIC DESIGN

David Saltman

...The codes and exact methods of making these designs are jealously guarded as Fritz Brenner's recipe for Sally Lunn. It stays within the family. You may be interested to know that one of the largest sects of Sufis, commonly called the "Naqshbandi School," is also called, among the Arabs, "The Designers." They are particularly occupied with this business of encoding information into rugs, calligraphy and architecture.

I'm told that there is a certain mosque, in Central Asia, where anyone who enters, irrespective of race or culture, immediately bursts out crying! *It has something to do with the architectural dimensions of the place, and their relationship to human physiology.*

CARAVANSERAI

ISFAHAN

He shows us a piece of paper ruled off into small squares. In each square is a number.

"This is a unique kind of 'magic square,' " he explains, "and it is the key to my work of designing.

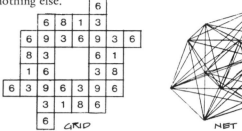

1	2	3	4	5	6	7	8	9
2	4	6	8	1	3	5	7	9
3	6	9	3	6	9	3	6	9
4	8	3	7	2	6	1	5	9
5	1	6	2	7	3	8	4	9
6	3	9	6	3	9	6	3	9
7	5	3	1	8	6	4	2	9
8	7	6	5	4	3	2	1	9
9	9	9	9	9	9	9	9	9

'MAGIC SQUARE'

"I have the task of designing this archway to transmit certain information and to give a certain feeling," he says with a gleam in his dark eyes. "Let us call the information and the feeling 'six.' "

We watch closely as he abstracts from the magic square all the rows and columns which contain a "six." He comes up with a kind of number grid, which represents "six" and nothing else.

					6		
		6	8	1	3		
	6	9	3	6	9	3	6
	8	3			6	1	
	1	6			3	8	
6	3	9	6	3	9	6	
		3	1	8	6		
		6					

GRID

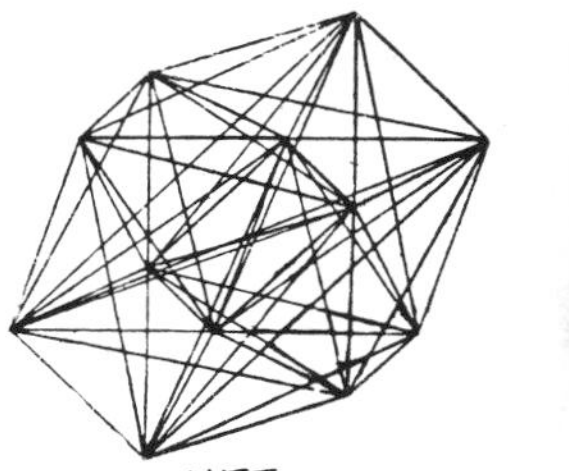

NET

Then he takes a fine pen and skillfully joins all the points of the squares of the "number six grid," forming a kind of net.

He overlays several "number six nets" at different angles to one another, simplifies the design, and comes up with this:

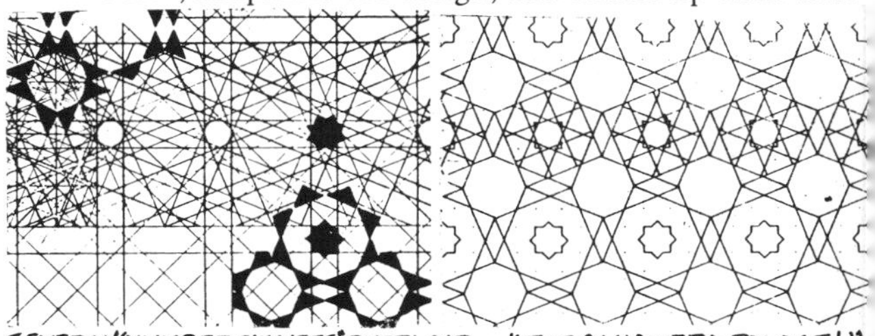

SEVERAL 'NUMBER SIX NETS' OVERLAID, ... "THE ALHAMBRA PALACE!"

He looks at us triumphantly. "And now, you see, it is a matter of building. I put one net onto another, until the lines grow thick...and Allah willing, we will end up with the Alhambra Palace!"

Depending on the particular code the designer is using, you could get hexagons within hexagons, octagons within octagons, or whatever.

The method of simplification is really elegant, in my opinion. When you actually do this business of laying the nets over one another, you find that many of the lines just coalesce into a black blob. The blobs stay in the final design. You also get the effect where a 20-sided figure approaches a circle, which may be rendered into a circle in the end, or may just stay a 20-sided figure. Sometimes everything within one of these circles is erased, and the line segments left over are connected with one another according to yet another numerical code. There are also certain geometric combinations which are simply not used, for whatever esoteric reasons, and when one of these turns up, the crucial lines are erased and the segments connected in some unusual way. In this what they call "lawful otherwise", according to the legends, is hidden the real secret information these designs are supposed to convey.

I had seen this kind of Islamic pattern often, but had found it impossible to follow. You trace out one line and it seems to meander and sashay along, like a wandering dervish, with no particular order or meaning. But taken as a whole, all these seemingly random lines somehow work together. The result is startling, and a perfect map of the Arab mind.

As I stood there in the Arabian twilight watching this designer work I got the definite feeling that a message was being delivered, to us personally, through the thick fog of the past. This man was a codemaster, a telegrapher, an artist in an artificial language. "It is a matter of design, and design within design...."

If you are interested in buying an Oriental rug from the source — and saving several thousand dollars — it would be worthwhile to learn how to count in Arabic. In fact, the more Arabic or Persian you know, the cheaper your rug will be.

The best introductory book on Arabic is called *Arabic Made Easy* by Mouncef Saheb-Ettaba. The publisher is McKay. All other books on Arabic that I've seen are completely impossible to understand.

An excellent introduction to Oriental rugs of all kinds is a little book called *Oriental Rugs in Colour* by Preben Liebetrau. Reprinted courtesy *Rolling Stone.*

LATH HOUSE AT THE FARM, U.C. SANTA CRUZ.

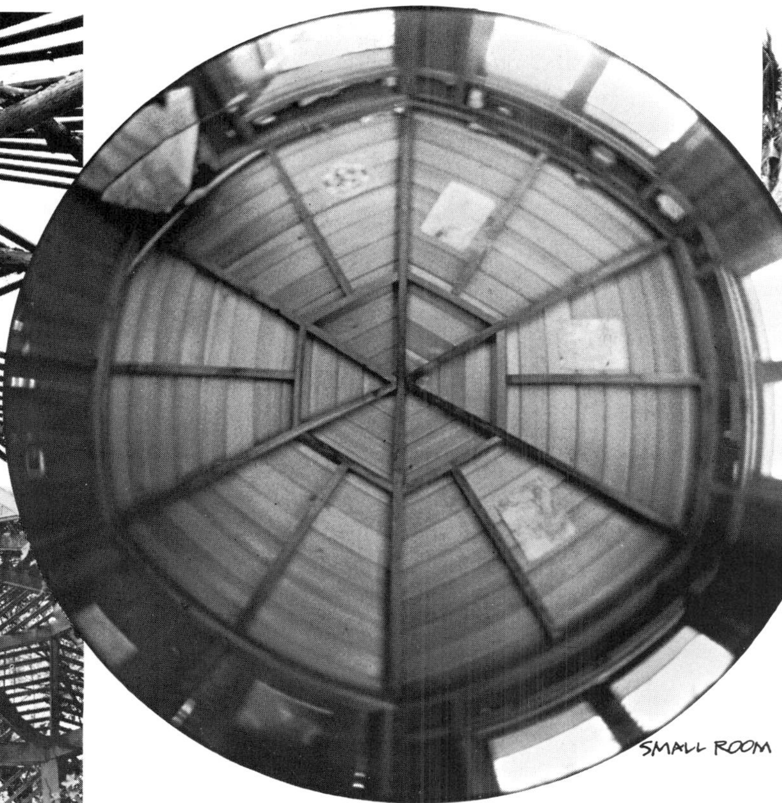

SMALL ROOM

SAMOA

DUTCH BARN

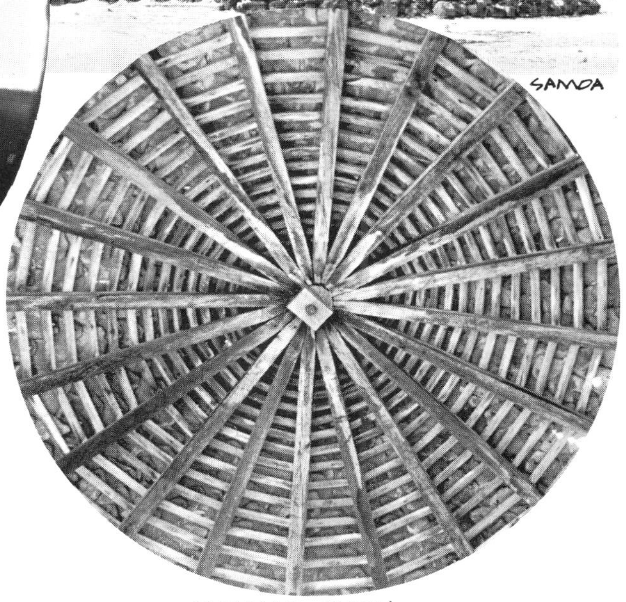

ENGLISH BARN

POLYGONS

Polygonal buildings are an alternative to building a dome, if you are interested in a rounded floor plan. The roofs are easier to build; they have vertical walls so windows and doors are easy to install, and extra rooms can be added on.

ANANDA

Ananda is a community following the spiritual teachings of Paramahansa Yogananda. In *Domebook 2* Alan Schmidt describes the domes they built as their first buildings. Since then, they have built a variety of buildings, some described here.

In the past four years, many structures at Ananda besides domes have been built. Perhaps the simplest and most attractive of these are the eight and nine sided houses. Combining the circular shape of a dome, the straight walls of a conventional house, and the pointed roof of a tipi, one finds a unique and inviting structure. The flowing, unifying feeling of a dome is not lost, and joining this with the upward surge expressed by tipis, the result is a blend of harmony and direction. The advantages of these structures are many. Since we deal only with straight, vertical wall construction, building is relatively easy. (Four of us built two nonagons in 3½ months. This included going to a Forestry Service tree farm and cutting and skinning both foundation and rafter poles.) The roof design, except for the center hub, is simple and easy to waterproof. Between the Farm and the Retreat, there are five 8-sided homes and three 9-sided ones. All are water tight and require only the usual home maintenance....

We (I use this term loosely, all credit goes to my good friend Jaya) designed the 20' octagon and the 24' nonagon as permanent structures for single persons or couples. The smaller octagon, with approximately 290 sq. ft., is well suited for one person. The nonagon is an efficient space for two, with roughly 405 sq. ft. of floor space. Lofts in these homes add space, and enhance the interior design....

Walls

The walls were standard construction. Because the roof pushes down and out, there is great stress on the walls. Outward stress from the roof will be transferred to the top plate of the walls, so all the plates should be tied to one another to form a solid continuous ring. To strengthen the walls, we used 3 3/8" bolts per corner. The weakest spot in the walls are where the corner studs and top framing meet. Here we used triangular sheet metal plates to add strength. For still more support you could use rectangular metal plates on both sides of the 2 x 4 studs joining the top plate.

Roof

In designing the roof, we wanted a method by which we could eliminate the need of a center pole. Our first successful method was to cut a four foot diameter circle of ¾" plywood and attach all the rafters to it. Those rafters coming from the corners of the house passed over the edge of the plywood and butted together at the peak. Holes were drilled through each of these rafter poles at three points along the pole between where it touches the plywood and the peak. Holes were also drilled in the plywood directly below the holes in the poles. 5/8" metal strap was then passed through the holes and tightened securely with a metal strapping tool. This tied all of the rafters together via the piece of plywood and held the roof together as a unit. Two disadvantages of this method: you need a strapping tool and the metal strap stretches and can break more readily than a bolt or metal brackets.

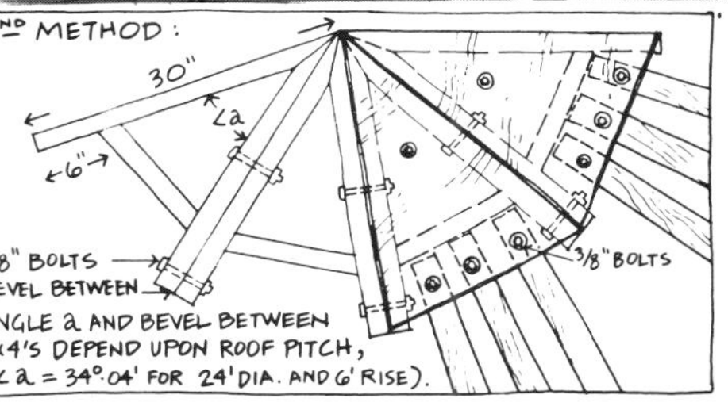

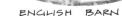

Our second method involved a center piece which looks like a nine sided coolie hat. We cut 2 x 4's to form nine (for a nine sided house) individual triangles. Onto each side of these triangles were nailed or screwed ¾" plywood triangles. We found it easier to cut the plywood into the proper size triangles and then make the 2 x 4 frame to fit rather than vice versa. We beveled the sides of the 2 x 4 where adjacent triangles met. Before attaching the plywood, the triangle frames are bolted together with 3/8" machine bolts to form one complete unit. The plywood is then attached to both sides. After securing another bolt through the center of each triangle to hold the "sandwich" together, it is ready to lift into place.

Positioning the plywood hat on those buildings not having a center post was surprisingly easy. First, we stapled about 16' of line (for those homes having a 16' peak) and plumb bob to the center of the hat. Then, we positioned 3 poles — forming a tripod — into the hat and lifted it into place. With one person per pole, and a fourth one relating the plumb to the center of the building, it was only a matter of minutes to place it in position. Once on center, we used 3/8" x 8" lag screws to secure the poles to our top framing plates. Then we merely inserted the remaining 24 poles into our plywood hat, and bolted each into place. We found it best to place our largest diameter poles on the nine corners, with the smaller ones in between. This will prevent a lot of difficulty when nailing your roofing in place....

Costs

Building costs for the octagon ran about $1000. This included 250 sq. ft. of deck and 72 sq. ft. of loft. The nonagon cost just under $1400, with 90 sq. ft. of loft and 300 sq. ft. of decking. We were able to purchase our lumber at about $100/1000 bd. ft....

Parting Thoughts

All the octagons and nonagons we've built have varied somewhat in design. Some have 12' roofs, others have 16' roofs, still others have offset center poles or 16' walls sloping to 14' and then to 8'. This variety in design has allowed us to keep the basic structure and then improvise as we like....

Hari Om,
Bryan

DOME SHINGLED WITH ALUMINUM PHOTO OFFSET PLATES

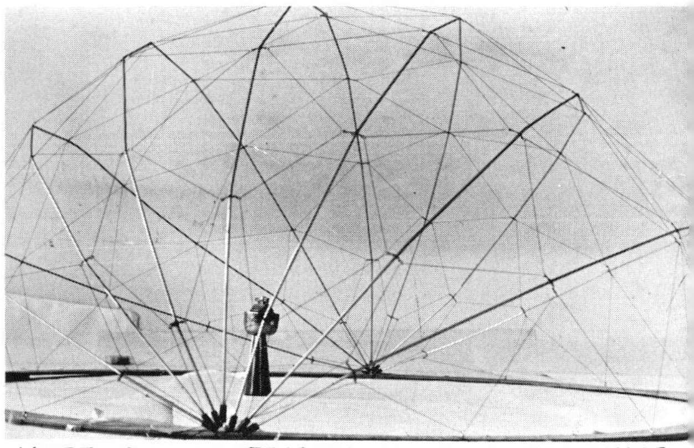

MODEL SHOWING TENSION-COMPRESSION ARCHES

ZARCH

My *Zarch,* constructed this winter in Sharon, Connecticut, is a 20 foot diameter, 5/8-sphere elongated by an 8 foot center section.

The foundation is pole construction using twelve 12" diameter cedar posts. The deck of 1 x 6 tongue and groove sub-flooring is supported by a 2 x 6 radial beam and joist system.

I made nine arches containing a total 78 segments. Each segment of the arches was 4 feet long (planning ahead for a plywood skin.) The ends of each segment were cut to fit an angle of 158°. For symmetry I made every angle in the arch the same (158°) and reinforced the gussets with glue and plenty of nails.

Seven hundred feet of double twist barbless fence wire tuned with a fence wire stretcher and held with staples provide the tension required to position and stabilize the arches. A few 2 x 4 spreaders between the arches were added to relieve my personal feelings of insecurity.

The skin, 162 plywood triangles, were cut from only 11 different size patterns. I shingled the plywood with aluminum photo offset plates, nailed developed side out so it is readable. I was able to obtain most of the plates at no cost and they have worked well. I only have a few minor leaks around the temporary plastic windows.

My *Zarch* is 14 feet high, 28 feet long, and 20 feet wide. A loft spans the northern third of the interior and forms two partially partitioned rooms beneath it. The area of the floor space is 543 sq. feet. The volume is 4,186 cu. feet. No electricity or running water and I heated it this winter with an Ashley wood stove.

This area of Connecticut is covered by strict building codes but I was able to get a permit from a tolerant yet slightly confused building inspector.

FLOOR FRAMING

Here is a variation on a dome shelter. The basic structural strength is the double triangle with a common base. The base (A) is rigid (wood), the sides (1,2,3,4) are flexible (cables). The base is placed in compression when the sides are pulled in tension. Vertices (B and C) are angular units which can readily change positions to alter designs.

A great many other shapes are possible. When joined together the basic triangular units have this shape on a flat plane:

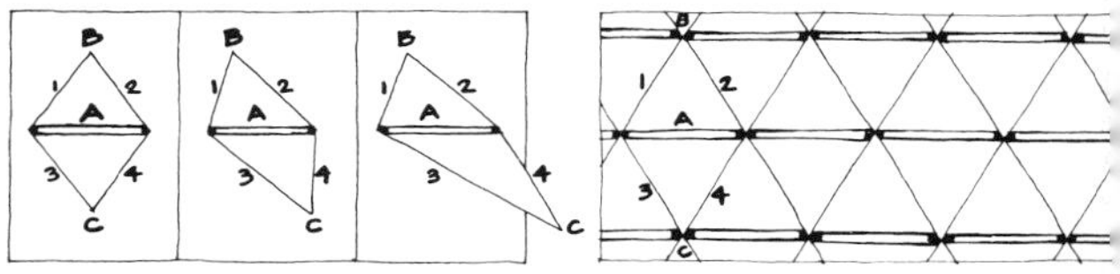

The skeleton is light weight, easy and quick to construct. The rigid compression units are merely fastened end to end by gussets to form an arch. The arch is raised and temporarily aligned. When all the arches are up, string your tension lines between the angular units and your *Zarch* is up. The tension and compression acting together keep the structure intact.

The biggest advantage the *Zarch* has over the geodesic is that it can form almost any aesthetically pleasing curved shape. New designs can be made by just varying the length of the tension and compression units. Crazy and asymmetrical roof lines can occur.

While the Albert B. Moore Associates have exclusive rights on the *Zarch* patent for commercial production, they do not object to an individual using this material for construction of his or her own shelter.

Geoffrey Bornemann
Falls Village, Conn.

The construction of this fiberglass dome, built by Jon Lazell, was described to us by one of his sons. Jon first welded a frame of steel water pipe, then infilled the pipe formed triangles with slabs of foam carved to a shallow, outward arching double-curve. He then applied fiberglass over the sealed foam, which was removed after the fiberglass cured. The dome, built in Southern California where temperatures and rainfall are mild, has developed leaks even though the skin is basically monolithic.

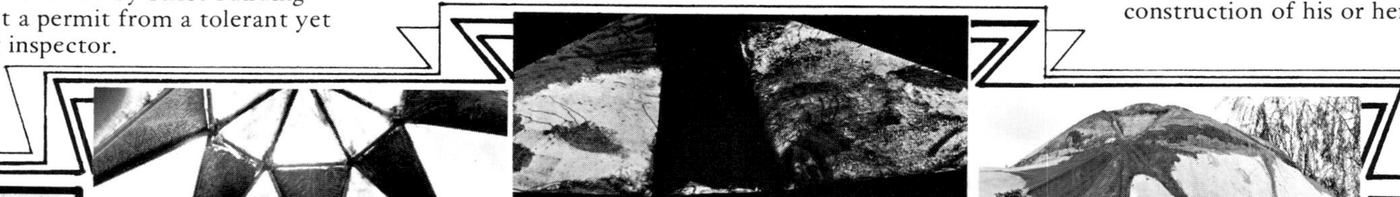

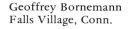

TET TRUSS

KENNER DECK

Bob Easton

Vhen Hugh Kenner asked me to help him build an conomical dome work studio on a very steep, rregular (35° slope) part of his back yard, he greed to let me cook up a new approach to a latform upon which to build the dome. It had een suggested that the dome sit directly on the lope; but the ground irregularity made a round oundation layout impractical. Also, from xperience I know that post and beam platforms n steep slopes are difficult to lay out, require igh batter boards, transits, and high vertical osts are made rigid only by extensive diagonal racing.

he tetrahedron-octahedron space frame truss as eveloped by Alexander Graham Bell always seemed o have some potential practical application to uilding. It occurred to me to "blow-up" the truss o 10' long members; that the total triangulation nd the angles involved would work very well on he extreme slope. Some basic experiments with odels (with Bill Ridenour's help) revealed the asic slope to structure relationship shown in the rawing below. The platform is a 20' diameter exagon, the two structural leg units are etrahedrons.

HOOKER DECK

Some more model work showed that extending the truss in all directions could strengthen the structure enough so the interior members could be eliminated to make a room under the dome platform (the vertical members slope but the headroom is 8'-6"). But most important, starting with the two bottom footings, the structure would determine exact footing locations as we progressed up the slope. The first members up would be light, structural beams would be laminated thereon later.

BELL'S TRUSS

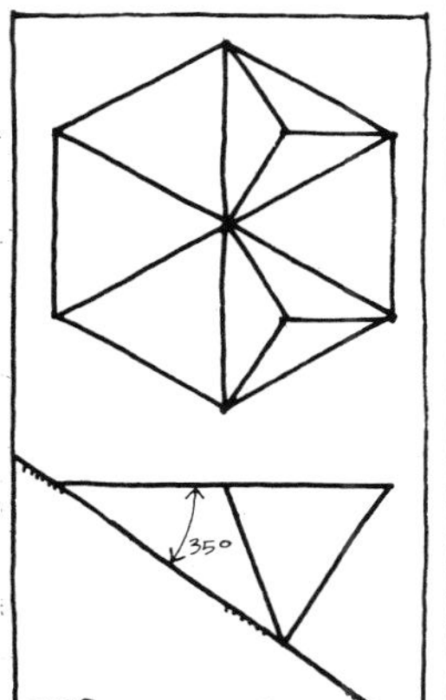

35° TO SLOPE

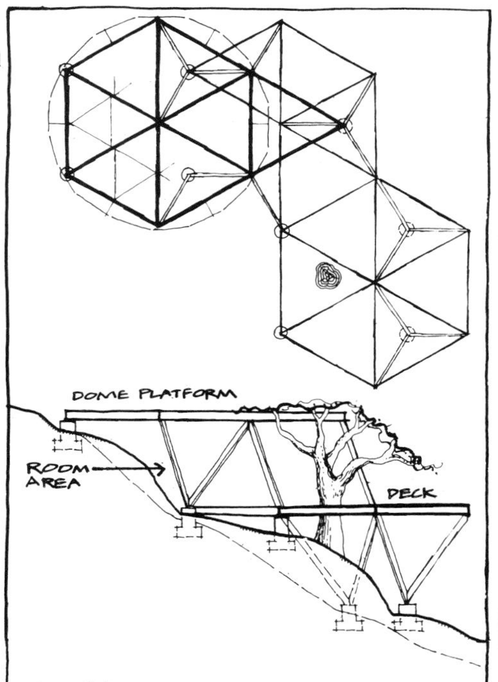

KENNER DECK

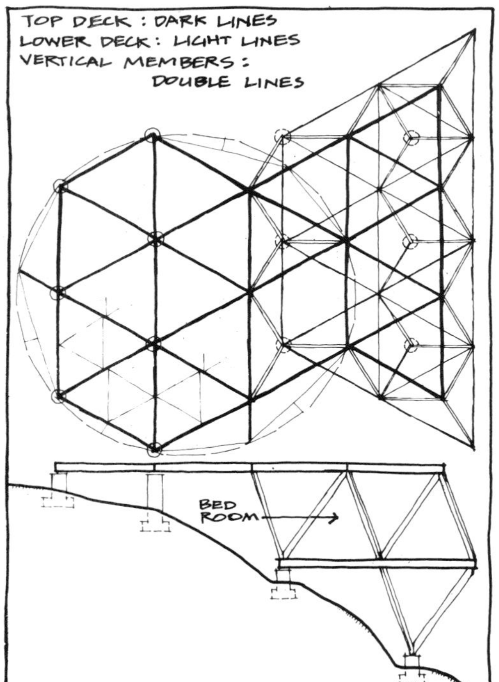

HOOKER DECK

Hugh was excited by the idea, and with Hugh, his son Michael, and Gilbert Jackson, we built the platform to expectations but never completed the dome because of Hugh and family's move to the east coast. However, a second platform is now finished. Carey Smoot and Jim Hooker used it to build Jim's house in Topanga Canyon near Los Angeles. They report the truss costs are equal to those of conventional framing even though Los Angeles County building code requirements caused joint design and member sizes in excess of the engineered design of the original platform.

These two experiments indicate the truss works structurally, and could be expanded vertically to build interesting structures on steep slopes without excavation or long posts. We built with wood because of its beauty, low cost and ease of working, but steel could be a practical alternative at least for the under floor framing. However, I urge people not to build it without an engineer's help, and top quality materials (many of the members are in tension). Write to: Bob Easton, P.O. Box 4811, Santa Barbara, Ca., 93108 for more information if you want to experiment with these ideas.

FROM HUGH BROWN'S TREEHOUSE

How about a triangulated paraboloid dome? Should be stronger load bearing than spherical if parabola axis vertical, also, two parabolas facing each other gives reflective focus to focus quality of ellipse, difference being parabolas reflect twice, ellipse once. I like idea of skinning with burlap (mentioned in "egg" in elliptical domes) & foam – how about using painted canvas? Some people do, & make boats from it – just need tough paint. This might be my first choice if it would hold up o.k. What about 2 layers of canvas, inside one sort of loose, filled with rock wool, plastic foam beads, excelsior, spanish moss or some other insulation? Also seems there ought to be a way to really open the whole dome up with lots & lots of openable windows for days of breeze & glorious weather: perhaps top-hinging a lot of the point-down triangles or something. What about a pod made of not plywood but poles pulled across up & over, & skinned somehow: see construction of shelter by Congo Pygmies also: backpacked foam dome I like: this is sort of what they do. I would think it would be profitable to investigate minimal area surfaces in general: "soap bubble shapes"; which should be fairly strong. There is a little piece of math that is very useful with these shapes: "Calculus of Variations" I think it is. Catenary of "Hyperboloid of Revolution." It is the minimal surface area between 2 perpendicularly parallel circles, coaxial, if the circles are not too far apart relative to their diameters. It would be interesting to build one somewhere with fairly stable boundary planes – say in a very steep ravine or between 2 skyscrapers or something. How about getting about 1,000,000 old bottles together & laying them up like bricks but end-on, mouths inward – small mouth bottles corked – air inside them would insulate, cement them together using geodesic frame, & up against plywood movable support as you go up – should make a very interesting effect (light & other) inside: any dump should be able to produce enough bottles, or intercept them at recycling station at glass factory. Sort of like cheap glass bricks, would be a stone dome of mostly visual holes – thick glass like coke bottles might be best....

What about domes (I'll go into town tumaro, & get some model materials, perhaps I'll find out why not) built up by triangulation, but not according to geodesic formulae – say, attach 1 end of a radial length string to center of curvature of dome to get vertices on sphere surface, and use whatever length boards you can find, 3 through 8 or 9 triangles coming together at different vertices, perhaps, really free pattern. A lot of what I have been saying in this letter is that it seems that a lot of what your book is about is stuck in arbuckyef formal formulas. Certain sections (ferro cement, pod, zomes) of considerable interest otherwise, but basically this thing ought to bust itself wide open in all sorts of directions. P. 80 "And it is in window patterns that our domes are perhaps most unique." Well, shit. Anyhow, however, these things seem to have a lot of little & not so little construction problems to be worked out, & you have made a lot of very solid progress. & on it goes. But nemm ind allat – nohow....

Can't really say where it says in your book, if it does, about dome's making you more at contact with outside & less isolated & rectilinear construction negating outside world or some such. I would say waitaminitnotsofast – to me domed shapes, with all structure perpendicular to center tend to be quite insulative spatially; unless a lot of dome is window, which makes that part of dome partially disappear – bedroom in my house here in Honduras is A-frame (approximately) with minimal vertical members (wires) supporting long horizontal palm leaves, gives very strong visual projection to view out end of room, which is of bay & mountains beyond, eye is drawn to connect with whatever is out there. Would seem to me, that geodesic-type dome would tend to make you more aware of peripheral vision occurances, & being in a place with things present all around rather than attention focused on a small area in front. Life in general in rural tropics does this over life in urban U.S.; i.e. here you are much more aware of things above & below & lateral & behind. So for this reason I would think domes would be good for getting your head in with nature, but not because 90° – base structures are more isolating – they aren't if you build so they aren't. (Many ways to do this.) But most 90° –base structures are in nature – negating areas – cities – it may be a question of association. Beware of sentimental bullshit. It would seem, though, that a basically curved structure without definite focal points or optical barriers would be more given to induce groovy states of meditation & mental floating, which is a lot of fun, & often produces good feelings and vibrations and ideas, also & general peace internal, yielding external, and understanding, or is this just all sentimental bullshit? I don't know, but I sure have a good time with my mind....

ZOMES

ZOME BUILT IN PLACITAS, NEW MEXICO BY JOHN MARTIN.

ZOME ADOBE IN PLACITAS. LEE & ALICE JOHNSON.

ZOME AT LAMA

ZOME AT LAMA

STEVE BAER

Lloyd: Building your own zome, if you were going to do it over again, would you do it the same way? What do you think about the aluminum, would a wood frame have worked as well?

Steve: Oh yeah, wood frame and composition shingles. I probably would have done it with aluminum again, just to work with it once, because I've been fascinated with it so long. Used on a large number of buildings it might bring fairly substantial savings since it's exterior interior finish, it's insulation, it's structural, it's all there. It's easy to put windows and doors in. I think that the honeycomb core aluminum skin sandwich panel is something that could make sense on a project of, you know, over a hundred houses.

What about Zomeworks. In the future, are you going to do mostly solar heating?

What we'd like to do is combine solar heating with zomes when possible and keep working in both fields. Because you can get such beautiful results with zomes if you can get the money to build them right. I just don't think there's any future for zomes as home built, one at a time things, because you can't get a crew together to build them with the expertise, and do a good job.

A lot of what's coming out in our book is discouraging people from trying to build domes by themselves.

I would add to that, discourage them. And the people who go on and do it anyway, well maybe they'll get a hell of a lot out of it. But I sure wouldn't encourage people because you usually end up with something that leaks, and is poorly designed. Like people write me and say we're going to Arizona or Wisconsin to start a commune and we want to use zomes. And I say don't use zomes, go to where you're going and drive around and find some local buildings you really dig and build them just like the ones you dig there and people will know how to do it and how to help you do it.

Instead of coming in with a preconceived notion.

No, you can't do that.

Yeah, I learned that the hard way.

But if you can get a team of people together who build domes or zomes, etc. God, they can really do beautiful stuff after a little experience.

Do you think that they're going to cost as much as a conventional building?

I think they're going to cost as much as a professional custom built house. However, if they are produced in numbers for a development or something they could be somewhat cheaper than a conventional house, if they were done on a big scale. Which is obvious because there is a savings in materials, etc.

OK. What about the water heater?

We're making modular panels for water heaters, they're 4 ft. sq. It's an anti-freeze system that blows through collectors and then into heat exchanges with an 82 gallon conventional water tank. When we finish making some improvements we'll be able to ship them out....

Now what's the best thing for someone who's interested in solar heating to do? Either they're building a new house or want to hook up a solar heater to an existing one.

They should contact us and we can advise them ho to do it or we can actually do a design. But it's important we be contacted as early as possible.

What if some guy writes from Oregon, can you give him solar advice?

We can, and quite quickly, too. We have climatological data for the entire U.S. and we can get pretty close. We'd probably charge a minimum of maybe $25 for an hour or two of simple design. Then for a complete analysis, like we have five consulting jobs now, we usually get a couple of hundred bucks. But the earlier they talk to us the better; otherwise they get set on some plan that doesn't take solar heating into account....

What do you think is the best book on solar heating?

Still Farrington Daniels (*Direct Use of the Sun's Energy, Yale University Press, 1964*).

Holly told me you were writing a book and that chapters have been published in an underground paper.

Yes, it's cookbook style and about 8 or 10 chapter have been published in the *Tribal Messenger* in Albuquerque. But I just haven't had the time to polish it up for a book...

Zomeworks Corporation, P.O. Box 712, Albuquerque, N.M. 87103.

ZOME

, SKYLID AND DAY CHAHROUDI

LARGE DOORS IN DOWN POSITION

ZOME FROM WINDMILL TOWER

The house consists of 10 exploded rhombic dodecahedra stretched and fused to form the different sized rooms...is heated by the four drum walls, four skylights and two small wood stoves.

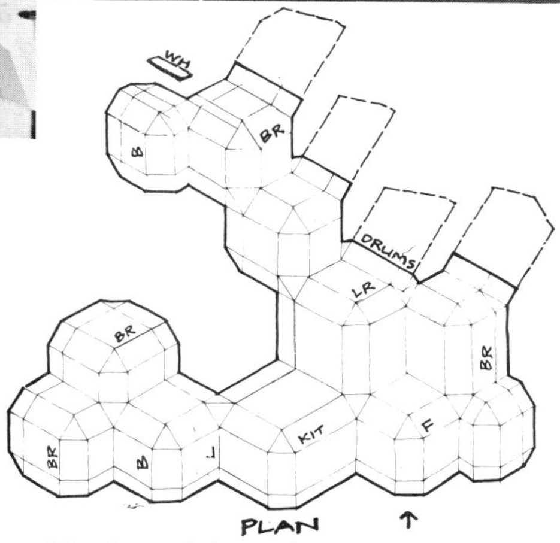

PLAN

SOLAR WATER HEATER

Steve Baer

ıe Zome went up during the fall and winter of '71 - '72. is constructed of 3" thick urecomb paper honeycomb re panels with urethane foam filling the spaces between e honeycomb. The skins are both .024" anodized uminum bonded to the honeycomb with contact cement.

ıe panels are joined together by aluminum strips pop veted to both sides. The strips are sealed with silicone alant. We have had some leaks where the zomes connect t none elsewhere. The honeycomb core panels are nice work with — they are light, very strong and have terior and interior skins already there. We expected e inside skin to be unpleasantly metallic but the satin ıish anodize is very nice — the exterior is a bit bright d harsh.

indows, skylights and doors are easy to install. You nply cut an opening in the panel with a skill saw and ıck them in. The panels are so strong that you can open the structure almost anywhere.

ıe retaining wall around the house was built by Clark chert who is working on a mosaic transformation to ver part of it. The wall is slanted at 54 ¾° — the angle the side of an equilateral pyramid.

ıe inside walls are made of adobe. The cabinets are made rough cut lumber and plywood. The adobe work and binet work were done by Richard Kallweit and Dick enry.

ıe house consists of 10 exploded rhombic dodecahedra retched and fused to form the different sized rooms. ıe zome is heated by the four drum walls, four skylights d two small wood stoves. The drum walls consist of gallon drums filled with water and placed behind a uth window in front of which are large doors made of neycomb panels. The doors are open during the day to llect sun and closed at night to conserve heat. The drum alls are extraordinarily efficient solar heat collectors nce the collecting surface is never above 100° F and e direct sun is augmented by reflection from the open ors.

ı clear days during December the drum walls collected arly 1400 BTU's per sq. ft. of glass.

There are no doors in the house. We planned doors, but now don't feel we want them. The geometry of the zomes gives a great deal of privacy without doors.

The two small pot bellied wood stoves are in the children's rooms at the north end of the house which we use for heating during extended periods of cloudy weather. During the winter the house usually stayed between 63° and 70° F without any fires, but did swing to such extremes as 56° F and 75° F. The house warms slowly during sunny weather and cools slowly during cloudy weather. This is because there is a great deal of thermal mass in the house, material that will not change temperature rapidly. The thermal mass is made up of drums filled with water, the adobe walls that divide the rooms from each other and the slab floor.

During the summer we use the thermal mass to great advantage by opening windows and top vents at night and closing them during the day. The house is then cooled at night and remains cool during the day.

In the winter the temperature will drop two or three degrees per day during cloudy weather. We wear sweaters when it gets chilly. All winter we burned less than a cord of wood in the wood stoves and in Albuquerque it was an exceptionally cold winter.

Three of the four skylights have skylids underneath them. The skylid© is a solar heating device we manufacture which opens during the day to let sunlight in and closes at night to prevent heat loss. It operates on temperature differentials and requires no outside energy. We plan to install more of these in the house. The skylid has a manual override that allows you to close them in the summer to prevent excessive heating.

The house has approximately 2000 ft.^{2} of floor area and only 400 ft.^{2} of solar collector which is about ½ the conventional ratio of collector to floor area. The excellent performance of the system is the result of the high efficiency of the drum walls.

The hot water for our bathroom is supplied by a solar water heater. We are now installing another water heater to heat the water for the kitchen also. The water for our house and the solar heated house next door is pumped by the windmill. Our consumption of electricity, which we buy from the power company, averages about 5 KWH/day.

We enjoy living in the house very much (what builder will say otherwise?) In the winter it is bright with sunshine and in the summer cool and shady — the high zome ceilings are very nice and the geometry of 120° angles is both interesting and efficient.

We welcome inquiries about solar energy. We do design, fabrication and consulting and also sell various publications concerning solar energy.

WOODEN DOME

Even after some years disillusionment building various domes I tried one more. This one was an attempt to blend the dome concept and wood construction. A final dome experiment. In some ways it worked, but in most it didn't. A description of the construction follows, because if someone is intent on having a dome, the framework and window details here are better than anything in *Domebook 2.* However, because this dome photographs well I'd like to point out its drawbacks before giving construction details. If you're thinking of building a dome home, I hope you'll think about the following things I've learned the hard way:

1. Building a dome, you are locked into abstract math, the computer that worked out measurements and angles. The frame is relatively quick and light, but the rest is hard. Putting in doors, windows, the kitchen, trying to fit 90° furniture against changing-angled walls: time consuming, frustrating. You can't innovate, improvise as with perpendicular walls. The polyhedron you've started with is pure form, unyielding to changes.

2. In the midst of cutting up 10' 2x3's for struts, 14-16' 1x8's for panelling I realized these short pieces would not be salvageable, unless for another dome. Wood can be used over and over, but each time it's shortened you diminish choices of the next builder.

3. I used shakes for the exterior because I found redwood logs on a beach and split them. However, although they look good, it was difficult and wasteful: shakes are made for flat surfaces, fitting them on a building with over 100 triangular planes and changing angles is maddening and difficult. Then, as you get up around the top of the dome you must cut shakes into pie shapes to fit the tightening circle: wasteful. In addition the slope at the top is too shallow for shakes to keep wind or water out.

4. I couldn't use salvaged doors and windows for entry and light, had to buy new plexiglas. Now I want more doors opening out, but just don't have the energy to fight the shell and weaken it in framing them in.

5. What generally excites people at first about a dome is the openness, the purity, yet this is precisely what stops me from making subdivisions, which would look awkward. Thus I ended up with a one-room house in which all sounds, smells and vibes are on top of everyone. O.K. for one person, or someone not needing privacy. After 4 years living in domes, the excitement of moonlight through overhead windows has worn off. Now we'd rather step out to look at the skies, and often yearn for a low-ceilinged wooden room, fireplace with bed nearby.

6. I'm afraid the dome won't last long, because the entire building is roof, it gets full weather treatment over entire surface, unlike a building with vertical walls and roof acting as umbrella for rain and sun. Domes require more maintenance and have a much shorter life span than rectilinear buildings.

Is it all bad? No, although it has many difficulties as a home, it makes a wonderful space for working in, a studio (or a large space to be spanned, such as an auditorium or theater). In fact, we're putting together this book in the dome, so it does have its functions. However, if you are considering building a dome, I urge you to read through the 32 pages of dome building information in this book.

FLOOR FRAMING

Foundation: approx 12" diameter "Baxter Poles", the green treated poles which are supposed to last for 75 years. I think next time I'd use concrete sonotubes instead as you know they'll never rot.

Flooring: 2x8" dry tongue and groove white pine, layed in radial pattern. Bang it in with sledge hammer from the end for tight fit. To darken it I used a weed burner flame thrower (scary device), then wirebrushed. Opposite effect of stain: hard grain dark, soft wood lighter. Much work, maybe a full week in all, but beautiful. Improves with age. No varnish, sealer or wax. Just wood.

HUB

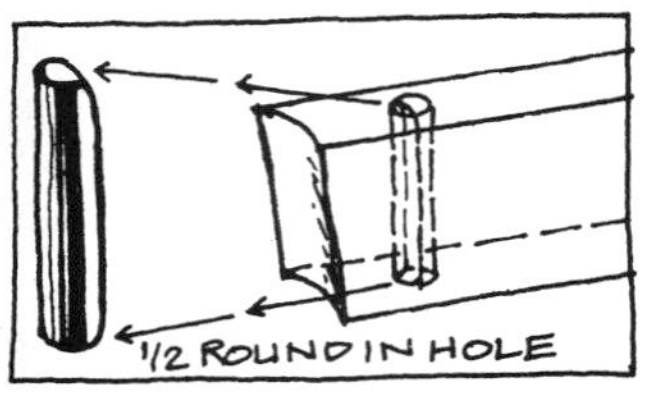

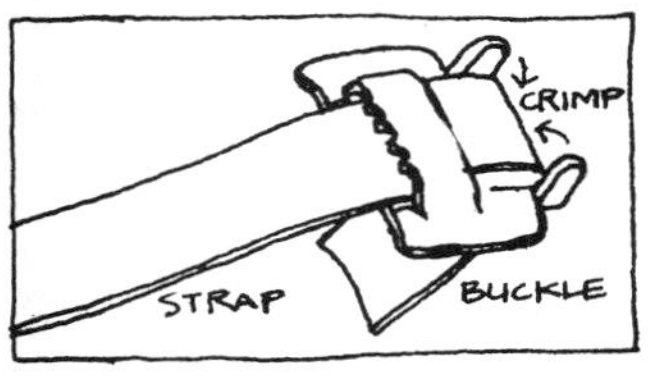

Pipe: any diameter available from Kilsby Tube Supply, San Leandr Calif. However 2 7/8" OD water pipe is cheaper, available anywher

I followed the experience of Jeff Morse and used a Band-it strapper and buckles in this dome, which is stronger than the strapping we used in *Domebook 2.* Band-it Co., 4799 Dahlia St., Denver, Colo. 80216. Strapping turns out to be expensive. If I had it to do again I'd probably use the plywood hub from *Domebook 2,* maybe with wooden dowels instead of bolts — An all-wooden structure.

Strap: bought 3 rolls about 30 lbs. each, had about ½ roll left over. So used about 75 lbs. ½" .025 strap. $97.00 worth.

Buckles: about $50; *tubing:* about $95.

Struts: This assumes an understanding of the information in the Pacific Dome section of *Domebook 2* (pp. 21-22). Table is for 4-frequency alternate hemisphere, 2"x3" struts.

Strut	No. each	Chord factor	x	Radius	=	Chord length	-	Subtract hub	+	Add for rounding operation	Strut length	Axial angle
A	30	.25318	x	180"	=	45 9/16"	-	2 7/8"*	+	3/8"**	43 1/16"**	82¾°
B	30	.29524	x	"	=	53 5/32	-	3 1/8	+	"	58 13/32	81½°
C	60	.29453	x	"	=	53	-	"	+	"	58 9/32	81½°
D	90	.31287	x	"	=	56 5/16	-	"	+	"	53 9/16	81°
E	30	.32492	x	"	=	58 1/2	-	"	+	"	55 3/4	80½°
F	30	.29859	x	"	=	53 3/4	-	"	+	"	51	81½°

Essential notes on this table:

* I rounded strut ends to fit perfectly around pipe. If you don't do this; adjust accordingly.

** These lengths are based on 3 1/8" outside diameter tubing, for all but the pentagon vertexes, where I used a 2 7/8" OD tubing. I wanted 2x3 struts to pack tightly.

Windows: here is what I did with my windows:

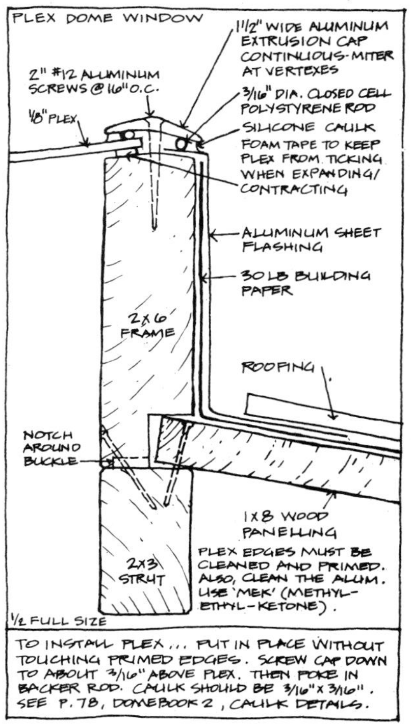

ROUNDING STRUT END

The 2x6 is to get the windows up above shakes. If you use plex, the silicone sealant (everything cleaned and primed) with aluminum extrusion is the best method. However it's a very difficult, painstaking process. As careful as we were I still have a few leaks, had to get up and caulk about three times last winter.

I used 1/8" plexiglas from Swedlow, Inc. in L.A. With some other people building a dome, we bought a 51" wide roll of 2nd quality plexiglas, about 2400 lbs., cost about 35 cents per sq. ft. I got some aluminum extrusions from Jeff Morse and followed his window details (*¾ Dome* in *Domebook 2*) with a few changes: I used a back rod for spacing, and put some foam tape between the plex and wood so it wouldn't tick during expansion-contraction.

A friend who manufactures skylights loaned us his large machine shop for a night and we were able to use a carbide tipped blade especially designed for plexiglas to cut all our window pieces. Without the use of this shop we would have had much trouble handling and cutting this much plastic.

Plexiglas is the best plastic window material, but you must be meticulous in measuring, cutting, cleaning and caulking. Its disadvantages compared to glass is it scratches easily, collects dust and probably will turn yellow in ten years. Working with it you must keep in mind that it expands and contracts tremendously wit temperature changes. Thus give it room, never put two pieces of plex next to each other, or to another material where it will bind when expanding: if you must bolt through it, use oversized holes. Jeff Morse discovered that you can use silicone sealant for a hinge.

HYPAR DOMES

Peter Calthorpe

A lot of work has been done with the hyperbolic paraboloid (The *hypar* is a minimal surface defined by straight lines between edges which are out of plane.) in cement and thin shell concrete, but it seemed constricted compared to the possibilities of such a free form material. In wood the form takes on a warmth and the sense of a boat hull losing its abstract and hard associations. On one hand the thin diaphragm of the hypar, one layer of common one-inch siding and a layer of 3/8" plywood, spanning thirty ft. with no joists and minimal edge members which receive no bending loads seems very economical. Yet the cost of clear dry one-inch lumber compensates for savings on joist material and the motivations become mainly aesthetic.

The hypar roofs work well with pole construction, which is a very satisfying system in every way. A good way to plumb poles is to hang two plumbobs off the top at right angles from the center of the pole and line the pole up with both of the lines. Because the diaphragm intersects the edge members at varying angles the beams have to be shaped to that constantly changing angle. We did this with a draw knife and a power planer. When the edge beams are all in place temporary joists are put in parallel to the edges. One by four rough sawn redwood with a tongue and groove was then placed on a diagonal and curved over the joists. The plywood is then glued over the siding and stapled on to hold the shape of the hypar and carry the tension. A tie rod holds the two diagonal poles together and the whole thing acts somewhat like a bow and string, the curve giving all the strength. It was amazing to find the diaphragm stronger in its final form after we removed the joists. There is a remarkable lightness and organic feeling to the roof; natural curves and flexible strength making it a thin wooden shell.

Unlike the normal hypars which are balanced with two corners up and two down, these were tilted to give a horizontal perimeter. This caused some asymmetry and less balanced loads but no significant problems. Alan Strain, one of the first people to experiment with wooden hypars, built a beautiful home with balanced roofs and filled in the height at the perimeter corners with glass. There is an extremely helpful section in the *Douglas Fir Use* book on this type of construction.

This structure was an attempt to use domes to create a complete house. Previously we had used domes as single use spaces; the sleeping units at Pacific, exhibition pavilions, children's play rooms, and classroom structures. The problem was to segregate the interior without destroying the beauty and unity of the form. One solution is to cluster domes for differing functions but this seemed to sacrifice the material economy of the dome by magnifying the amount of exterior surfaces (and the accompanying sealing problems).

I fear however that the solution we tried involved an equally frustrating sacrifice, that of structural stability. The idea of splitting the dome and juxtaposing the sections in different ways to allow vertical faces and a natural interior division was very attractive. The vertical faces allowed simple doors and windows (in domes) as well as very interesting shapes and contrasts. Much of the beauty of the house is in the forms created by the vertical faces. And the divisions of space resulting from the different dome sections seems effective and not damaging to the flowing type of spaces. However, the structural aspects of the complete dome were lost and compensation was steel arches to strengthen the perimeter of each section. (Expensive and hard to work with.)

The exterior skin is urethane foam with a special rubber based coating for protection and a color coat of latex. The foam had the unexpected and pleasing quality of reproducing the surface it was sprayed on, so the lines of the interior redwood paneling and even some of the grain shows and gives an interesting texture. The paneling inside is definitely the most beautiful interior for domes and the interior is always the most beautiful aspect.

The dome sections themselves are squashed elliptical 'zafu' quarters stood up, with a six freq. octahedronal breakdown.

The amount of work involved and the problems encountered convince me that domes are more difficult to build and make work than conventional structures. But ease is not the only criterion for design of a house. A fascinating idea at this point would be the combination of dome sections with rectilinear structures to perhaps integrate the best of both.

RED ROCKERS

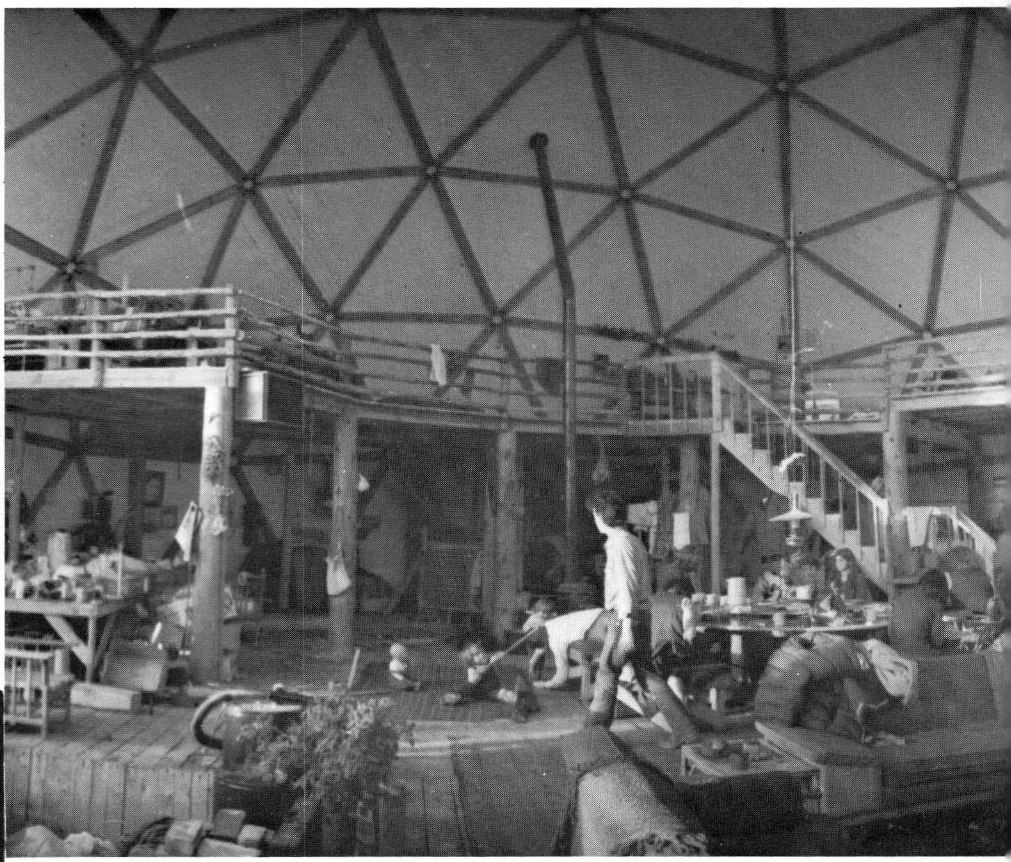

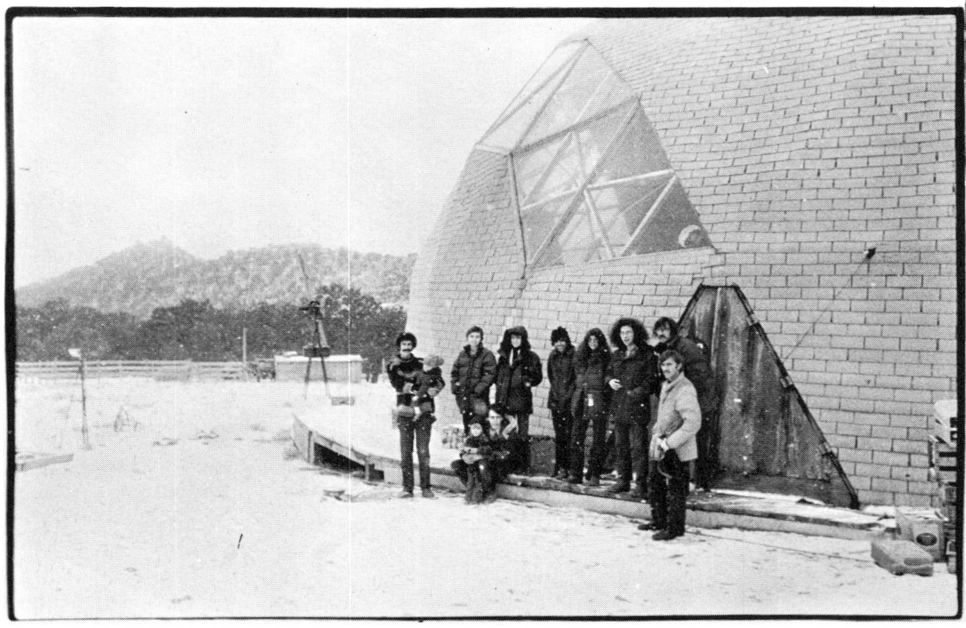

We (about 11 men & women) built our 60 foot dome three years ago for about $2500. What we wanted was a place to all live together in order to change our lives and ourselves.

We wanted to create a structure that didn't remind us of anything — a new kind of space in which to create new selves. We also needed a space that was large enough to house a lot of people and their trips — a space that was voluminous enough to assume different shapes as we changed and our needs changed.

The dome was the most beautiful in the very beginning — empty. From the Red Rocks (our favorite lookout point and the landmark on our land from which we took our name) it looked like a huge cut wooden gem. Then WE moved in and took over. I think there were thirteen of us the first night we slept in the dome and we were overcome with the massiveness of the space and the weird accoustics. We went immediately about the business of tacking up our pictures of Meher Baba and George Jackson. One week after we moved into the dome we had a huge Thanksgiving party (now an annual event) and there were 180 people holding hands in the pre-dinner circle.

The first winter we all slept in a circle along the wall of the dome. By the second winter we had added a kid's room (we started having kids) and a big mezzanine style sleeping platform that extends about three quarters of the way around the dome's circumference. The platform improved our lives a great deal because it is nice and warm there in the wintertime and most of the beds have a beautiful view through the arc of windows that serves to heat the dome on sunny winter days. Two babies have been born on the platform, one on the main floor, and last summer we delivered one in a tipi right behind the dome. We have spent a lot of time and energy getting our kitchen together. We built a big brick range with lots of gas burners and a cast iron griddle that we use every morning for pancakes, french toast and eggs from our chickens. We built counter tops and cabinets with tongue & groove oak and pine that we salvaged from an ancient ghost town dance hall nearby.

When we first put in our arc of windows we covered it with vinyl. Last spring we replaced the vinyl with plexiglass. We used fancy silicone sealant for the windows which was very expensive and totally ineffective. Last week we re-sealed the arc windows with asphalt sealant and right now we are (incredibly) without leaks. Plexiglass is o.k. but it is very expensive and scratches too easily.

In the summertime most of the Red Rockers move out of the dome into tipis or temporary shelters. This summer many of us are moving out and won't be moving back. We plan on building four small houses this summer. These will be sleeping spaces — small shelters designed for one, two or three people, probably without kitchen facilities. After three years of living in a heap, most of us have decided that in order to keep becoming new people, to keep growing and changing, we need more privacy. We are still a communal family that wishes to hold together but we need a new kind of shelter for this period in our growth — shelters that will be places to make love in and argue in, to play music and write and worry and think and decide what the next step is to be.

The dome will continue to be a center for us. We will continue to eat together here and meet together and hang out. The platform will be converted into a sewing area, general crafts area, etc.

WHAT WE LOVE ABOUT OUR DOME...
IT'S BIG
IT'S A WONDERFUL PLACE FOR PARTIES, MEETINGS, AND WEDDINGS
KIDS LOVE IT!
LOTS OF ROOM TO CRUISE AROUND AND GET INTO TROUBLE
LOTS OF PLACES TO CRAWL INTO AND HIDE
IT'S WARM
ONE ASHLEY STOVE HEATS THE WHOLE THING

WHAT WE HATE ABOUT OUR DOME

IT LEAKS
AROUND THE WINDOWS

T ATTRACTS FLIES !
(SOME KIND OF GIANT WINDOWSHADE SHOULD SOLVE THESE)

NO MORE
ARRY McGUIRE
RECORDS !
I DON'T WANT TO WAIT TILL EVERYONE'S ASLEEP
WHO TOOK MY EMKO ?
DID YOU HEAR THE ONE ABOUT THE GURU'S DAUGHTER ?
IF ONLY I WERE POPULAR
SIGH
GRUNT !
CHAUVINIST PIG !
WHARAA
WHERE'S THE TIGER BALM ?
TOO MUCH NOISE AND RAK-A-RAK
- HARDLY ANY PRIVACY

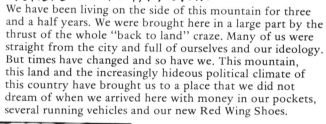

WHERE WE'RE AT NOW...

We have been living on the side of this mountain for three and a half years. We were brought here in a large part by the thrust of the whole "back to land" craze. Many of us were straight from the city and full of ourselves and our ideology. But times have changed and so have we. This mountain, this land and the increasingly hideous political climate of this country have brought us to a place that we did not dream of when we arrived here with money in our pockets, several running vehicles and our new Red Wing Shoes.

We decided to build a big dome before we ever saw the land we were to build it on. We bought lumber at the lumber mill and shingles at the shingle store. Now, with far less money and far more sense we think of building our sleeping shelters out of WHAT IS HERE. One shelter will be built out of stone (which is plentiful here) others will be made from rammed earth, adobe bricks, etc. Floorboards, doors, moulding etc. can be gleaned from old abandoned buildings. Glass is easily obtainable in these parts from friendly commercial nursery owners who are currently replacing all the glass on their structures with corrugated fiberglass. Lumber is prohibitive. Even here where it is plentiful it runs 12 cents a board foot for crappy unseasoned stuff. Used lumber is seasoned lumber. We have just finished building a garage/shop. The floor, sub-floor, and roof are all built with free second hand materials. The lumber that we did buy for the building was extremely low-grade but made fine siding because we tongue & grooved it ourselves. (Simply a matter of buying an inexpensive moulding head attachment for our table saw). The fancy sliding garage doors, which are our pride and joy, we bought for $20 from a garage in town that was "modernizing."

We have learned that with less money and more scavenging we are building better and better structures. This seems to hold true in all areas of our lives. In terms of food, shelter, health care, education, communication: the less we are dependent on the mainstream culture for materials, goods and information the more high quality these aspects of our lives become.

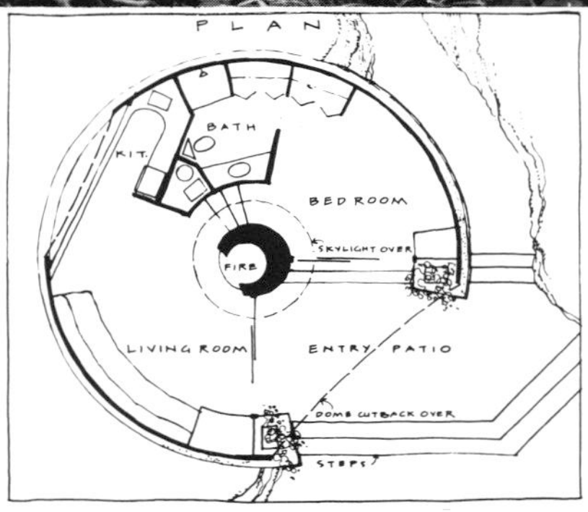

IVY DOME

This concrete dome, covered with ivy and blending beautifully with surrounding land was built by Burton and Katherine Wilson, and designed by John Watson. It's a rare accomplishment to have a home enhance rather than mar the landscape. Here is Burton's description of building the dome:

Enclosed are some photos of the dome as it is today. Just couldn't get back any farther for pictures because the brush has grown right up to it.

The slab is standard with stirrups on the outer edge and extra re-enforcement in the center for the fireplace column. Built a "rubble" wall around the perimeter. This is a standard way here to build water tanks, etc. Just build a form (I think ours was 14" thick) place rocks in it and pour concrete over that, forcing it between the rocks. This wall is three feet tall and on the top of it we placed half inch diameter steel pins that protruded about four inches upright above the top surface of the wall. These were located at equal distances (about 3' apart) and were used to place the pipes on that would form the arch.

The pipes were one inch water pipe that we had formed to an arch at a steel yard. These were placed on the pins and laid on the fireplace column that had been left at 13' for this purpose. Then laid an 8' diameter hoop (also formed at the steel yard) on the top of the pipes and welded it to each one. Now with an arc-torch cut off the inside spokes as they were not needed — it was now free standing. To hold the pipes in position we laid lengths of half inch cold rolled steel on them and welded it to each pipe. Laid four inch fencing on this and tied it on with wires.

On the under side tied sheets of expanded metal plaster lath. Next cut rope in to one inch lengths, shredded this and mixed it with mortar. We now *scratch coated* this on to the lath. The rope had been added to enable the mortar to stay on the lath and not just fall off. When it was set we sprayed it with gunite (the way swimming pools are made) inside and out. Waterproofed it by mopping it with tar over glass-cloth. On top we laid a four inch thick layer of concrete and vermiculite, held in place with chicken screen until it set. The vermiculite was used to make the concrete lighter with an insulation against the sun. It is the finished roof and has never been touched since. We used plexiglass for the windows in the sky-light around the fireplace column.

I don't think that I would make any changes in it were I building it again. We might have 'over built' a bit on the steel web but didn't know and couldn't find any one that could help us as problems arose....

Burton Wilson
Austin, Texas

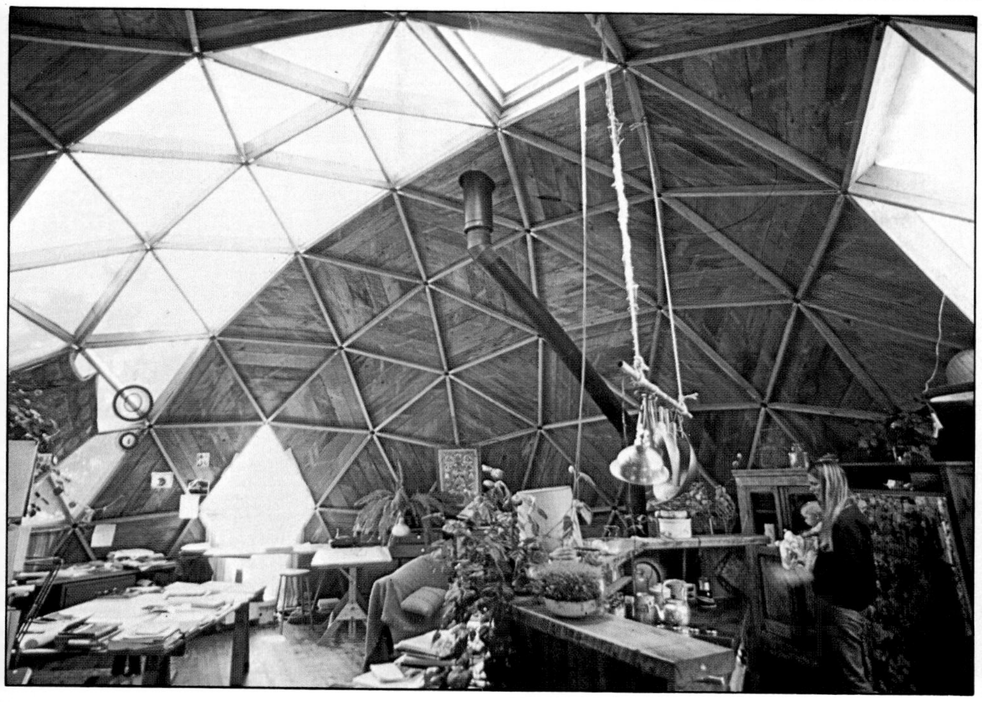

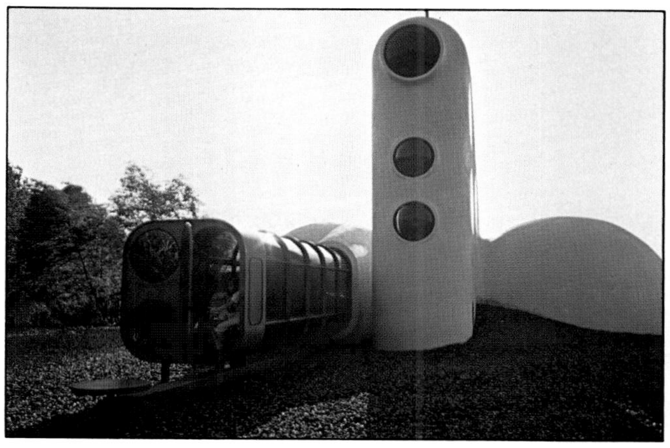

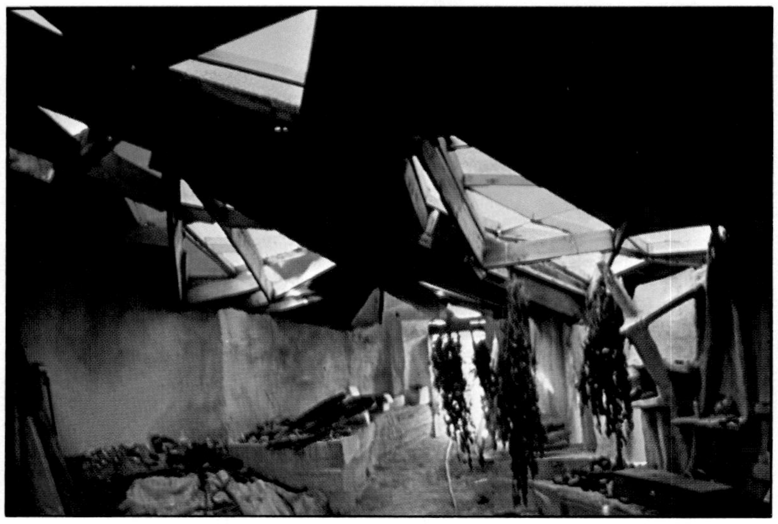

...like I found it in 2001, where the dude had finally split out of the satellite and was heading towards Jupiter, just as he was coming in, what they had done was they had used different types of film, infra-red for one, and just taken a plane and flown over Grand Canyon at a high speed, low, what it created you know, is in some respects synonymous to what the house is, you know, and certainly our cell structure in our body is synonymous with that...

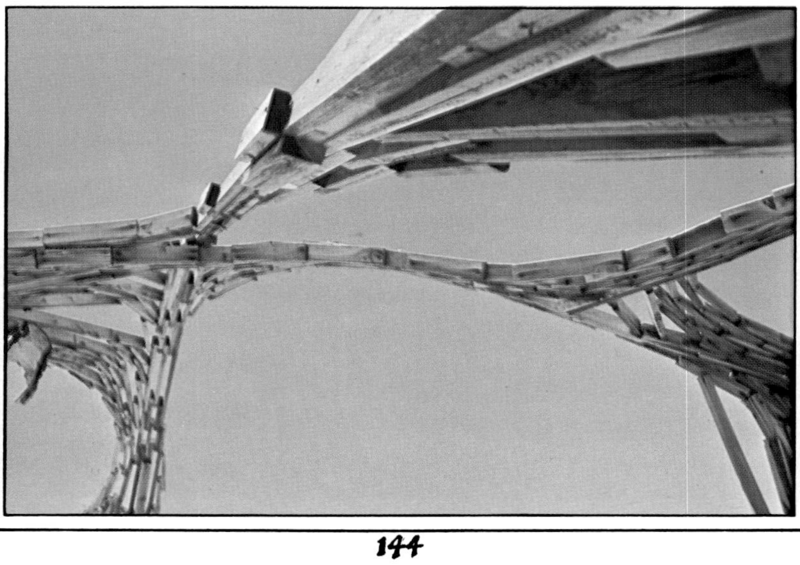

FREE FORM

...letting the light create the illusion that if you started looking at something or if you were staying in a room long enough, or however, you would start moving through it, it would take you on to other things.

...it's like three dimensional sculpturing, you know, we just got into building a house out here that's like jewelry....

...Building this house was more of like feeling where you went as you started working with it, you know, the material and just playing it from there. Visual sort of feel, there was no preplanning....

ROOF FRAMING

THE ROOF

Bob De Buck and Jerry Thorman made jewelry in Truchas, New Mexico, saved some money, got some land farther south and started to build. After several years of highly imaginative and outrageous work they ended up with this 5,000 sq. ft. free form house. Now, four years later, Bob has moved on to Oregon, and Jerry has built a smaller house nearby using many of the unique principles employed in their first adventure. The house is for sale right now, anyone interested can write: Free Form, P.O. Box 3255, Albuquerque, N.M. Photos can only partially convey what it feels like to be inside. On the next page is an interview with Bob De Buck. Below are Bob Easton's impressions.

RUTH AND ASA; BELOW, BOB

Day Chahroudi invited us down from Colorado to see the house, he was living there in an extra room. We called Steve Baer from Santa Fe, to ask directions, and he said something like, turn off the freeway, drive two miles, look to the right and you'll see what looks like a huge junk pile... when we saw it, we couldn't believe it; the place from a distance was totally outrageous, but as we got close, its magic began.

I've been conditioned to admire craftsmanship, order (studs 16" on center, plumbed). I've built domes, which required accuracy to 1/32 of an inch. I've always thought that order and competence were synonymous with craftsmanship, that structure grew from discipline, but on the other hand, I never believed I was uptight or over-trained. I've always improvised and tried to build imaginatively. As we walked up to the house I saw Day's head pop up out of a hole in the roof: I was deconditioned in an instant. Bob and Jerry had sacrificed every sacred cow of architecture and building.

The place is a maze of wandering spaces (you never know exactly where you are) gathered around a central shaft of four or five poured concrete columns that form a flue for the fireplace and Bob's shop furnace. The roof beams radiate out from these columns, most are old bridge timbers bought from Steve Baer, but the truss spanning 45' over Bob's shop is made of car wheel rims welded with pipe. Bob says they built the roof, then decided where to put the walls. The roof framing between the beams and the walls are built with scrap lumber and plywood scavenged from apartment projects.They would gather scrap until they had a large enough of a pile to do a room...and then new rooms would appear over the weekend (the roof now covers 5,000 square feet of area).

The *organic* effect they wanted was accomplished by plastering the framing. The wood is covered with tar paper and chicken wire and plastered with 2 coats of gypsum plaster. The roof? Sealing problems? The average rainfall in their part of New Mexico is only 6-10 inches per year, so the problem can never be *critical*...how could a million joints ever be sealed; what does it matter if there are a few leaks, nothing else in this house fits conventional wisdom...the floors all slope, there isn't a straight up and down plumb wall in the whole house, each door is a different parallelogram, even the kitchen shelves slope. The colored skylights cast intense light across the curving walls and floor...we woke from our night's sleep bathed in pools of color from the morning sun.

But there are ideas here beyond the appearance of the house. Bob and Jerry found and used scrap lumber, materials and nails that otherwise would have been burned or dumped. The laminated short 2x4's used to build rigid flowing curves is an idea that could be developed by people interested in free-form structures. The idea of the roof as a tent, with walls independent structurally from the roof, able to be moved. But maybe the best idea here is the pure outrageous freedom of the builders, an audacious exploration past all the rules, conceptions, boundaries of even what passes today as the frontiers of building research.

The place was unfinished when we were there, walls and skylights were open. Basically, it's a huge timber and plaster tent, the roof and walls *stretch* between the ground and beams.

I've never experienced anything like the freedom of the place...the plan and structure, the color (I never thought I would see polychromatic plexiglass ever used so it looked *right*) and the excitement of being in the changing light. Life in a desert camp...

BED ROOM

JERRY'S HOUSE. BUILT AFTER THEY DID BOB'S PLACE. THE TWO LOWER LEFT COLOR PHOTOS ARE OF JERRY'S.

Jewelry can be like a man creating a dream. An illusion of completeness, of fulfilling, exasperation. Especially in art. And specially in sculpturing. Because it is so three dimensional. And this is just related into a larger piece of jewelry, or a larger piece of sculpture, but you just happen to live in it, is what it amounts to.

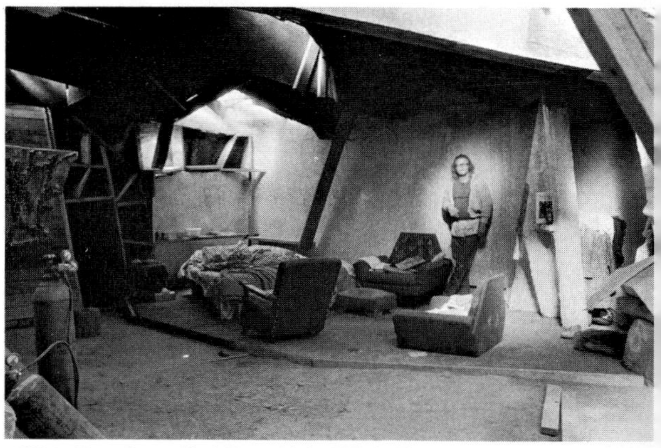

LEFT: ROOF TOP, ABOVE: JERRY, BELOW: KITCHEN SINK

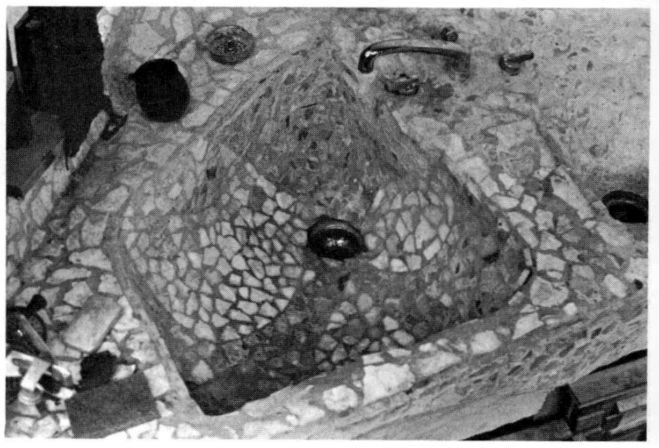

WHEEL RIM TRUSS SPANS 45' OVER BOB'S SHOP. 2" PIPE WELDED TOP AND BOTTOM.

BOB SAYS: TOOLS NOT TO HAVE: STRAIGHTEDGE, SQUARE, LEVEL, PLUMB...

Lloyd: What about the adobes, you bought a bunch of adobe blocks, and the wind, the rain destroyed them? Were you going to build the house out of adobe?

Bob: Yeah, we bought 10,000 adobes and we lost about 8,000 of them in a rainstorm. We were going to use them for the walls, a lot of the back-up walls.

L: Did you have a house like you built in mind at that time?

B: No. Building this house was more of like feeling where you went as you started working with it, you know, the material and just playing it from there. Visual sort of feel, there was no preplanning. Today if I started working on it I couldn't tell you exactly where it was going to go...it's like three dimensional sculpturing, you know, we just got into building a house out here that's like jewelry. A lot of it is like sculpturing. Using a three dimensional media, which is sculpturing, and by using those three dimensions, creating a fourth dimension of like motion, movement, through the sculpture...letting the light create the illusion that if you started looking at something or if you were staying in a room long enough, or however, you would start moving through it, it would take you on to other things.

L: And how does that relate to your jewelry?

B: Jewelry can be like a man creating a dream. An illusion of completeness, of fulfilling, exasperation. Especially in art. And specially in sculpturing. Because it is so three dimensional. And this is just related into a larger piece of jewelry, or a larger piece of sculpture, but you just happen to live in it, is what it amounts to.

L: Were there other influences besides the jewelry?

B: OK, let me put it this way, the inspiration, like as we move along through it, like I found it in *2001*, where the dude had finally split out of the satellite and was heading towards Jupiter, just as he was coming in, what they had done was they had used different types of film, infra-red for one, and just taken a plane and flown over Grand Canyon at a high speed, low, what it created you know, is in some respects synonymous to what the house is, you know, and certainly our cell structure in our body is synonymous with that...

L: When you guys would go out to work on it, did you work in spurts? Did you just work when you were feeling good? Or did you go out there each day without having anything in mind of what you were going to do?

B: No, when I say that we didn't have anything in mind that mostly pertains to the skeleton of the building. But there's daily projects you have

BOB AND JERRY BUILT TWO SCRAP 2x4 SCULPTURES AFTER BUILDING THEIR HOUSES. ONE IS USED AS A BARN, AND GOAT PEN, THE OTHER AS A SHELTER.

...But today, I could go back and start working on a piece of that house that I started 4 years ago. And it will change. There are places in that house where there's two or three walls. The first one didn't work out, the second one didn't work out, it just didn't relate to the rest of the house. You work in splotches, the whole house sort of develops all as one...

to plan out, you have to calculate what's going to work. But today, I could go back and start working on a piece of that house that I started 4 years ago. And it will change. There are places in that house where there's two or three walls. The first one didn't work out, the second one didn't work out, it just didn't relate to the rest of the house. You work in *splotches*, the whole house sort of develops all as one whereas if we had confined ourselves to one particular area, it would have been sort of a hodge podge mess because as we worked out from that one point, we ourselves would have grown and our ideas would have developed...

L: What about the process and materials?

B: The process has been developing now, it's almost a system, in the sense that you can do whatever you want to, and there is a way to make the building waterproof and to keep it warm... tarpaper is not necessary. At least on the inside of the house. And expanded metal lath is about the best material to work with, because you can form it and shape it, it's a great sculpturing medium. And you can just really get after it. Structalite, which is gypsum plaster, and perlite is a wonderful material. Take a bag, throw it into the wheelbarrow, add some water, mix it up and you can just start plastering. Real good bonding abilities and it has a nice finish, it also insulates to a degree. Structalite I think is American Plaster's trade name for it and Litemix is U.S. Gypsum's.

L: Litemix is a gypsum and a perlite, huh?

B: Right.

L: I think I got most of this from you when I was there. Have you guys run across anything new?

B: Well, we got into the Dow Latexes.

L: Yeah, what about that stuff?

B: That is really super stuff. That's where it's at in the future. That's where concrete is heading. It's unavoidable. It's just too good a material.

L: It's a kind of flexible concrete?

B: Yeah, you use it only on the outside, like in skinning the roof. It's very impressive stuff. Day's been working on some new materials. There's new stuff called Hypalon out that they use for tank linings that costs 18 cents a square foot. You could just skin a roof with it. And chicken wire and plaster it. Plaster is so flexible that you can actually sculpture it. That's why I prefer plaster to anything else.

L: Hypalon is that rubber stuff, you mean. That's been around a long time but they're just manufacturing it again now, I guess.

B: Well now like Latex, they were using that back in 1932...

WALL AND ROOF FRAMING AT BOB'S.

TRADITIONAL YURT

by Aron Faegre

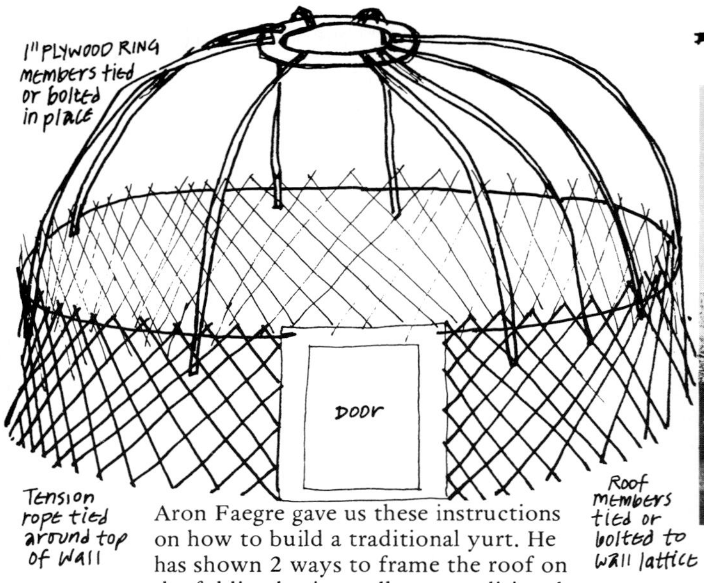

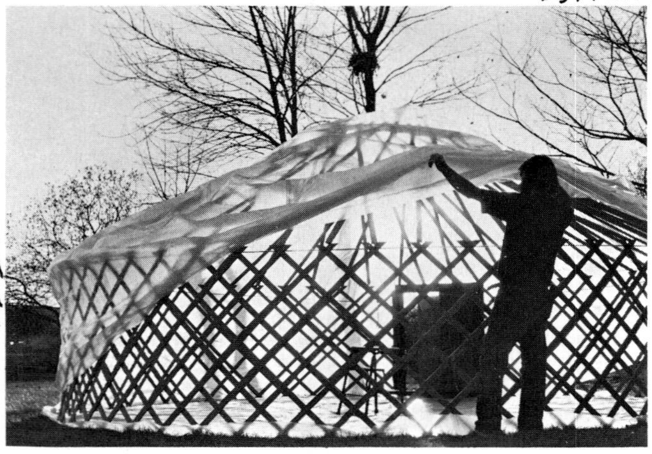

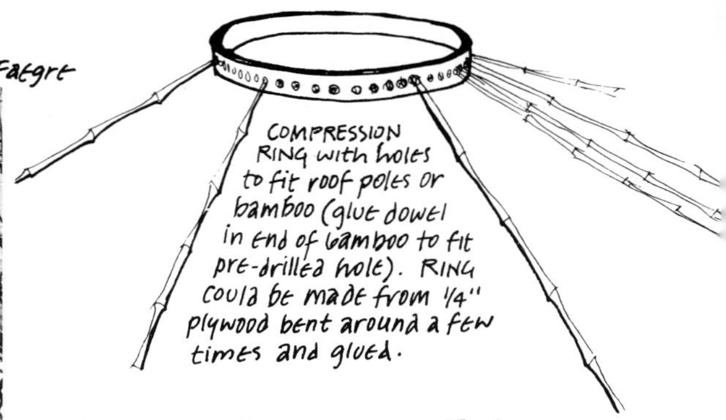

COMPRESSION RING with holes to fit roof poles or bamboo (glue dowel in end of bamboo to fit pre-drilled hole). RING could be made from 1/4" plywood bent around a few times and glued.

Aron Faegre gave us these instructions on how to build a traditional yurt. He has shown 2 ways to frame the roof on the folding lattice wall, one traditional, the other a folding lattice of his own design.

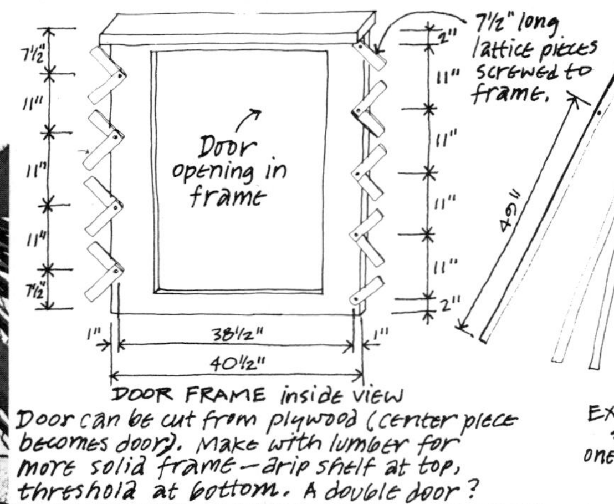

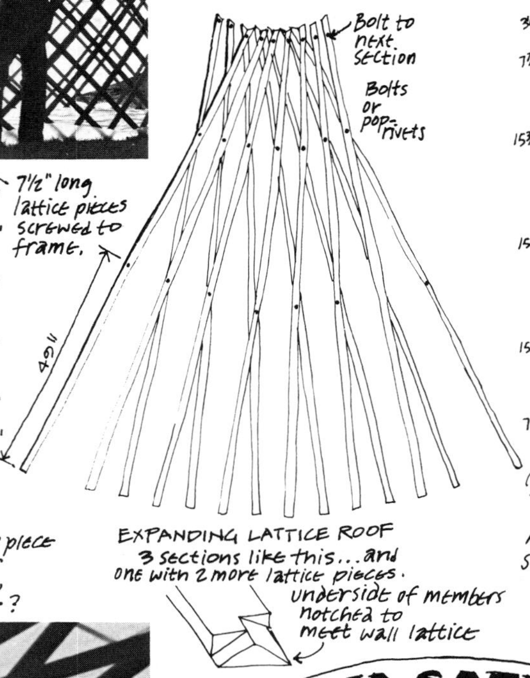

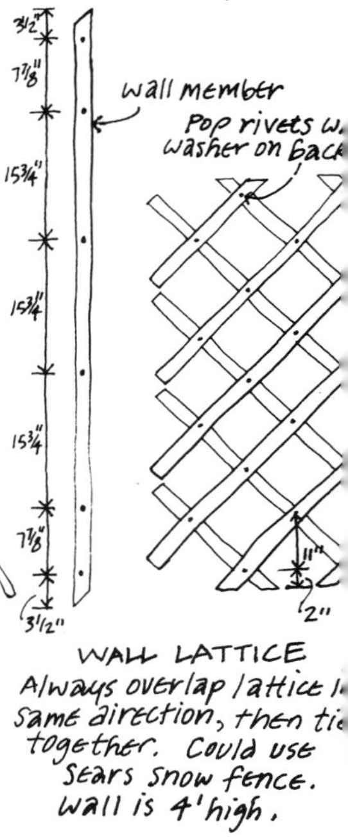

LATTICE ROOF WALL LATTICE ROOF

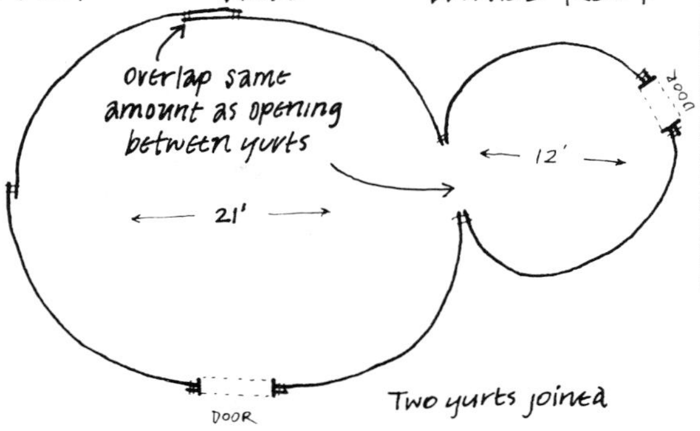

It is interesting that the nomadic tribes never used this method of connection. I believe it illustrates the power of the symbolic aspects of a house. To the nomads the inside of the yurt was divided into quadrants and areas that had different social and religious uses. The quadrants and areas were ordered by the two important points of the yurt: the doorway between inside and outside, and the center above which was the smokehole-skylight. These divisions reflected the ordering of the house life of the nomads. To join two yurts together by a passageway would destroy this order — would disrupt their life. For us non-Mongolians, several rooms may be more practical (and symbolically meaningful) in our house life.

The power of the smoke hole (nowadays the Mongolians run a stovepipe up through the hole and glass in the rest) is astounding. Gary Snyder's "Through the Smoke Hole" in *In the Back Country* recreates the myth that is so natural to the yurt:

> "There is another world above this one; or outside of this one; the way to it is thru the smoke of this one, & the hole that smoke goes through. The ladder is the way through the smoke hole; the ladder holds up, some say, the world above; it might have been a tree or pole; I think it is merely a way."

And then through the myth, as through the smoke hole, we are reborn:

> "*wha, wha, wha* flying
> *in* and *out* thru the smoke hole
> plain men
> come out of the ground."

NORTH STAR

...Second summer here in Cape Breton shows I stumbled into something extraordinary with this half-assed building I made last summer. All I wanted was to get out of the rain, so one whole wall is screen; what's wonderful is that I'm comfortable in all kinds of summer weather — & they got 'em all here; snow on June 18, heat wave a week later. To regulate temperature I can (see sketch):

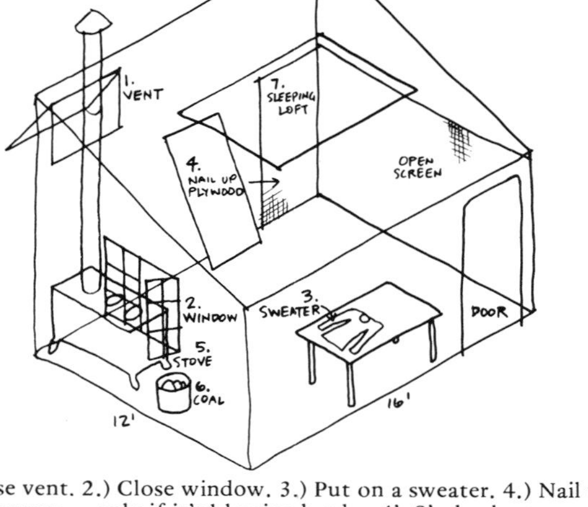

1.) Close vent. 2.) Close window. 3.) Put on a sweater. 4.) Nail up corner screen — only if it's blowing hard — 4'x8' ply sheet — no corner draft. 5.) Start wood stove 6.) Put coal in stove. 7.) Move into sleeping loft.

From here at the table, or in hammock hanging right in front of screen, the view is one spectacular wall-to-wall mural — not picture window at all. I live outside, except for bugs, rain, heat, & wind. And cold is negotiable. This seems an ideal kind of building for California.

This morning I was pondering on foundation structure for the house on the hill, where there's gale winds routinely, arrived at concrete pier & grade-beam (I guess it's called) with an adapted — laminated — mortise & tenon scheme. The problem was: how to make heavy timbers that I could lift & assemble alone. Some as long as 16 ft. So, the basic material is rough-cut 2x6, nailed together in place to make 6x6, permitting joints like

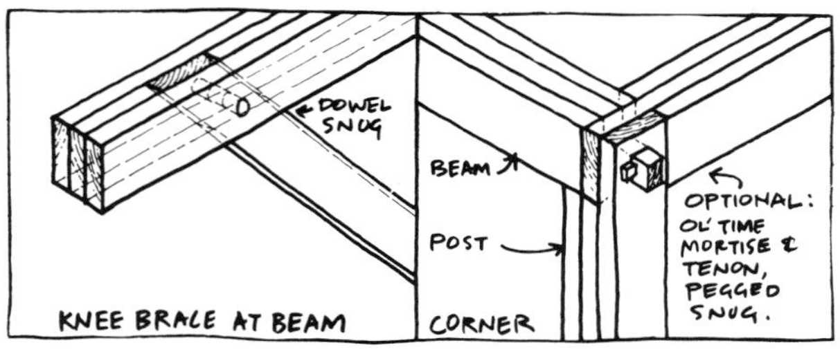

Last night I was on the mountain with a level & flashlight, lining up my house with the north star. I want to make a peak window & a hole in the upstairs floor, so you can stand in the middle of the downstairs front room (dark), look up from a certain place in the center of everything, and see Polaris.

I'd rather build a house than a book. But I'm surely glad you're building the book you are.

Stewart Brand

VIVA SATIVA

Greetings from Goa,

My huskey, Karma, & I moved to Earth People's Park, Vt., for its first froasty toehold in Planet Politics, and some winter sojourning; between Sept. 1970, and April 1971, we lived in a tarpaper tepee which cost $3.50 in hardware and four days of New York City scrounging. Unable to afford Thoreau's Middle Income expenditures, I checked around downtown building sites after completion. Much spin off — junk windows, nails, pipes and tarpaper. Trucked it merrily up up to the North East corner of Vermont and into the spruce and birch forests. It was already snowing in Zarvargo like ice mists. The timber smothered sound and suggested shelter at once! Lucked out and found half a tepee's worth of stripped poles axed and dried over the summer. I interspaced these with dead birch saplings choked out by the pushy pine people. Crisscrossed more saplings from the bottom of each pole to a point 7 ft. up on the next pole, for snow support. When encircled the effect was of the bottom strata of four frequency dome. This attractive shelter was brutally covered with plain tarpaper laid in long parabolic curves — shingled about twelve inches. Even the inner diamond patterned poles disappeared as an internal skin of tarpaper slid into place. The space between the two black membranes (pole width about 6 inches). I stuffed with green spruce tufts, pruned a little from many forestry citizens. A fire hazard if not changed seasonally. The superior ecologican sets the calendar in order! Made an arc over the sleeping section, and lined it with our beloved lightweight Aluminanized Mylar Space blanket. With a good oak, maple or birch fire, the arc produced a Dutch Oven effect. We were often without a stitch on -35° outside. A sauna like dash dive into the sharp blue electric snow bank. Zounds! Dug a deep firepit, and lined it with head-sized stones, this heated the floor which was covered with chips from the seldom idle woodpile (work ran behind schedule due to having to constantly chop fuel). Ran some scroundged ½ ft. piping underground from firepit to outdoors where a wind funnel was snow sculptured to it. A flattened juice can — gulloitin type valve regulated the fire mouth of the tube for enormous bellow blasts on latent coals. I put a 6 ft. tunnel on the entrance, with blankets hung at each end (light logs tied to the bottom of each blanket). Perfect doorway. Another blanket cut diagonally produced fine wind flaps. Another batch of spruce boughs piled with their curves bowed up in the middle filled my log bed frame wonderfully. Some good days were spent wetting a snowheap until it became a curved Art Noveau frozen latrine open face to the south — pleasant for day use. As greyhaired Adam Alexander observed, "Vermont if not *enlighting*, is at least *enwhitening*." Three windows faced east, south, and west. At night, they shone firebright at odd snow sloped angles like golden petals of the Secret Flower. Sweet Free Vermont. See you next year. Boom Shiva! Viva sativa! Stay tuned —

Walking Horse

The need for a larger structure with semi-separate areas for family use led to the development of the concentric-Yurt. However, the attractive division of circular space is difficult. As a solution we developed the idea of placing one Yurt within another.

The inner Yurt wall acts as a support for the increased roof span. The inner Yurt is raised to meet the larger roof, making a one-and-a-half story structure.

WOODEN YURT

by Bill Coperthwaite

The traditional Yurt, made of light poles and covered with thick felt, was a portable home which the nomads of inner Asia carried with them in search of suitable grazing for their herds. These nomadic Yurt builders appear to be the first to use the principle of the tension band in architecture. This development allowed the roof (or roof-wall) of a structure to be raised above the ground without the use of internal posts or truss work. This solved a basic architectural problem of eliminating negative space (the area where most tent structures, gables and A-frames meet the floor). The challenge was to have neither negative space, truss work nor posts blocking the interior of the dwelling. These ancient people made an ingenious discovery that, at once, gave their tent a positive wall angle, a clear inner space, a circular shape to fend off strong winds and provided minimal wall surface for heat loss and still have a portable shelter. This invention was a tension band — a simple belt of several ropes, made from the hair of yak, camel, goat or sheep, sewn side by side into a belt — used to encircle the building at the eaves and take the outward thrust of the roof. The hoops on a barrel work in the same way.

The modern Yurt takes from its predecessor the tension band, central skylight and low profile. With these elements in mind and using modern materials (such as milled lumber, glass and steel), the present Yurt has evolved over a ten year period. The design now includes outward sloping walls, an important departure from the Mongolian prototype. The sloping walls give increased rigidity and strength to the structure, and present a natural back rest as well as a feeling of greater spaciousness within.

The roof is accordian pleated or folded plate (called pie crust by the wags) for greater strength. A by-product of this design is the ring of triangular windows, fitted under the eaves. Although sufficient light comes through the central skylight, the quality of light entering the peripheral windows adds greatly to the attractiveness of the interior.

It is out of profound respect for the technical genius of these people that the name Yurt was chosen for this contemporary structure.

The quality of space in the Yurt is different from that which I have experienced in any other structure. From the outside, the Yurt is unimposing. With its low sod roof and wall of weathered pine, it blends easily into the landscape. The curves offer as little resistance to the eye as to the wind, adding to the impression of smallness. From without, the possibility of standing erect inside seems doubtful. Upon entering it is a surprise to find not only head room, but spaciousness. The eye tends to follow the upward, expanding slope of the wall. The radially extending roof lines meet the wall at a ring of light which helps to carry the eye beyond, giving a feeling of greater expanse.

Proceeds from the sale of Yurt plans and the holding of Yurt workshops go to the support of the Yurt Foundation — a non-profit foundation dedicated to the collecting of knowledge about simpler living from the cultures of the world — believing that by blending this knowledge with selected contributions from modern society, we can design a more beautiful and humane culture.

Plans for the standard Yurt cost $3.50; Concentric Yurt, $5.00. These can be ordered from the designer:

Wm. S. Coperthwaite, Director
Yurt Foundation
Bucks Harbor, Maine 04618

Extracts from Interview with Bill Coperthwaite by Bruce Williamson, *Mother Earth News*

You see, one of my goals in design — whether for a house or a pair of shoes — is to involve the machine less and the person's own capabilities more and more. That doesn't mean throwing out all machines or spending two weeks making a needle ... such an approach can get ridiculous. But it does mean that whenever I can find a reasonably simple way to eliminate special material or tools in making a yurt, I'll do it... not in order to produce a quick result — like a molded plastic house — but in order to let the builder use his own personal skills.

More specifically, the customs of small rural groups provide us with alternatives that we're not aware of because we're conditioned to think in terms of standardized methods. The use of locally available materials is a good example. The Finns, for instance, believe that lilac and the small mountain ash are two of the best woods in the world for making rake teeth. Because modern industry uses white oak for that purpose — to be sure of an adequate supply of material — we in this society don't realize that there are these other possibilities ... when in reality we could make a rake out of that bush in the front yard that would be better than the commercial product, would cost nothing, and would involve us in its fashioning.

I think that our society went through a long period of putting *things* down, saying that one should rise above the material level ... and of course, it's possible to want objects in a miserly way, to need a collection of them as proof of one's worth. But there's another sense in which things are important because we respect and understand them ... because we've come into an intimate relationship with them.

If a certain dish gets broken we feel badly ... and if the dish doesn't matter to us at all then it's time to get some better dishes ... ones we like.

We need to teach children to value the spirit within inanimate objects ... the beauty that we see when we know who made a particular item or when we know the way it was formed or when we know how it works. We need to emphasize the interrelationship of all things, not only in a practical way, but out of respect for the skill that created an article. This attitude takes for granted a life intimately surrounded by things made by friends.

It's extremely important to gather this kind of information while it's still available. Since folk knowledge isn't considered essential by the mass culture, the old skills are dying out. The Canadian Eskimo who knows how to harpoon a seal through a hole in the ice or make a kayak in a certain way is disappearing ... and in many cases we can't reconstruct what he knew.

Traditional knowledge is the product of thousands of generations of handing down from father to son, mother to daughter ... and once that chain is broken we have to start all over again, which can be pretty much impossible. But if we can find and learn certain kinds of knowledge while they're still being transmitted, then we can become part of the chain and pass on what we know to other people throughout the world.

Crafts are important in this society because many people grow up lacking self-confidence, and the development of hand skills is a good way to promote a feeling of self-worth. Traditionally, though, we think of crafts as belonging to the specialists and their apprentices. An advantage of working with handcrafts of different cultures is that I'm stimulated to design techniques that enable more people to participate in these skills.

I've also found that — when I'm among people like the Eskimos to whom manual skills are common — crafts become an important means of communication. The fact that I've developed the ability to create with my hands not only establishes that I'm sincere in wanting to learn from people of another background, but drastically reduces the time it normally takes to overcome stereotypes and barriers ... the language barrier, for one.

HOGAN & TEA HOUSE

TIPI-SNAIL SHELL

This house was conceived as an extension of the kind of space that a tipi gives — round, uplifting space — yet with strength and protection enough for a Massachussettes winter and with window space to look out of. Using whole trees for structural members gave the space the natural feel we wanted it to have as well as giving us free materials. A local logger agreed to let us fell and haul off these "sticks" or dead, standing trees. They were pine, some 40' long. They were stripped of bark and thoroughly poisoned before we used them to prevent their carrying insects into the other wood in the house. The center-post is a 96 year old hemlock, also standing dead, that we cut off the land/the land the house is on. We cut it down to 24 feet long, leaving stubs of the branches on to climb up to the loft. Michael inscribed a poem on it.

The basic structure of the house can be described as follows: First we built concrete piers and a small cinderblock basement (to get the plumbing underground — minus 30 degrees in Jan. — and for a cold cellar for vegetables). Then we erected the "tipi poles" on top of the piers with the obliging help of the surrounding trees, two large block and tackles, and many people heaving on ropes. The poles were each lag-screwed to the center post through a 2"x12"x¼" steel strap bent to the proper angle in Woolman Hill's forage. This took some considerable angle-carving where all the poles sat on top of the center post, something we came to be quite proficient at by the time we finished this house. Once the poles were erected, we shaped the edge of a 2"x6" to fit on each pole for hip rafters and framed the triangles of the roof and the floor quite conventionally. Old wood and windows from the Blinns' barn and other places were used as much as possible and since the house has been finished the Potters (the folks living in it) found an old slate roof they are putting on and are building a chimney (shown on the floor plan) of cinder-block-type chimney tiles and covering it with fieldstone. Jim made stained glass windows for the house in which he used mica from the surrounding hills.

A nice process was the "house-book," a spiral notebook in which drawings, calculations, estimates, questions for local wisdom, answers, detail sketches, etc. were kept. It made for a nice fluid continuity in the design-building process. "Method-of-the-week" awards were given for solutions to problems, compound angle cuts, nailing in impossible places....

Andy Shapiro and Doug Glasser
Alexandria, Va.

The photos are of a hole I made/built as an environment for meditating. It is 6 feet deep and 7 feet in diameter. Entrance and exiting are accomplished by jumping in and pulling oneself out. All the wood used is either cut from nearby *Eucalyptus Globulus* or is recycled from previous projects, tools used were shovel, hammer, and hand saw. The hole and the teahouse are hidden away in the hills of Los Angeles (undoubtedly illegal).

Leonard Koren
Beverly Hills, Calif.

$65 TIPI

On an island off the west coast of Canada there is a comfortable, cozy, creative shelter built by its owners for a total of just sixty-five dollars.

The basic construction is four trimmed fir trees stacked as a pyramid. They are attached at the top to one another by old steel logging cable well nailed into place. Other, smaller poles form the lateral supports for the heavy, handsplit cedar shingles that comprise the exterior.

All windows and doors were salvaged from abandoned cabins and houses. What little sawn lumber was used was obtained by an arrangement with a sawmill about 25 miles away. In exchange for a truck load of rough logs they obtained a percentage of the lumber cut from these logs.

The floor of the dwelling is of flat stones with a hard clay mix in the joints. A scattering of handwoven rugs and reed mats makes the floor both attractive and usable.

The stone fireplace and wood stove were built from native rock gathered along the beaches. There are no utilities except for gravity flow spring water piped to a faucet in the kitchen. Wood provides heat and kerosene for lamps. A privy, whose location is changed periodically, is "outback."

Total expenditure:

Insulation	$50.00
Nails and miscellaneous	15.00
Total	$65.00

Because the house can be made larger or smaller, square or rectangular or even round, exact measurements and specifications are not given here...merely the rough sketches showing the basic construction and finish details....

Bill Kaysing

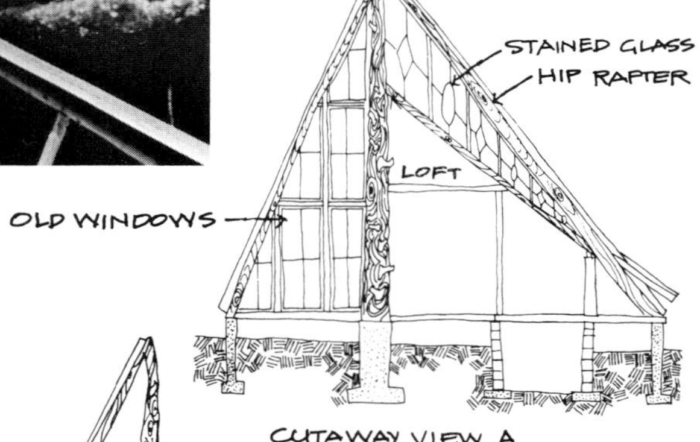

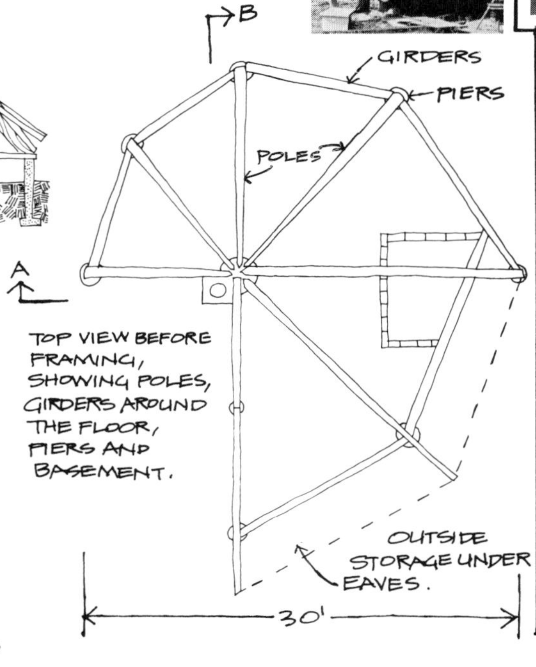

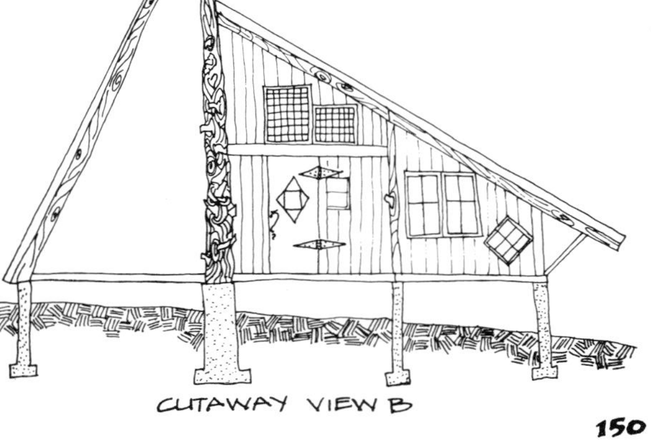

SOD IGLU

by Ole Wik

When Europeans came to Arctic North America, they brought with them the ideas of log cabins and frame houses, which came to displace a variety of native-style dwellings. In many cases the trade-off brought advantages of dryness, airiness and permanence at the expense of heat economy. Instead of heating small, semi-subterranean dwellings with seal oil lamps, for example, the coastal Eskimos moved into frame cabins and had to start scrambling for driftwood to feed the hungry sheet-iron stoves. Later oil replaced wood as space heating fuel, and the people had to take wage employment to meet their fuel bills.

A good answer to cold-country housing is the modified sod iglu, which incorporates the best features of old and new. Basically, a sod iglu is a semi-underground structure built of light poles or rough-sawn planks leaned against a sturdy log framework and then covered with polyethylene sheeting, dirt, moss and, eventually, snow. The underground siting and the heavy dirt banking provide a heat sink which buffers the room against the extreme cold on the outside. Consider that, with a log or frame wall, you can heat the place all week when it is -40°F outside, and at the end of that week the outside air is still -40°. With a sod iglu, the dirt on the other side of your wall will not be anywhere near -40° to start with (due to the heat of the earth itself), and after a week of heating it will be well warmed. Heat from the room is stored in the daytime and released at night, thus maintaining a very even temperature through the 24 hours. (Anybody who has slept in mid-winter in a log cabin and had their cup of water freeze by the bedside will appreciate this.)

The heat economy of a sod iglu is not hard to understand. In order to calculate the heat flow through the walls of a house, you multiply the area of the walls, times the heat flow per unit area per unit time per degree of inside-outside temperature differential, times the difference in indoor-outdoor temperatures. To reduce heat flow, you can concentrate on reducing the first factor (make the cabin smaller). Or, you can reduce the second factor (by using better insulation). These are the conventional approaches. The sod iglu concentrates on the third factor: the temperature differential across the wall is reduced by placing a mass of dirt there, which absorbs heat and warms up. The wall can then be almost dispensed with; we have used nothing more than the vinyl sheeting supported by widely-spaced poles.

The materials in this structure cost about $50, which covers the poly sheeting, spikes, nails and window material and staples. Everything else — logs, poles, dirt, moss, snow — came from the land. The only tools needed were a shovel, bow saw, axe, drawknife, brace and bit, hammer, level and staple gun. My wife, Manya, and I built the place in just about one month, from the first shovelful of dirt to the first night inside.

Over the years we have built four dwellings on this principle, and our neighbors have built another ten. Years of cumulative experience in temperatures ranging from 90° to -70° have amply demonstrated the superiority of the iglu's heat economy to that of nearby log cabins and frame houses. If you are building for cold country and if you have a site that is reasonably well drained, take time to consider the possibility of building a sod iglu.

The illustrations show three steps in the construction of our latest sod iglu. Photo above shows the basic frame. Four upright piles are sunk 3' into the bottom of the pit, which is itself 2½' deep. (We wanted to go 3' deep on the pit, but our site is near a creek and the water table was too high.) Two logs span the long distance between the pilings (which, incidentally, outline a rectangle 9' X 12'). Two more logs span the short distance, and then the ridge pole is laid on these. Simple square notches are used throughout. The logs are all butted in such a way as to take the compressive forces of the dirt load to be applied later. The joints in the frame are fastened with 10" metal spikes.

Due to the long window in front, the dirt load on the back wall was plainly going to be greater than that in front, so we ran diagonal log braces from the top of the rear pilings to the region of the base of the front pilings, anchoring them by butting against a log floor joist that lay against the base of the front pilings. Likewise the side opposite the door carries a proportionately greater load, so a brace runs from right to left, beginning below the window. These braces are all-important, as we learned by experience; without them an iglu gradually leans away from the greater loads, and eventually has to be abandoned.

In left photo the pole walls have been applied. Here is one real beauty of this type of construction. Fitting the logs for a good, tight log cabin is an exacting, time-consuming art. With an iglu you can leave cracks up to 6" wide and simply seal them with the vapor barrier that is applied over the frame. In this particular iglu we placed the poles close together, mainly for appearance. We also peeled them very carefully, so that they would be as light in color as possible.

The poles need not be prime trees — in fact, a forest that is entirely cleaned out of log-cabin-type prime logs can yield hundreds of smaller trees suited for iglu construction. Many of these can be culled from stands that are really too thick anyway — crowded stands tend to be tall and thin and straight. If you get a pole with a twist in it, you can usually match it with another that has a similar bend, and work out the little differences as the wall progresses.

The walls have a slant of about 4 - 4½" per foot of height. This is important, as then gravity keeps the dirt and moss held firmly against the wall. The floor measurement is thus greater than the frame measurement; our 9' X 12' frame corresponds to a floor size of about 14' X 18'. Note that no fancy measurement is needed for the wall poles. Put them in place, hammer them down into the dirt as far as they will go, nail them to the top log, and simply saw off the excess. The whole iglu, in fact, is a low-tolerance operation — you don't have to worry a lot about precision and fit.

In the right photo the roof has not yet been added, but the principle is exactly the same. Notice that a pole has been added as a brace between the ridge pole and the log above the window. This is to carry the force from the back wall to the front without stressing the roof poles.

The center photo shows the completed structure. After the roof was added (notice the ends of the poles overhanging the window), the whole frame was covered with black polyethylene sheeting (6 mil thick). Then dirt was backfilled onto the entire structure. Near the top of the walls, where the dirt banking is thinnest, we put a layer of moss against the plastic before backfilling. On the roof we put two layers of moss, for an aggregate thickness of 10". This is good insulation, and in winter we have an additional two feet of snow. Snow is almost entirely dead air, which is the principle of most insulations; use it as much as you can.

For windows we used clear vinyl sheeting stapled to poles tacked vertically in the window opening. In midwinter we put removable shutters over these on the outside; they consist of two layers of ordinary clear polyethylene sheeting. These let in adequate light but you cannot see through them very well, and we are still working on better window arrangements. But the plastic works well enough, and is cheap and easy to use. When we leave the iglu for a week or more in winter, we bank snow onto the windows and door, thus sealing the whole place in a blanket of snow. Once we were gone for 2½ weeks at Christmastime, and although the outdoor temperatures reached -45°, when we returned the tinned fruit was not even frozen.

We built our floor from split poplar logs, using an axe, drawknife and plane for the smoothing. The result is very pleasing, but the labor involved was enormous. You can get by for a winter with a sand floor (we've done it), but if you can afford plywood or planks you'll really have a first-class structure.

Finally, remember that there is no reason why you have to stick with poles if you are contemplating a sod iglu. You can set up a chainsaw mill and rough-cut planks for walls and ceiling, or you can use plywood. We once built a rigid-frame plywood structure with dirt insulation for the sloping walls, moss for the roof, and commercial fiberglass batts for the vertical end walls. Or, you may want to consider placing some other sort of structure partly underground to get the heat-bank effect. No matter what type of house you have, consider banking it with snow in the wintertime.

We built our iglu with the long, deep Alaska winter in mind, and we had a warm, comfortable winter. When spring came, the snow insulation removed itself, and we discovered a further advantage: now, when the days are hot in mid-summer, the earth is cooler than the outside air, and we go into the iglu to get away from the heat.

WILL WOOD

JOINT AT DECK SEAT

TUB UNDER HUGE BOULDER
PAINTINGS BY JIM WALKER

Will Wood is building a *Shangri-la* of small structures, gardens, terraces and stonework among the boulders in the mountains near Southern California. He saw Maybeck's and Gaudi's work in books, then in person, and with this inspiration, began to build. He never studied architecture or building, he used an innate sense of craftmanship and the advice and help of others (especially Jim Walker's) to learn to build. He bought Mexican adobe and redwood to build his house with, but now his preoccupation with stonework reflects his awareness that you should build with what's around you.

by KEN KERN

I believe, with as much certainty as ever before, in the viability of amateur-constructed owner-built housing. Yet after two decades of experience with owner-builders I have witnessed a few examples of outstanding building success and not a few unfortunate failures. Failures, that is, in terms of unsatisfied personal building experiences or unsatisfying structural or esthetic building results.

The failures interest me most — not because they are more numerous, but because they invariably stem from some minor misjudgement which would have been avoidable had correct advice been offered and accepted at an opportune time. A book comprising examples of owner-built failures would seem to have as much practical value to prospective home builders as illustrations of neat and slick projects.

Experience has shown me that the greatest majority of owner-builder projects fail where little or no regard is given to one or another of three different aspects:

1) Personal Evaluation. Abilities as a do-it-yourself homebuilder can be fairly well determined by a self-scoring test. The test I use (send to me for a free copy — Sierra Rt., Oakhurst, Cal. 93644) indicates what types of jobs one can successfully handle and what types should be hired out. It also gives insight into whether one would really enjoy building one's own home. This test can go far in revealing probable success or failure.

2) Simplified Building. I have always been at a loss to determine why it is that unskilled owner-builders invariably attempt to build complicated structures. The less ability they have to draw from, the more structurally involved the project becomes. Too many people hold the view that a building is only beautiful to the degree that it is complex, whereas the opposite usually holds true. The best of all of man's creations has been the simple one — simple in design and simple in method of execution.

My attraction to curvilinear walls lies in the fact that curvilinear structures can be built from a central radius pole. Corners (as in the traditional log cabin) can be unduly complicated to build. Single-pitch shed-type roofs can be attractive, easier to build, require less material, and offer better solar and ventilation controls than the traditional gable roof.

3) Building Priority. Simple structure will be more universally acceptable once the shelter priority is placed in proper perspective. In my judgement human housing commands far too much of our energy and concern. People who produce their own food and fuel requirements, and who hand-craft their own furnishings, and who maintain the tools and equipment required to keep their home production operating efficiently usually view human housin more realistically: subordinate, perhaps, or integral with the garden, barn, and shop facilities. In reality the garden, barn and shop become an extension of "house" thereby requiring less formal home and homestead development. In this context I am reminded of the homestead that anarchist-philosopher J. William Lloyd built for himself in the Los Angeles foothills, 1930.

MYTHS & ROOTS

FRANK LLOYD WRIGHT HOUSE, OAK PARK ILL. 1889

HOUSE BY BRUCE PRICE, TUXEDO PARK, N.Y. 1886

From *The Shingle Style.*

Many architects and designers do not acknowledge their sources. Much that passes for invention or innovation is actually misappropriated (and often highly publicized) work of others. This tends to inhibit architecture students or designers, as the achievements of the "masters" appear stunning when out of the context of their historical background. Every creative person has roots.

CROSS SECTION OF OLD ST. PETER'S BASILICA, ROME

In *Planning for Good Acoustics* Hope Bagenal explains why the acoustical conditions of...(St. Peter's)...must by their very nature lead to a definite kind of music. When the priest wished to address the congregation he could not use his ordinary speaking voice. If it were powerful enough to be heard throughout the church, each syllable would reverberate for so long that an overlapping of whole words would occur and the sermon would become a confused and meaningless jumble. It therefore became necessary to employ a more rhythmic manner of speaking, to recite or intone. In large churches with a marked reverberation there is frequently what is termed a "sympathetic note" — that is to say "a region of pitch in which tone is apparently reinforced." If the reciting note of the priest was close to the "sympathetic note" of the church — and Hope Bagenal tells us that probably both of them were, then as now, somewhere near A or A flat — the sonorous Latin vowels would be carried full-toned to the entire congregation. A Latin prayer or one of the psalms from the Old Testament could be intoned in a slow and solemn rhythm, carefully adjusted to the time of reverberation.

The priest began on the reciting note and then let his voice fall away in a cadence, going up and down so that the main syllables were distinctly heard and then died away while the others followed them as modulations. In this way the confusion caused by overlapping was eliminated. The text became a song which lived in the church and in a soul-stirring manner turned the great edifice into a musical experience. Such, for instance, are the Gregorian chants which were especially composed for the old basilica of S. Peter in Rome.

Experiencing Architecture

ARCHITECT & JOINER

In the later years of the eighteenth century certain architects had entered into the profession from the carpenter's bench. One such subscribes himself "Architect and Joiner". There emerged, though, a host of more or less able professional architects, but they took their cue from their patrons, and in a highly artificial age they made architecture an affair of rules, so much so that only a pedant could understand it. At this point the craftsman became a mere tool, his functions being limited to carrying out the instructions of the learned architect. Lost was the spirit of the past, where the craftsman was able to put his heart into his work and he and the architect alike could appreciate their mutual dependence.

From *Short History of the Building Crafts*

LONG HAIR, MASONIC LODGES, SEEDS OF ARCHITECTURE

he great gothic cathedrals of Europe were not built by architects" as we know them today, at least not of the rofessional, sedentary variety. They were built by masons nspired by newly acquired knowledge of the buildings of he Middle East, and generated by competitive bishops in hat is known as the Cathedral Crusade. The skills acquired y masons during these years led to the establishment of he masonic lodges and is part of the history of architecture. in those days, the men responsible for the great found headwork and handwork perfectly reconcilable and there was as yet no divorce between esign and construction, architect and craftsman.

The story of the cathedrals, the men who built them, is told in *Master Builders of the Middle Ages,* by David Jacobs, and is summarized briefly here. The two quotes below are from this book.

the last years of the 11th century the knights of the y Roman Empire resolved to drive the Moslems from Holy Land. This led to the First Crusade. Returning ctorious to Europe, the crusaders brought eyewitness counts of the magnificent Byzantine church, the Haga phia of Constantinople, and of the graceful, delicate rchitecture of the Moslems.

> *Perhaps more importantly, the crusaders brought home a new spirit. They had gone off to war crying "God wills it!" It became the cry of an era and gave the impression that all events — the most brutal war, the abuse of the peasantry, the laborious building of a church — were inevitable, divinely inspired, irreversible. All causes were crusades. The spirit needed time to take hold and grow and make its way into medieval thinking, but the First Crusade had given it the first push.*

Gradually, as Europeans heard of the pointed and slender columns of eastern architecture which seemed to transform stone into soaring, heaven-reaching cathedrals, there began n Europe a new type of crusade: competitive church building. Bishops competed with zeal to have the grandest, mightiest churches.

Previous to this time, the mason was uneducated and lliterate, selected from the peasantry for his strong back. Each monastery or nobleman had a resident builder whose ask was to follow his lord's instructions in assembly of materials and construction. But as competition increased, Europe's bishops unwittingly gave the masons a most valuable tool: knowledge. As the churchmen encouraged he emergence of the Gothic style, with its technical complexity, masons were sent out to study other churches and gradually became more sophisticated. The bishops became increasingly dependent upon the mason, and some masons who showed individual talent were allowed to operate independently for profit.

ERMAN SCULPTOR ADAM KRAFFT A. 1475. CARVED HIMSELF AS MASON.

CARVING OF MAN WITH TOOTHACHE, WELLS.

> *As soon as he realized how important he was becoming, the mason quickly asserted and flaunted his independence. He asked for higher pay, and he got it. He dressed garishly, favoring showy silks and satins and huge capes of bright solid colors, lined with gaily patterned prints. He let his hair grow long and cultivated an unruly beard — at a time when short-cropped hair and a clean-shaven face were signs of piety and self-sacrifice. His speech was peppered with curses; he was irreverent; he contradicted the bishop's orders for contradiction's sake.*

By 1230, the Church decided it had had enough. The bishops of France issued a decree: the masons would have to change their ways, and as a start they were to cut their long hair and shave their beards. In a few weeks' time, the masons issued a counter decree: not only did they refuse, but unless the bishops lifed the ultimatum, they'd burn down every cathedral in France!

The bishops capitulated, and the Church, which had prior to this time controlled monarchs, could not handle its upstart builders.

During these times the masons formed their *lodges.* Travelling masons stayed there as they went about to study other edifices or to inspect a quarry for suitable stone. The churches were built from models, which were then destroyed, and in the lodges masons exchanged information amongst themselves.

THE VAULTS OF THE GOTHIC CATHEDRALS WERE JUST THE CEILINGS. HUGE TIMBER TRUSSES SPANNED OVER THE VAULTS TO SUPPORT ROOF. THE GOTHIC VAULT.

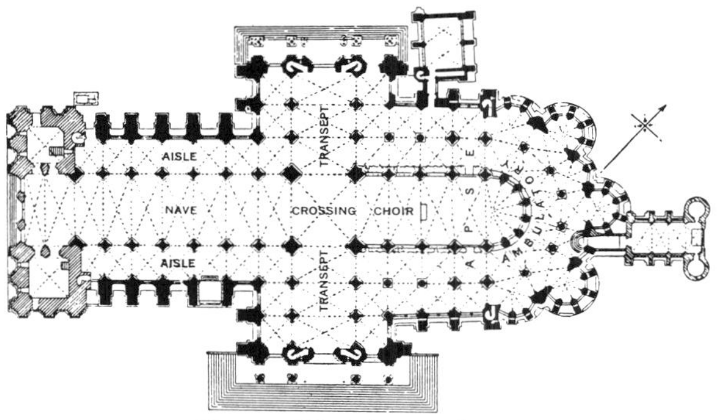

PLAN OF CHARTES

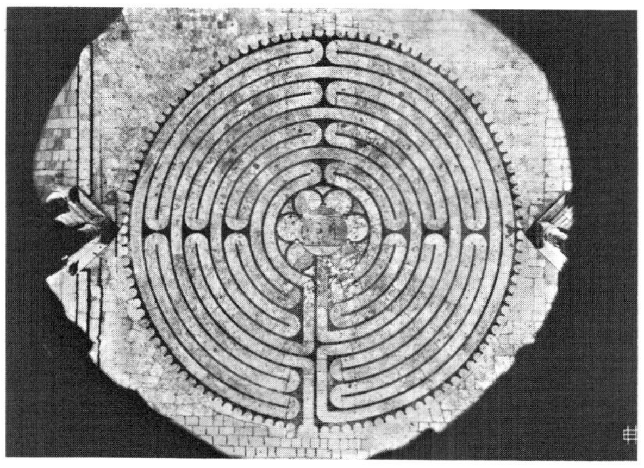

Masons competed to leave their personal signature most prominently on their works. Grand prize must go to the mason of Chartres cathedral, which had a mosaic labyrinth on the floor in the center of the church. According to a precedent supposedly set at Solomon's Temple in Jerusalem, the precise center of the labyrinth was the spot to which all pilgrimages were made. At Chartres, after the pilgrim made his way on his knees for 768 feet across the tile labyrinth to the center, he arrived at the mason's signature.

DESIGNER-BUILDERS

Master Builders of the Middle Ages does not speculate on precisely how architecture derived from these conditions, but it does not seem unreasonable to assume that the invention of the printing press had something to do with it. Where before, buildings were constructed from models, and secrets kept in the lodges amongst masons, the printing press may have made available this knowledge to the gentlemen and aristocrats. Val Agnoli thinks that the printing press showed drawings and etchings of Rome, the grand tour of Greece, and that the wealthy families sent their young men there to study, and they came back architects. It became a gentlemen's art. Talking about it Val said:

> *It's still a gentlemen's art. You still need a bankroll. You have to come to a firm with a bankroll or influential friends, certainly not with ability as a designer or as an engineer. There are a million guys who can draw and design and it's mainly a cultural, social outlet for aristocracy...those cathedrals were built from scale models, not drawings, not blueprints — that's something that comes later.*

CABINS, LAKE TAHOE, CAL. 1923. BERNARD MAYBECK

Bernard Maybeck, early California architect, is known for his imaginative building that balanced traditional design with his own unique innovations. (In 1907 he built a house in Berkeley with walls of gunny sacks dipped in a frothy cement mix. The house is in good shape and is still inhabited.)...

...Ivan Melvin, recalling Maybeck, said, "In his casket he wore his cap. He didn't look much different than he did during the depression when he called me over and said: 'You haven't anything to do and I haven't anything to do so we'll just build a house. We'll lay it out today.' We walked over his land and we found a spot and we laid it out. He wanted me to build a house too. 'Take some land, as little as you want or as much and pay me whenever you like.' He just sold the land by the foot.

"He was always trying something out," Melvin continued. "Once we tore down an electric stove in 1923 and built the rings into a tile counter and put the oven in the wall. He built a house once without any metal hinges. He built a house for his son Wallen with double concrete walls and put rice hulls between them. He could draw the prettiest picture of a stairway, sitting there with the drawing board on his lap, and I'd ask him what size posts to put in and he'd say, 'Six by sixes. Put in a few.' But if he took a notion to draw details he drew them perfect.

"He was a good man. He was a real common man. And he knew what he was looking for."

From *Five California Architects.*

It is rare today to find a man who both builds and designs. Yet this is the ideal situation, for the designer-builder is involved in the entire process of creating a building from its concept to its realization. It is only recently in history that there has been a divorce between conception of a building and its construction, between headwork and handwork.

In the following six pages are conversations with three friends who are designer-builders. They are remarkably open and candid, perhaps because a builder's work is done in the open, for all to see.

HOUSE IN ROSS VALLEY, 1906

BERNARD MAYBECK, 1862-1957

TOWER FOLLOWS SHAPE OF WINDSWEPT TREES.

VAL AGNOLI

Val Agnoli is an architect and a master builder, a rare combination in these specialized times. He has built gracefully curved, sculptural houses, as well as simple buildings, and designs for other people as well as himself. His latest project is a simple rectilinear house built of used 4x4's, 2" tongue and groove siding, and asphalt tab shingles.

Lloyd: One thing that interests me is that you're going from curves, very sculptural curves, and now into something much simpler. Is it because you're just moving around trying different things?

Val: I don't know. You feel like you can get almost as much done with straight lines. This thing here, I'm building this way because I want a simple tranquil building. And it's kind of surprising to see how much you can do with just a square little house, you know. Well, little things, like the way this drops down here. I mean, it's not relying on circular forms to give it its value. But I think I'd like to do a wild, a really wild building again. Yeah, really. I mean completely.

Well, how about your own house?

Well, that's a conservative building. It's just an A-frame turned around into elliptical form.

How long did it take you to build?

Not too long. Year and a half, maybe. That was a real ego trip, you know, it was being forced into a role of what I could do as a designer. It was a pretentious move. Had to show your stuff, you know. And it had to be realistic enough so that it was still worth some money when it was done.

How long ago did you do that?

Four years, five years ago.

And the tower came after that?

Yes.

The guy who built the tower, was he meticulous?

Yes. Super, super. And that's what influenced me a lot in the building of it, the design of it was loose. A lot more perforations, a lot more overhang, a lot more balcony, in and out. More transparent. Then as we became closer and closer working on it, we became real buddies, and I picked up on a lot of his vibes, a lot of his direction because I wanted to please him and I respected him. He's a very smart fellow. The trouble is, we ended up making the building tougher and harder and stronger and more fortress-like. Lost all the easy parts to look at, you know. And consequently I felt, personally, I was — at the end of that building I was a fucking paranoic wreck. You know, I was thinking the way he was thinking. And I really got angry with myself for being psyched out that way.

I think that's a transition, just from going around talking to experienced builders, doing complex things then getting back into simple things. Maybe you go back and forth.

You need a taste for adventure, you know. And challenge. It's as exciting as hell. I mean you can hardly catch your breath sometimes when you're fooling around with a real wild design. You don't know what the hell's happening. Like an animal in new territory. It's real intense and exciting. But all of a sudden you say, what the hell am I doing here?

How did you begin building? Your family?

Yeah, my father is a carpenter in New England. Cape Cod houses. Formula houses.

Have you been a boat builder, too?

Yes, but not a real first-class boat builder like... I could build a Monterey hull, I guess, but I haven't. I've been fooling around with rudimentary stuff.

How long have you been housebuilding?

Long time. I guess I built my first house when I was 18. Then I left that to go to school and got an architecture degree. I was also working on big projects like bridges over the Connecticut River. Working as a carpenter. And the St. Lawrence Seaway. When I got to the St. Lawrence Seaway, I decided it was so impersonal and so tough, to can it and go to school. I went to college very late and I had enough practical experience where it really held me back in design. I wasn't able to be naive about lines. I knew what the hell was going to happen when I tried to build it, so I really had to fight against my better judgment.

Bob says you should never draw anything unless you've picked it up. You should never draw any kind of piece of wood unless you know what it feels like.

That's what the enemy of architecture is. It's always been its weakest point, is that it uses paper. And that's its strongest point, too. Because if you stay right here, you can't — and you have a pile of lumber — you just can't visualize that. The pile of lumber will drag. The lumber drags and the gravity drags you back down to reality and you can't soar. So an architect is an unrealistic guy who rises up, then he gets pulled down two pegs by the guy who has to build it, but you're still pretty high up in the air when you finish. The dome...it's a beautiful technical thing. But it's so absolute, so strong, that — Jesus, as a western culture, we don't need any more of this rigid philosophy or manifested form to keep reminding us.

Right, because once you start with it, you're locked in.

Soleri and Fuller are geniuses but the motherfuckers are pushing us right into a locked-in system.

It strikes me often that their kind of thinking is an aristocratic fascism. They're both from very wealthy families, they think in terms of one man laying it all out.

Not once did these guys ever conceive of the fact that maybe you can't live in a fucking hole in the ground or domed city. Maybe what they need to do is walk out into the middle of that field to be sane and valid again. I heard Fuller speak about 15 years ago and he made so much sense he frightened the shit out of me. I don't know why. Everything he said was yeah, yeah, right, and that's true, that's true. But he frightens me, I don't know why.

Well, I listened because I had no education and I had never heard the stuff he was talking about before.

Well, when you talk about a family, a residence, a place — it's the only place that a man can go to have some peace and quiet, in his own little space when he slides that door shut, you know. And that shouldn't be the epitome of rigidity and ... I mean, just this thing over the top of you, you know, this network over the top of you, this grid...it's just as bad or worse, in fact, than the old grid they laid on cities. That was a controlled mania and that's what I like about Boston so much, are all of those fucked-up streets going everywhere.

I thought of that in great preference to Soleri's ideas. You know, just let the city grow, unmanaged, and come back 200 years later and dig up the streets for the underground systems but don't have one man lay it out all ahead of time.

Tokyo is that way, too. Tokyo developed out of a feudal city and so streets are going around this way and there's no grid. They don't all go to the center obviously like Washington or Paris. And that's why it's so hard to get around, but that's why Japanese streets are so beautiful....

I went to this shelter conference in September in L.A. and Soleri was the feature lecturer. The Barrio Planners who were the people from that area in L.A. gave a talk and hardly anybody went to hear them. And one guy got up with an eyepatch, great big guy, Chicano, and said "Who is this Poblari," ah, you know, he says "Who says we should live in these anthills? What is all this dome stuff? I want a house with a yard with a barbecue and a fence around it, and who is this guy who says we should live like this." And of course, nobody was listening to him.

No, because there's no percentages in listening to somebody like that. There is, though, in kneeling before the shrine and nodding to the fellow next to you, "Isn't this great." Well, I don't know. I envision Fuller getting to the moon and putting a dome completely around it. I think that's where his head is. It's religion.

Yeah, that's true. Conceptualization is the thing we've been talking about, the difference between the concept and the reality, and that's the white man's way of doing things, it seems. You get this idea and you don't go out building a house from the materials you have to work with or the site, you go out from an idea. You jam everything into this preconception and, you know...

Jesus Christ, I just remembered. One of our projects in school at Oklahoma with Bruce Goff was — the client had so many thousand railroad ties and this was beginning design and he wanted you to utilize those in the design. And I had just come from working on the St. Lawrence Seaway (laughs) so I ripped them all into 2 by 4's and laminated them together. He was so fucking heartbroken. I said I'm not going to work with those fucking railroad ties. I made it conform to something I knew how to use...

Houses being built now are way too big, too hard to heat, use too much materials.

Sure, that man can do what he's going to do in a third of that area with a roof that leaks a little bit. It doesn't make any difference, water doesn't hurt. If this roof leaks, in this house, it really doesn't make that much difference, a little bit, you know. A lot, of course, and it starts dry rotting. So you gotta be able to live with that little bit of nature coming in on you. And we can't do it. Westerners can't do it.

In California you need so much less space, you can live half outdoors.

You could use planting. For example, you can have an intermediate zone that's planting and kind of like a Eucalyptus trellis with something on it... much of the time you could use that as a building. Like this kitchen. What I want to do is let this kitchen — the breeze is down here. I want to flow out here like the same floorplan, so you can have a table out here and you could sit here and eat, peel vegetables under this kind of shade here...And when the weather gets bad you abandon this outdoor stuff and you come into a small tight space and you get on each other's nerves. But at least the guy and his wife who are going to build it can afford to build the goddamned thing. They don't have to go into debt and ruin their whole lives to pay for the crappy house. So this is too big. This is 600 square feet.

600 SQ FT HOUSE

Val then showed me the kitchen, which is a corner of the house with concrete slab floor. (Rest of the house is wood floor, and you step down into the kitchen.) He's going to put 1x1 duckboards over the concrete. He'll be able to hose it out. For cooking he'll build a charcoal fire outside and bring it in for cooking after coals are glowing. I asked him about wood or charcoal cooking on hot days.

You know, I have a great aunt who lives in Italy in the Alps and she has what they call a *fasces.* You know the sign for the Fascist movement in Italy was a band of little twigs bound together. It's called a *fasces.* They'd go into the woods and pick up the droppings from all the trees and they have them in the kitchen, about that long, little twigs. And she has a little cast-iron stove there but it's shallow. It doesn't have a deep fire box. Takes the thing off, breaks a couple in there, lights it and puts a pan right on it. She doesn't build a big fire, you know, just a little fire like that and it's done.

...like a eucalyptus trellis with something on it...much of the time you could use that as a building...so you can have a table out here and you could sit here and eat, peel vegetables under this kind of shade here...

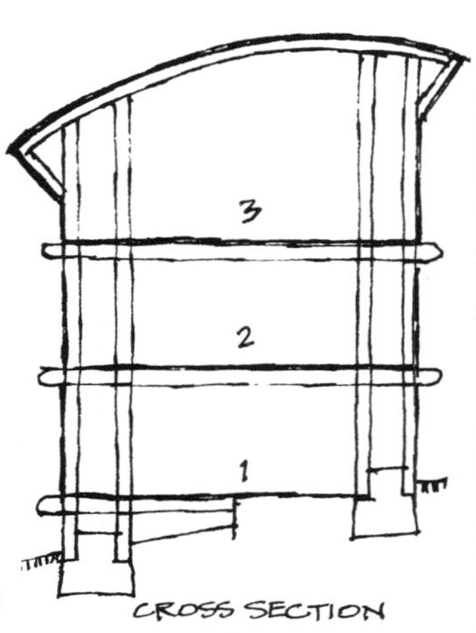

CROSS SECTION

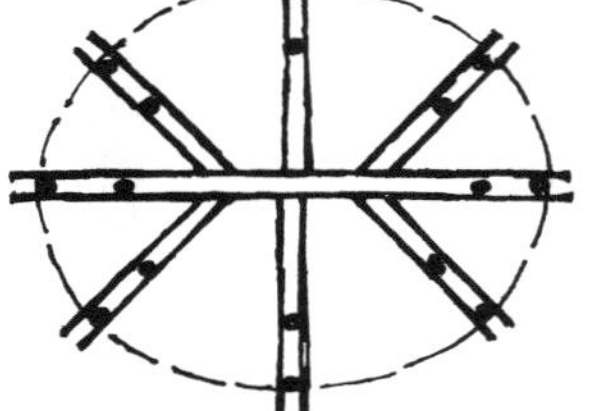

FLOOR FRAMING PLAN
TOWER

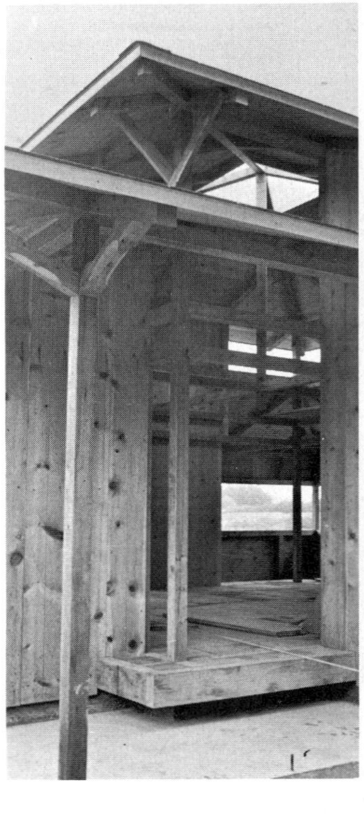

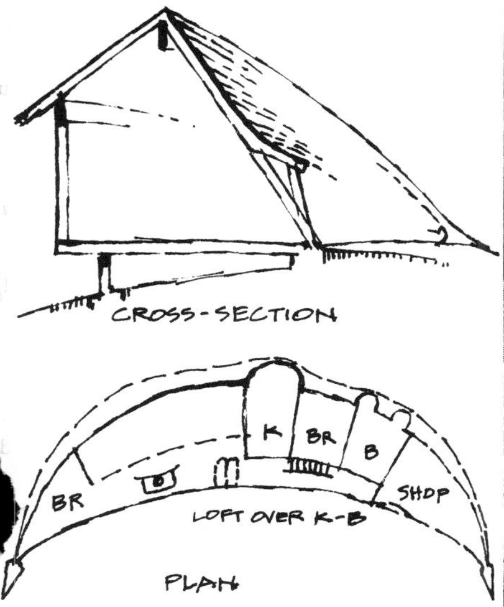

Val's own house. The roof is framed with 2x4 rafters on two foot centers, with 1x4 stripping spaced 4" apart over the rafters. The underside of the shingle roof is exposed between the tripping. The curving ridge and plate beams are minated 5/8" and ¾" boards, respectively.

You need a taste for adventure, you know. And challenge. It's as exciting as hell. I mean you can hardly catch your breath sometimes when you're fooling around with a real wild design. You don't know what the hell's happening. Like an animal in new territory. It's intense and exciting. But all of a sudden you say, what the hell am I doing here?

Color pictures on p. 143

...when we went up to the high lands we built things of volcanic rock; when we came down into the cedars we built with cedar logs; when we came down more we did adobe houses. What I'm trying to say is, when you got to the desert you dug holes, when you got out along the bluffs you did cave dwellings, wherever you are, take what you've got to build with.

The most important thing is...flowing with materials

DOUG MADSEN

On a steep California hillside, overlooking a vast expanse of ocean is the unique world of houses, shacks, pools, lake and beautiful garden built by Doug Madsen. Doug says the most important thing to him in building is to *flow with the materials.* As we started our talk, Doug got out his family album to show me different shelters in the history of his family:

Doug: My grandmother, who was a Danish aristocrat, said this is the way they lived in Denmark in the ninth century, and it's very interesting what they did. They found a hole and then they just bored into it and they rocked it and then in the front they put a chimney. *Turning to a picture of a log cabin:* this was on our sheep ranch. My father built this; mother lived in Paris and we had a house in Beverly Hills. My mother went to see my father at his sheep ranch and I was born there. It was 35 miles by 10 miles...when we went up to the high lands we built things of volcanic rock; when we came down into the cedars we built with cedar logs; when we came down more we did adobe houses. What I'm trying to say is, when you got to the desert you dug holes, when you got out along the bluffs you did cave dwellings, wherever you are, take what you've got to build with.

Lloyd: Sure, where trees grow, wood is good there because it's damp, when you're in the desert, use adobe because it's dry and you need the insulation.

D: If you've got grass, make a grass shack. I think our lives have got to go back to something like that. *Turning pages of the album:* this is where we went to in Idaho. We fattened lambs. This is in the summertime. Of course mother rode side saddle. This is when I was going to Europe every year. And we had big houses then, lots of servants... When they got into brick they got into elegant living. Now I want to show you what type house the cowboys and their wives lived in. This is our Utah life. *Turns to sod house.* Look at this beautiful house. They built whole villages that way. What did the cowboy have to have? He had to have his guns, his chaps, something to keep him warm in....

Turns to photo of tent: Imagine living in either one of those would be a delight. Now see they're getting to be rich. This is an old piece of canvas, the only luxury they had here. This is the beginning of a ranch near Miles City, Montana. While tents provided shelter, the men had already knocked together one building. From their store-bought lumber. Even at this hand-to-mouth stage the family had managed to acquire one luxury. This is how they lived. Now they're starting to get rich. Go build a ranch house, shows a sign of the owner's success, in goes a side porch, a second story plus a mechanical hay barn and mowing machine in the front yard, but the real luxury is a large quantity of window glass. If you look at this you get into mansions again. The things that really built this country and made the people really happy, were these small things. To me, this is where we have to go.

L: I think it is going to move back into that direction, people doing things for themselves. Probably that other building in there, that simple frame farm building might be the first step to moving back to a simpler kind of architecture.

D: They had something like ten children and they lived in that with the pigs, with the chickens, with the hay, with everything. And they didn't live in the house much because they were all outside working. And this is what I think even in the big city, I'm outside 99% of the time, I come in very seldom. And I built this house and I find that it's too big for me. I'm rattling around in it. Too much space. My little house over there is much more in keeping.

L: When you came down here, what was the first thing you built?

D: I lived in my car, I lived in a tent the first year and I built myself a one-room studio in the back. The planning commission said I had to have more to live in. So then I started building something larger and then I got myself into a bigger place than I really wanted to have, because of them. I mean what do you want to do in life anyway? Scrub and clean and take care of a house or do you want to have time to do meditation? Clothes are the same thing. I remember spending hours with clothes and hours with houses. I just think if we could do something about a house with a fireplace, let's get outside. We're bathing altogether too much; do the French sponge bath type thing more. Have more time for meditation.

L: People have gotten back to that way of thinking, builders and architects have gone just too far in the direction of building big places, too ambitious. All your places here seem to open up pretty well, so you just sort of drift outside and that seems important in building. Knowing it could have saved me a lot of mistakes if I had known this stuff years ago.

D: Personally, I think you should never use a bulldozer. You should never move the land.

L: You're just building around rocks and things here, aren't you? What would you advise someone to do who hasn't built anything, doesn't have any building skills.

D: Right now I'm working on something kind of exciting. I would make modular frames. I would make these frames 4x8.

L: You'd build them in place?

D: First I'd build a slab, then I would pour these frames right where I wanted them to be put up. Like this. I'd put anything I had down here first. Like if I had leaves or pebbles or shells, then I'd put my iron through this thing, put my cement and on top of it I'd put pebbles and when I'm finished I'd stand it up. I'd have whatever I needed here, a 2x4, 4x4, put a top across here.

L: What about your windows?

D: Put a window in there and tilt it up.

L: What would you do for a roof?

D: Depends on where I am. In country like this, I don't like flat roofs. I'd get any old piece of wood and make shingles out of it.

L: Why don't you like flat roofs?

D: They take too heavy a lumber.

L: As opposed to a triangular roof you mean? Just a plain gable.

D: Yeah. It gives you height, a skylight in there.

L: When you built these places, had you drawn up a plan?

D: I'll tell you something funny about me. Everything I do comes in dreams.

L: You get a general concept in a dream?

D: I don't know much about anything. I get a problem in the night, I wake up the next morning, it's solved. You have to kind of ask for help. Almost have to say, not *my* will be done, but *thy* will be done. And something else happens, magic happens. You go to town and you find the exact 2x4 you want. A lot of magic in it.

L: Do you think you can build the way you want for somebody else?

D: Yes. I've done a lot of them.

L: Do you find it easier to do it for yourself?

D: Nope. I find if I like the person, like LaVerne, I did the gallery up there. Put the gallery in the second floor. It's a matter of harmony with a person. It's a matter of teaching them to like old wood.

L: How about women as opposed to men. Do you find that they have a sensitivity?

D: I worked with Lolly Facett. We built about four houses up there out of nothing. Then Virginia Varda. Her father left her with an old barn in Monterey. She and I turned it into a magnificent carriage house. Really elegant.

L: I've come to the conclusion that men tend to think of buildings from the outside, the standard male builder, ego trip, architect, you know. And women tend to think of a house from the inside.

D: I must be male and female, I think of both. All your books on building generally will show a picture of the building from the outside which shows you how it looks to birds flying by but not how it is to live in it. I found that an interesting thing. In other words, instead of making a concrete lined pond you just make a lake. That's the kind of thing we had in our house in Utah. The mud on your toes is exciting. We've lost all of that contact. It's like fucking, you know, so many people have lost all contact and really getting into enjoying sex. They get into a pattern or position and that's the end of it. And you'd be surprised at people. Scared to death of sex. Scared to touch anyone. ...

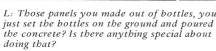

L: Those panels you made out of bottles, you just set the bottles on the ground and poured the concrete? Is there anything special about doing that?

D: I just find cement one of the most exciting things to work with. Most important is the frame and putting plenty of steel reinforcement in the frame. A 4x8 in redwood. 2x4's and reinforcing steel bars. Then putting clear, thick glass and putting windows in. And you put bits of tile and shelves and whatever you want in. And all you do is just flip it up.

L: Your colors? You got your colors from some special place?

D: Color: What I try to do, someone wants a purple bathroom, I can do it because I know where purple rocks are. I know where green rocks are, I know where yellow rocks are, I know where red rocks are. OK. Now, if they want a wall, 4x8, I can dye the cement the color of whatever I want. And I either use an analogous or complimentary hue.

L: Are those commonly available, those pigments?

D: Yes, you just buy those cement dyes. And there are so many different types that are so beautiful.

L: Every builder goes through the process of building one house after another, he goes through an evolution, and you're always refining, you work out trips. I tend to find myself simplifying.

D: That's right. Less and less and less.

L: Yeah, and not even worrying about it being so outstandingly innovative anymore. That's not that important to me, that the house be a masterpiece.

D: The most important thing to me is, Lloyd, I almost think of *flowing* with materials...Like Lolly Facett. I built with Lolly. I'd say, Lolly, give me a 2x4, kind of pretty, 8 ft. long. She'd go find it. And she'd know what I meant by a pretty 2x4. Scroungy 2x4 that doesn't need any strength but would be interesting. Or a 2x4 for structure.

L: Well, when you built, like this place here, you had to do the floor, so you had to have that sill in, by the time you had the floor in, did you know the windows were going to be about like that. You had that planned?

D: Yeah. This all came to me. See, first I was going to dig back in there and probably have maybe a 10x10 cave and do a platform here. When I started digging back, that rock was going to slip on me and I couldn't hold it up and I saw that so I thought I'd better cement that in a hurry. So that left my cave out, so I thought, well, my deck out here is going to have... gorgeous, you've never seen anything so beautiful...and moonlight night was just this deck here. And then I built this, the bedroom. And bath and kitchen, they're real tiny, well then I wanted a bigger kitchen, bathroom. So, again, it's this thing of flowing with what happens today. Then, all of a sudden the whole thing came to me. Why not have my studio upstairs, and go up there and work, and come down here? Now, I would never do that again. I'd work in my studio. I've got so much shit up there now.

L: You'd live in your studio, you mean?

D: I wouldn't be so ambitious. I'm doing 10 different types of things and I realize what I'm doing, of course after my analysis. See what happened, when I was married, then we had the child, then I divorced. There was a sense in the whole thing of failure, marriage failed, relationship to my wife failed, relationship to my child failed. But being Cancer rising in my chart, I've got to have a home, and I'd never really had a home. Mother had these big mansions by the time I came along, with lots of servants, but we never had a home. And I never had a female that would cook, we had servants that would cook, but I never had a female that could cook, and fuck, and go to the beach with and in the garden with, it was always some blond elegant model that went to the opera with and you know, I never had this all around woman. I build a house for a slut. And for an elegant person, and she doesn't come around. Or if she does come around I kick her out. Then I go build another house and it's like a perpetual pigeon making nests. I know this about myself now. Because in Paris, I built myself this, I bought a chateau. I fixed it up gorgeous. I was going with (blank's) daughter. It was going to be magnificent. But she never turned me on sexually. Socially, riding horses, everything that matters except sex. And I ended up going to the can cans and picking up the floozies and having my sex. So I had my life in these different segments, see. So this could be

one of my frustrations in building. Such a dilettante. I have to do a little sculpture, a little painting, a little of this and a little of that and a little of this, because I've had all those experiences and I have to keep them going. I know this, I can't simplify myself. I see a beautiful piece of wood, I want to do some sculpture, if I see a beautiful piece of material, I want to design a dress. If I see a horse, I want to go back to riding. Painting and sculpting and houses. And I see a piece of wood that turns me on and I build a house with it. Now I'm sure if a woman came in here, she'd do it different. Houses the way they are, and labor the way it is, and servantless, and gardenerless, and helpless. So many days I have to take my own horses, I have to milk my own goats, take care of the chickens. So you're gonna have to be completely self-sufficient. The future man, I think, has to be. He has to do more mechanical things. Electricity, plumbing, masonry. Guy with all the money in the world and his big cadillac, and can't fix his toilet. Know what I mean? And this is what I really call a total experience.

...go onto a piece of property, wherever you're going to build, sit there, as long as you can. Meditate. Watch where the sun is when it comes up, watch when the sun goes down. Watch where the winds are. If you can, don't build for a year. Find out what spring has for that spot. What summer's like. What winter's like. Find out what you respond to. Where the winds come from, if you want fragrances you have to plant flowers...

L: Doing everything yourself, you mean?

D: Then you don't fuck up. You know where the plumbing is going, the electricity is going, it's in your mind and you can carry it through. You can tell the plumber this and this and this and argue with him.

L: That was a good thing that I got by building down here, by going out and putting in your water supply, you know exactly where your water comes from, you dig your septic tank, you know where the sewage goes, you compost the garden, you know where the food is going, you eat the food out of the garden so you become aware of the cycle, you understand your relationship. The closer you are into a city, the more alienated you are from that.

D: Also your road. When it rains, you know where to put your road. If it's too hot you know how to put shade on your plants. And also, you can see where the moss is on the trees. If you want a tree that likes lots of heat, you put it amongst rocks, and the rocks reflect heat. You can take it ten feet away and plant a fig tree and it won't take because it needs more heat. But you see, the more you're in tune with this, which I call total environment, then the house expresses total environment, or you. Now you may build here, where you don't need the windows all open, if you build two hundred feet above here, you may have to have more air coming through, it's hotter. 500 feet, it's another thing, it may go in that canyon over there and in that canyon it's cold.

L: Unfortunately, there's no way for somebody to learn those things very fast.

D: The most important thing, I believe this, go onto a piece of property, wherever you're going to build, sit there, as long as you can. Meditate. Watch where the sun is when it comes up, watch when the sun goes down. Watch where the winds are. If you can, don't build for a year. Find out what spring has for that spot. What summer's like. What winter's like. Find out what you respond to. Where the winds come from, if you want fragrances you have to plant flowers. See what I'm trying to say. Like I sat here for months, meditating. I was 20 years building this.

Born in France, and now living in the mountains behind a small Southern California coastal town, Robert Venable is building a boat, but he is best known for his small houses and cabinet work. He, Jeanine and I are good friends; we live in a house Robert rebuilt. The following Sunday morning's conversation was recorded when Lloyd was visiting us...

Bob Easton

...Now what I'm referring to is when the house is not a trip. The house is just a garment. You don't spend all your life knitting a sweater. You need the sweater, you knit it, you put it on...and now you look at the world....

ROBERT VENABLE

Bob: I mean, there's something in the Domebook that's continuing to interest people. I don't think the domes. I think it's one of the few books out that shows the potential of personal freedom. Domes or no domes. It could have been igloos.

Robert: No, no, no. When there is a form of exterior pressure like this, which tend to strengthen the collective — the collective, there is always a particular reaction, most unconscious, from many people. I have seen this with cars. Many people like to have a four-wheel-drive vehicle. Quite sedate people. And as the road facilities increase, more and more people wish to have four-wheel-drive vehicle. They are selling now more four-wheel-drive vehicles than they were selling twenty years ago when there was indeed bad road conditions. This is only a psychological type of reaction...So they are selling more four-wheel-drive vehicles in the same way — for the same reason they are selling — that your Domebook is...

B: Yeah.

R:Now all this is particularly — is generally unconscious, and a lot of people then are satisfied or feel — will accept a particular form of garrisoned society, let us say, as long as they have an escape mechanism which is the four-wheel-drive vehicle.

B: It's younger people who are frustrated now and looking for a way to go, you see. Now, they're not buying the four-wheel...

R: No, no...let's forget about the four-wheel-drive vehicles. I was just mentioning this as a ...

B: This next book has got to be a turn-on, to inspire people to carry these aspects of freedom into other areas of their lives, you see, because very few people are going to be able to build.

Lloyd: You know, if we give them choices, almost like mandalas for builders and designers, this iron age cottage, you know, you can't go out and build it and live in it now, but just look at it and see how these guys built from what they had to build with, not from an abstract concept, like a geodesic dome. Here they had poles and rock and this is what they did. We don't want to propose going in any one direction, but instead, here's all this stuff that people have done, beautiful old things, crazy new things, crazy old things....

R: Yeah, because the freedom that we are talking about right now, at least, the meaning of it is, after all, a self-decision process, helping people to a self-decision. And one of the means of self-decision is the liberty of being free in designing one's own dwelling, home.

B: ...it's archetypal — you know, it's something that's common to all men.

R: Yeah, it's pretty basic, and of course it may be argued that in primitive society nobody would build their hut on the same level, but this is mostly because of — for exterior weather conditions and soil and whatever material is available.

B: Yeah, they had to respond to their material.

R: Yes. Also one might mention that whatever divergence of view a man might have with his tribe, he at the time didn't feel to manifest it by changing his dwelling, his house. I mean, he had more of a theater of choice....

L: Well, I've continually made the mistake of building from a preconceived notion, from a concept, and then end up spending a year having to build it. I think it's dangerous in building, it's not like doing a painting where if you screw up or it takes you too long, it's not such a big thing. But a building....

R: Yes, yes, it's a statement. And not too many people have a chance to be making a statement and a lot of people are afraid...all you can do is encourage the timorous, the timid, to be himself. That's *all* you can do. That's enough. The vast majority of people are still children who want to suck at the mother's breast, and that is 99%, this is an *incredible* large number. And to them it doesn't make any difference the color of the milk. They have no idea....So, there is a few who naturally seek independence and among these few there is quite a large percentage who are a little timid or timorous about it and have a sense of maybe not being good boys, and this book is to encourage those ones — to be bad boys. (*Laughter*) And that's all, you know.

Later, talking about whether domes are a good use of building materials or not...

B: Maybe it's not waste, maybe it's...an economy of materials, a wisdom in the use of materials.

L: Well, we may have inspired a lot of waste, I mean, it might not take much plywood to make a dome, but if it only lasts for 10 years, then it is a waste, you know, where it should last for maybe a hundred years. Now I worry about that.

R: I disagree. Precisely one of the saddest things that people are still hanging on is the spirit of the fortress, the spirit of building a house that's going to enslave them, instead of living their life with a light building material that is deciduous. So that one built for one particular experience of life now is to me the preferred thing. There is no reason at all to build for a century, there is no reason at all any more to transmit to your child, your estate. That is still present in the back of the mind of many people. This is the child, this is the child security....

B: ...Well, like a lot of the domes are just badly built. And what I'm talking about is not from that psychological thing, but almost from the craftsman's point of view, that if you're going to build something build it well so that someone else can use it after you leave it. Not in terms of passing it down as an inheritance....

L: The idea of portable shelter always sounds good, but if you're going to have plumbing, wiring, a septic tank, road or path, it's just not sensible to move the building. If you have a floor — there's no portable floor — then I think you should build it to last.

R: I don't want to give a new feeling of guilt to anyone and the new feeling of guilt is the idea of waste, for instance, and the idea of being non-ecological. That is the new sin. And I would not pass that upon nobody. If there is some people who want to build a fortress, fine. All I'm saying is that to encourage those who want to live temporary and that's all. And ten years in the economy I was talking to you about a month ago which is not necessarily an economy of money, but an economy of flight, an economy of action, an economy of energy. A centralization not outside of it, I mean, if one man builds something for 10 years, that's fine. Any building at all of any kind lasts twenty-five or thirty years. Anything. Well, we live in a speed up period, twenty-five years is the equivalent of a century 50 years ago. So what is meant by long time or temporary is relative. So to me, I know nothing worse than the father passing his property on along with his sin to the son. Nothing worse than that. You build a shelter, a home, or whatever, for twenty years for any kind of thing, is a lifetime.

L: Yeah. I just think in terms of the way the building ages, you know. I mean, if the building gets worse with age, then you've got a liability. You know, there are buildings that get nicer, that sort of soften and blend in, like this building here. But, fiberglass domes get worse. Plastics get uglier. And if you're walking through the woods whaam, there you see this thing and it's got dust all over it and it's kind of shiny underneath the dust and the windows are yellow, that worries me. That's where I think about permanency as opposed to....

R: Watch out once again, Lloyd, remember that permanency is a female, permanency is a source of laws. Everything that is permanent are tend to become a lawmaker. A tradition maker. So let's not insist too much on the word permanency. There are a few words to avoid and permanency is one of them.

B: Right, but let's define what we mean better. Now what I'm talking about is when you tie your tent over there, over your work bench and your saw, you tie it such a way that the wind's not going to blow it off.

R: No, because I've got two ounce of brain. That's all, and I....

B: That's all I'm talking about. The guy who builds the plastic dome is not tying his tent right. That's all I'm trying to say. Now that doesn't have anything to do with permanency, it doesn't have anything to do with tradition, and it doesn't have anything to do with guilt.

R: I wish you would never mention the word *waste* or the word *ecology*. I wish you would only be concerned with the *life element* and that's all.

B: I'm not holy, you're not holy.

L: We're not holy in the terms of guilt and ecology. The reason I do all of this is because I want there to be beautiful places like this to live inside and for people to look at as they walk around amidst the increasing ugliness.

R: I would be interested by seeing beauty that is reflected by people who have a sense of theirself. That's what I'm interested in because after all, that's what the earth is for, I'm here to say. The earth is there to be exploited. And just because it has been badly exploited doesn't mean that ultimately we are not going to exploit it at all anymore. Because it is at our service. It's there for us to use. And let's not talk about it. About waste. Let's talk about the spirit of man. And that's all....

L: ...I'm also trying to tell people not to be so uptight about the building inspectors. I see people really changing their buildings for the sake of the building inspector. You take in this piece of paper to the building inspector and six months later you're supposed to have built that same thing in spite of changes in your head while building....

R: This control or action by mother/father state is a sacrilege going around. I was so much conditioned by it during the year I was building here...When I went to Europe, I got a little farm and stayed there about year and I proceeded to make a bathroom inside and one day I was working on the bathroom and my *god dammit*...it took me quite a long time to realize that I was in Europe and that was all right. I thought about going to the authorities and saying may I build a bathroom.

B: Do you know why you didn't have to get a permit in Europe? I would guess the reason you didn't is because you didn't build something in Europe unless you knew how to do it. Is that true or not true? Is that a result of the guild....

R: No, it has nothing to do with that. It is simply where I was building which is farm country, which is a small village....Rigor of the control system hadn't reached there. Like it was in California 20 years ago where rigors were restricted to the downtown area. Now it is everywhere. The whole country is a garrisoned country and many part of Europe have still not reached that point....

B: I was thinking, that the way you get past the building department is to become a good builder. To do it right. To do it well. Now not necessarily to code, but to build well.

R: Who is going to decide? Here now we have the hang-up with the authority once again. And your whole book is to free the people from the hang-up of authority.

Jeanine: Then there are people who would take advantage of that and create unsafe...exploit the people who are going to be living in them. The building inspector had a function. I mean the whole idea of controls originated because of things that had happened before which was exploitation of people by bad builders.

R: Let me make another point. Every piece of property in this country is government owned. In fact, we live in a communistic country, or a communistic regime is based on the old belief of private property. The old belief of private property still what keeps the whole thing going, it's an illusion. Actually, your house or anything is part of the national gross income. It does not belong to you. It is very important that every piece of property is at its most value as possible because it increase the national gross income. That's exactly what it is. On top of this, the mother bureaucrat government look over us like a mother hen over its brood. We are supposed to be crook and sinners and we are supposed to tattle on other bad stuff and others are supposed to be so foolish that they can't make a judgement of themselves, so we have to be watched, so we are deprived of our right to risk and to our own judgement. That's communistic thing.

B: And if we have to license everything we do, it's a way of like getting a little approval and they can keep...tabs on us.

J: That all sounds fine, but what really happens, how many people are going to educate themselves? I agree. That would be wonderful if everyone could build what he wanted. But it's just not realistic.

R: Look, this is just what I was going to add. This is the same kind of vicious circle, once you start something you just keep adding. If someone decide he's impotent and incompetent, and is going to delegate the right to govern his own life to the establishment. This is going to increase more and more and more and more. It just seems to me that such basic things as looking after a house anybody can do. Once the idea of delegation has taken place, then it increases all the time, the more it increase, the more there is control, the more....

J: But it's the people who....

R: Let them suffer to begin with...then they themselves will establish...when we buy a car we are being cheated. *Royally*. By the establishment. And we haven't any choice at all because the whole country is, the whole car system is made to deteriorate.

J: How do you get the people individually to have the desire and the will to control their own lives is the essence, is the problem, because it's going to happen no matter...because you allow...this...

B: We're not allowing anything, we don't have any power.

R: He's right.

J: I don't say you, I mean everyone, all of us. For example, the men who built the car industry abused their freedom. The people only got headlights and other safety factors through legislation.

R: No, no, no, I can't go on that trip. The more safety there is, the more safety there is needed. The less people are alert, the less people are alive, the more....Safety, security, those are enemy words for me. Obviously there is not much risk and danger involved into a house, and yet I have heard building inspector look at a house and mention that they were inspecting that it might endanger the life of the next buyer. Now, I mean, this is talking like the priest of God. What do you mean? *Endanger* the life of the next buyer. *Wow*! Let people discover that themselves....

J: I don't trust the big electric companies that come out and wire my house. I don't trust some 18 year old, either, coming in and wiring my house, who has no knowledge but thinks he does. I'm not saying I want the building inspector. I'm saying I want confidence...competence.

R: Why do you want competence?

J: Because I think, in a way, we're very naive, it sounds paranoid, but I've had tragedies in my life and I know we're naive as to the things we allow to go on and by just being unaware...there are a lot of people who have had their houses burned down. A lot of incompetent things have gone on....

R: I don't want to talk about that. That is so beside the point. I mean this is talking like a bureaucrat. And this bureaucrat talk. *Safety, security, happiness*. Ah, happiness, that is very important.

B: Yes, you have your house burn down with a child in it and you talk to that man. That's not all that rare.

R: Houses burn down all the time.

J: I mean like the tract houses that have the plumbing system blow up....

STUDIO BUILT BY ROBERT

JEANINE AND CHANDRA

R: I really don't care about that.

L: Well, she's talking about the rules that do apply to the slipshod contractor and anybody who is building is concerned with how those rules apply to doing what you want to do and your own freedom.

B: What we're saying is that rules come after the fact. It's a way of living and it's got nothing to do with things after the fact or the person writing down the notes to music, we're talking about the music. And when you talk about the bureaucrat you're talking of the written notes after the music is played. That's not what we're concerned with.

R: When it started when I mentioned the hypocrisy of...that's precisely the kind of discussion to avoid, am I right? Immediately we got into it.

B: Let everybody else talk about those aspects, we're not talking about them.

R: Talking like this just show that we're trying to talk freely as possible or to act freely as possible. Wherever the word security or safety is concerned, immediately, the freedom just evaporate.

B: Maybe there is a correlation between danger and freedom....

L: Absolutely. Freedom is dangerous.

R: But once again, let's not even state it.

J. Oh, shit, that's fine, you're sitting here so safe and so middle-class.

R: What feelings you express are legitimate, not legitimate...understandable but it takes a long time to answer you properly, that has also to do something with the breaking of a family, the fact that within the family there is no more head of family, we have been made single, everbody is separate, and therefore for the nostalgia of basic security has to rely on the state....

J: No, I don't agree with that. The answer is that it takes individual knowledge and responsibility for everyone to free himself.

R: Those are things I don't even want to discuss. Because this is it. Our particular society has devoured the family. With centralization, a particular type of centralized society that is mass media, standardization cannot tolerate the power at the base upon which this society evolved to begin with. See there is constant paradox. It destroys support, it destroys base. And in order to remain active it has to become more and more tight, more and more fascist, more and more control. There used to be, and I'm telling *you (sharply, to Jeanine)* there used to be a time when, back to king day, back to feudal day, when it was pyramid type of thing, it depended on a strong, solid base. That is to say, the family was protected as being a source of power. And there was such thing as the head of a family and there was estate or peasantry or the absolute inviolability, — please listen....

J: I want you all to shout really loud right into that tape recorder....

R: I am explaining it, why it has been....

J: Your superiority is *unsufferable!* (gets up and leaves table.)

R: Christ sake, man, I mean....

B: Well, I mean, it's exactly what you're saying, the family has disintegrated. Now everybody is an individual. Jeanine wants her individuality. She wants to sit here as an equal at the table.

R: And at the same time there is a desire for protection, and it used to be that the head of a family was a man, no matter how bad or good, who furnished the protection. Now that no longer exists. The only entity, the only abstraction is the state with its laws and regulations and there is among the people who are in need of protection and that's the 99%, they want control. And there is control without responsibility. There used to be a time when the head of the family could be judged by his friends. That's no longer. But really it is the fact that our particular form of civilization today cannot tolerate power anywhere else but at the top, not at the base. This agitate the whole thing hence the desire for people to have a four-wheel-drive vehicle, once again, give them this feeling of being, I mean, it's such a pathetic illusion: it never goes anywhere, there is no road, and now they are making national park with trail for four-wheel-drive vehicle, I mean if that is not a merry-go-round, I would like to know what it is. And pretty soon they will make those roads safe. Safe for four-wheel-drive. You pay your 25 cents and get on the wooden horses....

Later, talking about the earlier remarks about building from a concept...

B: If we need a metaphor to build today, let's upgrade the level of our metaphor. Let it be an example of experience. In other words, we don't need the concept anymore. What we need is to open our eyes to what we have right around us. We don't need to carry an oppressive concept in our mind which dictates our actions. We live too much in the superficiality of our minds.

R: Sometimes we preserve the superficiality because we do not dare to dig into something that's going to be solid and stable and anchor us and force us to look upon ourself, so we treat a house with the same kind of superficiality that we treat other things, the quick passing...a house is no longer a home since it is a very superficial type of thing and a lot of people move into those prefab house accepting the superficiality, satisfied by it, not wishing to get into something which cling to their skin, which make them too much aware of it. One of your goal, as I understand it, is to give those people the restoration of the idea of the home. And the home is a place of darkness, is a place of birth, is a place of death, is a place where you can observe the outside from the inside, is a place where you can be objective, is a place where you can be free and have all your fancy and all your imagination. It's a place of privacy. It's a place of darkness in the sense that you can dwell within yourself without being too much ashamed or letting go your inhibitions. It's a place of privacy, it's a place where you lick your wound. That's what a home is. And a home that is superficial is not a home. So it does appear to me that indeed, the idea of the house as a home can be, is well worth to show it....

B: All right now, your idea for a house that you've told me about before presupposes a temperate climate....

R: It's something that is personal and it do suppose a temperate or semi-arid climate, such as Mediterranean climate. Which we have here. As far as this country is concerned, I would imagine that most of the western, that is to say Nevada, Arizona, New Mexico, California....Phil Briton, one day, we did talk about houses, he asked me what is your idea of house and I took a piece of chalk and I drew on the concrete, we were standing outside just like this (gestures) which is indeed my idea of the house. So the idea at first would be... those are walls...can change this...the wind blows

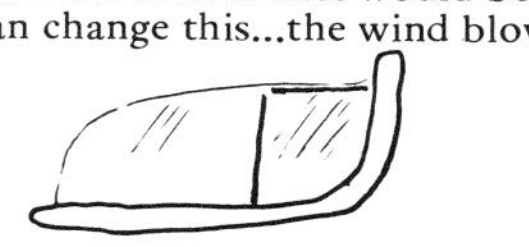

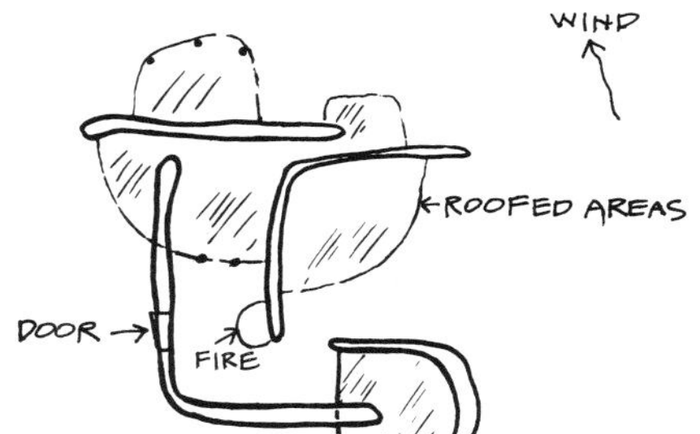

from this direction...this is already quite elaborate... ah, you may decide to put a roof someplace...so if you roof that up....

L: Would you build the walls then decide about the roof?

R: Why not, Or which ever way...now that roof can go over the wall or underneath, it wouldn't make much difference.

L: Now if you were going to build something like that would you draw it before you started to build it? If you had a site? And how would you make the walls? What would you make them out of, if you were going to build it right now, say....

R: Probably from a brick, adobe or ferro cement. There could be door and opening here if one wished. Now the idea is that it's this outdoor indoor type of thing. Some of those places could be unclosed, permanently or semi-permanently with cotton or with things but it has to be so you live half outdoor, half indoor all the time. It rains and that's fine, the whole thing can be tiled. And there can be a big fire pit someplace. You live in a house and you experience with your body all those change of season. Also you dress yourself. Most Americans wear light cotton suit, where in Europe it is more the tweed, English tweed and Mohair, heavily woven....

L: Well, but the Japanese carry the heater around with them, so the idea is that you heat person and not the house. The Victorian house, you heat the big, high rooms; instead put on a sweater and turn the thermostat down 20 degrees.

R: But, I like this because one could decorate himself with clothing. Instead of warming the house you warm yourself, you warm yourself with very colorful...decorate yourself instead of decorating the house.

B: In the Tibetan monasteries, the heating there is tea, they drink tea....

L: Or strip, if it's hot.

R: Right, I mean this is one place you can walk naked all the time. So that's my idea of the house.

L: Do you think it matters how long it takes you to build?

R: It would not. There is hardly any foundation at all, that would defy the building code, I mean, this is to be free, make a foundation for it and I envision Spanish style on the floor but it can be whatever it wants. And all of a sudden you decide to have a — because your friend drop by all the time and you put a lean-to where they shack up with sleeping bag.

B: It's not a concept type of house.

L: It could be so simple.

R: That's exactly it. So the idea that you clothe yourself, and being outside. Sometime, I like it. *(Robert now lives mostly outside.)* There is no doubt. Sometimes in the evening and it doesn't make any difference if it's a nice evening or if it is rain, but I light the fire in the hibachi and I'm looking around and I see all those houses shut with light inside, and without any false ego on my part, I really can't understand it. Why do they shut themselves in? Well, I did that for four years. Like everybody else. But why do they shut themselves in. It's ridiculous. *Bling* goes the door. I can hear the door closing. I can hear door opening, somebody step outside, *bling* go the door. Okay, fine, when he come back in, the door open and *bling*, close... It's always close, tight, shut. Okay, now this kind of house, the roof can be held with post every 2,3 feet, and panel puts in if one want to on a chilly day. But surprisingly, I wish — the only luxury I wish I had, that I could warm up the coffee in the morning in my bedroom, instead of having to step outside. That's the only luxury I deal with that I could improve on, but I have no room to put the heater.

L: Do you do all your cooking outside?

R: Yes.

L: How long have you been living there?

R: A year and a half. But I have been camping for the past three year....

B: A lot of people now — let's say, the people working on solar heat and wind energy, are middle-class American youth, because they still preserve the idea of the insulated house.

R: Well, this kind of house *(we're sitting inside, around a table)* lend itself to the form of intellectual conversation that we delude ourself into having. When you are outside, it's not so conducive to this intellectualization in which we like to indulge.

Imagine now if we would be outside around...there would be a fire going, and it's a little cold so we would have sweater on and would be sitting on benches or on whatever and around the fire. And if it drizzle a little bit, then it drizzle a little bit. If it drizzle too much, we would move to the lean-to, but still we would be outside. The only part that would be enclosed would be bedroom and the john, even if one want to enclose that....At the same time, since we are Western, since we are sophisticated...intellectualization will take place there. But it will take place entirely differently....

RAMP AT STUDIO

RIDGE PLATES

BOAT AND CAMP

And the home is a place of darkness, is a place of birth, is a place of death, is a place where you can observe the outside from the inside, is a place where you can be objective, is a place where you can be free and have all your fancy and all your imagination. It's a place of privacy. It's a place of darkness in the sense that you can dwell within yourself without being too much ashamed or letting go your inhibitions. It's a place of privacy, it's a place where you lick your wound. That's what a home is. And a home that is superficial is not a home....

ROBERT

L: Well, what are you going to do now. Are you going to take off in your boat?

R: Oh, yes. For this boat is also a very particular trip because the particular frame of mind in which I was when I started the boat, I slowly changed. And there's all kind of changed things that take place. I'm faced now with living now the particular kind of life that I anticipated living on the boat, an easy type of life. So sometimes I wonder, is it necessary to go on with the boat. Also, the boat... it's a product of an action, of doing, and I have created something which do not possess me, but even some of where I am. Some of my mind. And it's also quite interesting because it would be in a sense, a little bit like a woman give birth to a baby, but in an entirely different way since a woman give birth to a baby totally independent of herself, she is entirely submissive to her process.

She's making no choices and she's submissive to the...she's only a vehicle. She's the lot upon which the house is built. She's totally submissive to the process, for lack of a better word, for the process which take place inside. When I build the boat, I am the process. I am just like all this intricate mechanism. I organize and produce living, total mystery, I am this process. And that blow my mind because I don't know what I am doing. Also because I have chosen, almost unconsciously, unorthodox type of life on this boat. There is a very orthodox hull but after that everything depart from it. So I run into problem I have to solve, problem that wouldn't have arise with such clarity if I had chosen a conventional design....

B: It's not a houseboat but it's not a sailboat either, like it's a boat for living on, really, isn't it?

R: It's a boat. It's a sailboat on which I intend to live. I remember when I was a kid I saw an old man on a boat, and he must have been about 70 and he was mending something, and he was on a boat, the boat was in drydock and it was a kind of dilapidated old boat and I envied that old man. He was there, he was oblivious of — whatever. That was his life. So that would be a little bit of what would happen, although I don't make any pattern. That's the idea. A place. A place I own. I also want a place of privacy. I don't want to go particularly in the desert island or anything like this. I'm not that interested in it. I want to *master* the working of sails. I think there is a great pleasure in this, in sailing well and making the best of the boat and in exploring the sea. And I am seasick, and I never *(laughter)*. I have no masochism, but it's — it's a nice way to be going places. I want to go in London. I want to go in big city; I'm interested in big city. I'm interested in desert and big city. If I had to choose, a choice of living, it's either desert or big city. And I had it, but we somehow compromise in choosing the small hick town and that was a very big mistake. So I want to go in China, I mean, I want to go in London and I want you to go there with your boat, you are already at the measure of self-assurance. You can look already at this country from a point of self-assurance and therefore that allow you to be less emotional, therefore more objective. Also you know that you can go anyplace, you know that you are distinct and instead of being only emotional about the idea of self-distinction — I am not like anybody — and you become frustrated and you become a psychopath like so many other people by knowing that you are not like anybody; distinct from them and never doing nothing about it. This, even if it carries with it a little bit of vanity, to me, it's a good type of thing. I am distinct from the rest, the other guys.

L: If you're manifesting it.

R: Well, so it gives me an indispensable — a known manifestation of self — and it just allows me to look with more objectivity and more equanimity also upon the rest of the world. I mean, gentleness come from strength only, not from weakness. It has never come from weakness, only from strength. Love come from strength, not from weakness. Now the love of the weak and the meek, they can shove it. I don't want it. I know it's anything but love. Only the strong can offer strength and gentleness and equanimity. If they don't they're not strong. So this is the kind of thing, that it's a home, a means of privacy and of traveling. And also really, in life I have nothing better to do. I mean already I think what the fuck really am I going to do. I can do nothing without kind of blowing it up, so I might as well go ahead this way. There is a good idea also, perhaps, the idea of the river boat — such as the Missisippi River, you know. A few years ago, when I married Margaret — we were married in New Orleans and we wanted to look for a place. And I said, "Hey, I know. One day I am going to build a river boat and it will sail all up and down the Mississippi River," and she turned to me and she said, "Are you crazy?" *(Laughter.)*

WOODLANDS

by Ken Kern

Woodland management practices in this country are not unlike agricultural practices: both are currently based on a *ground-line* philosophy. That is, the harvest, the cutting, the profit are secured at ground-line. Monoculture is the accepted ground-line agricultural practice, and the same mentality results in clear-cutting and even-growth production in woodland management. A top-of-the-ground harvest orientation even twists the minds of trained foresters who advocate such abhorrent practices as control-burning and crop-dusting.

Bad forestry is in some respects more serious to a nation's welfare than bad farming. For one thing, there is more to lose: the world contains twice as much forest area as it contains land under cultivation...a third of the earth's surface is classified as forest soil. Originally in the U.S., before the white man came, over one-half of our two billion acres was in forest. Forests have been decimated in places like England for so many centuries that even the forest terminology has atrophied...forests become known as "woods".

Every seriously operated homestead should contain within its bounds some percentage of woodland acres. Besides the obvious "ground-line" value of timber, firewood, fenceposts and pulpwood, a woodland helps to control wind and water erosion. Experiments in Wisconsin showed that soil losses were one hundred times as great from pastured land as from unpastured woodland, and water losses were sixty times as great. Protection and shelter from the wind can be an essential attribute of proper woodland planning. And the food and shelter offered to wildlife from a woodland can help to maintain one's homestead as a balanced, complex association of plants and animals—which includes in its natural harmony birds, bacteria, insects and fungi—rather than *attempting* to eliminate them, something which never succeeds and only results in their lopsided development.

Woodlands can also influence the microclimate. Winds carry moisture that is lifted by the sun from large bodies of water. This moisture-laden air will move indefinitely or until it reaches a woodland. Trees transpire...cooling the air, spraying the sky and multiplying the cloud cover. This *upward* rainfall rises until it meets the moisture-laden air and then drops down as precipitation. Trees located on hills and mountains offer the best obstruction to clouds and thereby increase the rainfall.

Traditionally, woodlands are located on land classed as non-agricultural because of inaccessibility, steepness or poor soil. The ground-line "growth factory" concept keeps most farmers from planting trees and shrubs in rich, deep soils. No matter. Conifers are best planted on eroded, compacted, humus-lacking soils that perhaps once supported hardwood trees. In time the conifers will build up the soil to the point

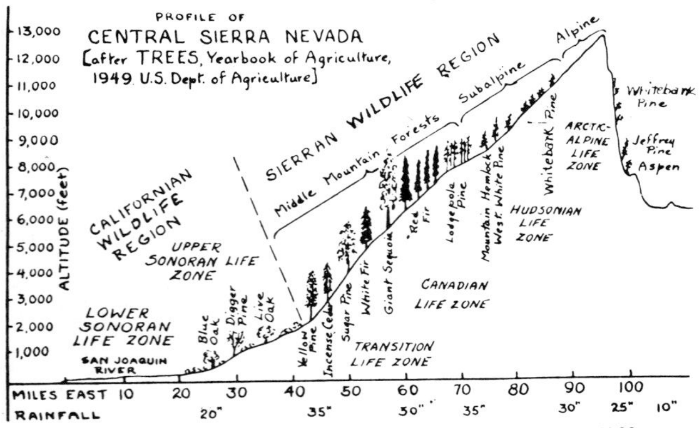

where hardwoods can become re-established. Trees differ widely in their soil and moisture requirements: where yellow poplar requires a deep and moist soil, black locust will thrive on a soil deep or shallow, moist or dry. Alders survive in wet, undrained soil while willows require wet, drained soil. Eucalyptus will dry up a swamp, but will also drain ground water on a water-needy land.

The inherent capacity of a tree species to withstand shade becomes a second factor that determines a woodland type. Again, the range of tolerance is very wide as accompanying charts indicate. Sugar maple seedlings require only two-percent full sunshine, and loblolly pine seedlings require nearly full sunlight to grow satisfactorily. Seedlings are generally more tolerant of shade than mature trees, especially if they are grown in a good exposure and on good soil. Shade givers grown near seedlings also provide valuable physical protection.

Someone once discovered that trees have one or another of three different-shaped root systems: a *spear* shape, as found in the oak which taps minerals from great depths; a *heart* shape, as found in the birch tree which lifts huge quantities of water; and a *flat* shape, which is designed for the support required by such trees as the Sitka spruce. Now it becomes obvious that a plantation of a single tree species would offer fierce competition at the same root level for food, moisture or support. For this reason a homesteader should make certain that his woodland contains a mixed variety of trees—different evergreens and different deciduous trees—with a wide variance in age and growth. A mixed woodland is far less subject to insect damage mainly because mixed woods supply a mixed humus. A loose, crumbly layering of mixed leaves is certainly a better stimulus to germination and to growth than a dense, impermeable layer of a single variety of needles or leaves. Witness the fatal Dutch elm disease which totally wiped out mono-planted roadside elm tree plantings all over the East. A mixed woodland also provides a varied supply of materials for use and for sale. One naturally aims at producing the highest grade woodland products...such as black walnut furniture wood, hardwood veneer, or saw logs, poles and pilings. Low-grade products such as fuelwood, pulpwood or railroad ties can be mostly a by-product or culls from high-grade products. Each of these main woodland products will be discussed in detail below.

High-grading wood products from one's own woodland is only possible where the homesteader does his own harvesting, as contrasted to selling "stumpage" (standing timber) to a contractor. Homestead-sized woodlands only attract the small "gyppo" contractor, who is notoriously destructive in his timber removal operations...destructive, that is, to the remaining woodlands that aren't stolen by him. The sale of stumpage brings only about 15% of the timber market value. About 30% of the operation is allocated to felling and to bucking the trees. Another 40% is taken up in yarding and in hauling, with the balance in profit. There is no part of this timber harvest operation that cannot be handled—even single-handed—by the homesteader. In due course one learns that a tall, straight, well-tapered southern yellow pine or Douglas fir will bring more money sold as a pole than sold as a saw log; large, high-grade logs are best sold as veneer timber. Lower-grade timber can be high-graded, too, and sold as saw logs or finally as pulpwood, fenceposts or cordwood. While contemplating this whole woodland operation, keep in mind two important factors: First, the work is performed during "slack time", in the winter months when homestead chores are minimal. Second, the homesteader can supplement his income by utilizing to the best advantage the resources that he has at hand. They are, in order of availability, *land, labor, management* and *tools.* Unlike other members of our society, the homesteader generally does not have access to wages, rent, interest or unearned profit.

Before delving further into the economics of woodland management, something should be said of woodlands as *sheltering* devices. My interest in windbreaks was first aroused years ago when I read about the Lake States Forest experiments at Holdrege, Nebraska. In these experiments exact fuel requirements were recorded in two identical test houses...one exposed to the winds and one protected by a nominal windbreak. With both houses maintained at a constant 70-degree inside temperature, *the one having windbreak protection required 30% less fuel.* It was also found that animals in a tree-protected yard gained 35% more pounds during a mild winter!

There is a distinct but thorough science to shelterbelt planting involving tree varieties, planting layout and tree spacing. For example, poplars are often used as windbreaks because they are able to withstand high wind pressure, but poplars are great soil robbers. So it becomes prudent to alternate poplars with a complementary variety such as an alder tree. The alder brings nitrogen into the soil...being one of the few non-legumes which have this property. And for reasons mentioned earlier, it is important to combine coniferous and deciduous trees. A dense monoculture of conifers may successfully break the force of prevailing winter winds, but at the same time they may obstruct the flow of cold air, thus impeding natural air drainage. The winter-bared branches of deciduous trees will not obstruct this important air movement.

The most effective windbreak is in the form of an "L", with the point to the prevailing winter winds. This layout is best both for preventing evaporation of soil moisture (thus raising soil temperature) and for catching and preventing drifting snow around walks and buildings. Snow in the shade of shelter trees melts slowly in the spring, conserving moisture for row and sod crops.

Row and sod crops benefit in other indirect ways by planting windbreak-woodlands in close proximity. The traditional treeless, field-crop pest is, of course, the sparrow. Its birds of prey are the owl and the sparrow hawk, which cannot function without the protection and the vantage point offered by woodlands. Birds are attracted to woodlands, and this is important to the *control* (not ground-line *eradication*) of insects. The tomtit is known to consume 80 pounds of agriculturally injurious catepillars in a summer; the starling destroys harmful larvae. The thrushes, jays and starlings are all valuable tree-seed planters. Wherever possible plant shrub fencing—such as multiflora roses—where fencing is required, to encourage bird habitations.

Some form of fencing—shrub, wood or wire—should definitely be installed around the homestead woodlands. Grazing a woods may help reduce fire hazard, but animals (especially sheep and goats) destroy young shoots and seedlings, compact the soil and, in general, disturb the physical structure of a woods....

The demand for saw logs is greater than for any other timber product, and a well-managed homestead woodland produces a minimum of 500 board feet of lumber for each acre each year. The key word here is "well-managed". According to the Department of Agriculture, an optimum-managed woodland and a good growing stock produces three times as much wood as the average untended woodland.

I can list three rules essential to the maintenance of optimal woodland conditions: (1) choose the best species, (2) keep the stand at optimum density and (3) correctly cut and prune. These factors will now be discussed in detail.

A conifer is preferred for construction because it is a softer wood and easier to work. Especially valuable, building-wise, are the pines, spruce, hemlock and Douglas fir. In checking over current nationwide lumber prices one notes that hemlock and beech have a very low stumpage value, whereas birch and maple have a very high value. There is an even greater price range in lumber *grade*: select white pine sells for $300 a thousand board feet, while number 4 common grade sells for $100.

A high lumber grade is best obtained by maintaining the woods at proper density. The younger the trees the greater the density. When forests are planted from scratch about 1,000 trees per acre are usually set out...the mature crop should contain about 200 trees. One should maintain a uniform rate of growth in a correctly managed woodland. If the trees are sparsely planted they will grow too fast. The faster a tree grows the more taper it will have. And, of course, the greater the taper the larger the knots. Lumber without knots brings three or four times more money. When a timber stand is crowded while the trees are young the lower branches naturally die and break off, thereby reducing knots. Conversely, a timber stand should not be allowed to grow too slowly...that is, in too crowded conditions. When a conifer or hardwood grows slowly the wood is light and weak and is, thus, a poor building material where structural strength is required.

Correct pruning, thinning and cutting practices constitute the final factor for maintaining optimum woodland conditions. Woodland thinning is done to reduce the density...maintaining a fast-growing diameter and slow-growing height. It is better to make moderate thinnings at frequent intervals than heavy thinnings infrequently. And pruning and thinning should be done in early spring, just before the growing season begins.

Some thought needs to be given to the question of when to harvest a saw log. A four-foot-diameter log contains more board feet of lumber, but it is not economical to grow a tree to such a large size. Small trees increase in board-foot volume much more rapidly than large trees: as a tree grows from 10 to 11 inches in diameter the board-foot volume increases 33%; from 20 to 21 inches the volume increases only 10%.

Before attempting to fell a mature tree, first consider the damage that may occur to surrounding trees on its way down. Keep in mind the slope of the ground, the lean of the tree, wind movement and final positioning for removal from the woods. An undercut is first made to guide the direction of fall. On a conifer the undercut chunk should be about one-fourth the diameter of the stump; a hardwood tree should be cut one-third the diameter. The main cut is

Relative Diameter Growth

Rapid	Moderate	Slow
European Larch	Douglas Fir	Cedar
Loblolly Pine	Ponderosa Pine	Hemlock
Aspen	Redwood	Longleaf Pine
Black Locust	Spruce	Beech
Cotton Wood	Black Walnut	Oak-Black & White
Willow	Elm	Sugar Maple

made on the opposite side of the undercut, slightly above its base.

Fuelwood is one of the neat cottage industry by-products of a well-managed woodland. From thinnings and prunings alone, about one cord of new wood is realized per acre per year. A cord of hardwood weighs two tons (twice the weight of soft wood) and, if dry, will give as much heat as 200 gallons of fuel oil or one ton of the very best anthracite coal....

This article is one of the chapters in Ken Kern's just-published *The Owner-Built Homestead*, available for $5.00 from the author, Sierra Route, Oakhurst, Calif., 93644. The chapter contains several drawings, which were not included here due to space limitations.

The maple grove that is planted by a young man may be enjoyed by him through more than half of an ordinary lifetime. With proper care it will perpetuate itself through a long course of years, and for aught we know (if the youn growth is protected) forever. It will occupy broken ground that could not otherwise be cultivated, and the timber, when taken out at greatest maturity, has a value which is gaining every year, aside from the annual revenue to be derived from the sap. The maple adorns and beautifies perhaps more than any other of our native forest trees...Th sugar season comes at a time when farm labor is least employed, and the occupation presents amenities beyond those which any other form of farm labor can afford.

Franklin Hough, Report on Forestry, 1884